GUIDELINES

ANALYSES AND OTHER TOOLS

To Benjamin W. Niebel (1918–1999) who reminded me to remember productivity while designing work for health and safety.

About the Authors

Andris Freivalds is professor of industrial engineering at the Pennsylvania State University. He obtained a B.S. degree in science engineering and M.S. and Ph.D. degrees in bioengineering from the University of Michigan. After biomechanics research at the Aerospace Medical Research Laboratory at Wright Patterson AFB, Ohio, and at the University of Nottingham, U.K., as a Fulbright Scholar, he shifted his focus to the reduction of work-related musculoskeletal injuries in U.S. industries. With the assistance of the Commonwealth of Pennsylvania in the development of the Center for Cumulative Trauma Disorders (CTD), he has provided ergonomic and engineering design services to over 75 Pennsylvania companies in controlling their workplace hazards and improving overall productivity. As part of these efforts, he was recognized by the Institute of Industrial Engineers with its Technical Innovation Award in 1995. He is a Certified Professional Ergonomist and a Fellow of the Ergonomics Society.

Benjamin Niebel received B.S., M.E., and P.E. degrees in industrial engineering from the Pennsylvania State University. After heading the industrial engineering department at Lord Manufacturing Company, he became interested in engineering education. He returned to Penn State as a faculty member in 1947 and then served as the head of the Department of Industrial Engineering from 1955 to 1979. During this time he wrote many textbooks including the original *Motion and Time Study* in 1955 and consulted regularly with many industries. For his outstanding service to both the profession and Penn State, he was awarded the Frank and Lillian Gilbreth Award by the Institute of Industrial Engineers in 1976, the Outstanding Engineering Alumnus Award by the College of Engineering in 1989, and the Penn State Engineering Society Distinguished Service Award in 1992. He passed away at the distinguished age of 80 while on a consulting trip for industry.

McGraw-Hill
Higher Education

NIEBEL'S METHODS, STANDARDS, & WORK DESIGN, TWELFTH EDITION

1 2 3 4 5 6 7 8 9 0 DOC/DOC 0 9 8

ISBN 978-0-07-337631-8
MHID 0-07-337631-0

Global Publisher: *Raghothaman Srinivasan*
Director of Development: *Kristine Tibbetts*
Developmental Editor: *Lorraine K. Buczek*
Project Coordinator: *Melissa M, Leick*
Lead Production Supervisor: *Sandy Ludovissy*
Designer: *John Albert Joran*
Associate Design Coordinator: *Brenda A. Rolwes*
Cover Designer: *Studio Montage, St. Louis, Missouri*
(USE) Cover Image: *(Hands holding PDA, Conceptual image of PDA's):* © *Comstock/PunchStock; (Executive people in a factory, Stopwatch, Worker overseeing production):* © *Royalty-Free/Corbis; (Man looking at a laptop):* © *Liquidlibrary/PictureQuest; (Laptop with cell phone, Hammer):* ©*Getty Images RF; (Doctor and nurse):* © *Stockbyte/PunchStock*
Senior Photo Research Coordinator: *Lori Hancock*
Compositor: *Carlisle Publishing Services*
Typeface: *10.5/12 Times*
Printer: *R. R. Donnelley Crawfordsville, IN*

Library of Congress Cataloging-in-Publication Data

Niebel, Benjamin W.
    Niebel's methods, standards, & work design / Andris Freivalds. — 12th ed.
      p. cm.
    Includes index.
    ISBN 978-0-07-337631-8—ISBN 0-07-33763I—0(hardcopy: alk. paper) 1. Work design. 2. Human engineering. 3. Time study. 4. Motion study. I. Freivalds, Andris. II. Title.
T60.8.N54 2009
658.5'4-dc22
                            2007046684

www.mhhe.com

Niebel's Methods, Standards, and Work Design

Twelfth Edition

Andris Freivalds
Professor of Industrial Engineering
The Pennsylvania State University

Benjamin W. Niebel
Late Professor of Industrial Engineering
The Pennsylvania State University

Boston   Burr Ridge, IL   Dubuque, IA   New York   San Francisco   St. Louis
Bangkok   Bogotá   Caracas   Kuala Lumpur   Lisbon   London   Madrid   Mexico City
Milan   Montreal   New Delhi   Santiago   Seoul   Singapore   Sydney   Taipei   Toronto

Brief Contents

Table of Contents

Preface

BACKGROUND

Faced with increasing competition from all parts of the world, almost every industry, business, and service organization is restructuring itself to operate more effectively. As downsizing and outsourcing become more common, these organizations must increase the intensity of cost reduction and quality improvement efforts while working with reduced labor forces. Cost-effectiveness and product reliability without excess capacity are the keys to successful activity in all areas of business, industry, and government and are the end result of methods engineering, equitable time standards, and efficient work design.

Also, as machines and equipment grow increasingly complex and semiautomated if not fully automated, it is increasingly important to study both the manual components and the cognitive aspects of work as well as the safety of the operations. The operator must perceive and interpret large amounts of information, make critical decisions, and control these machines both quickly and accurately. In recent years, jobs have shifted gradually from manufacturing to the service sector. In both sectors, there is increasingly less emphasis on gross physical activity and a greater emphasis on information processing and decision making, especially via computers and associated modern technology. The same efficiency and work design tools are the keys to productivity improvement in any industry, business, or service organization, whether in a bank, a hospital, a department store, a railroad, or the postal system. Furthermore, success in a given product line or service leads to new products and innovations. It is this accumulation of successes that drives hiring and the growth of an economy.

The reader should be careful not to be swayed or intimidated by the latest jargon offered as a cure-all for an enterprise's lack of competitiveness. Often these fads sideline the sound engineering and management procedures that, when properly utilized, represent the key to continued success. Today we hear a good deal about reengineering and use of cross-functional teams as business leaders reduce cost, inventory, cycle time, and nonvalue activities. However, experience in the past few years has proved that cutting people from the payroll just for the sake of automating their jobs is not always the wise procedure. The authors, with many years of experience in over 100 industries, strongly recommend sound methods engineering, realistic standards, and good work design as the keys to success in both manufacturing and service industries.

WHY THIS BOOK WAS WRITTEN

The objectives of the twelfth edition have remained the same as for the eleventh: to provide a practical, up-to-date college textbook describing engineering methods to measure, analyze, and design manual work. The importance of ergonomics and work design as part of methods engineering is emphasized, not only to increase productivity, but also to improve worker health and safety and thus company bottom-line costs. Far too often, industrial engineers have focused solely on increasing productivity through methods changes and job simplification, resulting in overly repetitive jobs for the operators and increased incidence rates of musculoskeletal injuries. Any cost reductions obtained are more than offset by the increased medical and worker's compensation costs, especially considering today's ever-escalating health care costs.

WHAT'S NEW IN THE TWELFTH EDITION

A new Chapter 8 on workplace and systems safety has been added that includes material on accident causation models, accident prevention, quantitative analyses, and general hazard control. This then completes the knowledge that a basic industrial engineer should have for managing a production line or a service center. Old Chapters 10 and 11 on ratings and allowances were combined as support materials to the new Chapter 10 on time study. Chapter 13 was expanded to include more material on BasicMOST.

Approximately 10 to 15 percent more examples, problems, and case studies have been added. The twelfth edition still provides a continued reliance on work design, work measurement, facilities layout, and various flow process charts for students entering the industrial engineering profession and serves as a practical, up-to-date source of reference material for the practicing engineer and manager.

HOW THIS BOOK DIFFERS FROM OTHERS

Most textbooks on the market deal strictly either with the traditional elements of motion and time study or with human factors and ergonomics. Few textbooks integrate both topics into one book or, for that matter, one course. In this day and age, the industrial engineer needs to consider both productivity issues and their effects on the health and safety of the worker simultaneously. Few of the books on the market are formatted for use in the classroom setting. This text includes additional questions, problems, and sample laboratory exercises to assist the educator. Finally, no text provides the extensive amount of online student and instructor resources, electronic forms, current information, and changes as this edition does.

ORGANIZATION OF THE TEXT AND COURSE MATERIAL

The twelfth edition is laid out to provide roughly one chapter of material per week of a semester-long introductory course. Although there are a total of 18 chapters,

Chapter 1 is short and introductory, much of Chapter 7 on cognitive work design and Chapter 8 on safety may be covered in other courses, and Chapter 15 on standards for indirect and expense work may not need to be covered in an introductory course, all of which leaves only 15 chapters to be covered in the semester.

A typical semester plan, chapter by chapter, using the first lecture number, might be as follows:

Chapter	Lectures	Coverage
1	1	Quick introduction on the importance of productivity and work design, with a bit of historical perspective.
2	3–6	A few tools from each area (Pareto analysis, job analysis/worksite guide, flow process charts, worker–machine charts) with some quantitative analysis on worker–machine interactions. Line balancing and PERT may be covered in other courses.
3	4	Operation analysis with an example for each step.
4	4	Full, but can gloss over basic muscle physiology and energy expenditure.
5	4	Full.
6	3–4	Basics on illumination, noise, temperature; other topics as desired may be covered in another course.
7	0–4	Coverage depends on instructor's interest; may be covered in another course.
8	0–5	Coverage depends on instructor's interest; may be covered in another course.
9	3–5	Three tools: value engineering, cost-benefit analysis, and crossover charts; job analysis and evaluation, and interaction with workers. Other tools may be covered in other classes.
10	3	Basics of time study.
11	3–5	One form of rating; first half of the allowances that are well established.
12	1–3	Coverage of standard data and formulas depends on instructor's interest.
13	4–7	Only one predetermined time system in depth; the second may be covered in another course.
14	2–3	Work sampling.
15	0–3	Coverage of indirect and expense labor standards depends on instructor's interest.
16	2–3	Overview and costing.
17	3–4	Day work and standard hour plan.
18	3–4	Learning curves, motivation, and people skills.

The recommended plan covers 43 lectures, with two periods for examinations. Some instructors may wish to spend more time on any given chapter, for which additional material is supplied, for example, work design (Chapters 4 to 7), and less time on traditional work measurement (Chapters 8 to 16), or vice versa. The text allows for this flexibility.

Similarly, if all the material is used (the second lecture number), there is enough material for one lecture course and one course with a lab, as is done at Penn State University. Both courses have been developed with appropriate materials such that they can be presented completely online. For an example of an online course using this text, go to *www.engr.psu.edu/cde/courses/ie327/index.html*

SUPPLEMENTARY MATERIAL AND ONLINE SUPPORT

The twelfth edition of this text continues to focus on the ubiquitous use of PCs as well as the Internet to establish standards, conceptualize possibilities, evaluate costs, and disseminate information. A website, hosted by the publisher at *www.mhhe.com/niebel-freivalds*, furthers that objective by providing the educator with various online resources, such as an updated instructor's manual. Design-Tools version 4.1.1, a ready-to-use software program for ergonomics analysis and work measurement, appears on the site as well. A special new feature of Design-Tools is the addition of QuikTS, a time study data collection program, and Quik-Samp, a work sampling program. The program may be downloaded via hot synch to a Palm device (m105 or higher) and used to collect time study data. The data are then uploaded directly to the time study form on DesignTools for easy and accurate calculation of standard time.

The book's website also links to a website hosted by the author at *www2.ie.psu.edu/Freivalds/courses/ie327new/index.html* which provides instructors with online background material, including electronic versions of the forms used in the textbook. Student resources include practice exams and solutions. Up-to-date information on any errors found or corrections needed in this new edition appear on this site as well. Suggestions received from individuals at universities, colleges, technical institutes, industries, and labor organizations that regularly use this text have helped materially in the preparation of this twelfth edition. Further suggestions are welcome, especially if any errors are noticed. Please simply respond to the OOPS! button on the website or by email to *axf@psu.edu*

ACKNOWLEDGMENTS

I wish to acknowledge the late Ben Niebel for providing me with the opportunity to contribute to his well-respected textbook. I hope the additions and modifications will match his standards and continue to serve future industrial engineers as they enter their careers. Thanks to Dr. Dongjoon Kong, University of Tennessee, for devoting so much of his time at Penn State to programming DesignTools. Thanks also to the following reviewers for their invaluable input:

David R. Clark, *Kettering University*
Luis Rene Contreras, *University of Texas, El Paso*
Jerry Davis, *Auburn University*
Corinne MacDonald, *Dalhousie University*
Gary Mirka, *Iowa State University*
Durward K. Sobek, *Montana State University*
Harvey Wolfe, *University of Pittsburgh*

Finally, I wish to express my gratitude to Dace for her patience and support.

Andris Freivalds

Methods, Standards, and Work Design: Introduction

KEY POINTS

- Increasing productivity drives U.S. industry.
- Worker health and safety are just as important as productivity.
- Methods engineering simplifies work.
- Work design fits work to the operator.
- Time study measures work and sets standards.

1.1 PRODUCTIVITY IMPORTANCE

Certain changes continually taking place in the industrial and business environment must be considered both economically and practically. These include the globalization of both the market and the producer, the growth of the service sector, the computerization of all facets of an enterprise, and the ever-expanding applications of the Internet and Web. The only way a business or enterprise can grow and increase its profitability is by increasing its productivity. Productivity improvement refers to the increase in output per work-hour or time expended. The United States has long enjoyed the world's highest productivity. Over the last 100 years, productivity in the United States has increased approximately 4 percent per year. However, in the last decade, the U.S. rate of productivity improvement has been exceeded by that of Japan, Korea, and Germany, and it may soon be challenged by China.

The fundamental tools that result in increased productivity include methods, time study standards (frequently referred to as work measurement), and work design. Of the total cost of the typical metal products manufacturing enterprise, 12 percent is direct labor, 45 percent is direct material, and 43 percent is overhead. All aspects of a business or industry—sales, finance, production, engineering, cost, maintenance, and management—provide fertile areas for the application of methods, standards, and work design. Too often, people consider only production, when other aspects of the enterprise could also profit from the application of productivity tools. In sales, for example, modern information retrieval methods usually result in more reliable information and greater sales at less cost.

Today, most U.S. businesses and industries are, by necessity, restructuring themselves by downsizing, to operate more effectively in an increasingly competitive world. With greater intensity than ever before, they are addressing cost reduction and quality improvement through productivity improvement. They are also critically examining all nonvalue business components, those that do not contribute to their profitability.

Since the production area within manufacturing industries utilizes the greatest number of engineers in methods, standards, and work design efforts, this text will treat that field in greater detail than any other. However, examples from other areas of the manufacturing industry, such as maintenance, transportation, sales, and management, as well as the service industry, will be provided.

Traditional areas of opportunity for students enrolled in engineering, industrial management, business administration, industrial psychology, and labor–management relations are (1) work measurement, (2) work methods and design, (3) production engineering, (4) manufacturing analysis and control, (5) facilities planning, (6) wage administration, (7) ergonomics and safety, (8) production and inventory control, and (9) quality control. However, these areas of opportunity are not confined to manufacturing industries. They exist, and are equally important, in such enterprises as department stores, hotels, educational institutions, hospitals, banks, airlines, insurance offices, military service centers, government agencies, and retirement complexes. Today, in the United States, only about 10 percent of the total labor force is employed in manufacturing industries. The remaining 90 percent is engaged in service industries or staff-related positions. As the United States becomes ever more service-industry-oriented, the philosophies and techniques of methods, standards, and work design must also be utilized in the service sector. Wherever people, materials, and facilities interact to obtain some objective, productivity can be improved through the intelligent application of methods, standards, and work design.

The production area of an industry is key to success. Here materials are requisitioned and controlled; the sequence of operations, inspections, and methods is determined; tools are ordered; time values are assigned; work is scheduled, dispatched, and followed up; and customers are kept satisfied with quality products delivered on time.

Similarly, the methods, standards, and work design activity is the key part of the production group. Here more than in any other place, people determine whether a product is going to be produced on a competitive basis, through efficient workstations, tooling, and worker and machine relationships. Here is where they are creative in improving existing methods and products and maintaining good labor relations through fair labor standards.

The objective of the manufacturing manager is to produce a quality product, on schedule, at the lowest possible cost, with a minimum of capital investment and a maximum of employee satisfaction. The focus of the reliability and quality control manager is to maintain engineering specifications and satisfy customers with the product's quality level and reliability over its expected life. The production control manager is principally interested in establishing and maintaining production schedules with due regard for both customer needs and the favorable economics obtainable with careful scheduling. The maintenance manager is primarily concerned with minimizing facility downtime due to unscheduled breakdowns and repairs. Figure 1.1 illustrates the relationship of all these areas and the influence of methods, standards, and work design on overall production.

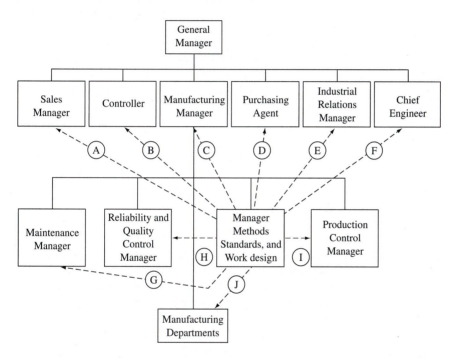

A—Cost is largely determined by manufacturing methods.
B—Time standards are the bases of standard costs.
C—Standards (direct and indirect) provide the bases for measuring the performance of production
    departments.
D—Time is a common denominator for comparing competitive equipment and supplies.
E—Good labor relations are maintained with equitable standards and a safe work environment.
F—Methods work design and processes strongly influence product designs.
G—Standards provide the bases for preventive maintenance.
H—Standards enforce quality.
 I—Scheduling is based on time standards.
 J—Methods, standards, and work design provide how the work is to be done and how long it will take.

Figure 1.1
Typical organization chart showing the influence of methods, standards, and work design on the operation of the enterprise.

1.2 METHODS AND STANDARDS SCOPE

Methods engineering includes designing, creating, and selecting the best manufacturing methods, processes, tools, equipment, and skills to manufacture a product based on the specifications that have been developed by the product engineering section. When the best method interfaces with the best skills available, an efficient worker–machine relationship exists. Once the complete method has been established, a standard time for the product must be determined. Furthermore there is the responsibility to see that (1) predetermined standards are met; (2) workers are adequately compensated for their output, skills, responsibilities, and experience; and (3) workers have a feeling of satisfaction from the work that they do.

The overall procedure includes defining the problem; breaking the job down into operations; analyzing each operation to determine the most economical manufacturing procedures for the quantity involved, with due regard for operator safety and job interest; applying proper time values; and then following through to ensure that the prescribed method is put into operation. Figure 1.2 illustrates

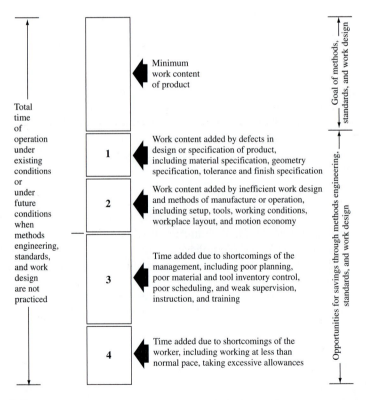

Figure 1.2
Opportunities for savings through the applications of methods engineering and time study.

the opportunities for reducing the standard manufacturing time through the application of methods engineering and time study.

METHODS ENGINEERING

The terms *operation analysis, work design, work simplification,* and *methods engineering* and *corporate reengineering* are frequently used synonymously. In most cases, the person is referring to a technique for increasing the production per unit of time or decreasing the cost per unit output—in other words, productivity improvement. However, methods engineering, as defined in this text, entails analyses at two different times during the history of a product. First, the methods engineer is responsible for designing and developing the various work centers where the product will be produced. Second, that engineer must continually restudy the work centers to find a better way to produce the product and/or improve its quality.

In recent years, this second analysis has been called corporate reengineering. In this regard, we recognize that a business must introduce changes if it is to continue profitable operation. Thus, it may be desirable to introduce changes outside the manufacturing area. Often, profit margins may be enhanced through positive changes in such areas as accounting, inventory management, materials requirements planning, logistics, and human resource management. Information automation can provide dramatic rewards in all these areas. The more thorough the methods study during the planning stages, the less the necessity for additional methods studies during the life of the product.

Methods engineering implies the utilization of technological capability. Primarily because of methods engineering, improvements in productivity are never-ending. The productivity differential resulting from technological innovation can be of such magnitude that developed countries will always be able to maintain competitiveness with low-wage developing countries. Research and development (R&D) leading to new technology is therefore essential to methods engineering. The 10 countries with the highest R&D expenditures per worker, as reported by the United Nations Industrial Development Organization (1985), are the United States, Switzerland, Sweden, Netherlands, Germany, Norway, France, Israel, Belgium, and Japan. These countries are among the leaders in productivity. As long as they continue to emphasize research and development, methods engineering through technological innovation will be instrumental in their ability to provide high-level goods and services.

Methods engineers use a systematic procedure to develop a work center, produce a product, or provide a service (see Figure 1.3). This procedure is outlined here, and it summarizes the flow of the text. Each step is detailed in a later chapter. Note that steps 6 and 7 are not strictly part of a methods study, but are necessary in a fully functioning work center.

1. *Select the project.* Typically, the projects selected represent either new products or existing products that have a high cost of manufacture and a low profit. Also, products that are currently experiencing difficulties in

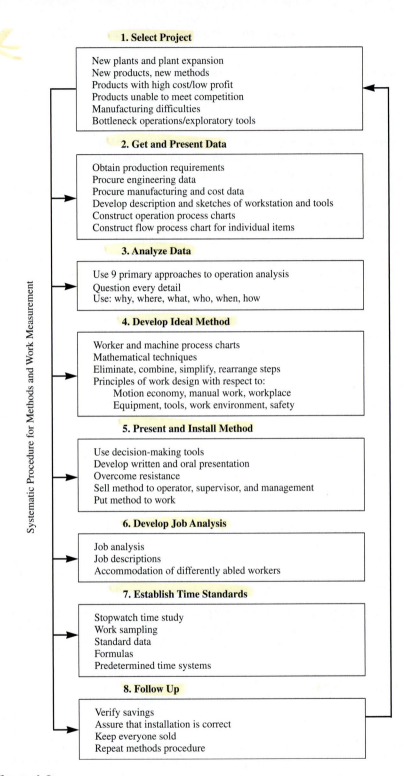

Systematic Procedure for Methods and Work Measurement

1. Select Project

New plants and plant expansion
New products, new methods
Products with high cost/low profit
Products unable to meet competition
Manufacturing difficulties
Bottleneck operations/exploratory tools

2. Get and Present Data

Obtain production requirements
Procure engineering data
Procure manufacturing and cost data
Develop description and sketches of workstation and tools
Construct operation process charts
Construct flow process chart for individual items

3. Analyze Data

Use 9 primary approaches to operation analysis
Question every detail
Use: why, where, what, who, when, how

4. Develop Ideal Method

Worker and machine process charts
Mathematical techniques
Eliminate, combine, simplify, rearrange steps
Principles of work design with respect to:
　　Motion economy, manual work, workplace
　　Equipment, tools, work environment, safety

5. Present and Install Method

Use decision-making tools
Develop written and oral presentation
Overcome resistance
Sell method to operator, supervisor, and management
Put method to work

6. Develop Job Analysis

Job analysis
Job descriptions
Accommodation of differently abled workers

7. Establish Time Standards

Stopwatch time study
Work sampling
Standard data
Formulas
Predetermined time systems

8. Follow Up

Verify savings
Assure that installation is correct
Keep everyone sold
Repeat methods procedure

Figure 1.3
The principal steps in a methods engineering program.

maintaining quality and are having problems meeting competition are logical projects for methods engineering. (See Chapter 2 for more details.)

2. *Get and present the data.* Assemble all the important facts relating to the product or service. These include drawings and specifications, quantity requirements, delivery requirements, and projections of the anticipated life of the product or service. Once all important information has been acquired, record it in an orderly form for study and analysis. The development of process charts at this point is very helpful. (See Chapter 2 for more details.)

3. *Analyze the data.* Utilize the primary approaches to operations analysis to decide which alternative will result in the best product or service. These primary approaches include purpose of operation, design of part, tolerances and specifications, materials, process of manufacture, setup and tools, working conditions, material handling, plant layout, and work design. (See Chapter 3 for more details.)

4. *Develop the ideal method.* Select the best procedure for each operation, inspection, and transportation by considering the various constraints associated with each alternative, including productivity, ergonomics, and health and safety implications. (See Chapters 3 to 7 for more details.)

5. *Present and install the method.* Explain the proposed method in detail to those responsible for its operation and maintenance. Consider all details of the work center, to ensure that the proposed method will provide the results anticipated. (See Chapter 8 for more details.)

6. *Develop a job analysis.* Conduct a job analysis of the installed method to ensure that the operators are adequately selected, trained, and rewarded. (See Chapter 8 for more details.)

7. *Establish time standards.* Establish a fair and equitable standard for the installed method. (See Chapters 9 to 15 for more details.)

8. *Follow up the method.* At regular intervals, audit the installed method to determine if the anticipated productivity and quality are being realized, whether costs were correctly projected, and whether further improvements can be made. (See Chapter 16 for more details.)

In summary, methods engineering is the systematic close scrutiny of all direct and indirect operations to find improvements that make work easier to perform, in terms of worker health and safety, and also allow work to be done in less time with less investment per unit (i.e., greater profitability).

WORK DESIGN

As part of developing or maintaining the new method, the principles of work design must be used to fit the task and workstation ergonomically to the human operator. Unfortunately, work design is typically forgotten in the quest for increased productivity. Far too often, overly simplified procedures result in machinelike repetitive jobs for the operators, leading to increased rates of work-related musculoskeletal disorders. Any productivity increases and reduced costs are more than offset by the increased medical and workers' compensation costs, especially considering today's ever-escalating health-care trends. Thus, it is necessary for the methods engineer to incorporate the principles of work design into any new method, so that it not only will be more productive but also will be safe and injury-free for the operator. (Refer to Chapters 4 to 7.)

STANDARDS

Standards are the end result of time study or work measurement. This technique establishes a time standard allowed to perform a given task, based on measurements of the work content of the prescribed method, with due consideration for fatigue and for personal and unavoidable delays. Time study analysts use several techniques to establish a standard: a stopwatch time study, computerized data collection, standard data, predetermined time systems, work sampling, and estimates based on historical data. Each technique is applicable to certain conditions. Time study analysts must know when to use a given technique and must then use that technique judiciously and correctly.

The resulting standards are used to implement a wage payment scheme. In many companies, particularly in smaller enterprises, the wage payment activity is performed by the same group responsible for the methods and standards work. Also, the wage payment activity is performed in concert with those responsible for conducting job analyses and job evaluations, so that these closely related activities function smoothly.

Production control, plant layout, purchasing, cost accounting and control, and process and product design are additional areas closely related to both the methods and standards functions. To operate effectively, all these areas depend on time and cost data, facts, and operational procedures from the methods and standards department. These relationships are briefly discussed in Chapter 16.

OBJECTIVES OF METHODS, STANDARDS, AND WORK DESIGN

The principal objectives of methods, standards, and work design are (1) to increase productivity and product reliability safely and (2) to lower unit cost, thus allowing more quality goods and services to be produced for more people. The ability to produce more for less will result in more jobs for more people for a greater number of hours per year. Only through the intelligent application of the principles of methods, standards, and work design can producers of goods and services increase, while, at the same time, the purchasing potential of all consumers grows. Through these principles, unemployment and relief rolls can be minimized, thus reducing the spiraling cost of economic support to nonproducers.

Corollaries to the principal objectives are as follows:

1. Minimize the time required to perform tasks.
2. Continually improve the quality and reliability of products and services.

3. Conserve resources and minimize cost by specifying the most appropriate direct and indirect materials for the production of goods and services.
4. Consider the cost and availability of power.
5. Maximize the safety, health, and well-being of all employees.
6. Produce with an increasing concern for protecting the environment.
7. Follow a humane program of management that results in job interest and satisfaction for each employee.

1.3 HISTORICAL DEVELOPMENTS

THE WORK OF TAYLOR

Frederick W. Taylor is generally conceded to be the founder of modern time study in this country. However, time studies were conducted in Europe many years before Taylor's time. In 1760, Jean Rodolphe Perronet, a French engineer, made extensive time studies on the manufacture of No. 6 common pins, while 60 years later, an English economist, Charles W. Babbage, conducted time studies on the manufacture of No. 11 common pins.

Taylor began his time study work in 1881 while associated with the Midvale Steel Company in Philadelphia. Although born in a wealthy family, he disdained his upbringing and started out serving as an apprentice. After 12 years' work, he evolved a system based on the "task." Taylor proposed that the work of each employee be planned out by the management at least one day in advance. Workers were to receive complete written instructions describing their tasks in detail and noting the means to accomplish them. Each job was to have a standard time, determined by time studies made by experts. In the timing process, Taylor advocated breaking up the work assignment into small divisions of effort known as "elements." Experts were to time these individually and use their collective values to determine the allowed time for the task.

Taylor's early presentations of his findings were received without enthusiasm, because many of the engineers interpreted his findings to be a new piece-rate system rather than a technique for analyzing work and improving methods. Both management and employees were skeptical of piece rates, because many standards were either typically based on the supervisor's guess or inflated by bosses to protect the performance of their departments.

In June 1903, at the Saratoga meeting of the American Society of Mechanical Engineers (ASME), Taylor presented his famous paper "Shop Management," which included the elements of scientific management: time study, standardization of all tools and tasks, use of a planning department, use of slide rules and similar timesaving implements, instruction cards for workers, bonuses for successful performance, differential rates, mnemonic systems for classifying products, routing systems, and modern cost systems. Taylor's techniques were well received by many factory managers, and by 1917, of 113 plants that had installed "scientific management," 59 considered their installations completely successful, 20 partly successful, and 34 failures (Thompson, 1917).

In 1898, while at the Bethlehem Steel Company (he had resigned his post at Midvale), Taylor carried out the pig-iron experiment that came to be one of the most celebrated demonstrations of his principles. He established the correct method, along with financial incentives, and workers carrying 92-lb pigs of iron up a ramp onto a freight car were able to increase their productivity from an average of 12.5 tons/day to between 47 and 48 tons/day. This work was performed with an increase in the daily rate of $1.15 to $1.85. Taylor claimed that workmen performed at the higher rate "without bringing on a strike among the men, without any quarrel with the men and were happier and better contented."

Another of Taylor's Bethlehem Steel studies that gained fame was the shoveling experiment. Workers who shoveled at Bethlehem owned their own shovels and would use the same one for any job—lifting heavy iron ore to lifting light rice coal. After considerable study, Taylor designed shovels to fit the different loads: short-handled shovels for iron ore, long-handled scoops for light rice coal. As a result, productivity increased, and the cost of handling materials decreased from 8 cents/ton to 3 cents/ton.

Another of Taylor's well-known contributions was the discovery of the Taylor–White process of heat treatment for tool steel. Studying self-hardening steels, he developed a means of hardening a chrome–tungsten steel alloy without rendering it brittle, by heating it close to its melting point. The resulting "high-speed steel" more than doubled machine cutting productivity and remains in use today all over the world. Later, he developed the Taylor equation for cutting metal.

Not as well known as his engineering contributions is the fact that in 1881, he was a U.S. tennis doubles champion. Here he used an odd-looking racket he had designed with a spoon curved handle. Taylor died of pneumonia in 1915, at the age of 59. For more information on this multitalented individual, the authors recommend Kanigel's biography (1997).

In the early 1900s, the country was going through an unprecedented inflationary period. The word efficiency became passé, and most businesses and industries were looking for new ideas that would improve their performance. The railroad industry also felt the need to increase shipping rates substantially to cover general cost increases. Louis Brandeis, who at that time represented the eastern business associations, contended that the railroads did not deserve, or in fact need, the increase because they had been remiss in not introducing the new "science of management" into their industry. Brandeis claimed that the railroad companies could save $1 million/day by introducing the techniques advocated by Taylor. Thus, Brandeis and the Eastern Rate Case (as the hearing came to be known) first introduced Taylor's concepts as "scientific management."

At this time, many people without the qualifications of Taylor, Barth, Merrick, and other early pioneers, were eager to make names for themselves in this new field. They established themselves as "efficiency experts" and endeavored to install scientific management programs in industry. They soon encountered a natural resistance to change from employees, and since they were not equipped to handle problems of human relations, they met with great difficulty. Anxious to make a good showing and equipped with only a pseudoscientific knowledge, they generally established rates that were too difficult to meet. Situations became so acute that some managers were obliged to discontinue the whole program in order to continue operation.

In other instances, factory managers would allow the establishment of time standards by the supervisors, but this was seldom satisfactory. Once standards were established, many factory managers of that time, interested primarily in the reduction of labor costs, would unscrupulously cut rates if some employee made what the employer felt was too much money. The result was harder work at the same, and sometimes less, take-home pay. Naturally, violent worker reaction resulted.

These developments spread in spite of the many favorable installations started by Taylor. At the Watertown Arsenal, labor objected to such an extent to the new time study system that in 1910 the Interstate Commerce Commission (ICC) started an investigation of time study. Several derogatory reports on the subject influenced Congress to add a rider to the government appropriations bill in 1913, stipulating that no part of the appropriation should be made available for the pay of any person engaged in time study work. This restriction applied to the government-operated plants where government funds were used to pay the employees.

It wasn't until 1947 that the House of Representatives passed a bill that rescinded the prohibition against using stopwatches and the use of time study. It is of interest that even today the use of the stopwatch is still prohibited by unions in some railroad repair facilities. It is also interesting to note that Taylorism is very much alive today, in contemporary assembly lines, in lawyer's bills that are calculated in fractional hours, and in documentation of hospital costs for patients.

MOTION STUDY AND THE WORK OF THE GILBRETHS

Frank and Lilian Gilbreth were the founders of the modern *motion-study technique,* which may be defined as the study of the body motions used in performing an operation, to improve the operation by eliminating unnecessary motions, simplifying necessary motions, and then establishing the most favorable motion sequence for maximum efficiency. Frank Gilbreth originally introduced his ideas and philosophies into the bricklayer's trade in which he was employed. After introducing methods improvements through motion study, including an adjustable scaffold that he had invented, as well as operator training, he was able to increase the average number of bricks laid to 350 per worker per hour. Prior to Gilbreth's studies, 120 bricks per hour was considered a satisfactory rate of performance for a bricklayer.

More than anyone else, the Gilbreths were responsible for industry's recognition of the importance of a detailed study of body motions to increase production, reduce fatigue, and instruct operators in the best method of performing an operation. They developed the technique of filming motions to study them, in a technique known as micromotion study. The study of movements through the aid of the slow-motion moving picture is by no means confined to industrial applications.

In addition, the Gilbreths developed the cyclegraphic and chronocyclegraphic analysis techniques for studying the motion paths made by an operator.

The cyclegraphic method involves attaching a small electric lightbulb to the finger or hand or part of the body being studied and then photographing the motion while the operator is performing the operation. The resulting picture gives a permanent record of the motion pattern employed and can be analyzed for possible improvement. The chronocyclegraph is similar to the cyclegraph, but its electric circuit is interrupted regularly, causing the light to flash. Thus, instead of showing solid lines of the motion patterns, the resulting photograph shows short dashes of light spaced in proportion to the speed of the body motion being photographed. Consequently, with the chronocyclegraph it is possible to compute velocity, acceleration, and deceleration, as well as to study body motions. The world of sports has found this analysis tool, updated to video, invaluable as a training tool to show the development of form and skill.

As an interesting side note, the reader may wish to read about the extreme lengths to which Frank Gilbreth went to achieve maximum efficiency even in his personal life. His eldest son and daughter recount vignettes of their father shaving with razors simultaneously in both hands or using various communication signals to assemble all the children, of which there were 12. Hence the title of their book *Cheaper by the Dozen* (Gilbreth and Gilbreth, 1948)! After Frank's relatively early death at the age of 55, Lillian, who had received a Ph.D. in psychology and had been a more than equal collaborator, continued on her own, advancing the concept of work simplification especially for the physically handicapped. She passed away in 1972 at the distinguished age of 93 (Gilbreth, 1988).

EARLY CONTEMPORARIES

Carl G. Barth, an associate of Frederick W. Taylor, developed a production slide rule for determining the most efficient combinations of speeds and feeds for cutting metals of various hardnesses, considering the depth of cut, size of tool, and life of the tool. He is also noted for his work in determining allowances. He investigated the number of foot-pounds of work a worker could do in a day. He then developed a rule that equated a certain push or pull on a worker's arms with the amount of weight that worker could handle for a certain percentage of the day.

Harrington Emerson applied scientific methods to work on the Santa Fe Railroad and wrote a book, *Twelve Principles of Efficiency,* in which he made an effort to inform management of procedures for efficient operation. He reorganized the company, integrated its shop procedures, installed standard costs and a bonus plan, and transferred its accounting work to Hollerith tabulating machines. This effort resulted in annual savings in excess of $1.5 million and the recognition of his approach, termed *efficiency engineering.*

In 1917, Henry Laurence Gantt developed simple graphs that would measure performance while visually showing projected schedules. This production control tool was enthusiastically adopted by the shipbuilding industry during World War I. For the first time, this tool made it possible to compare actual performance against the original plan, and to adjust daily schedules in accordance with capacity, backlog, and customer requirements. Gantt is also known for his invention of a wage payment system that rewarded workers for above-standard performance, eliminated any penalty for failure, and offered the boss a bonus for every worker who performed above standard. Gantt emphasized human relations and promoted scientific management as more than an inhuman "speedup" of labor.

Motion and time study received added stimulus during World War II when Franklin D. Roosevelt, through the U.S. Department of Labor, advocated establishing standards for increasing production. The stated policy advocated greater pay for greater output but without an increase in unit labor costs, incentive schemes to be collectively bargained between labor and management, and the use of time study or past records to set production standards.

EMERGENCE OF WORK DESIGN

Work design is a relatively new science that deals with designing the task, workstation, and working environment to fit the human operator better. In the United States, it is more typically known as *human factors,* while internationally it is better known as *ergonomics,* which is derived from the Greek words for work (*erg*) and laws (*nomos*).

In the United States, after the initial work of Taylor and the Gilbreths, the selection and training of military personnel during World War I and the industrial psychology experiments of the Harvard Graduate School at Western Electric (see the Hawthorne studies in Chapter 9) were important contributions to the work design area. In Europe, during and after World War I, the British Industrial Fatigue Board performed numerous studies on human performance under various conditions. These were later extended to heat stress and other conditions by the British Admiralty and Medical Research Council.

World War II, with the complexity of military equipment and aircraft, led to the development of the U.S. military engineering psychology laboratories and a real growth of the profession. The start of the race to space with the launch of *Sputnik* in 1957 only accelerated the growth of human factors, especially in the aerospace and military sectors. From the 1970s on, the growth has shifted to the industrial sector and, more recently, into computer equipment, user-friendly software, and the office environment. Other driving forces for the growth in human factors are the rise in product liability and personal injury litigation cases and also, unfortunately, tragic, large-scale technological disasters, such as the nuclear incident at Three-Mile Island and the gas leak at the Union Carbide Plant in Bhopal, India. Obviously, the growth of computers and technology will keep human factors specialists and ergonomists busy designing better workplaces and products and improving the quality of life and work for many years to come.

ORGANIZATIONS

Since 1911, there has been an organized effort to keep industry abreast of the latest developments in the techniques inaugurated by Taylor and Gilbreth. Technical organizations have contributed much toward bringing the science of time study, work design, and methods engineering up to present-day standards. In 1915, the

Taylor Society was founded to promote the science of management, while in 1917 the Society of Industrial Engineers was organized by those interested in production methods. The American Management Association (AMA) traces its origins to 1913, when a group of training managers formed the National Association of Corporate Schools. Its various divisions sponsor courses and publications on productivity improvement, work measurement, wage incentives, work simplification, and clerical standards. Together with the American Society of Mechanical Engineers (ASME), AMA annually presents the Gantt Memorial Medal for the most distinguished contribution to industrial management as a service to the community.

The Society for the Advancement of Management (SAM) was formed in 1936 by the merger of the Society of Industrial Engineers and the Taylor Society. This organization emphasized the importance of time study and methods and wage payment. Industry has used SAM's time study rating films over a long period of years. SAM annually offers the Taylor key for the outstanding contribution to the advancement of the science of management and the Gilbreth medal for noteworthy achievement in the field of motion, skill, and fatigue study. In 1972, SAM combined forces with AMA.

The Institute of Industrial Engineers (IIE) was founded in 1948 with the purposes of maintaining the practice of industrial engineering on a professional level; fostering a high degree of integrity among the members of the industrial engineering profession; encouraging and assisting education and research in areas of interest to industrial engineers; promoting the interchange of ideas and information among members of the industrial engineering profession (e.g., publishing the journal *IIE Transactions*); serving the public interest by identifying persons qualified to practice as industrial engineers; and promoting the professional registration of industrial engineers. IIE's Society of Work Science (the result of merging the Work Measurement and Ergonomics Divisions in 1994) keeps the membership up to date on all facets of this area of work. This society annually gives the Phil Carroll Award and M. M. Ayoub Award for achievement in work measurement and ergonomics, respectively.

In the area of work design, the first professional organization, the Ergonomics Research Society, was founded in the United Kingdom in 1949. It started the first professional journal, *Ergonomics,* in 1957. The U.S. professional organization The Human Factors and Ergonomics Society was founded in 1957. In the 1960s, there was rapid growth in the society, with membership increasing from 500 to 3,000. Currently, there are well over 5,000 members organized in 20 different technical groups. Their primary goals are to (1) define and support human factors/ergonomics as a scientific discipline and in practice, with the exchange of technical information among members; (2) educate and inform business, industry, and government about human factors/ergonomics; and (3) promote human factors/ergonomics as a means for bettering the quality of life. The society also publishes an archival journal, *Human Factors,* and holds annual conferences where members can meet and exchange ideas.

With the proliferation of national professional societies, an umbrella organization, the International Ergonomics Association, was founded in 1959 to coordinate ergonomic activities at an international level. At present, there are 42 individual societies encompassing over 15,000 members worldwide.

PRESENT TRENDS

Practitioners of methods, standards, and work design have come to realize that such factors as gender, age, health and well-being, physical size and strength, aptitude, training attitudes, job satisfaction, and motivation response have a direct bearing on productivity. Furthermore, present-day analysts recognize that workers object, and rightfully so, to being treated as machines. Workers dislike and fear a purely scientific approach and inherently dislike any change from their present way of operation. Even management frequently rejects worthwhile methods innovations because of a reluctance to change.

Workers tend to fear methods and time study, for they see that the results are an increase in productivity. To them, this means less work and consequently less pay. They must be sold on the fact that they, as consumers, benefit from lower costs, and that broader markets result from lower costs, meaning more work for more people for more weeks of the year.

Some fears of time study today are due to unpleasant experiences with efficiency experts. To many workers, motion and time study is synonymous with the speedup of work and the use of incentives to spur employees to higher levels of output. If the new levels established were normal production, the workers were forced to still greater exertions to maintain their previous earning power. In the past, shortsighted and unscrupulous managers did resort to this practice.

Even today, some unions oppose the establishment of standards by measurement, the development of hourly base rates by job evaluation, and the application of incentive wage payment. These unions believe that the time allowed to perform a task and the amount that an employee should be paid represent issues that should be resolved by collective bargaining arrangements.

Today's practitioners must use the "humane" approach. They must be well versed in the study of human behavior and accomplished in the art of communication. They must also be good listeners, respecting the ideas and thinking of others, particularly the worker at the bench. They must give credit where credit is due. In fact, they should habitually give the other person credit, even if there is some question of that person deserving it. Also, practitioners of motion and time study should always remember to use the questioning attitude emphasized by the Gilbreths, Taylor, and the other pioneers in the field. The idea that there is "always a better way" needs to be continually pursued in the development of new methods that improve productivity, quality, delivery, worker safety, and worker well-being.

Today, there is a greater intrusion by the government in the regulation of methods, standards, and work design. For example, military equipment contractors and subcontractors are under increased pressure to document direct labor standards as a result of MIL-STD 1567A (released 1975; revised 1983 and

1987). Any firm awarded a contract exceeding $1 million is subject to MIL-STD 1567A, which requires a work measurement plan and procedures, a plan to establish and maintain engineered standards of known accuracy and traceability, a plan for methods improvement in conjunction with standards, a plan for the use of the standards as an input to budgeting, estimating, planning, and performance evaluation, and detailed documentation for all these plans. However, this requirement was cancelled in 1995.

Similarly, in the area of work design, Congress passed the OSHAct establishing the National Institute for Occupational Safety and Health (NIOSH), a research agency for developing guidelines and standards for worker health and safety, and the Occupational Safety and Health Administration (OSHA), an enforcement agency to maintain these standards. With the sudden increase in repetitive-motion injuries in the food processing industry, OSHA established the Ergonomics Program Management Guidelines for Meatpacking Plants in 1990. Similar guidelines for general industry slowly evolved into a final OSHA Ergonomics Standard, signed into law by President Clinton in 2001. However, the measure was rescinded soon afterward by Congress.

With increasing numbers of individuals with different abilities, Congress passed the Americans with Disabilities Act (ADA) in 1990. This regulation has a major impact on all employers with 15 or more employees, affecting such employment practices as recruiting, hiring, promotions, training, laying off, firing, allowing leaves, and assigning jobs.

While work measurement once concentrated on direct labor, methods and standard development have increasingly been used for indirect labor. This trend will continue as the number of traditional manufacturing jobs decreases and the number of service jobs increases in the United States. The use of computerized techniques will also continue to grow, for example, predetermined time systems such as MOST. Many companies have also developed time study and work sampling software, using electronic data collectors for compiling the information required.

Table 1.1 illustrates the progress made in methods, standards, and work design.

SUMMARY

Industry, business, and government are in agreement that the untapped potential for increasing productivity is the best hope for dealing with inflation and competition. The principal key to increased productivity is a continuing application of the principles of methods, standards, and work design. Only in this way can greater output from people and machines be realized. The U.S. government has pledged itself to an increasingly paternalistic philosophy of providing for the disadvantaged—housing for the poor, medical care for the aged, jobs for minorities, and so on. To accommodate the spiraling costs of labor and government taxes and still stay in business, we must get more from our productive elements—people and machines.

Table 1.1 Progress Made in Connection with Methods, Standards, and Work Design

Year	Event
1760	Perronet makes time studies on No. 6 common pins.
1820	Charles W. Babbage makes time studies on No. 11 common pins.
1832	Charles W. Babbage publishes *On the Economy of Machinery and Manufactures.*
1881	Frederick W. Taylor begins his time study work.
1901	Henry L. Gantt develops the task and bonus wage system.
1903	Taylor presents paper on shop management to ASME.
1906	Taylor publishes paper *On the Art of Cutting Metals.*
1910	Interstate Commerce Commission starts an investigation of time study. Gilbreth publishes *Motion Study.* Gantt publishes *Work, Wages, and Profits.*
1911	Taylor publishes text *The Principles of Scientific Management.*
1912	Society to Promote the Science of Management is organized. Emerson estimates $1 million per day can be saved if eastern railroads apply scientific management.
1913	Emerson publishes *The Twelve Principles of Efficiency.* Congress adds rider to government appropriation bill stipulating that no part of this appropriation should be made available for the pay of any person engaged in time study work. Henry Ford unveils the first moving assembly line in Detroit.
1915	Taylor Society is formed to replace the Society to Promote the Science of Management.
1917	Frank B. and Lillian M. Gilbreth publish *Applied Motion Study.*
1923	American Management Association is formed.
1927	Elton Mayo begins Hawthorne study at Western Electric Company's plant in Hawthorne, IL.
1933	Ralph M. Barnes receives the first Ph.D. granted in the United States in the field of industrial engineering from Cornell University. His thesis leads to the publication of "Motion and Time Study."
1936	Society for the Advancement of Management is organized.
1945	Department of Labor advocates establishing standards to improve productivity of supplies for the war effort.
1947	Bill is passed allowing the War Department to use time study.
1948	The Institute of Industrial Engineers is founded in Columbus, Ohio. Eiji Toyoda and Taichi Ohno at Toyota Motor Company pioneer the concept of lean production.
1949	Prohibition against using stopwatches is dropped from appropriation language. The Ergonomics Research Society (now The Ergonomics Society) is founded in the United Kingdom.
1957	The Human Factors and Ergonomics Society is founded in the United States. E. J. McCormick publishes *Human Factors Engineering.*
1959	International Ergonomics Association is founded to coordinate ergonomics activities worldwide.
1970	Congress passes the OSHAct, establishing the Occupational Safety and Health Administration.
1972	Society for the Advancement of Management combines with the American Management Association.
1975	MIL-STD 1567 (USAF), Work Measurement, is released.
1981	NIOSH lifting guidelines are first introduced.
1986	MIL-STD 1567A, Work Measurement Guidance Appendix, is finalized.
1988	ANSI/HFS Standard 100-1988 for Human Factors Engineering of Visual Display Terminal Workstations is released.

Table 1.1 (*continued*)

Year	Event
1990	Americans with Disabilities Act (ADA) is passed by Congress. Ergonomics Program Management Guidelines for Meatpacking Plants are established by OSHA. This serves as a model for developing an OSHA ergonomics standard.
1991	NIOSH lifting guidelines are revised.
1995	Draft ANSI Z-365 Standard for Control of Work-Related Cumulative Trauma Disorders is released.
1995	MIL-STD 1567A Work Measurement is canceled.
2001	OSHA Ergonomics Standard signed into law but rescinded soon afterward by Congress.
2006	50th Anniversary of the Human Factors and Ergonomics Society.

QUESTIONS

1. What is another name for *time study?*
2. What is the principal objective of methods engineering?
3. List the eight steps in applying methods engineering.
4. Where were time studies originally made and who conducted them?
5. Explain Frederick W. Taylor's principles of scientific management.
6. What is meant by motion study, and who are the founders of the motion-study technique?
7. Was the skepticism of management and labor toward rates established by "efficiency experts" understandable? Why?
8. Which organizations are concerned with advancing the ideas of Taylor and the Gilbreths?
9. What psychological reaction is characteristic of workers when methods changes are suggested?
10. Explain the importance of the humanistic approach in methods and time study work.
11. How are time study and methods engineering related?
12. Why is work design an important element of methods study?
13. What important events have contributed to the need for ergonomics?

REFERENCES

Barnes, Ralph M. *Motion and Time Study: Design and Measurement of Work.* 7th ed. New York: John Wiley & Sons, 1980.

Eastman Kodak Co., Human Factors Section. *Ergonomic Design for People at Work.* New York: Van Nostrand Reinhold, 1983.

Gilbreth, F., and L. Gilbreth. *Cheaper by the Dozen.* New York: T. W. Crowell, 1948.

Gilbreth, L. M. *As I Remember: An Autobiography.* Norcross, GA: Engineering & Management Press, 1988.

Kanigel, R. *One Best Way.* New York: Viking, 1997.

Konz, S., and S. Johnson. *Work Design.* 5th ed. Scottsdale, AZ: Holcomb Hathaway, 2000.

Mundell, Marvin E. *Motion and Time Study: Improving Productivity.* 5th ed. Englewood Cliffs, NJ: Prentice-Hall, 1978.

Nadler, Gerald. "The Role and Scope of Industrial Engineering." In *Handbook of Industrial Engineering,* 2d ed. Ed. Gavriel Salvendy. New York: John Wiley & Sons, 1992.

Niebel, Benjamin W. *A History of Industrial Engineering at Penn State.* University Park, PA: University Press, 1992.

Salvendy, G., ed. *Handbook of Human Factors.* New York: John Wiley & Sons, 1987.

Saunders, Byron W. "The Industrial Engineering Profession." In *Handbook of Industrial Engineering.* Ed. Gavriel Salvendy. New York: John Wiley & Sons, 1982.

Taylor, F. W. *The Principles of Scientific Management.* New York: Harper, 1911.

Thompson, C. Bertrand. *The Taylor System of Scientific Management.* Chicago: A. W. Shaw, 1917.

United Nations Industrial Development Organization. *Industry in the 1980s: Structural Change and Interdependence.* New York: United Nations, 1985.

WEBSITES

The Ergonomics Society—http://www.ergonomics.org.uk/
Human Factors and Ergonomics Society—http://hfes.org/
Institute of Industrial Engineers—http://www.iienet.org/
International Ergonomics Association—http://www.iea.cc/
OSHA—http://www.osha.gov/

Problem-Solving Tools

KEY POINTS

- *Select the project* with the exploratory tools: Pareto analyses, fish diagrams, Gantt charts, PERT charts, and job/worksite analysis guides.

- *Get and present data* with the recording tools: operation, flow, worker/machine, and gang process charts, and flow diagrams.

- *Develop the ideal method* with quantitative tools: worker/machine relationships with synchronous and random servicing and line balancing calculations.

A good methods engineering program will follow an orderly process, starting from the selection of the project and ending with the implementation of the project (see Figure 1.3). The first, and perhaps most crucial, step—whether designing a new work center or improving an existing operation—is the identification of the problem in a clear and logical form. Just as the machinist uses tools such as micrometers and calipers to facilitate performance, so the methods engineer uses appropriate tools to do a better job in a shorter time. A variety of such problem-solving tools are available, and each tool has specific applications.

The first five tools are primarily used in the first step of methods analysis, *select the project*. Pareto analysis and fish diagrams evolved from Japanese quality circles of the early 1960s (see Chapter 18) and were quite successful in improving quality and reducing costs in their manufacturing processes. Gantt and PERT charts emerged during the 1940s in response to a need for better project planning and control of complex military projects. They can also be very useful in identifying problems in an industrial setting.

Typically, project selection is based on three considerations: economic (probably the most important), technical, and human. Economic considerations may involve new products, for which standards have not been implemented, or existing products that have a high cost of manufacturing. Problems could be large amounts of scrap or rework, excessive material handling, in terms of either cost or distance, or simply "bottleneck" operations. Technical considerations may include processing techniques that need to be improved, quality control problems due to the method, or product performance problems compared to the competition. Human considerations may involve highly repetitive jobs, leading to work-related musculoskeletal injuries, high-accident-rate jobs, excessively fatiguing jobs, or jobs about which workers constantly complain.

The first four exploratory tools are most typically used in the analyst's office. The fifth tool, job/worksite analysis guide, helps identify problems within a particular area, department, or worksite and is best developed as part of a physical walk-through and on-site observations. The guide provides a subjective identification of key worker, task, environmental, or administrative factors that may cause potential problems. It also indicates appropriate tools for further, more quantitative evaluations. Use of the job/worksite analysis guide should be a necessary first step before extensive quantitative data are collected on the present method.

The next five tools are used to record the present method, and they comprise the second step of methods analysis, *get and present the data*. Pertinent factual information—such as the production quantity, delivery schedules, operational times, facilities, machine capacities, special materials, and special tools—may have an important bearing on the solution of the problem, and such information needs to be recorded. (The data are also useful in the third step of methods analysis, *analyze the data*.)

The final three tools are more useful as a quantitative approach in the fourth step of methods analysis, *develop the ideal method*. Once the facts are presented clearly and accurately, they are examined critically, so that the most practical, economical, and effective method can be defined and installed. They should therefore be used in conjunction with the operational analysis techniques described in Chapter 3. Note that most of the tools from all four groupings can easily be utilized in the operational analysis phase of development.

2.1 EXPLORATORY TOOLS

PARETO ANALYSIS

Problem areas can be defined by a technique developed by the economist Vilfredo Pareto to explain the concentration of wealth. In *Pareto analysis*, items of interest are identified and measured on a common scale and then are ordered in descending order, as a cumulative distribution. Typically 20 percent of the ranked items account for 80 percent or more of the total activity; consequently, the technique is sometimes called the *80-20 rule*. For example, 80 percent of the total inventory is found in only 20 percent of the inventory items, or 20 percent of the jobs account for approximately 80 percent of the accidents (Figure 2.1), or 20 percent of the jobs account for 80 percent of the workers' compensation costs. Conceptually, the methods analyst concentrates the greatest effort on the few jobs that produce most of the problems. In many cases, the Pareto distribution can be transformed to a straight line using a lognormal transformation, from which further quantitative analyses can be performed (Herron, 1976).

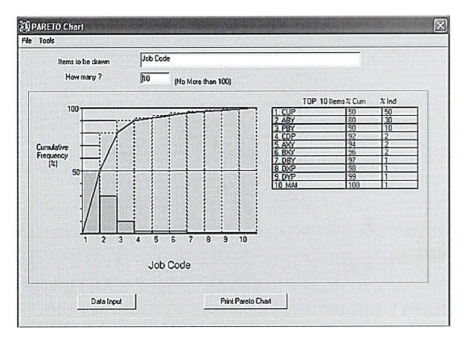

Figure 2.1 Pareto distribution of industrial accidents.
Twenty percent of job codes (CUP and ABY) cause approximately 80 percent of accidents.

FISH DIAGRAMS

Fish diagrams, also known as *cause-and-effect diagrams*, were developed by Ishikawa in the early 1950s while he was working on a quality control project for Kawasaki Steel Company. The method consists of defining an occurrence of a typically undesirable event or problem, that is, the *effect*, as the "fish head" and then identifying contributing factors, that is, the *causes*, as "fish bones" attached to a backbone and the fish head. The principal causes are typically subdivided into five or six major categories—the human, machines, methods, materials, environmental, administrative—each of which is further subdivided into subcauses. The process is continued until all possible causes are listed. A good diagram will have several levels of bones and will provide a very good overview of a problem and its contributing factors. The factors are then critically analyzed in terms of their probable contribution to the overall problem. Hopefully, this process will also tend to identify potential solutions. An example of a fish diagram used to identify operator health complaints in a cutoff operation is shown in Figure 2.2.

Fish diagrams have worked quite successfully in Japanese quality circles, where input is expected from all levels of workers and managers. Such diagrams may prove to be less successful in U.S. industry, where the cooperation between labor and management may be less effective in producing the desired solutions and outcomes (Cole, 1979).

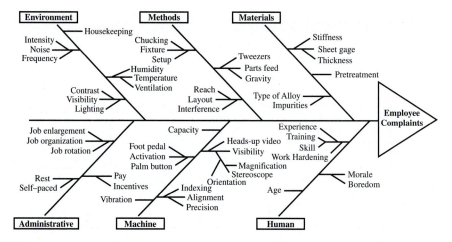

Figure 2.2 Fish diagram for operator health complaints in a cutoff operation.

GANTT CHART

The *Gantt chart* was probably the first project planning and control technique to emerge during the 1940s in response to the need to manage complex defense projects and systems better. A Gantt chart simply shows the anticipated completion times for various project activities as bars plotted against time on the horizontal axis (Figure 2.3a). Actual completion times are shown by shading the bars appropriately. If a vertical line is drawn through a given date, you can easily determine which project components are ahead of or behind schedule. For example, in Figure 2.3a, by the end of the third month, mock-up work is behind schedule. A Gantt chart forces the project planner to develop a plan ahead of time and provides a quick snapshot of the progress of the project at any given time. Unfortunately, it does not always completely describe the interaction between different project activities. More analytical techniques, such as PERT charts, are required for that purpose.

The Gantt chart can also be utilized for sequencing machine activity on the plant floor. The machine-based chart can include repair or maintenance activity by crossing out the time period in which the planned downtime will occur. For example, in the job shop in Figure 2.3b, in the middle of the month, lathe work is behind schedule, while production on the punch press is ahead of schedule.

PERT CHARTING

PERT stands for Program Evaluation and Review Technique. A *PERT chart,* also referred to as a network diagram or *critical path method,* is a planning and control tool that graphically portrays the optimum way to attain some predetermined objective, generally in terms of time. This technique was employed by the U.S.

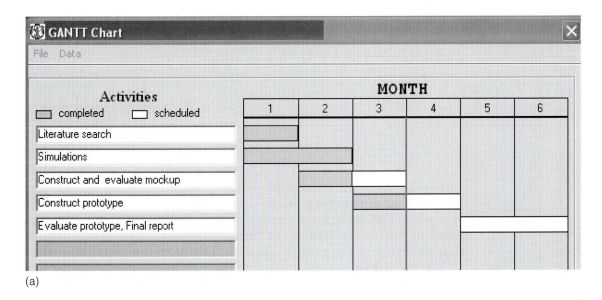

(a)

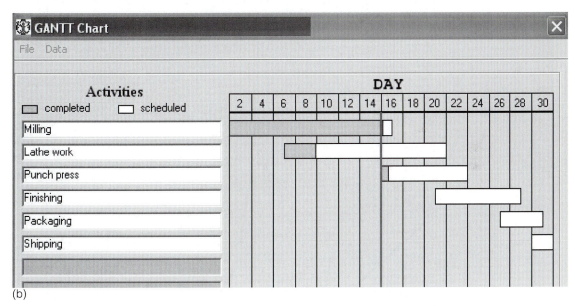

(b)

Figure 2.3 Example of (a) project-based Gantt chart and (b) machine- or process-based Gantt chart.

military in the design of such processes as the development of the Polaris missile and the operation of control systems in nuclear-powered submarines. Methods analysts usually use PERT charting to improve scheduling through cost reduction or customer satisfaction.

In using PERT for scheduling, analysts generally provide two or three time estimates for each activity. For example, if three time estimates are used, they are based on the following questions:

1. How much time is required to complete a specific activity if everything works out ideally (optimistic estimate)?
2. Under average conditions, what would be the most likely duration of this activity?
3. What is the time required to complete this activity if almost everything goes wrong (pessimistic estimate)?

With these estimates, the analyst can develop a probability distribution of the time required to perform the activity.

On a PERT chart, events (represented by nodes) are positions in time that show the start and completion of a particular operation or group of operations. Each operation or group of operations in a department is defined as an activity and is called an *arc*. Each arc has an attached number representing the time (days, weeks, months) needed to complete the activity. Activities that utilize no time or cost yet are necessary to maintain a correct sequence are called *dummy activities* and are shown as dotted lines (activity *H* in Fig 2.4.).

Dummy activities are typically used to indicate precedence or dependencies, because, under the rules, no two activities can be identified by the same nodes; that is, each activity has a unique set of nodes.

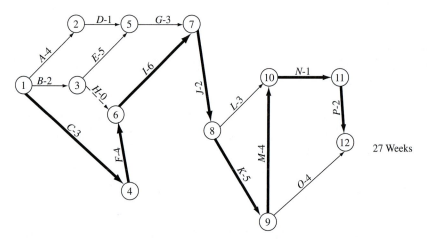

Figure 2.4   Network showing critical path (heavy line).
Circled numbers are nodes that represent the beginning and ending of activities which are represented as lines. The values above each line represent the normal duration of that activity in weeks.

The minimum time needed to complete the entire project corresponds to the longest path from the initial node to the final node. Termed the *critical path* in Figure 2.4, the minimum time needed to complete the project is the longest path from node 1 to node 12. While there is always one such path through any project, more than one path can reflect the minimum time needed to complete the project.

Activities not on the critical path have a certain time flexibility. This time flexibility, or freedom, is referred to as *float* and is defined as the amount of time that a noncritical activity can be lengthened without delaying the project's completion date. This implies that when the intent is to reduce the project completion time, termed *crashing*, it is better to concentrate on activities that lie on the critical path than on those on other pathways.

Although the critical path can be found through trial and error there is a formal procedure for uniquely finding the critical path using various time concepts. These are (1) the earliest starting (ES) for each activity such that all precedence relationships are upheld and (2) the earliest finish (EF) for that activity, which is the earliest start plus the estimated time for that activity, or

$$EF_{ij} = ES_{ij} + t_{ij}$$

where i and j are the nodes.

These times are typically found by a forward pass through the network, as shown in the network table in Table 2.1. Note that for an activity that has two predecessor activities, the earliest start is computed as the maximum of the previous earliest furnishes

$$ES_{ij} = \max (EF_{ij})$$

Similar to the earliest start and finish times are the latest start (LS) and latest finish (LF) times which are found through a backward pass through the

Table 2.1 Network Diagram

Activity	Nodes	ES	EF	LS	LF	Float
A	(1, 2)	0	4	5	9	4
B	(1, 3)	0	2	3	5	3
C	(1, 4)	0	3	0	3	0
D	(2, 5)	4	5	9	10	5
E	(3, 5)	2	7	5	10	3
F	(4, 6)	3	7	3	7	0
G	(5, 7)	7	10	10	13	3
H (Dummy)	(3, 6)	2	2	7	7	0
I	(6, 7)	7	13	7	13	0
J	(7, 8)	13	15	13	15	0
K	(8, 9)	15	20	15	20	0
L	(8, 10)	15	18	21	24	3
M	(9, 10)	20	24	20	24	0
N	(10, 11)	24	25	24	25	0
O	(9, 12)	20	24	23	27	3
P	(11, 12)	25	27	25	27	0

network. The latest start time is the latest time an activity can start without delaying the project. It is found by subtracting the activity time from the latest finish time.

$$LS_{ij} = LF_{ij} - t_{ij}$$

Where two or more activities emanate from one node, the latest finish time is the minimum of the latest start times of the emanating activities.

$$LF_{ij} = \min\ (LS_{ij})$$

The network table for the network diagram in Fig 2.5 is given in Table 2.1. Float is formally defined as

$$Float = LS - ES$$

or

$$Float = LF - EF$$

Note that all activities with float equal to zero define the critical path, which for this example, is 27 weeks.

Several methods can be used to shorten a project's duration, and the cost of various alternatives can be estimated. For example, Table 2.2 identifies the normal times and costs as well as crash times and crash costs that would occur in the project shown in Fig 2.4 were shortened. Using this table, and the network diagram, and assuming that a linear relationship exists between time and the cost per week, various alternatives shown in Table 2.3 can be computed.

Note that at 19 weeks, a second critical path is developed through nodes 1, 3, 5, and 7, and any further crashing would need to consider both paths.

JOB/WORKSITE ANALYSIS GUIDE

The *job/worksite analysis guide* (see Figure 2.5) identifies problems within a particular area, department, or worksite. Before collecting quantitative data, the

Table 2.2 Cost and Time Values to Perform a Variety of Activities under Normal and Crash Conditions

Activities	Nodes	Normal Weeks	Normal $	Crash Weeks	Crash $	Cost per Week $
A	(1, 2)	4	4,000	2	6,000	1,000
B	(1, 3)	2	1,200	1	2,500	1,300
C	(1, 4)	3	3,600	2	4,800	1,200
D	(2, 5)	1	1,000	0.5	1,800	1,600
E	(3, 5)	5	6,000	3	8,000	1,000
F	(4, 6)	4	3,200	3	5,000	1,800
G	(5, 7)	3	3,000	2	5,000	2,000
H	(3, 6)	0	0	0	0	—
I	(6, 7)	6	7,200	4	8,400	600
J	(7, 8)	2	1,600	1	2,000	400
K	(8, 9)	5	3,000	3	4,000	500
L	(8, 10)	3	3,000	2	4,000	1,000
M	(9, 10)	4	1,600	3	2,000	400
N	(10, 11)	1	700	1	700	—
O	(9, 12)	4	4,400	2	6,000	800
P	(11, 12)	2	1,600	1	2,400	800

Table 2.3 Times and Costs for Various Alternatives for the Network Shown in Figure 2.4 and Data Given in Table 2-2

Schedule (weeks)	Least Expensive Alternative	Total Weeks Gained	Total Added Cost ($)
27	Normal duration of project	0	0
26	Crash activity M (or J) by one week for an added cost of $400	1	400
25	Crash activity J (or M) by one week for an additional cost of $400	2	800
24	Crash activity K by one week for an additional cost of $500	3	1,300
23	Crash activity K by another week for an additional cost of $500	4	1,800
22	Crash activity I by one week for an additional cost of $600	5	2,400
21	Crash activity I by another week for an additional cost of $600	5	3,000
20	Crash activity P by one week for an additional cost of $800	7	3,800
19	Crash activity C by one week for an additional cost of $1,200	8	5,000

analyst first walks through the area and observes the worker, the task, the workplace, and the surrounding working environment. In addition, the analyst identifies any administrative factors that may affect the worker's behavior or performance. All these factors provide an overall perspective of the situation and help guide the analyst in using other, more quantitative tools for collecting and analyzing data. The example in Figure 2.5 shows the application of the job/worksite analysis guide to a hot-end operation in a television manufacturing facility. Key concerns include the lifting of heavy loads, heat stress, and noise exposure.

2.2 RECORDING AND ANALYSIS TOOLS

OPERATION PROCESS CHART

The *operation process chart* shows the chronological sequence of all operations, inspections, time allowances, and materials used in a manufacturing or business process, from the arrival of raw material to the packaging of the finished product. The chart depicts the entrance of all components and subassemblies to the main assembly. Just as a blueprint displays such design details as fits, tolerances, and specifications, the operation process chart gives manufacturing and business details at a glance.

Two symbols are used in constructing the operation process chart: a small circle denotes an operation, and a small square denotes an inspection. An operation

Job/Worksite Analysis Guide		
Job/Worksite: HOT END **Analyst:** A F **Date:** 1-27-		
Description: INSERTING STEM INTO FUNNEL		
Worker Factors		
Name: **Age:** 42 **Gender:** (M) F **Height:** 6' **Weight:** 180		
Motivation: High Medium (Low) **Job Satisfaction:** High Medium (Low)		
Education Level: Some HS (HS) College **Fitness Level:** High (Medium) Low		
Personal Protective Equipment: (Safety Glasses) Hard Hat Safety Shoes (Ear Plugs) Other GLOVES, SLEEVES		

Task Factors		**Refer To:**
What happens? How do parts flow in/out? FUNNEL FROM BELT TO INSERTING MACHINE, THEN SEALER THEN BACK TO BELT		*Flow Process Charts*
What kinds of motions are involved? REPETITIVE LIFTING, WALKING, GRASPING		*Video Analysis, Principles of Motion Economy*
Are there any jigs/fixtures? Automation? YES, TO POSITION FUNNEL + BASIC PROCESS, NONE FOR LIFTING		
Are any tools being used? NO		*Tool Evaluation Checklist*
Is the workplace laid out well? Any long reaches? SOME WALKING & REACHING		*Workstation Evaluation Checklist*
Are there awkward finger/wrist motions? How frequent? NO		*CTD Risk Index*
Is there any lifting? YES, HEAVY GLASS FUNNELS		*NIOSH Lifting Analysis, UM2D Model*
Is the worker fatigued? Physical workload? SOME , SOME		*Heart Rate Analysis, Work-rest Allowances*
Is there any sensory input, information processing, decision making, or mental workload? MINIMAL		*Cognitive Work Evaluation Checklist Display Design and GUI Checklists*
How long is each cycle? What is the standard time? ~ 1½ MIN		*Time Study, MTM-2 Checklist*

Work Environment Factors		***Work Environment Checklist***
Is the illumination acceptable? Is there glare? YES , NO		*IESNA Recommended Values*
Is the noise level acceptable? NO, EAR PLUGS REQUIRED		*OSHA Levels*
Is there heat stress? YES!.!		*WBGT*
Is there vibration? NO		*ISO Standards*

Administrative Factors		Remarks:
Are there wage incentives? NO		
Is there job rotation? Job enlargement? YES , NO		
Is training or work hardening provided? YES		
What are the overall management policies? ?		

Figure 2.5 Job/worksite analysis guide for a hot-end job in a television manufacturing facility.

takes place when a part being studied is intentionally transformed or when it is being studied or planned prior to productive work being performed on it. An inspection takes place when the part is being examined to determine its conformity to a standard. Note that some analysts prefer to outline only the operations, calling the result an *outline process chart.*

Before beginning the actual construction of the operation process chart, analysts identify the chart with a title, Operation Process Chart, and other information, such as the part number, drawing number, process description, present or proposed method, date, and name of the person doing the charting. Additional information may include such items as the chart number, plant, building, and department.

Vertical lines indicate the general flow of the process as work is accomplished, while horizontal lines feeding into the vertical flow lines indicate material, either purchased or worked on during the process. Parts are shown as entering a vertical line for assembly or leaving a vertical line for disassembly. Materials that are disassembled or extracted are represented by horizontal material lines drawn to the right of the vertical flow line, while assembly materials are shown as horizontal lines drawn to the left of the vertical flow line.

In general, the operation process chart is constructed such that vertical flow lines and horizontal material lines do not cross. If it becomes necessary to cross a vertical and a horizontal line, use conventional practice to show that no juncture occurs; that is, draw a small semicircle in the horizontal line at the point where the vertical line crosses it (see Figure 2.6).

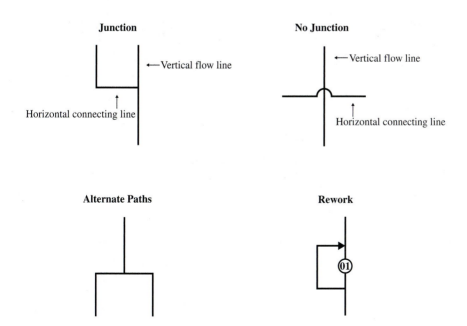

Figure 2.6 Flowcharting conventions.

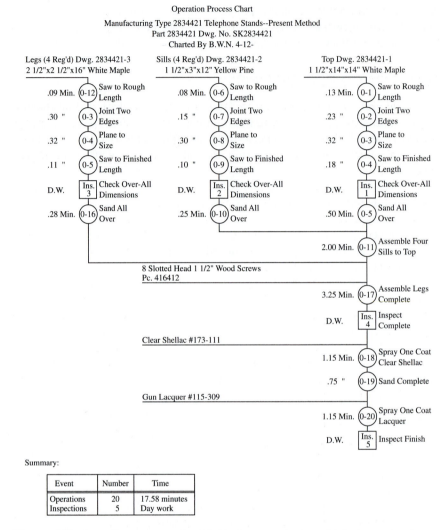

Operation Process Chart

Manufacturing Type 2834421 Telephone Stands--Present Method
Part 2834421 Dwg. No. SK2834421
Charted By B.W.N. 4-12-

Figure 2.7 Operation process chart illustrating manufacture of telephone stands.

Time values, based on either estimates or actual measurements, may be assigned to each operation and inspection. A typical completed operation process chart illustrating the manufacture of telephone stands is shown in Figure 2.7.

The completed operation process chart helps analysts visualize the present method, with all its details, so that new and better procedures may be devised. It shows analysts what effect a change on a given operation will have on the preceding and subsequent operations. It is not unusual to realize a 30 percent reduction in performance time by utilizing the principles of operations analysis (see Chapter 3) in conjunction with the operation process chart, which inevitably suggests possibilities for improvement. Also since each step is shown in its proper chronological sequence, the chart in itself is an ideal plant layout. Consequently, methods analysts find this tool extremely helpful in developing new layouts and improving existing ones.

FLOW PROCESS CHART

In general, the *flow process chart* contains considerably greater detail than the operation process chart. Consequently, it is not usually applied to entire assemblies but rather for each component of an assembly. The flow process chart is especially valuable in recording nonproduction hidden costs, such as distances traveled, delays, and temporary storages. Once these nonproduction periods are highlighted, analysts can take steps to minimize them and hence their costs.

In addition to recording operations and inspections, flow process charts show all the moves and storage delays encountered by an item as it goes through the plant. Flow process charts therefore need several symbols in addition to the operation and inspection symbols used in operation process charts. A small arrow signifies transportation, which can be defined as moving an object from one place to another, except when the movement takes place during the normal course of an operation or inspection. A large capital D indicates a delay, which occurs when a part is not immediately permitted to be processed at the next workstation. An equilateral triangle standing on its vertex signifies a storage, which occurs when a part is held and protected against unauthorized removal. These five symbols (see Figure 2.8) are the standard set of process chart symbols (ASME, 1974). Several other nonstandard symbols may sometimes be utilized for clerical or paperwork operations and for combined operations, as shown in Figure 2.9.

Two types of flowcharts are currently in general use: product or material (see Figure 2.10, preparation of direct mail advertising) and operative or person (see Figure 2.11, service personnel inspecting LUX field units). The product chart provides the details of the events involving a product or a material, and the operative flowchart details how a person performs an operational sequence.

Like the operation process chart, the flow process chart is identified by a title, Flow Process Chart, and accompanying information that usually includes the part number, drawing number, process description, present or proposed method, date, and name of the person doing the charting. Additional data that may be valuable for completely identifying the job being charted include the plant, building, or department; chart number; quantity; and cost.

For each event in the process, the analyst writes a description of the event, circles the appropriate process chart symbol, and indicates the times for processes or delays and distances for transportations. The analyst then connects succeeding event symbols with a vertical line. The right-hand column provides space for the analyst to enter comments or make recommendations for potential changes.

To determine the distance moved, the analyst need not measure each move accurately with a tape or a 6-ft rule. A sufficiently correct figure usually results by counting the number of columns that the material moves past and then multiplying

Figure 2.8   The ASME standard set of process chart symbols.

this number, less 1, by the span. Moves of 5 ft or less are usually not recorded; however, they may be if the analyst feels that they materially affect the overall cost of the method being plotted.

All delay and storage times must be included on the chart. The longer a part stays in storage or is delayed, the more cost it accumulates and the longer the customer must wait for delivery. It is therefore important to know how much time a part spends at each delay or storage. The most economical method of determining the duration of delays and storages is to mark several parts with chalk, indicating the exact time they went into a storage or were delayed. Then check the section periodically to see when the marked parts are brought back into production. By taking a number of cases, recording the elapsed time, and then averaging the results, analysts can obtain sufficiently accurate time values.

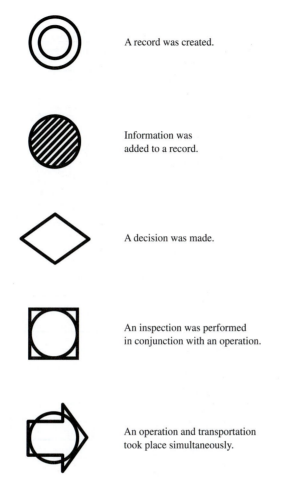

A record was created.

Information was
added to a record.

A decision was made.

An inspection was performed
in conjunction with an operation.

An operation and transportation
took place simultaneously.

Figure 2.9 Nonstandard process chart symbols.

The flow process chart, like the operation process chart, is not an end in and
of itself; it is merely a means to an end. This tool facilitates the elimination or
reduction of the hidden costs of a component. Since the flowchart clearly shows
all transportations, delays, and storages, the information it provides can lead to a
reduction of both the quantity and duration of these elements. Also, since distances are recorded on the flow process chart, the chart is exceptionally valuable
in showing how the layout of a plant can be improved. These techniques are
described in greater detail in Chapter 3.

FLOW DIAGRAM

Although the flow process chart gives most of the pertinent information related
to a manufacturing process, it does not show a pictorial plan of the flow of work.
Sometimes this information is helpful in developing a new method. For example,

Flow Process Chart   Page 1 of 1

Location: Dorben Ad Agency		Summary			
Activity: Preparing Direct Mail Ads		Event	Present	Proposed	Savings
Date: 1-26-98		Operation	4		
Operator: J.S.	Analyst: A.F.	Transport	4		
Circle appropriate Method and Type:		Delay	4		
Method: (Present) Proposed		Inspection	0		
Type: Worker (Material) Machine		Storage	2		
Remarks:		Time (min)			
		Distance (ft)	340		
		Cost			

Event Description	Symbol		Time (In Minutes)	Distance (In Feet)	Method Recommendation
stock room	○ ◇ D □ ▽				
to collating room	○ ◇ D □ ▽			100	
collating rack by type	○ ◇ D □ ▽				
collate 4 sheets	○ ◇ D □ ▽				
stack	○ ◇ D □ ▽				
to folding room	○ ◇ D □ ▽			20	
jog, fold, crease	○ ◇ D □ ▽				
stack	○ ◇ D □ ▽				
to angle stapler	○ ◇ D □ ▽			20	
staple	○ ◇ D □ ▽				
stack	○ ◇ D □ ▽				
to mail room	○ ◇ D □ ▽			200	
addressing	○ ◇ D □ ▽				
mailbag	○ ◇ D □ ▽				
	○ ◇ D □ ▽				
	○ ◇ D □ ▽				
	○ ◇ D □ ▽				
	○ ◇ D □ ▽				
	○ ◇ D □ ▽				

(handwritten note: OPERATIONS IN ORDER)

Figure 2.10   Flow process chart (material) for preparation of direct mail advertising.

before a transportation can be shortened, the analyst needs to see or visualize where room can be made to add a facility so that the transportation distance can be shortened. Likewise, it is helpful to visualize potential temporary and permanent storage areas, inspection stations, and work points.

The best way to provide this information is to take an existing drawing of the plant areas involved and then sketch in the flow lines, indicating the movement of the material from one activity to the next. A pictorial representation of the layout of floors and buildings, showing the locations of all activities on the flow

Flow Process Chart

Location: Dorben Co.	Summary			
Activity: Field Inspection of LUX	**Event**	**Present**	**Proposed**	**Savings**
Date: 4-17-97	Operation	7		
Operator: T.Smith — Analyst: R. Ruhf	Transport	6		
Circle appropriate Method and Type:	Delay	2		
Method: (Present) Proposed	Inspection	6		
Type: (Worker) Material Machine	Storage	0		
Remarks:	Time (min)	32.60		
	Distance (ft)	375		
	Cost			

Event Description	Symbol	Time (In Minutes)	Distance (In Feet)	Method Recommendation
Leave vehicle, walk to front door, ring bell.	O ⊘ D □ ▽	1.00	75	Call home in advance to reduce waiting delays.
Wait, enter home.	O ⊘ D □ ▽			
Walk to field reservoir.	O ⊘ D □ ▽	.25	25	
Disconnect field reservoir from unit.	⊘ ◇ D □ ▽	.35		
Inspect for dents, cracks in shroud, cracked glass or missing hardware.	O ◇ D □ ▽	1.25		This can be done while walking back to vehicle.
Clean unit with approved cleaner and disinfectant.	⊘ ◇ D □ ▽	2.25		This can be done more effectively at vehicle.
Return to vehicle with empty tank.	O ⊘ D □ ▽	1.00	75	
Unlock vehicle, place empty tank in fixture and connect hardware.	⊘ ◇ D □ ▽	1.75		
Open valve; begin fill.	⊘ ◇ D □ ▽	.25		
Wait for tank to fill.	O ◇ D □ ▽	12.00		Clean unit while being filled.
Check humidifier for proper function.	O ◇ D □ ▽	.5		Eliminate. No need to do this twice.
Check pressure (indicator).	O ◇ D □ ▽	.2		
Check reservoir contents (indicator).	O ◇ D □ ▽	.2		
Return to patient with filled tank.	O ⊘ D □ ▽	1.10	100	
Hook up filled tank.	⊘ ◇ D □ ▽	1.00		
Check humidifier for proper function.	O ◇ D □ ▽	.75		
Wait for patient to remove nasal cannula or face mask.	O ◇ D □ ▽	2.00		
Install new nasal cannula or face mask.	⊘ ◇ D □ ▽	2.50		
Check flows with patient.	O ◇ D □ ▽	2.25		
Affix a dated, initialed inspection sticker.	⊘ ◇ D □ ▽	1.00		Perform this while unit being filled.
Return to vehicle.	O ⊘ D □ ▽	1.00	100	

Figure 2.11 Flow process chart (worker) for field inspection of LUX.

process chart, is a *flow diagram*. When constructing a flow diagram, analysts identify each activity by symbols and numbers corresponding to those appearing on the flow process chart. The direction of flow is indicated by placing small arrows periodically along the flow lines. Different colors can be used to indicate flow lines for more than one part.

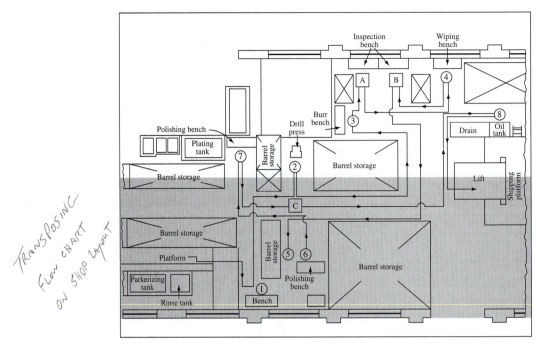

Figure 2.12   Flow diagram of the old layout of a group of operations on the Garand rifle.

(Shaded section of plant represents the total floor space needed for the revised layout [Figure 2.13]. This represented a 40 percent savings in floor space.)

Figure 2.12 illustrates a flow diagram made in conjunction with a flow process chart to improve the production of the Garand (M1) rifle at Springfield Armory. This pictorial representation, together with the flow process chart, resulted in savings that increased production from 500 rifle barrels per shift to 3,600—with the same number of employees. Figure 2.13 illustrates the flow diagram of the revised layout.

The flow diagram is a helpful supplement to the flow process chart because it indicates backtracking and possible traffic congestion areas, and it facilitates developing an ideal plant layout.

WORKER AND MACHINE PROCESS CHARTS

The *worker and machine process chart* is used to study, analyze, and improve one workstation at a time. The chart shows the exact time relationship between the working cycle of the person and the operating cycle of the machine. These facts can lead to a fuller utilization of both worker and machine time, and a better balance of the work cycle.

Many machine tools are either completely automatic (the automatic screw machine) or semiautomatic (the turret lathe). With these types of facilities, the

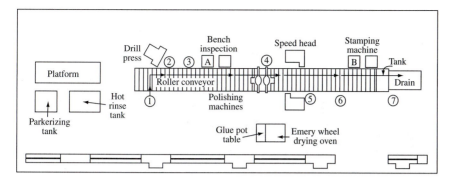

Figure 2.13 Flow diagram of the revised layout of a group of operations on the M1 Garand rifle.

Specific work station vs entire shop

operator is often idle for a portion of the cycle. The utilization of this idle time can increase operator earnings and improve production efficiency.

The practice of having one employee operate more than one machine is known as *machine coupling*. Because organized labor may resist this concept, the best way to sell machine coupling is to demonstrate the opportunity for added earnings. Since machine coupling increases the percentage of "effort time" during the operating cycle, greater incentive earnings are possible if a company is on an incentive wage payment plan. Also, higher base rates result when machine coupling is practiced, since the operator has greater responsibility and can exercise greater mental and physical effort.

When constructing the worker and machine process chart, the analyst must first identify the chart with a title such as Worker and Machine Process Chart. Additional identifying information would include the part number, drawing number, operation description, present or proposed method, date, and name of the person doing the charting.

Since workers and machine charts are always drawn to scale, the analyst selects a distance in inches to conform with a unit of time such that the chart can be neatly arranged. The longer the cycle time of the operation being charted, the shorter the distance per decimal minute of time. Once exact values have been established for the distance, in inches per unit of time, the chart is begun. The left side shows the operations and time for the worker, and the right side shows the working time and the idle time of the machine or machines. A solid line drawn vertically represents the employee's working time. A break in the vertical work–time line signifies idle time. Likewise, a solid vertical line under each machine heading indicates machine operating time, and a break in the vertical machine line designates idle machine time. A dotted line under the machine column indicates loading and unloading machine time, during which the machine is neither idle nor productive (see Figure 2.14).

The analyst charts all elements of occupied and idle time for both the worker and the machine through the termination of the cycle. The bottom of the chart

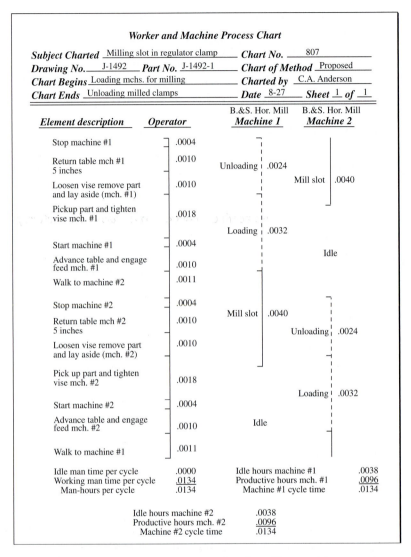

Figure 2.14   Worker and machine process chart for milling machine operation.

shows the employee's total working time and total idle time, as well as the total working time and idle time of each machine. The productive time plus the idle time of the worker must equal the productive time plus the idle time of each machine that the worker operates.

Accurate elemental time values are necessary before the worker and machine chart can be constructed. These time values should represent standard times that include an acceptable allowance for fatigue, unavoidable delays, and personal delays (see Chapter 11 for more details). The analyst should never use overall stopwatch readings in the construction of the chart.

The completed worker and machine process chart clearly shows the areas in which both idle machine time and idle worker time occur. These areas are generally a good place to start in effecting improvements. However, the analyst must also compare the cost of the idle machine with that of the idle worker. It is only when total cost is considered that the analyst can safely recommend one method over another. Economical considerations are presented in the next section.

GANG PROCESS CHARTS

The *gang process chart* is, in a sense, an adaptation of the worker and machine chart. A worker and machine process chart helps determine the most economical number of machines one worker can operate. However, several processes and facilities are of such magnitude that instead of one worker operating several machines, several workers are needed to operate one machine effectively. The gang process chart shows the exact relationship between the idle and operating cycles of the machine and the idle and operating times per cycle of the workers who service that machine. This chart reveals the possibilities for improvement by reducing both idle operator time and idle machine time.

Figure 2.15 illustrates a gang process chart for a process in which a large number of idle work-hours exist, up to 18.4 h per 8-h shift. The chart also shows that the company is employing two more operators than are needed. By relocating some of the controls of the process, the company was able to reassign the elements of work so that four, rather than six, workers could effectively operate the extrusion press. A better operation of the same process is shown on the gang process chart in Figure 2.16. The savings of 16 h per shift was easily developed through the use of this chart.

2.3 QUANTITATIVE TOOLS, WORKER AND MACHINE RELATIONSHIPS

Although the worker and machine process chart can illustrate the number of facilities that can be assigned to an operator, this can often be computed in much less time through the development of a mathematical model. A worker and machine relationship is usually one of three types: (1) synchronous servicing, (2) completely random servicing, and (3) a combination of synchronous and random servicing.

SYNCHRONOUS SERVICING

Assigning more than one machine to an operator seldom results in the ideal case where both the worker and the machine are occupied during the whole cycle. Such ideal cases are referred to as *synchronous servicing,* and the number of machines to be assigned can be computed as

$$n = \frac{l + m}{l}$$

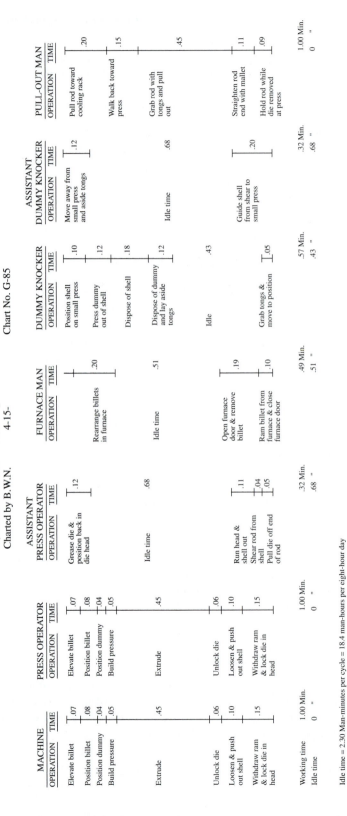

Figure 2.15 Gang process chart of the present method of operation of a hydraulic extrusion process.

GANG PROCESS CHART—PROPOSED METHOD

Hydraulic Extrusion Press Dept. II 　　　　　Bellefonte, Pa. Plant

Charted by B.W.N 　　　　4-15 　　　　　　Chart G-85

MACHINE OPERATION	TIME	MACHINE OPERATION	TIME	ASSISTANT PRESS OPERATOR OPERATION	TIME	DUMMY KNOCKER OPERATION	TIME	PULL-OUT MAN OPERATION	TIME
Elevate billet	.07	Elevate billet	.07	Grease die & position back in die head	.12	Position shell on small press	.10	Pull rod toward cooling rack	.20
				Walk to furnace	.05	Press dummy out of shell	.12		
Position billet	.08	Position billet	.08					Walk back toward press	.15
Position dummy	.04	Position dummy	.04	Rearrange billets in furnace	.20	Dispose of shell	.18		
Build pressure	.05	Build pressure	.05						
Extrude	.45	Extrude	.45	Return to press	.05	Dispose of dummy and lay aside tongs	.12	Grab rod with tongs and pull out	.45
				Idle time	.09				
				Open furnace door & remove billet	.19	Idle time	.23		
				Ram billet from furnace & close furnace door	.10				
Unlock die	.06	Unlock die	.06			Grab tongs & move to position	.05		
Loosen & push out shell	.10	Loosen & push out shell	.10	Run head & shell	.11	Guide shell from shear to small press	.20	Straighten rod end with mallet	.11
Withdraw ram & lock die in head	.15	Withdraw ram & lock die in head	.15	Shear rod from shell	.04			Hold rod while die removed at press	.09
				Pull die off end of rod	.05				

Working time 1.00 Min. 　　1.00 Min. 　　　.91 Min. 　　　.77 Min. 　　　1.00 Min.
Idle time 　　0 　　　　　　0 　　　　　.09 Min. 　　　.23 Min. 　　　　0

Figure 2.16 Gang process chart of the proposed method of operation of a hydraulic extrusion process.

where n = number of machines the operator is assigned

　l = total operator loading and unloading (servicing) time per machine

　m = total machine running time (automatic power feed)

For example, assume a total cycle time of 4 min to produce a product, as measured from the start of the unloading of the previously completed product to the end of the machine cycle time. Operator servicing, which includes both the unloading of the completed product and the loading of the raw materials, is 1 min, while the cycle time of the automatic machine cycle is 3 min. Synchronous servicing would result in the assignment of

$$n = \frac{1 + 3}{1} = 4 \text{ machines}$$

Graphically, this assignment would appear as shown in Figure 2.17, with the operator moving to the second machine once the first machine is serviced. By the time the fourth machine is serviced, the operator would need to return to the first machine to service it, since that machine's automatic cycle would have just ended.

If the number of machines in this example is increased, machine interference takes place, and we have a situation in which one or more of the facilities sit idle for a portion of the work cycle. If the number of machines is reduced to

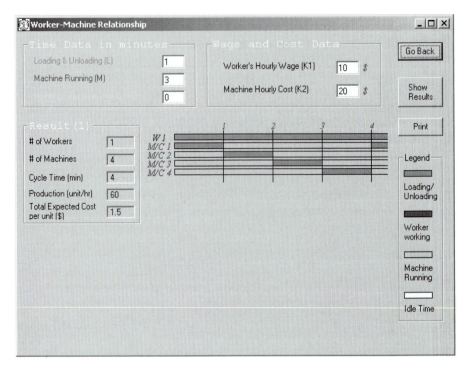

Figure 2.17 Synchronous servicing assignment for one operator and four machines.

some figure less than 4, then the operator is idle for a portion of the cycle. In such cases, the minimum total cost per piece usually represents the criterion for optimum operation.

An additional complication occurs because of less than ideal conditions. The operator may need to walk between machines or clean and adjust the machines. This worker time also needs to be accounted for based on the cost of each idle machine and the hourly rate of the operator.

The number of machines that the operator should be assigned under realistic conditions can be reestimated by the lowest whole number from the revised equation:

$$n_1 \leq \frac{l + m}{l + w}$$

where n_1 = lowest whole number

w = total worker time (not directly interacting with the machine, typically walking time to the next machine)

The cycle time with the operator servicing n_1 machines is $l + m$, since in this case, the operator is not busy the whole cycle, yet the facilities are occupied during the entire cycle.

Using n_1, we can compute the total expected cost (TEC) as follows:

$$\text{TEC}_{n_1} = \frac{K_1(l + m) + n_1 K_2(l + m)}{n_1} \tag{1}$$

$$= \frac{(l + m)(K_1 + n_1 K_2)}{n_1}$$

where   TEC = total expected cost in dollars per unit of production from one machine

      K_1 = operator rate, in dollars per unit of time

      K_2 = cost of machine, in dollars per unit of time

After this cost is computed, a cost should be calculated with $n_1 + 1$ machines assigned to the operator. In this case, the cycle time is governed by the working cycle of the operator, since there is some idle machine time. The cycle time is now $(n_1 + 1)(l + w)$. Let $n_2 = n_1 + 1$. Then the total expected cost with n_2 facilities is

$$\text{TEC}_{n_2} = \frac{(K_1)(n_2)(l + w) + (K_2)(n_2)(n_2)(l + w)}{n_2} \tag{2}$$

$$= (l + w)(K_1 + n_2 K_2)$$

The number of machines assigned depends on whether n_1 or n_2 gives the lowest total expected cost per piece.

Synchronous Servicing   EXAMPLE 2.1

It takes an operator 1 min to service a machine and 0.1 min to walk to the next machine. Each machine runs automatically for 3 min, the operator earns $10.00/h and the machines cost $20.00/h to run. How many machines can the operator service?

The optimum number of machines that the operator can service is

$$n = (l + m)/(l + w) = (1 + 3)/(1 + 0.1) = 3.6$$

The number being fractional, leaves two choices. The operator can be assigned 3 machines (option 1), in which case the operator will be idle some portion of the time. Or the operator can be assigned 4 machines (option 2), in which case the machines will be idle some portion of the time. The best choice may be based on the economics of the situation, that is, the lowest cost per unit.

In option 1, the expected production cost from Equation (1) is (divided by 60 so as to convert hours to minutes)

$$\text{TEC}_3 = (l + m)(K_1 + n_1 K_2)/n_1 = (1 + 3)(10 + 3 \times 20)(3/60) = \$1.556/\text{unit}$$

An alternate approach is to calculate the production rate R per hour:

$$R = \frac{60}{l + m} \times n_1$$

The production rate is based on the machines being the limiting factor (i.e., the worker is idle at times) and the machines producing on one unit per machine per 4.0-min total cycle (1.0-min service time, 3.0. min of machine time). With 3 machines running 60 min/h, the production rate is

$$R = \frac{60}{1 + 3} \times 3 = 45 \text{ units/h}$$

The expected cost is then the cost of the labor and machines divided by the production rate:

$$TEC_3 = (K_1 + n_1K_2)/R = (10 + 3 \times 20)/45 = \$1.556/\text{unit}$$

In option 2, the expected production cost from Equation (2) is

$$TEC_4 = (1 + w)(K_1 + n_2K_2) = (1 + 0.1)(10 + 4 \times 20)/60 = \$1.65/\text{unit}$$

In the alternate approach, the production rate is based on the worker being the limiting factor (i.e., the machines are idle at times). Since the worker can produce one unit per 1.1-min cycle (1.0. min service time and 0.1-min walk time), the production rate (R) per hour for the alternate approach is

$$R = \frac{60}{1 + w} = \frac{60}{1.1} = 54.54 \text{ units/h}$$

The expected cost is then the cost of the labor and machines divided by the production rate:

$$TEC_4 = (K_1 + n_1K_2)/R = (10 + 4 \times 20)/54.54 = \$1.65/\text{unit}$$

Based on lowest cost, the setup with three machines is best. However, if there is market demand at a good sales price, profits can be maximized using a four-machine setup. Note also that in this example, with a walk time of 0.1 min, production decreases from the ideal of 60 units/min (see figure on page 48).

Note the effect of reducing loading/unloading time from 1. min to 0.9 min, a relatively small amount. The optimum number of machines that the operator can service is now

$$n = (l + m)/(l + w) = (0.9 + 3)/(0.9 + 0.1) = 3.9$$

Although the number is still fractional, it is very close to 4, the realistic amount. If the operator is assigned three machines (option 1), operator will be idle a greater portion of the time, increasing from 0.7 to 0.9 min or to almost 25% of the time. The expected production cost from Equation (1) is (with 60 needed to convert hours to minutes)

$$TEC_3 = (l + m)(K_1 + n_1K_2)/n_1 = (0.9 + 3)(10 + 3 \times 20)(3/60)$$

$$= \$1.517/\text{unit}$$

The alternate approach yields a production rate

$$R = \frac{60}{l + m} \times n_1 = \frac{60}{3.9} \times 3 = 46.15 \text{ units/h}$$

The expected cost is the cost of the labor and machines divided by the production rate:

$$TEC_3 = (K_1 + n_1K_2)/R = (10 + 3 \times 20)/46.5 = \$1.517/\text{unit}$$

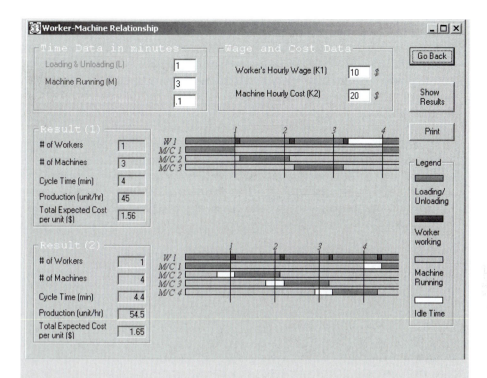

If the worker is assigned the more realistic number of 4 machines (option 2), the more costly machine idle time decreases from 0.4 to 0.1 min. The expected production cost from Equation (2) is

$$\text{TEC}_4 = (l + w)(K_1 + n_2K_2) = (0.9 + 0.1)(10 + 4 \times 20)/60 = \$1.50/\text{unit}$$

The alternate approach yields a production rate R per hour of

$$R = \frac{60}{l + w} = \frac{60}{1.0} = 60 \text{ units/h}$$

The expected cost is the cost of the labor and machines divided by the production rate:

$$\text{TEC}_4 = (K_1 + n_1K_2)/R = (10 + 4 \times 20)/60 = \$1.50/\text{unit}$$

Based on lowest cost and minimum idle time, the setup with 4 machines is now best. Note that a 10% decrease in loading/unloading time (from 1 to 0.9 min) yielded several positive improvements:

- A 10% increase in production (60 compared to 54.54 units/h)
- A reduction of idle time from 0.7 min for the operator (17.5% of cycle time) in the first scenario to 0.1 min for the machines in the second scenario
- A 3.6% decrease in unit costs from $1.556 to $1.50 per unit

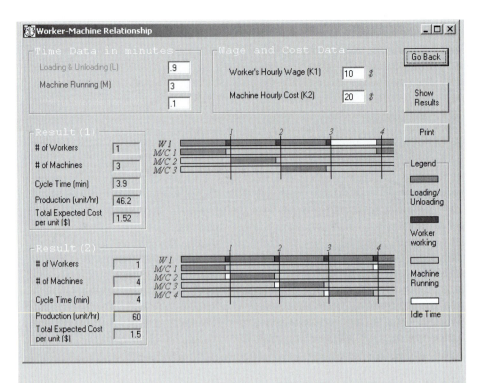

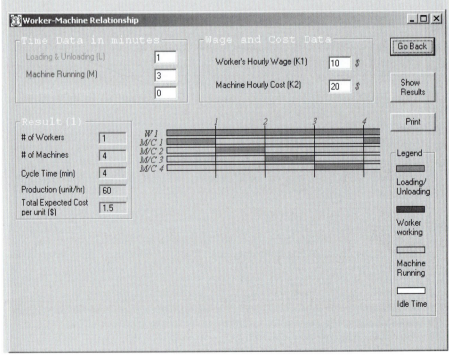

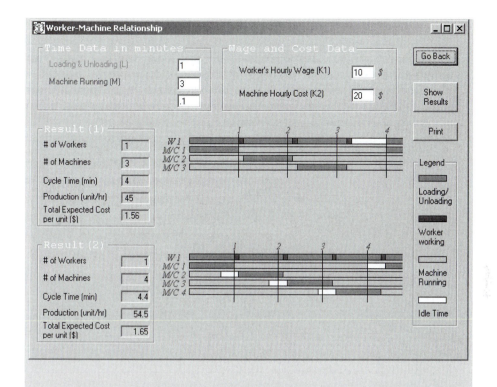

This demonstrates the importance of decreasing loading or setup time, to be discussed in greater detail in Chapter 3. Note also that decreasing the walk time by a comparable amount (0.1 min which in this case eliminates it completely) results in the ideal case shown below or in Figure 2.17 with the same unit cost of $1.50.

RANDOM SERVICING

Completely *random servicing* situations are those cases in which it is not known when a facility needs to be serviced or how long servicing takes. Mean values are usually known or can be determined; with these averages, the laws of probability can provide a useful tool in determining the number of machines to assign a single operator.

The successive terms of the binomial expansion give a useful approximation of the probability of 0, 1, 2, 3, . . ., n machines down (where n is relatively small), assuming that each machine is down at random times during the day and that the probability of downtime is p and the probability of runtime is $q = 1 - p$. Each term of the binomial expansion can be expressed as a probability of m (out of n) machines down:

$$P(m \ of \ n) = \frac{n!}{m!(n-m)!}p^m q^{n-m}$$

As an example, let us determine the minimum proportion of machine time lost for various numbers of turret lathes assigned to an operator where the average machine runs unattended 60 percent of the time. Operator attention time (machine is down or requires servicing) at irregular intervals is 40 percent on average. The analyst estimates that three turret lathes should be assigned per operator on this class of work. Under this arrangement, the probabilities of m (out of n) machines down would be as follows:

Machines down m	Probability
0	$\dfrac{3!}{0!(3-0)!}(0.4^0)\,(0.6^3) = (1)(1)(0.216) = 0.216$
1	$\dfrac{3!}{1!(3-1)!}(0.4^1)\,(0.6^2) = (3)\,(0.4)\,(0.36) = 0.432$
2	$\dfrac{3!}{2!(3-2)!}(0.4^2)\,(0.6^1) = (3)(0.16)(0.6) = 0.288$
3	$\dfrac{3!}{3!(3-3)!}(0.4^3)\,(0.6^0) = (1)\,(0.064)\,(1) = 0.064$

By using this approach, the proportion of time that some machines are down may be determined, and the resulting lost time of one operator per three machines may be readily computed. In this example, we have the following:

No. of machines down	Probability	Machine hours lost per 8-h day
0	0.216	0
1	0.432	0*
2	0.288	$(0.288)(8) = 2.304$
3	0.064	$(2)(0.064)(8) = 1.024$
	1.000	3.328

*Since only one machine is down at a time, the operator can be attending the down machine.

$$\text{Proportion of machine time lost} = \frac{3.328}{24.0} = 13.9 \text{ percent}$$

Similar computations can be made for more or less machine assignments to determine the assignment resulting in the least machine downtime. The most satisfactory assignment is usually the arrangement showing the least total expected cost per piece, while the total expected cost per piece for a given arrangement is computed by the expression

$$\text{TEC} = \frac{K_1 + nK_2}{R}$$

where K_1 = hourly rate of the operator
$\quad\quad K_2$ = hourly rate of the machine
$\quad\quad n$ = number of machines assigned
$\quad\quad R$ = rate of production, pieces from n machines per hour

The rate of production, in pieces per hour, from n machines is computed with the mean machine time required per piece, the average machine servicing time per piece, and the expected downtime or lost time per hour.

For example, under a five-machine assignment to one operator, an analyst determined that the machining time per piece was 0.82 h, the machine servicing time per piece was 0.17 h, and the machine downtime was an average of 0.11 h per machine per hour. Thus, each machine was available for production work only 0.89h each hour. The average time required to produce one piece per machine would be

$$\frac{0.82 + 0.17}{0.89} = 1.11$$

Therefore, the five machines would produce 4.5 pieces per hour. With an operator hourly rate of $12 and a machine hourly rate of $22, we have a total expected cost per piece of

$$\frac{\$12.00 + 5(\$22.00)}{4.5} = \$27.11$$

Random servicing | EXAMPLE 2.2

An operator is assigned to service three machines that have an expected downtime of 40 percent. When running, each machine can produce 60 units/h. The operator is paid $10.00/h, and a machine costs $60.00/h to run. Is it worth hiring another operator to keep the machines running?

Case A - One operator

Machines down m	Probability	Machine hours lost per 8-h day
0	$\frac{3!}{0!\,3!}(0.4)^0(0.6)^3 = 0.216$	0
1	$\frac{3!}{1!\,2!}(0.4)^1(0.6)^2 = 0.432$	0
2	$\frac{3!}{2!\,1!}(0.4)^2(0.6)^1 = 0.288$	$0.288 \times 8 = 2.304$
3	$\frac{3!}{3!\,0!}(0.4)^3(0.6)^0 = 0.064$	$0.064 \times 16 = 1.024$

Considering that a total of 3.328 production hours (2.304 + 1.024) are lost in an 8-h day, only 1,240.3 units (20.672 × 60) can be produced at an hourly average of 155.04. The unit cost is

$$\text{TEC} = (10 + 3 \times 60)/155.04 = \$1.23/\text{unit}$$

Case B - Two operators

Machines down m	Probability	Machine hours lost per 8-h day
0	$\dfrac{3!}{0!\ 3!}(0.4)^0(0.6)^3 = 0.216$	0
1	$\dfrac{3!}{1!\ 2!}(0.4)^1(0.6)^2 = 0.432$	0
2	$\dfrac{3!}{2!\ 1!}(0.4)^2(0.6)^1 = 0.288$	0
3	$\dfrac{3!}{3!\ 0!}(0.4)^3(0.6)^0 = 0.064$	$0.064 \times 8 = 0.512$

There is a considerable improvement from case A. Since only 0.512 production hour is lost in an 8-h day, production increases to 1,409.28 units (23.488 × 60) or an hourly average of 176.16. The unit cost is

$$\text{TEC} = (2 \times 10 + 3 \times 60)/176.16 = \$1.14/\text{unit}$$

Therefore it is more cost-efficient to hire another operator and keep the machines running.

Note that hiring a third operator to keep all three machines running all the time would not be cost-efficient. Although the total production increases marginally, the total cost increases more, and the unit cost becomes

$$\text{TEC} = (3 \times 10 + 3 \times 60)/180 = \$1.17/\text{unit}$$

COMPLEX RELATIONSHIPS

Combinations of synchronous and random servicing are perhaps the most common type of worker and machine relationships. Here the servicing time is relatively constant, but the machines are serviced randomly. Furthermore, the time between breakdowns is assumed to have a particular distribution. As the number of machines increases and the relationship between the operator and machines becomes more complex, machine interference and consequent delay times increase. In practice, machine interference predominantly occurs from 10 to 30 percent of the total working time, with extremes of up to 50 percent. Various approaches have been developed to deal with such situations.

One such approach assumes an expected workload for the operator based on the number of machines assigned and mean machine running times and mean

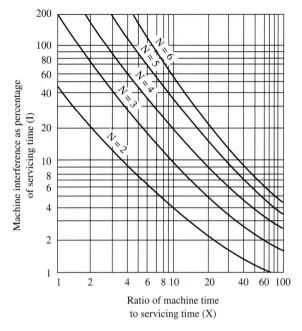

Figure 2.18
Machine interference as a percentage of servicing time when the number of machines assigned to one operator is six or less.

servicing times. For up to six machines, the use of the empirical curves illustrated in Figure 2.18 is recommended.

For seven or more machines, *Wright's formula* (Wright, Duvall, and Freeman, 1936) can be used:

$$I = 50\{\sqrt{[(1 + X - N)^2 + 2N]} - (1 + X - N)\}$$

where I = interference, expressed as a percentage of the mean servicing time
  X = ratio of mean machine running time to mean machine servicing time
  N = number of machine units assigned to one operator

An application of this formula is shown in Example 2.3.

Calculation of Machine Interference Time   **EXAMPLE 2.3**

In a quilling production, an operator is assigned 60 spindles. The mean machine running time per package, determined by stopwatch study, is 150 min. The standard mean servicing time per package, also developed by time study, is 3 min. The computation of the machine interference, expressed as a percentage of the mean operator attention time, is

$$I = 50\{\sqrt{[(1 + X - N)^2 + 2N]} - (1 + X - N)\}$$

$$= 50\left[\sqrt{\left(1 + \ +\frac{150}{3.00} - 60\right)^2 + 120} - \left(1 + \frac{150}{3.00} - 60\right)\right]$$

$$I = 50\left[\sqrt{(1 + 50 - 60^2) + 120} - (1 + 50 - 60)\right]$$

$$I = 1.159\%$$

Thus, we would have

Machine running time	150.0 min
Servicing time	3.0 min
Machine interference time	$11.6 \times 3.0 = 34.8$ min

Using queuing theory with the time between breakdowns assumed to have an exponential distribution, Ashcroft (1950) extended the above approach and developed tables to determine machine interference times. These are shown in Table A3-13 (Appendix 3) and provide values of machine running time and machine interference time for values of the service ratio k:

$$k = l/m$$

where l = servicing time
         m = machine running time

The total cycle time to produce one piece is

$$c = m + l + i$$

where c = total cycle time
         i = machine interference time

Note that the values of machine running time and machine interference time in Table A3-13 are given as a percentage of total cycle time. Also, any walking or worker time w should be included as part of servicing time. Example 2.4 demonstrates *Ashcroft's method* for calculating machine interference time.

EXAMPLE 2.4	Calculation of Machine Interference, Using Ashcroft's Method

With reference to Example 2.3:

$$k = l/m = 3/150 = 0.02$$

$$N = 60$$

From Table A3–13, Appendix 3, with exponential service time and $k = 0.02$ and $N = 60$, we have a machine interference time of 16.8 percent of the cycle time. We have $T_i = 0.168c$, where c is the cycle time to produce one unit per spindle. Then

$$c = m + l + i$$

$$c = 150 + 3.00 + 0.168c$$

$$0.832c = 153$$

$$c = 184 \text{ min}$$

and

$$T_i = 0.168c = 30.9 \text{ min}$$

The interference time computed by the equation (34.8 min, Example 2.3) closely agrees with that developed here by the queuing model. However, as n (the number of machines assigned) becomes smaller, the proportional difference between the two techniques increases.

LINE BALANCING

The problem of determining the ideal number of workers to be assigned to a production line is analogous to that of determining the number of workers to be assigned to a workstation; the gang process chart solves both problems. Perhaps the most elementary *line balancing* situation, yet one that is very often encountered, is one in which several operators, each performing consecutive operations, work as a unit. In such a situation, the rate of production is dependent on the slowest operator. For example, we may have a line of five operators assembling bonded rubber mountings prior to the curing process. The specific work assignments might be as follows: Operator 1, 0.52 min; operator 2, 0.48 min; operator 3, 0.65 min; operator 4, 0.41 min; operator 5, 0.55 min. Operator 3 establishes the pace, as is evidenced by the following:

Operator	Standard minutes to perform operation	Wait time based on slowest operator	Standard time (min)
1	0.52	0.13	0.65
2	0.48	0.17	0.65
3	0.65	—	0.65
4	0.41	0.24	0.65
5	0.55	0.10	0.65
Totals	2.61		3.25

The efficiency of this line can be computed as the ratio of the total actual standard minutes to the total allowed standard minutes, or

$$E = \frac{\sum_{1}^{5} SM}{\sum_{1}^{5} AM} \times 100 = \frac{2.61}{3.25} \times 100 = 80\%$$

where E = efficiency

　　SM = standard minutes per operation

　　AM = allowed standard minutes per operation

Details on standard times will be covered later in Chapter 9.

Some analysts prefer to consider percent idle time (%Idle):

$$\% \text{Idle} = 100 - E = 20\%$$

In a real-life situation similar to this example, the opportunity for significant savings exists. If an analyst can save 0.10 min on operator 3, the net savings per cycle is not 0.10 min, but 0.10×5, or 0.50, min.

Only in the most unusual situations would a line be perfectly balanced; that is, the standard minutes to perform an operation would be identical for each member of the team. The "standard minutes to perform an operation" is not really a standard. It is only a standard to the individual who established it. Thus, in our example, where operator 3 has a standard time of 0.65 min to perform the first operation, a different work measurement analyst might have allowed as little as 0.61 min, or as much as 0.69 min. The range of standards established by different work measurement analysts on the same operation might be even greater than the range suggested. The point is that whether the issued standard is 0.61, 0.65, or 0.69, the typical conscientious operator should have little difficulty in meeting the standard. In fact, the operator will probably better the standard in view of the performance of the operators on the line with less work content in their assignments. Those operators who have a wait time based on the output of the slowest operator are seldom observed as actually waiting. Instead, they reduce the tempo of their movements to utilize the number of standard minutes established by the slowest operator.

The number of operators needed for the required rate of production can be estimated by

$$N = R \times \Sigma \text{AM} = R \times \frac{\Sigma \text{SM}}{E}$$

where N = number of operators needed in the line

　　R = desired rate of production

For example, assume that we have a new design for which we are establishing an assembly line. Eight distinct operations are involved. The line must produce 700 units per day (or $700/480 = 1.458$ units/min), and since it is desirable to minimize storage, we do not want to produce many more than 700 units/day. The eight operations involve the following standard minutes based on existing standard data: Operation 1, 1.25 min; operation 2, 1.38 min; operation 3, 2.58 min; operation 4, 3.84 min; operation 5, 1.27 min; operation 6, 1.29 min; operation 7, 2.48 min; and operation 8, 1.28 min. To plan this assembly line for the most economical setup, we estimate the number of operators required for a given level of efficiency (ideally, 100 percent), as follows:

$$N = 1.458 \times (1.25 + 1.38 + 2.58 + 3.84 + 1.27 + 1.29 + 2.48$$

$$+ 1.28)/1.00 = 22.4$$

For a more realistic 95 percent efficiency, the number of operators becomes $22.4/0.95 = 23.6$.

Since it is impossible to have six-tenths of an operator, you would endeavor to set up the line utilizing 24 operators. An alternate approach would be to utilize part-time hourly workers.

Next, we estimate the number of operators to be utilized at each of the eight specific operations. Since 700 units of work are required a day, it will be necessary to produce 1 unit in about 0.685 min (480/700). We estimate the number of operators needed on each operation by dividing the number of minutes allowable to produce one piece into the standard minutes for each operation, as follows:

Operation	Standard minutes	Standard minutes Min/unit	No. of operators
Operation 1	1.25	1.83	2
Operation 2	1.38	2.02	2
Operation 3	2.58	3.77	4
Operation 4	3.84	5.62	6
Operation 5	1.27	1.86	2
Operation 6	1.29	1.88	2
Operation 7	2.48	3.62	4
Operation 8	1.28	1.87	2
Total	15.37		24

To identify the slowest operation, we divide the estimated number of operators into the standard minutes for each of the eight operations. The results are shown in the following table.

Operation 1	$1.25/2 = 0.625$
Operation 2	$1.38/2 = 0.690$
Operation 3	$2.58/4 = 0.645$
Operation 4	$3.84/6 = 0.640$
Operation 5	$1.27/2 = 0.635$
Operation 6	$1.29/2 = 0.645$
Operation 7	$2.48/4 = 0.620$
Operation 8	$1.28/2 = 0.640$

Thus, operation 2 determines the output from the line. In this case, it is

$$\frac{2 \text{ workers} \times 60 \text{ min}}{1.38 \text{ standard min}} = 87 \text{ pieces, or } 696 \text{ pieces/day}$$

If this rate of production is inadequate, we would need to increase the rate of production of operator 2. This can be accomplished by

1. Working one or both of the operators at the second operation overtime, thus accumulating a small inventory at this workstation.
2. Utilizing the services of a third part-time worker at the workstation of operation 2.
3. Reassigning some of the work of operation 2 to operation 1 or operation 3. (It would be preferable to assign more work to operation 1.)
4. Improving the method at operation 2 to diminish the cycle time of this operation.

In the preceding example, given a cycle time and operation times, an analyst can determine the number of operators needed for each operation to meet a desired production schedule. The production line work assignment problem can also be to minimize the number of workstations, given the desired cycle time; or, given the number of workstations, assign work elements to the workstations, within the restrictions established, to minimize the cycle time.

An important strategy in assembly line balancing is work element sharing. Two or more operators whose work cycle includes some idle time might share the work of another station, to make the entire line more efficient. For example, Figure 2.19 shows an assembly line involving six workstations. Station 1 has three work elements—A, B, and C—for a total of 45 seconds (s). Note that work elements B, D, and E cannot begin until A is completed and that B, D, and E can occur in any order. It may be possible to share element H between stations 2 and

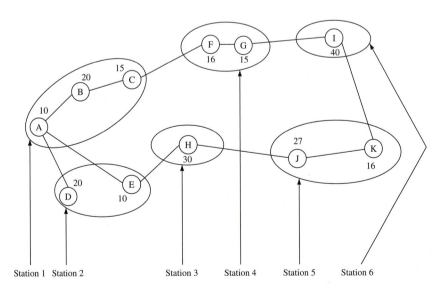

Figure 2.19   Assembly line involving six workstations.

4, with only a 1-s increase in cycle time (from 45 to 46 s), while saving 30 s per assembled unit. We should note that element sharing may result in an increase in material handling, since parts may have to be delivered to more than one location. In addition, element sharing may necessitate added costs for duplicate tooling.

A second possibility for improving the balance of an assembly line involves dividing a work element. Referring again to Figure 2.19, it may be possible to divide element H, rather than have one-half of the parts go to station 2 and the other half to station 4.

Many times, it is not economical to divide an element. An example would be driving home eight machine screws with a power screwdriver. Once the operator has located the part in a fixture, gained control of the power tool, and brought the tool to the work, it would usually be more advantageous to drive home all eight screws, rather than only a portion of them, leaving the rest for a different operator. Whenever elements can be divided, workstations may be better balanced as a result of the division.

The following procedure for solving an assembly line balancing problem is based on Helgeson and Birnie (1961). The method assumes the following:

1. Operators are not able to move from one workstation to another to help maintain a uniform workload.
2. The work elements that have been established are of such magnitude that further division would substantially decrease the efficiency of performing the work element. (Once established, the work elements should be identified with a code.)

The first step in the solution of the problem is to determine the sequence of individual work elements. The fewer the restrictions on the order in which the work elements can be done, the greater the probability that a favorable balance in the work assignments will be achieved. To determine the sequence of the work elements, the analyst determines the answer to the following question: What other work elements, if any, must be completed before this work element can be started? This question is applied to each element to establish a precedence chart for the production line under study (see Figure 2.20). Functional design, available production methods, floor space, and so on can all introduce constraints with respect to work element sequence.

A second consideration in the production line work assignment problem is zoning restraints. A zone represents a subdivision that may or may not be physically separated or identified from other zones in the system. Confining certain work elements to a given zone may be justified, to congregate similar jobs, working conditions, or pay rates. Or zoning restraints may help to identify physically specific stages of a component, such as keeping it in a certain position while performing certain work elements. As an example, all work elements related to one side of a component might be performed in a certain zone before the component is allowed to be turned over.

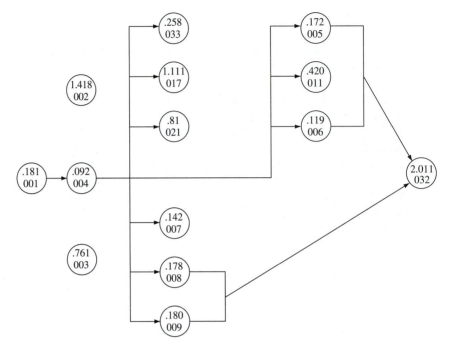

Figure 2.20   Partially completed precedence chart.
Note that work elements 002 and 003 may be done in any sequence with respect to any
of the other work elements and that 032 cannot be started until 005, 006, 008, and 009
have been completed. Note also that after 004 has been finished, we may start 033, 017,
021, 005, 011, 006, 007, 008, or 009.

Obviously, the more zoning restraints placed on the system, the fewer the
combinational possibilities available for investigation. The analyst begins by
making a sketch of the system and coding the applicable zones. Within each
zone, the work elements that may be done in that area are shown. The analyst
then estimates the production rate, using the expression

$$\text{Production per day} = \frac{\text{working min/day}}{\text{cycle time of system (min/unit)}}$$

where the cycle time of the system is the standard time of the limiting zone or
station.

Consider another assembly line with the following, the precedence chart:

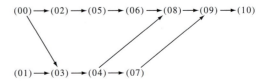

Estimated work unit time (minutes)	Work unit	Work unit										
		00	01	02	03	04	05	06	07	08	09	10
0.46	00			1	1	1	1	1	1	1	1	1
0.35	01				1	1			1	1	1	1
0.25	02						1	1		1	1	1
0.22	03					1			1	1	1	1
1.10	04								1	1	1	1
0.87	05							1		1	1	1
0.28	06									1	1	1
0.72	07										1	1
1.32	08										1	1
0.49	09											1
0.55	10											
6.61												

Figure 2.21 A precedence matrix used for a line balancing problem.

This precedence graph shows that work unit (00) must be completed before (02), (03), (05), (06), (04), (07), (08), (09), and (10); and work unit (01) must be completed before (03), (04), (07), (08), (09), and (10). Either (00) or (01) can be done first, or they can be done concurrently. Also, work unit (03) cannot be started until work units (00) and (01) are completed, and so on.

To describe these relationships, the precedence matrix illustrated in Figure 2.21 is established. Here the numeral 1 signifies a "must precede" relationship. For example, work unit (00) must precede work units (02), (03), (04), (05), (06), (07), (08), (09), and (10). Also, work unit (09) must precede only work unit (10).

Now, a *positional weight* must be computed for each work unit. This is done by computing the summation of the given work unit and all the work units that must follow it. Thus, the positional weight for work unit (00) would be

$$\Sigma\,00, 02, 03, 04, 05, 06, 07, 08, 09, 10$$
$$= 0.46 + 0.25 + 0.22 + 1.10 + 0.87 + 0.28 + 0.72 + 1.32$$
$$\quad + 0.49 + 0.55$$
$$= 6.26$$

Listing the positional weights in decreasing order of magnitude gives the following:

Unsorted work elements	Sorted work elements	Positional weight	Immediate predecessors
00	00	6.26	—
01	01	4.75	—
02	03	4.40	(00), (01)
03	04	4.18	(03)
04	02	3.76	(00)
05	05	3.51	(02)
06	06	2.64	(05)
07	08	2.36	(04), (06)
08	07	1.76	(04)
09	09	1.04	(07), (08)
10	10	0.55	(09)

Work elements must then be assigned to various workstations. This process is based on the positional weights (i.e., those work elements with the highest positional weights are assigned first) and the cycle time of the system. The work element with the highest positional weight is assigned to the first workstation. The unassigned time for this workstation is determined by subtracting the sum of the assigned work element times from the estimated cycle time. If there is adequate unassigned time, the work element with the next highest-positional weight is assigned, provided that the work elements in the "immediate predecessors" column have already been assigned. Once a workstation's allotted time has been filled, the analyst moves on to the next workstation, and the procedure continues until all the work elements have been assigned.

As an example, assume that the required production per 450-min shift is 300 units. The cycle time of the system is 450/300 = 1.50 min, and the final balanced line is shown in Table 2.4.

Under the arrangement illustrated, with six workstations, we have a cycle time of 1.32 min (workstation 4). This arrangement produces 450/1.32 = 341 units, which more than meets the daily requirement of 300.

However, with six workstations, we also have considerable idle time. The idle time per cycle is

$$\sum_{1}^{6} 0.04 + 0.22 + 0.17 + 0 + 0.11 + 0.77 = 1.31 \text{ min}$$

For more favorable balancing, the problem can be solved for cycle times of less than 1.50 min. This may result in more operators and greater production per day, which may have to be stored. Another possibility involves operating the line under a more efficient balancing for a limited number of hours per day.

A variety of commercially available software packages, as well as Design Tools, eliminate the drudgery of the calculations and perform these steps automatically.

Table 2–2 Balanced Assembly Line

Station	Work Element	Positional weight	Immediate predecessors	Work element time	Station time Cumulative	Station time Unassigned	Remarks*
1	00	6.26	—	0.46	.46	1.04	—
1	01	4.75	—	0.35	.81	0.69	—
1	03	4.40	(00), (01)	0.22	1.03	0.47	—
1	04	4.18	(03)	1.10	2.13	—	N.A.
1	02	3.76	(00)	0.25	1.28	0.22	—
1	05	3.56	(02)	0.87	2.05	—	N.A.
2	04	4.18	(03)	1.10	1.10	0.40	—
2	05	3.56	(02)	0.87	1.97	—	N.A.
3	05	3.56	(02)	0.87	0.87	0.63	—
3	06	2.64	(05)	0.28	1.15	0.35	—
3	08	2.36	(04), (06)	1.32	2.47	—	N.A.
4	08	2.36	(04), (06)	1.32	1.32	0.18	—
4	07	1.76	(04)	0.72	2.04	—	N.A.
5	07	1.76	(04)	0.72	0.72	0.78	—
5	09	1.04	(07), (08)	0.49	1.21	0.29	—
5	10	0.55	(09)	0.55	1.76	—	N.A.
6	10	0.55	(09)	0.55	0.55	0.95	—

*N.A. means not acceptable.

SUMMARY

The various charts presented in this chapter are valuable tools for presenting and solving problems. Just as several types of tools are available for a particular job, so several chart designs can help solve an engineering problem. Analysts should understand the specific functions of each process chart and choose the appropriate one for solving a specific problem and improving operations.

Pareto analyses and fish diagrams are used to select a critical operation and to identify the root causes and contributing factors leading to the problem. Gantt and PERT charts are project scheduling tools. The Gantt chart provides only a good overview, and the PERT chart quantifies the interactions between different activities. The job/worksite analysis guide is primarily used on a physical walk-through to identify key worker, task, environmental, and administrative factors that may cause potential problems. The operation process chart provides a good overview of the relationships between different operations and inspections on assemblies involving several components. The flow process chart provides more details for the analysis of manufacturing operations, to find hidden or indirect costs, such as delay time, storage costs, and material handling costs. The flow diagram is a useful supplement to the flow process chart in developing plant layouts. The worker/machine and gang process charts show machines or facilities in conjunction with the operator or operators, and are used to analyze idle operator time and idle machine time. Synchronous and random servicing calculations and line balancing techniques are used to develop more efficient operations through quantitative methods.

These 13 tools are very important for methods analysts. The charts are valuable descriptive and communicative aids for understanding a process and its related activities. Their correct use can aid in presenting and solving the problem, and in selling and installing the solution. Quantitative techniques can determine the optimum arrangement of operators and machines. Analysts should be acquainted with sufficient algebra and probability theory to develop a mathematical model that provides the best solution to the machine or facility problem. Thus, they are effective in presenting improved methods to management, training employees in the prescribed method, and focusing pertinent details, in conjunction with plant layout work.

QUESTIONS

1. What does the operation process chart show?
2. What symbols are used in constructing the operation process chart?
3. How does the operation process chart show materials introduced into the general flow?
4. How does the flow process chart differ from the operation process chart?
5. What is the principal purpose of the flow process chart?
6. What symbols are used in constructing the flow process chart?
7. Why is it necessary to construct process charts from direct observation, as opposed to information obtained from the foreman?
8. In the construction of the flow process chart, what method can be used to estimate distances moved?
9. How can delay times be determined in the construction of the flow process chart? Storage times?
10. When would you advocate using the flow diagram?
11. How can the flow of several different products be shown on the flow diagram?
12. What two flowchart symbols are used exclusively in the study of paperwork?
13. What are the limitations of the operation and flow process charts and the flow diagram?
14. Explain how PERT charting can save a company money.
15. What is the purpose of crashing?
16. When is it advisable to construct a worker and machine process chart?
17. What is machine coupling?
18. In what way does an operator benefit through machine coupling?
19. How does the gang process chart differ from the worker and machine process chart?
20. In a process plant, which of the following process charts has the greatest application: worker and machine, gang, operation, flow? Why?
21. What is the difference between synchronous and random servicing?
22. Reducing which of the three times—worker, machine, or loading—would have the greatest effect on increasing productivity? Why?

PROBLEMS

1. Based on the following crash cost table, what would be the minimum time to complete the project described by Figure 2.4, whose normal costs are shown in Table 2.2 What would be the added cost to complete the project within this time period?

	Crash Schedule	
	Weeks	$
A	2	7,000
B	1	2,500
C	2	5,000
D	0.5	2,000
E	4	6,000
F	3	5,000
G	2	6,000
H	0	0
I	4	7,600
J	1	2,200
K	4	4,500
L	2	2,200
M	3	3,000
N	1	700
O	2	6,000
P	1	3,000

2. The machining time per piece is 0.164 h, and the machine loading time is 0.038 h. With an operator rate of $12.80/h and a machine rate of $14/h, calculate the optimum number of machines for lowest cost per unit of output.

3. At Dorben Company, a worker is assigned to operate several machines. Each of these machines is down at random times during the day. A work sampling study indicates that, on average, the machines operate unattended 60 percent of the time. Operator attention time at irregular intervals averages 40 percent. If the machine rate is $20/h and the operator rate is $12/h, what would be the most favorable number of machines (from an economic standpoint) that should be operated by one operator?

4. The analyst in the Dorben Company wishes to assign a number of similar facilities to an operator, based on minimizing the cost per unit of output. A detailed study of the facilities reveals the following:

$$\text{Loading machine standard time} = 0.34 \text{ min}$$
$$\text{Unloading machine standard time} = 0.26 \text{ min}$$
$$\text{Walk time between two machines} = 0.06 \text{ min}$$
$$\text{Operator rate} = \$12.00/ \text{ h}$$
$$\text{Machine rate (both idle and working)} = \$18.00/ \text{ h}$$
$$\text{Power feed time} = 1.48 \text{ min}$$

How many of these machines should be assigned to each operator?

5. A study reveals that a group of three semiautomatic machines assigned to one operator operates unattended 80 percent of the time. Operator service time at irregular intervals averages 20 percent of the time on these three machines. What would be the estimated machine hours lost per 8-h day because of lack of an operator?

6. Based upon the following data, develop your recommended allocation of work and the number of workstations.

work unit	Estimated Work unit time (min)
0	0.76
1	1.24
2	0.84
3	2.07
4	1.47
5	2.40
6	0.62
7	2.16
8	4.75
9	0.65
10	1.45

The minimum required production per day is 90 assemblies. The following precedence matrix was developed by the analyst.

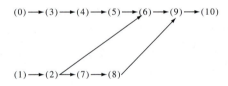

7. How many machines should be assigned to an operator for lowest-cost operations when

   a. Loading and unloading time on one machine is 1.41 min.

   b. Walking time to the next facility is 0.08 min.

   c. Machine time (power feed) is 4.34 min.

   d. Operator rate is $13.20/h.

   e. Machine rate is $18.00/h.

8. What proportion of machine time would be lost in operating four machines when each machine operates unattended 70 percent of the time and the operator attention time at irregular intervals averages 30 percent? Is this the best arrangement for minimizing the proportion of machine time lost?

9. In an assembly process involving six distinct operations, it is necessary to produce 250 units per 8-h day. The measured operation times are as follows:

   a. 7.56 min.

   b. 4.25 min.

    c.   12.11 min.

    d.   1.58 min.

    e.   3.72 min.

    f.   8.44 min.

    How many operators would be required at 80 percent efficiency? How many operators will be utilized at each of the six operations?

10. A study reveals the following steps in the assembly of a truss (small triangle of three small pieces within a large triangle of three larger pieces):

    Forklift delivers 2×4 pieces of pine from outside storage area (20 min).

    Bandsaw operator cuts six pieces to appropriate length (10 min).

    Assembler #1 gets three short pieces, bolts small triangle (5 min).

    Assembler #2 gets three long pieces, bolts large triangle (10 min).

    Assembler #3 gets one of each triangle and fastens into truss (20 min).

    Supervisor inspects complete truss and prepares for delivery (5 min).

    a.   Complete a flow process chart of the operation.

    b.   What are the %Idle time and production for an unbalanced, linear assembly line?

    c.   Balance the assembly line using appropriate workstations. What are the %Idle time and production now?

11. The current operation consists of the following elements:

    Operator removes pressed unit (0.2 min).

    Operator walks to inspection area, checks for defects (0.1 min).

    Operator files rough edges (0.2 min).

    Operator places unit on conveyor for further processing and returns to press (0.1 min).

    Operator cleans press die element with compressed air (0.3 min).

    Operator sprays lubricant into die (0.1 min).

    Operator places sheet metal into press, pushes START (0.2 min).

    Press cycles automatically for 1.2 min.

    Given that the operator is paid $10/h and that presses cost $15/hr to run, find and draw the worker-machine chart for the lowest-cost operation. What is the production? What is the unit cost?

12. Given the OSHA recordable (i.e., those that must be recorded on the OSHA 300 log and open to inspection) injuries shown on the next page, what can you conclude about the injuries? Which job code would you study first? If you had limited resources, where would you put them?

13. Exploratory analysis has identified the following job as a problem area. Complete a flow process chart (material type) for the following engine stripping, cleaning, and degreasing operation.

    Engines are stored in the old-engine storeroom. When needed, an engine is picked up by an electric hoist on a monorail, transported to the stripping bay, and unloaded onto an engine stand. There the operator strips the engine, putting the engine parts into the degreasing basket. The basket is transported to the degreaser, loaded into the degreaser, degreased, and then unloaded from the degreaser. The basket with degreased engine parts is then transported to the cleaning area, where the parts are simply dumped on the ground for drying. After several minutes of drying, the parts are lifted to the cleaning benches and cleaned. The cleaned parts are collected in special trays to await transport. The parts are loaded onto a trolley and transported to the inspection station. There they are slid from the trays onto the inspection benches.

| | Type of injury | | |
Job Code	Strain/sprain	CTD	Other
AM9	1	0	0
BTR	1	2	0
CUE	2	0	1
CUP	4	4	19
DAW	0	0	2
EST	0	0	2
FAO	3	1	1
FAR	3	1	3
FFB	1	0	1
FGL	1	0	1
FPY	1	2	0
FQT	0	0	3
FQ9	2	0	3
GFC	0	0	1
IPM	4	1	16
IPY	1	0	0
IP9	1	0	0
MPL	1	0	0
MST	0	0	0
MXM	1	0	2
MYB	1	1	3
WCU	1	0	1

14. Given the following operations and unit times in minutes (#1 = 1.5, #2 = 3, #3 = 1, #4 = 2, #5 = 4), balance the production line with the goal of producing 30 units/h.

15. The following activities and times (in minutes) were recorded for a mold operator:

- Removes molded piece from die   0.6
- Walks 10 ft to a workbench   0.2
- Boxes widget and places on conveyor   1.0
- Walks back to molder   0.2
- Blows out dirt from mold   0.4
- Sprays oil into mold, pushes "GO"   0.2
- Mold cycles automatically   3.0

The cycle then repeats itself. The operator is paid $10.00/h, and it costs $15.00/h to run the molder. What is the optimum number of machines that can be assigned to the operator to produce the widgets at lowest cost? Draw a worker-machine chart.

16. TOYCO produces toy shovels on a 20-ton press. The steps taken by the press operator to produce one shovel are:

   - Remove finished shovel and put on conveyor   0.1 min
   - Remove debris from the dies   0.2 min
   - Spray dies with oil   0.1 min
   - Check raw material (flat sheet) for defects   0.3 min
   - Place flat sheet into press   0.1 min
   - Press cycles automatically   1.0 min

   The operator is paid $10/h, and the press costs $100/h to run. Raw material in shovel costs $1.00, and it sells for $4.00. What is the optimum number of presses for one operator for lowest unit cost? Draw the worker-machine chart for this situation.

17. In the project shown below, activities are represented by arrows, and the number for each activity also indicates its normal duration (in days).

   a. Determine the critical path and the length of this project.

   b. Assume that each activity, except 1 and 2, can be crashed up to 2 days at a cost equal to the activity number. E.g., activity 6 normally takes 6 days, but could be crashed to 5 days for a cost of $6, or to 4 days for a total of $12. Determine the least-cost 26-day schedule. Show the activities that are crashed and the total crash costs.

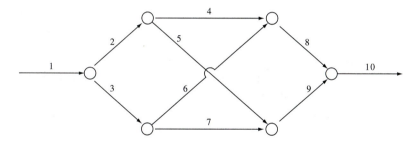

REFERENCES

Ashcroft, H. "The Productivity of Several Machines under the Care of One Operator." *Journal of the Royal Statistical Society B,* 12, no. 1 (1950), pp. 145–51.

ASME. *ASME Standard—Operation and Flow Process Charts, ANSI Y15.3-1974.* New York: American Society of Mechanical Engineers, 1974.

Baker, Kenneth R. *Elements of Sequencing and Scheduling.* Hannover, NJ: K. R. Baker, 1995.

Cole, R. *Work, Mobility, and Participation: A Comparative Study of American and Japanese Industry.* Berkeley, CA: University of California Press, 1979.

Helgeson, W. B., and D. P. Birnie, "Assembly Line Balancing Using Ranked Positional Weight Technique," *Journal of Industrial Engineering,* 12, no. 6, (1961) pp. 394–398.

Herron, D. "Industrial Engineering Applications of ABC Curves." *AIIE Transactions* 8, no. 2 (June 1976), pp. 210–18.

Moodie, C. "Assembly Line Balancing." In *Handbook of Industrial Engineering,* 2d ed., Ed. Gavriel Salvendy. New York: John Wiley & Sons, 1992, pp. 1446–1459.

Stecke, K. "Machine Interference." In *Handbook of Industrial Engineering,* 2d ed., Ed. Gavriel Salvendy. New York: John Wiley & Sons, 1992, pp. 1460–1494.

Takeji, K., and S. Sakamato, "Charting Techniques." In *Handbook of Industrial Engineering,* 2d ed., Ed. Gavriel Salvendy. New York: John Wiley & Sons, 1992, pp. 1415–1445.

Wright, W. R., W. G. Duvall, and H. A. Freeman. "Machine Interference." *Mechanical Engineering,* 58, no. 8 (August 1936), pp. 510–14.

SELECTED SOFTWARE

Design Tools (available from the McGraw-Hill text website at www.mhhe.com/niebel-freivalds). New York: McGraw-Hill, 2002.

Operation Analysis

- Use operation analysis to improve the method by asking *what*.
- Focus on the purpose of operation by asking *why*.
- Focus on design, materials, tolerances, processes, and tools by asking *how*.
- Focus on the operator and work design by asking *who*.
- Focus on the layout of the work by asking *where*.
- Focus on the sequence of manufacture by asking *when*.
- Always try to *simplify* by eliminating, combining, and rearranging operations.

Methods analysts use operation analysis to study all productive and nonproductive elements of an operation, to increase productivity per unit of time, and to reduce unit costs while maintaining or improving quality. When properly utilized, methods analysis develops a better method of doing the work by simplifying operational procedures and material handling and by utilizing equipment more effectively. Thus, firms are able to increase output and reduce unit cost; ensure quality and reduce defective workmanship; and facilitate operator enthusiasm by improving working conditions, minimizing operator fatigue, and permitting higher operator earnings.

Operation analysis is the third methods step, the one in which analysis takes place and the various components of the proposed method crystallize. It immediately follows obtaining and presenting facts using a variety of flow process charting tools presented in Chapter 2. The analyst should review each operation and inspection presented graphically on these charts and should ask a number of questions, the most important of which is *why*:

1. Why is this operation necessary?
2. Why is this operation performed in this manner?
3. Why are these tolerances this close?
4. Why has this material been specified?
5. Why has this class of operator been assigned to do the work?

The question why immediately suggests other questions, including how, who, where, and when. Thus, analysts might ask,

1. How can the operation be performed better?
2. Who can best perform the operation?
3. Where could the operation be performed at a lower cost or improved quality?
4. When should the operation be performed to yield the least amount of material handling?

For example, in the operation process chart shown in Figure 2.7, analysts might ask the questions listed in Table 3.1 to determine the practicability of the methods improvements indicated. Answering these questions helps initiate the elimination, combination, and simplification of the operations. Also in obtaining the answers to such questions, analysts become aware of other questions that may lead to improvement. Ideas seem to generate more ideas, and experienced analysts usually arrive at several improvement possibilities. Analysts must keep an open mind, so that previous disappointments do not discourage trying new ideas. Such opportunities for methods improvement usually appear in every plant, with consequent beneficial results.

Note that much of the information presented in Chapter 3 is currently used in a repackaged format termed *lean manufacturing*. Lean manufacturing originated with the Toyota Motor Corporation as a means of eliminating waste in the aftermath of the 1973 oil embargo and followed the footsteps of the Taylor system of scientific management but in much broader approach, targeting not only manufacturing costs, but also sales, administrative, and capital costs. Highlights of the Toyota Production System (TPS) included seven types of *muda* or waste (Shingo, 1987): (1) overproduction, (2) waiting for the next step, (3) unnecessary transportation, (4) inappropriate processing, (5) excess inventory, (6) unnecessary motion, and (7) defective products. The overlaps with traditional approaches are exemplified by the following: (1) waiting and transportation wastes are elements to be examined and eliminated within flow process charts analyses, (2) waste of motion summarizes the Gilbreths' lifelong work in motion study culminating the principles of work design and motion economy, (3) waste of overproduction and excess inventory are based on the additional storage requirements and material handling requirements to move items into and out of storage, and (4) waste in defective products is an obvious waste producing scrap or requiring rework.

A corollary to the seven mudas is the 5S system to reduce waste and optimize productivity by maintaining an orderly workplace and consistent methods. The 5S pillars are (1) *sort* (seiri), (2) *set in order* (seiton), (3) *shine* (seiso), (4) *standardize* (seiketsu) and (5) *sustain* (shitsuke). *Sort* focuses on removing all unnecessary items from the workplace and leaving only the bare essentials. *Set in order* arranges needed items so that they are easy to find and use. Once the clutter is removed, *shine* ensures further cleanliness and tidiness. Once the first three pillars

Table 3.1 Questions to Ask in the Manufacture of Telephone Stands

Question	Method Improvement
1. Can fixed lengths of 1½″ × 14″ white maple be purchased at no extra square footage cost?	Eliminate waste ends from lengths that are 14″.
2. Can purchased maple boards be secured with edges smooth and parallel?	Eliminate jointing of ends (operation 2).
3. Can boards be purchased to thickness size and have at least one side planed smooth? If so, how much extra will this cost?	Eliminate planing to size.
4. Why cannot two boards be stacked and sawed into 14″ sections simultaneously?	Reduce time of 0.18 (operation 4).
5. What percentage of rejects do we have at the first inspection station?	If the percentage is low, perhaps this inspection can be eliminated.
6. Why should the top of the table be sanded all over?	Eliminate sanding of one side of top and reduce time (operation 5).
7. Can fixed lengths of 1½″ × 3″ yellow pine be purchased at no extra square footage cost?	Eliminate waste ends from lengths that are not multiples of 12″
8. Can purchased yellow pine boards be secured with edges smooth and parallel?	Eliminate jointing of one edge.
9. Can sill boards be purchased to thickness size and have one side planed smooth? If so, how much extra will this cost?	Eliminate planing to size.
10. Why cannot two or more boards be stacked and sawed into 14″ sections simultaneously?	Reduce time of 0.10 (operation 9).
11. What percentage of rejects do we have at the first inspection of the sills?	If the percentage is low, perhaps this inspection can be eliminated.
12. Why is it necessary to sand the sills all over?	Eliminate some sanding and reduce time (operation 10).
13. Can fixed lengths of 2½″ × 2½″ white maple be purchased at no extra square footage cost?	Eliminate waste ends from lengths that are not multiples of 16″
14. Can a smaller size than 2½″ × 2½″ be used?	Reduce material cost.
15. Can purchased white maple boards be secured with edges smooth and parallel?	Eliminate jointing of edges.
16. Can leg boards be purchased to thickness size and have sides planed smooth? If so, how much extra will this cost?	Eliminate planing to size.
17. Why cannot two or more boards be stacked and sawed into 14″ sections simultaneously?	Reduce time (operation 15).
18. What percentage of rejects do we have at the first inspection of the legs?	If the percentage is low, perhaps this inspection can be eliminated.
19. Why is it necessary to sand the legs all over?	Eliminate some sanding and reduce time (operation 16).
20. Could a fixture facilitate assembly of the sills to the top?	Reduce assembly time (operation 11).
21. Can a sampling inspection be used on the first inspection of the assembly?	Reduce inspection time (operation 4).
22. Is it necessary to sand after one coat of shellac?	Eliminate operation 19.

have been implemented, *standardize* serves to maintain the order and consistent approach to housekeeping and the methods. Finally, *sustain* maintains the full 5S process on a regular basis.

3.1 OPERATION PURPOSE

This is probably the most important of the nine points of operation analysis. The best way to simplify an operation is to devise some way to get the same or better results at no additional cost. An analyst's cardinal rule is to try to eliminate or combine an operation before trying to improve it. In our experience, as much as 25 percent of the operations being performed can be eliminated if sufficient study is given to the design and process. This also corresponds closely to eliminating the muda of inappropriate processing.

Far too much unnecessary work is done today. In many instances, the task or the process should not be simplified or improved, but eliminated entirely. Eliminating an activity saves money on the installation of an improved method, and there is no interruption or delay because no improved method is being developed, tested, and installed. Operators need not be trained on the new method, and resistance to change is minimized when an unnecessary task or activity is eliminated. With respect to paperwork, before a form is developed for information transfer, analysts should ask, Is the form really needed? Today's computer-controlled systems should reduce the generation of forms and paperwork.

Unnecessary operations frequently result from improper planning when the job is first set up. Once a standard routine is established, it is difficult to change, even if such a change would eliminate a portion of the work and make the job easier. When new jobs are planned, the planner may include an extra operation if there is any possibility that the product would be rejected without that extra work. For example, in turning a steel shaft, if there is some question whether to take two or three cuts to maintain a 40-microinch finish, the planner invariably specifies three cuts, even though proper maintenance of the cutting tools, supplemented by ideal feeds and speeds, would allow the job to be done with two cuts.

Unnecessary operations often develop because of the improper performance of a previous operation. A second operation must be done to "touch up" or make acceptable the work done by the first operation. In one plant, for example, armatures were previously spray painted in a fixture, making it impossible to cover the bottom of the armature with paint because the fixture shielded the bottom from the spray blast. It was therefore necessary to touch up the armature bottoms after spray painting. A study of the job resulted in a redesigned fixture that held the armature and still allowed complete coverage. In addition, the new fixture permitted seven armatures to be spray painted simultaneously, while the old method called for spray painting one at a time. Thus, by considering that an unnecessary operation may have developed because of

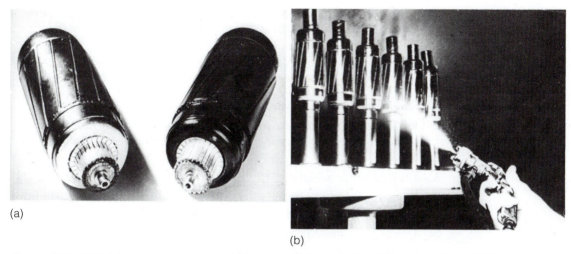

(a)

(b)

Figure 3.1 (a) Painted armature as removed from the old fixture (*left*) and as removed from the improved fixture (*right*). (b) Armature in spray-painting fixture allowing complete coverage of the armature bottom.

the improper performance of a previous operation, the analyst was able to eliminate the touch-up operation (see Figure 3.1).

As another example, in the manufacture of large gears, it was necessary to introduce a hand-scraping and lapping operation to remove waves in the teeth after they had been hobbed. An investigation disclosed that contraction and expansion, brought about by temperature changes in the course of the day, were responsible for the waviness in the teeth's surfaces. By enclosing the whole unit and installing an air-conditioning system within the enclosure, the company was able to maintain the proper temperature during the whole day. The waviness disappeared immediately, and it was no longer necessary to continue the hand-scraping and lapping operations.

To eliminate an operation, analysts should ask and answer the following question: Can an outside supplier perform the operation more economically? In one example, ball bearings purchased from an outside vendor had to be packed in grease prior to assembly. A study of bearing vendors revealed that "sealed-for-life" bearings could be purchased from another supplier at lower cost.

The examples given in this section highlight the need to establish the purpose of each operation before endeavoring to improve the operation. Once the necessity of the operation has been determined, the remaining nine steps to operation analysis should help to determine how it can be improved.

3.2 PART DESIGN

Methods engineers are often inclined to feel that once a design has been accepted, their only recourse is to plan its economical manufacture. While introducing even a slight design change may be difficult, a good methods analyst should still review every design for possible improvements. Designs can be changed, and if improvement is the result and the activity of the job is significant, then the change should be made.

To improve the design, analysts should keep in mind the following pointers for lower-cost designs on each component and each subassembly:

1. Reduce the number of parts by simplifying the design.
2. Reduce the number of operations and the length of travel in manufacturing by joining the parts better and by making the machining and assembly easier.
3. Utilize a better material.
4. Liberalize tolerances and rely on key operations for accuracy, rather than on series of closely held limits.
5. Design for manufacturability and assembly.

Note that the first two will help in reducing muda in unappropriate processing, unnecessary transportation, and excess inventory.

The General Electric Company summarized the ideas for developing minimum cost designs in Table 3.2.

The following examples of methods improvement resulted from considering a better material or process in an effort to improve the design. Conduit boxes were originally built of cast iron. The improved design, making a stronger, neater, lighter, and less expensive conduit box, was fabricated from sheet steel. A four-step process was used to bend a part into the desired shape (see Figure 3.2). This was inefficient and stressed the metal at the bends. The design was slightly altered so that the less expensive process of extruding could be utilized. The extruded sections were then cut to the desired length. In the redesigned process, three steps were eliminated.

Design simplification through the better joining of parts was used in assembling terminal clips to their mating conductors. The original practice required turning up the end of the clip to form a socket. The socket was filled with solder, and the wire conductor was then tinned, inserted into the solder-filled socket, and held there until the solder solidified. The altered design called for resistance welding the clip to the wire conductor, eliminating both the forming and dipping operations. The original part was designed with three components that had to be assembled (see Figure 3.3). A significantly less costly approach utilized a one-piece design which could be machined as a solid piece, eliminating two components and several operations.

Just as opportunities exist to improve productivity through better product design, similar opportunities exist to improve the design of forms (whether hard copy or electronic) used throughout an industry or business. Once a form is proved necessary, it should be studied to improve both the collection and flow of information. The following criteria apply to the development of forms:

1. Maintain simplicity in the form design, keeping the amount of necessary input information at a minimum.

Table 3.2 Methods for Minimum Cost Design

Castings
1. Eliminate dry sand (baked-sand) cores.
2. Minimize depth to obtain flatter castings.
3. Use minimum weight consistent with sufficient thickness to cast without chilling.
4. Choose simple forms.
5. Symmetrical forms produce uniform shrinkage.
6. Liberal radii—no sharp corners.
7. If surfaces are to be accurate with relation to each other, they should be in the same part of the pattern, if possible.
8. Locate parting lines so that they will not affect looks and utility, and need not be ground smooth.
9. Specify multiple patterns instead of single ones.
10. Metal patterns are preferable to wood.
11. Use permanent molds instead of metal patterns.

Moldings
1. Eliminate inserts from parts.
2. Design molds with smallest number of parts.
3. Use simple shapes.
4. Locate flash lines so that the flash does not need to be filed and polished.
5. Minimize weight.

Punchings
1. Use punched parts instead of molded, cast, machined, or fabricated parts.
2. "Nestable" punchings economize on material.
3. Holes requiring accurate relation to each other should be made by the same die.
4. Design to use coil stock.
5. Design punchings to have minimum sheared length and maximum die strength with fewest die moves.

Formed parts
1. Drawn parts instead of spun, welded, or forged parts.
2. Shallow draws if possible.
3. Liberal radii on corners.
4. Bent parts instead of drawn.
5. Parts formed of strip or wire instead of punched from sheet.

Fabricated parts
1. Self-tapping screws instead of standard screws.
2. Drive pins instead of standard screws.
3. Rivets instead of screws.
4. Hollow rivets instead of solid rivets.
5. Spot or projection welding instead of riveting.
6. Welding instead of brazing or soldering.
7. Use die castings or molded parts instead of fabricated construction requiring several parts.

Machined parts
1. Use rotary machining processes instead of shaping methods.
2. Use automatic or semiautomatic machining instead of hand-operated.
3. Reduce the number of shoulders.
4. Omit finishes where possible.
5. Use rough finish when satisfactory.
6. Dimension drawings from same point as used by factory in measuring and inspecting.
7. Use centerless grinding instead of between-center grinding.
8. Avoid tapers and formed contours.
9. Allow a radius or undercut at shoulders.

Screw machine parts
1. Eliminate second operation.
2. Use cold-rolled stock.
3. Design for header instead of screw machine.
4. Use rolled threads instead of cut threads.

Welded parts
1. Fabricated construction instead of castings or forgings.
2. Minimum sizes of welds.
3. Welds made in flat position rather than vertical or overhead.
4. Eliminate chamfering edges before welding.
5. Use "burnouts" (torch-cut contours) instead of machined contours.
6. Lay out parts to cut to best advantage from standard rectangular plates and avoid scrap.
7. Use intermittent instead of continuous weld.
8. Design for circular or straight-line welding to use automatic machines.

Treatments and finishes
1. Reduce baking time to minimum.
2. Use air drying instead of baking.
3. Use fewer or thinner coats.
4. Eliminate treatments and finishes entirely.

Assemblies
1. Make assemblies simple.
2. Make assemblies progressive.
3. Make only one assembly and eliminate trial assemblies.
4. Make component parts *right* in the first place so that fitting and adjusting will not be required in assembly.
   This means that drawings must be correct, with proper tolerances, and that parts must be made according to drawings.

General
1. Reduce number of parts.
2. Reduce number of operations.
3. Reduce length of travel in manufacturing.

Source: Adapted from *American Machinist*, reference sheets, 12th ed., New York: McGraw-Hill Publishing Co.

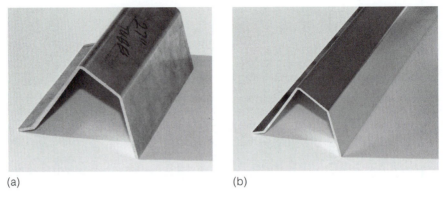

(a)   (b)

Figure 3.2   Part redesign to eliminate three steps
(a) A four-step process was used to bend this piece into the desired shape. This is inefficient and stresses the metal at the bends. (Courtesy of Alexandria Extrusion Company.) (b) This piece was extruded in one step and will later be cut to appropriate lengths. (Courtesy of Alexandria Extrusion Company.)

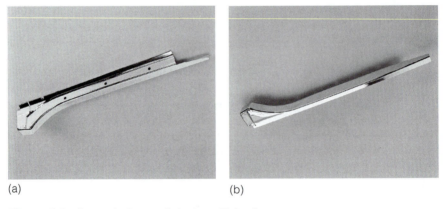

(a)   (b)

Figure 3.3   Part redesign to eliminate multiple pieces
(a) Original part was designed in three pieces and had to be assembled. (Courtesy of Alexandria Extrusion Company.) (b) Improved one-piece design can be machined as a solid piece. (Courtesy of the Minister Machine Company.)

2. Provide ample space for each bit of information, allowing for different input methods (writing, typewriter, word processor).
3. Sequence the information input in a logical pattern.
4. Color-code the form to facilitate distribution and routing.
5. Confine computer forms to one page.

3.3   TOLERANCES AND SPECIFICATIONS

The third of the nine points of operation analysis concerns tolerances and specifications that relate to the quality of the product, that is, its ability to satisfy given needs. While tolerances and specifications are always considered when reviewing the design, this is usually not sufficient; they should be considered independently of the other approaches to operation analysis.

Designers may have a tendency to incorporate specifications that are more rigid than necessary when developing a product. This can be due to a lack of knowledge about cost and the thought that it is necessary to specify closer tolerances and specifications than are actually needed to have the manufacturing departments produce to the actual required tolerance range.

Methods analysts should be well versed in the details of cost and should be fully aware of what unnecessarily close tolerances and/or rejects can do to the selling price. Figure 3.4 illustrates the pronounced relationship between the increased cost of tighter machining tolerances. If designers are being needlessly

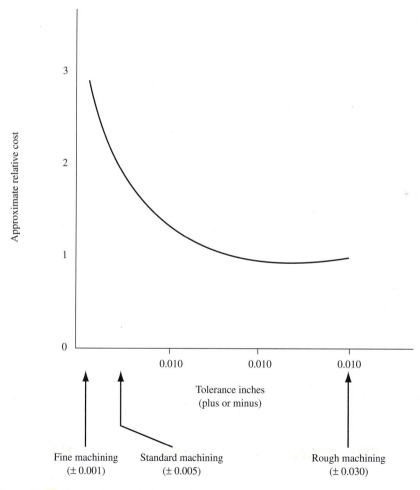

Figure 3.4 Approximate relationship between cost and machining tolerance.

tight in establishing tolerances and specifications, management should embark on a training program clearly presenting the economies of specifications. Developing quality products in a manner that actually reduces costs is a major tenet of the approach to quality instituted by Taguchi (1986). This approach involves combining engineering and statistical methods to achieve improvements in cost and quality by optimizing product design and manufacturing methods. This step corresponds to reducing the muda of inappropriate processing.

One manufacturer's drawings called for a 0.0005-in tolerance on a shoulder ring for a DC motor shaft. The original specifications called for a 1.8105 to 1.8110-in tolerance on the inside diameter. This close tolerance was deemed necessary because the shoulder ring was shrunk onto the motor shaft. Investigation revealed that a 0.003-in tolerance was adequate for the shrink fit. The drawing was immediately changed to specify a 1.809 to 1.812-in inside diameter. This change meant that a reaming operation was eliminated because someone questioned the absolute necessity of a close tolerance.

Analysts should also take into consideration the ideal inspection procedure. Inspection is a verification of quantity, quality, dimensions, and performance. Such inspections can usually be performed by a variety of techniques: spot inspection, lot-by-lot inspection, or 100 percent inspection. Spot inspection is a periodic check to ensure that established standards are being realized. For example, a nonprecision blanking and piercing operation set up on a punch press should have a spot inspection to ensure the maintenance of size and the absence of burrs. As the die begins to wear or as deficiencies in the material being worked begin to show up, the spot inspection would catch the trouble in time to make the necessary changes, without generating an appreciable number of rejects.

Lot-by-lot inspection is a sampling procedure in which a sample is examined to determine the quality of the production run or lot. The size of the sample depends on the allowable percentage of defective unity and the size of the production lot being checked. A 100 percent inspection involves inspecting every unit of production and rejecting the defective units. However, experience has shown that this type of inspection does not ensure a perfect product. The monotony of screening tends to create fatigue, thus lowering operator attention. The inspector may pass some defective parts, or reject good parts. Because a perfect product is not ensured under 100 percent inspection, acceptable quality may be realized by the considerably more economical methods of lot-by-lot or spot inspection.

For example, in one shop, a certain automatic polishing operation had a normal rejection quantity of 1 percent. Subjecting each lot of polished goods to 100 percent inspection would have been quite expensive. Management therefore decided, at an appreciable saving, to consider 1 percent the allowable percentage defective, even though this quantity of defective material would go through to plating and finishing, only to be thrown out in the final inspection before shipment.

By investigating tolerances and specifications and taking action when desirable, the company can reduce the costs of inspection, minimize scrap, diminish repair costs, and keep quality high. Also the company is addressing the muda of defective products.

3.4 MATERIAL

One of the first questions an engineer considers when designing a new product is, What material shall be used? Since choosing the correct material may be difficult because of the great variety available, it is often more practical to incorporate a better and more economical material into an existing design.

Methods analysts should consider the following possibilities for the direct and indirect materials utilized in a process:

1. Finding a less expensive and lighter material.
2. Finding materials that are easier to process.
3. Using materials more economically.
4. Using salvage materials.
5. Using supplies and tools more economically.
6. Standardizing materials.
7. Finding the best vendor from the standpoint of price and vendor stocking.

FINDING A LESS EXPENSIVE AND LIGHTER MATERIAL

Industry is continually developing new processes for producing and refining materials. Monthly publications summarize the approximate cost per pound of steel sheets, bars, and plates, and the cost of cast iron, cast steel, cast aluminum, cast bronze, thermoplastic and thermosetting resins, and other basic materials. These costs can be used as anchor points from which to judge the application of new materials. A material that was not competitive in price yesterday may be very competitive today.

One company used Micarta spacer bars between the windings of transformer coils. Separating the windings permitted the circulation of air between the windings. An investigation revealed that glass tubing could be substituted for the Micarta bars at a considerable savings. The glass tubing was less expensive, and it met service requirements better because the glass could withstand higher temperatures. Furthermore, the hollow tubing permitted greater air circulation than did the solid Micarta bars.

Another company also used a less expensive material that still met service requirements in the production of distribution transformers. Originally, a porcelain plate separated and held the wire leads coming out of the transformers. The company found that a fullerboard plate stood up just as well in service, yet was considerably less expensive. Today, one of the many types of plastic available would provide an even cheaper solution.

Another concern for manufacturers, especially today with high transportation costs due to continual increases in crude oil prices, is the weight of the product itself. Finding a lighter material or decreasing the amount of raw material used is of prime concern. A good example is shown by the changing nature of beverage cans (see Figure 3.5). All steel cans in the early 1970's weighed 1.94 oz (55 gr)

Figure 3.5 Decreasing weight of beverage cans
(a) All steel can from 1970 weighing 1.94 oz (55 gr) (b) Steel can with aluminum top and bottom from 1975 weighing 1.69 oz (48 gr) (c) All aluminum can from 1980 weighing 0.6 oz (17 gr) (d) All aluminum ribbed can from 1992 weighing 0.56 (16 gr) (Courtesy R. Voigt, Penn State)

(see Figure 3.5a). By replacing the top and bottom with aluminum discs, approximately 0.25 oz (7 gr) in weight savings could be achieved (see Figure 3.5b). Going to an all aluminum can decreased the total weight to 0.6 oz (17 gr) for considerable weight savings (see Figure 3.5c). However, the walls of the can became so thin that the walls easily crumpled. This was solved by creating ribbing in the walls (see Figure 3.5d).

Methods analysts should remember that items such as valves, relays, air cylinders, transformers, pipe fittings, bearings, couplings, chains, hinges, hardware, and motors can usually be purchased at less cost than they can be manufactured.

FINDING A MATERIAL THAT IS EASIER TO PROCESS

Some materials are usually more readily processed than others. Referring to handbook data on the physical properties usually helps analysts discern which material will react most favorably to the processes to which it must be subjected in its conversion from raw material to finished product. For example, machinability varies inversely with hardness, and hardness usually varies directly with strength.

Today the most versatile material is reinforced composites. Resin transfer molding can produce more complex parts advantageously from the standpoint of quality and production rate than most other metal and plastic forming procedures. Thus, by specifying a plastic made of reinforcing carbon fibers and epoxy, the analyst can substitute a composite for a metal part, at both a quality and a cost advantage. This step is also addressing the muda of inappropriate processing.

USING MATERIAL MORE ECONOMICALLY

The possibility of using material more economically is a fertile field for analysis. If the ratio of scrap material to that actually going into the product is high, then greater utilization should be examined. For example, if the material put into a plastic compression mold is preweighed, it may be possible to use only the exact amount required to fill the cavity; excessive flash can also be eliminated.

In another example, the production of stampings from sheet metal should utilize multiple dies carefully arranged to assure maximum use of material. Given consistent raw material and standard-sized dies, this typically is done through the use of CAD-assisted layout, yielding efficiencies exceeding 95% (i.e., less than 5% scrap). Similar approaches are utilized in the garment industry in the layout of patterns on cloth and the glass industry for the cutting of different sized windows. However, if the material is not consistent, then problems arise and the layout may still need to be performed by a human operator. The production of leather seats for automobiles requires layout of cutting dies on a tanned hide before entering a rolling press, which applies pressure on the dies to cut the leather in appropriate patterns. The operator needs to be highly skilled in handling variably sized cow hides full of imperfections from brands and barbed wire, especially to maximize the usage of quite expensive leather (see Figure 3.6).

Many world-class manufacturers are finding it not only desirable, but absolutely necessary, to take weight out of existing designs. For example, Ford engineers all looking a 40 percent weight reduction to achieve an 80 mi/gal fuel efficiency for the Taurus. This will require the cladding of stainless steel to high-strength aluminum to replace chrome-plated steel bumpers, as well as a much greater use of plastics and structural composites to replace ferrous components. Similar weight reduction is taking place on many other well-known products, such as washing machines, video cameras, VCRs, suitcases, and TV sets.

Today, powder coating is a proven technology that is replacing many other methods of metal finishing. Coating powders are finely divided particles of organic polymers (acrylic, epoxy, polyester, or blends) that usually contain pigments, fillers, and additives. Powder coating is the application of a suitable formulation to a substrate, which are then fused into a continuous film by the application of heat, forming a protective and decorative finish. In view of current environmental regulations affecting traditional metal finishing operations, such

Figure 3.6 Layout of cutting dies on a tanned hide before entering a rolling press (note the careful layout of the dies to maximize the use of the expensive leather).

as electroplating and wet painting, powder coating offers a safer and cleaner environment. The methodology can also provide a durable, attractive, cost-effective finish for metal surfaces used in many commercial products, such as wire shelving, control boxes, trailer hitches, water meters, handrails, boat racks, office partitions, and snow shovels.

USING SALVAGE MATERIALS

Materials can often be salvaged, rather than sold as scrap. By-products from an unworked portion or scrap section can sometimes offer real possibilities for savings. For example, one manufacturer of stainless steel cooling cabinets had 4- to 8-in-wide sections left as cuttings on the shear. An analysis identified electric light switchplate covers as a possible by-product. Another manufacturer, after salvaging the steel insert from defective bonded rubber ringer rolls, was able to utilize the hollow, cylindrical rubber rolls as bumpers for protecting moored motorboats and sailboats.

If it is not possible to develop a by-product, then scrap materials should be separated to obtain top scrap prices. Separate bins should be provided for tool steel, steel, brass, copper, and aluminum. Chip-haulers and floor sweepers should specifically be instructed to keep the scrap segregated. For electric lightbulbs, for example, the brass socket would be stored in one area, and after the glass bulb is broken and disposed of, the tungsten filament is removed and stored separately for greatest residual value. Many companies save wooden boxes from incoming shipments, and then saw the boards to standard lengths for use in making smaller boxes for outgoing shipments. This practice is usually economical, and it is now being followed by many large industries, as well as by service maintenance centers.

There are also a few interesting examples from the food industry. A manufacturer of tofu processes the beans, centrifuges out the edible protein material, and leaves behind tons of waste fiber. Rather than paying to haul it away to a landfill, the manufacturer gives it away to local farmers for hog feed, as long as they come and pick it up. Similarly, meatpackers utilize everything from a cow: hides, bones, even blood, all except the "moo."

USING SUPPLIES AND TOOLS FULLY

Management should encourage full use of all shop supplies. One manufacturer of dairy equipment introduced the policy that no new welding rod was to be distributed to workers without the return of old tips under 2 in long. The cost of welding rods was reduced immediately by more than 15 percent. Brazing or welding is usually the most economical way to repair expensive cutting tools, such as broaches, special form tools, and milling cutters. If it has been company practice to discard broken tools of this nature, the analyst should investigate the potential savings of a tool salvage program.

Analysts can also find a use for the unworn portions of grinding wheels, emery disks, and so forth. Also, items such as gloves and rags should not be discarded simply because they are soiled. Storing dirty items and then laundering them is less expensive than replacing them. Methods analysts can make a real contribution to a company by simply minimizing waste, one of the mudas in the TPS system.

STANDARDIZING MATERIALS

Methods analysts should always be alert to the possibility of standardizing materials. They must minimize the sizes, shapes, grades, and so on of each material utilized in the production and assembly processes. The typical economies resulting from reductions in the sizes and grades of the materials employed include the following:

- Purchase orders are used for larger amounts, which are almost always less expensive per unit.
- Inventories are smaller, since less material must be maintained as a reserve.
- Fewer entries need to be made in storage records.
- Fewer invoices need to be paid.
- Fewer spaces are needed to house materials in the storeroom.
- Sampling inspection reduces the total number of parts inspected.
- Fewer price quotations and purchase orders are needed.

The standardization of materials, like other methods improvement techniques, is a continuing process. It requires the continual cooperation of the design, production planning, and purchasing departments and fits in nicely with the 5S system.

FINDING THE BEST VENDOR

For the vast majority of materials, supplies, and parts, numerous suppliers will quote different prices, quality levels, delivery times, and willingness to hold inventories. It is usually the responsibility of the purchasing department to locate the most favorable supplier. However, the best supplier last year may not be the best one now. The methods analyst should encourage the purchasing department to rebid the highest-cost materials, supplies, and parts to obtain better prices and superior quality and to increase vendor stocking, where the vendors agree to hold inventories for their customers. It is not unusual for methods analysts to achieve a 10 percent reduction in the cost of materials and a 15 percent reduction in inventories by regularly pursuing this approach through their purchasing departments.

Perhaps the most important reason for continued Japanese success in the manufacturing sector is the *keiretsu*. This is a form of business and manufacturing organization that links businesses together. It can be thought of as a web of interlocking relationships among manufacturers—often between a large manufacturer and its principal suppliers. Thus, in Japan such companies as Hitachi and Toyota and other international competitors are able to acquire parts for their products from regular suppliers who produce to the quality called for and are continually looking for improvement so as to provide better prices for the firms in their network. Alert purchasing departments are often able to create relationships with suppliers comparable to the so-called production keiretsu.

3.5 MANUFACTURE SEQUENCE AND PROCESS

As manufacturing technology in the twenty-first century eliminates labor-intensive manufacturing in favor of capital-intensive procedures, the methods engineer will focus on multiaxis and multifunctioning machining and assembly. Modern equipment is capable of cutting at higher speeds on more accurate, rigid and flexible machines that utilize both advanced controls and tool materials. Programming functions permit in-process and postprocess gaging for tool sensing and compensation, resulting in dependable quality control.

The methods engineer must understand that the time utilized by the manufacturing process is divided into three steps: inventory control and planning, setup operations, and in-process manufacturing. Furthermore, it is not unusual to find that these procedures, in aggregate, are only about 30 percent efficient from the standpoint of process improvement.

To improve the manufacturing process, the analyst should consider (1) rearranging the operations; (2) mechanizing manual operations; (3) utilizing more efficient facilities on mechanical operations; (4) operating mechanical facilities more efficiently; (5) manufacturing near the net shape; and (6) using robots, all which address the muda of inappropriate processing.

REARRANGING OPERATIONS

Rearranging operations often results in savings. As an example, the flange of a motor conduit box required four drilled holes, one in each corner. Also, the base had to be smooth and flat. Originally, the operator began by grinding the base, then drilling the four holes using a drill jig. The drilling operation threw up burrs, which then had to be removed in another step. By rearranging the operation so that the holes were drilled first and the base then ground, analysts eliminated the deburring operation. The base-grinding operation automatically removed the burrs.

Combining operations usually reduces costs. For example, a manufacturer fabricated the fan motor support and the outlet box of its electric fans. After painting the parts separately, operators then riveted them together. By having the outlet box riveted to the fan motor support prior to painting, analysts effected an appreciable time savings for the painting operation. Similarly, using a more complex machine that combines several operations can reduce the time to produce the finished piece and increase productivity (see Figure 3.7). Although the machine may be more expensive, considerable savings are incurred by reduced labor costs.

In another example, the market for aluminum cylinder head castings is growing, and foundries are finding it cost-effective to go from the steel-mold casting process to the lost foam process. Lost foam is an investment casting

(a)

(b)

(c)

Figure 3.7 Combining operations to eliminate steps. Stock material shown in (b) is cut to size and threaded in one step on the Citizen CNC lathe shown in (a) to yield the finished piece shown in (c). (a) Citizen CNC lathe. (Courtesy of Jergens, Inc.) (b) Stock material. (Courtesy of Jergens, Inc.) (c) Finished piece. (Courtesy of Jergens, Inc.)

procedure that uses an expendable pattern of polystyrene foam surrounded by a thin ceramic shell. Steel-mold castings require considerable subsequent machining. In comparison, the lost foam process reduces the amount of machining and also eliminates the sand disposal costs usually associated with investment casting.

Before changing any operation, however, the analyst must consider possible detrimental effects on subsequent operations down the line. Reducing the cost of one operation could result in higher costs for other operations. For example, a change recommended in the manufacture of AC field coils resulted in higher costs and was therefore not practical. The field coils were made of heavy copper bands, which were formed and then insulated with mica tape. The mica tape was hand-wrapped on the already coiled parts. The company decided to machine wrap the copper bands prior to coiling. This did not prove practical, as the forming of the coils cracked the mica tape, necessitating time-consuming repairs prior to product acceptance.

MECHANIZING MANUAL OPERATIONS

Today, any practicing methods analyst should consider using special-purpose and automatic equipment and tooling, especially if production quantities are large. Notable among industry's latest offerings are program controlled, numerically controlled (NC), and computer controlled (CNC) machining and other equipment. These afford substantial savings in labor cost as well as the following advantages: reduced work-in-process inventory, less parts damage due to handling, less scrap, reduced floor space, and reduced production throughput time. For example (see Figure 3.8), whereas two operators are required for a manually operated machine tool, only one operator is required for a computer-controlled machine tool. Use of a robotic arm operating a fully automated machine tool would not even require the one operator, considerably reducing labor costs (albeit with higher initial capital costs).

Other automatic equipment includes automatic screw machines; multiple-spindle drilling, boring, and tapping machines; index-table machine tools; automatic casting equipment combining automatic sand-mold making, pouring, shakeout, and grinding; and automatic painting and plating finishing equipment. The use of power assembly tools, such as power nut- and screwdrivers, electric or air hammers, and mechanical feeders, is often more economical than the use of hand tools.

To illustrate, a company that produces specialty windows was using manual methods to press rails over both ends of plate window glass that had been covered with a synthetic rubber wrap. The plates of glass were held in position by two pads that were pneumatically squeezed together. The operator would pick up a rail and position it over the end of the window glass and then pick up a mallet and hammer the rail into position over the glass. The operation was slow and it resulted in considerable operator work-related musculoskeletal disorders.

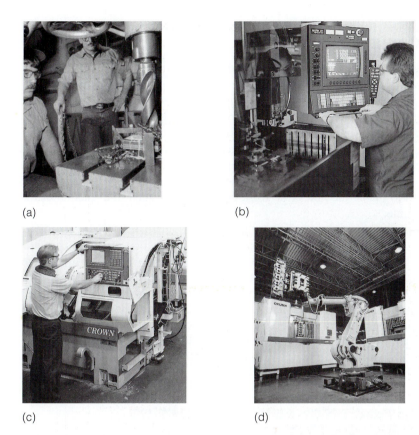

Figure 3.8 Mechanizing manual operations can reduce labor costs.
(a) Two operators are required for a manually operated machine tool. (b) A computer-controlled machine tool requires only one operator. (c) A state-of-the-art computer-controlled machine tool still requires one operator but performs more operations.
(d) A robotic arm operating a fully automated machine tool requires no operators.
[(a) © Yogi, Inc./CORBIS; (b) © Molly O'Bryon Welpott; (c) and (d) Courtesy of Okuma.]

Furthermore, scrap was high because of glass breakage due to pounding the rails over the glass. A new facility was designed that pneumatically squeezed rails onto the window glass over the synthetic rubber wrap. Operators enthusiastically accepted the new facility because the work was much easier to perform; health problems disappeared, productivity increased, and glass breakage dropped to near zero.

The application of mechanization applies not only to process operations, but also to paperwork. For example, bar coding applications can be invaluable to the operations analyst. Bar coding can rapidly and accurately enter a variety of data. Computers can then manipulate the data for some desired objective, such as counting and controlling inventory, routing specific items to or through a process, or identifying the state of completion and the operator currently working on each item in a work-in-process.

UTILIZING MORE EFFICIENT MECHANICAL FACILITIES

If an operation is done mechanically, there is always the possibility of a more efficient means of mechanization. At one company, for example, turbine blade roots were machined by using three separate milling operations. Both the cycle time and the costs were high. When external broaching was introduced, all three surfaces could be finished at once, for considerable time and cost savings. Another company overlooked the possibility of utilizing a press operation. This process is one of the fastest for forming and sizing processes. A stamped bracket had four holes that were drilled after the bracket was formed. By using a die designed to pierce the holes, the work could be performed in a fraction of the drilling time.

Work mechanization applies to more than just manual work. For example, one company in the food industry was checking the weight of various product lines with a balance. This equipment required the operator to note the weight visually, record the weight on a form, and subsequently perform several calculations. A methods engineering study resulted in the introduction of a statistical weight control system. Under the improved method, the operator weighs the product on a digital scale programmed to accept the product within a certain weight range. As the product is weighed, the weight information is transferred to a personal computer that compiles the information and prints the desired report.

OPERATING MECHANICAL FACILITIES MORE EFFICIENTLY

A good slogan for methods analysts is, "Design for two at a time." Usually multiple-die operation in presswork is more economical than single-stage operation. Again, multiple cavities in diecasting, molding, and similar processes are viable options when there is sufficient volume. On machine operations, analysts should be sure that proper feeds and speeds are used. They should investigate the grinding of cutting tools for maximum performance. They should check to see whether the cutting tools are properly mounted, whether the right lubricant is being used, and whether the machine tool is in good condition and is adequately maintained. Many machine tools are operated at a fraction of their possible output. Endeavoring to operate mechanical facilities more efficiently nearly always pays dividends.

MANUFACTURING NEAR THE NET SHAPE

Using a manufacturing process that produces components closer to the final shape can maximize material use, reduce scrap, minimize secondary processing such as final machining and finishing, and permit manufacturing with more environmentally friendly materials. For example, forming parts with powder metals (PM) instead of conventional casting or forging often provides the manufacture of near-net shapes for many components, resulting in dramatic economic savings as well as functional advantages. In the case of forged PM connecting rods, it has been reported that they have reduced the reciprocating mass of competing alternatives, resulting in less noise and vibration as well as major cost economies.

CONSIDERING THE USE OF ROBOTS

For cost and productivity reasons, it is advantageous today to consider the use of robots in many manufacturing areas (see Figure 3.9). For example, assembly areas include work that typically has a high direct labor cost, in some cases accounting for as much as one-half of the manufacturing cost of a product. The principal advantage of integrating a modern robot in the assembly process is its inherent flexibility. It can assemble multiple products on a single system and can be reprogrammed to handle various tasks with part variations. In addition, robotic assembly can provide consistently repeatable quality with predictable product output.

A robot's typical life is approximately 10 years. If it is well maintained and if it is used for moving small payloads, the life can be extended to up to 15 years. Consequently, a robot's depreciation cost can be relatively low. Also, if a given robot's size and configuration are appropriate, it can be used in a variety of operations. For example, a robot could be used to load a die-casting facility, load a quenching tank, load and unload a board drop-hammer forging operation, load a plate glass washing operation, and so on. In theory, a robot of the correct size and configuration can be programmed to do any job.

In addition to productivity advantages, robots also offer safety advantages. They can be used in work centers where there is danger to the worker because of the nature of the process. For example, in the die-casting process, there can be considerable danger due to hot metal splashing when the molten metal is injected into the die cavity. One of the original applications for robots was die casting. In one company, a five-axis robot developed by Unimation, Inc., serves a 600-ton microprocessor-controlled die-casting machine. In the operation, the robot moves into position when the die opens, grasps the casting by its slug, and clears it from the cavity. At the same time, it initiates automatic die-lubrication sprays. The robot displays the casting to infrared scanners, then signals the die-casting machine to accept another shot. The casting is deposited by the robot on an output station for trimming. Here an operator, remote from the die-casting machine, safely trims the casting preparatory to subsequent operations.

Automobile manufacturers have placed particular emphasis on the use of robots in welding. For example, at Nissan Motors, 95 percent of the welds on vehicles are made by robots; and Mitsubishi Motors reported that about 70 percent

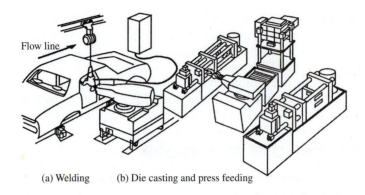

Flow line

(a) Welding   (b) Die casting and press feeding

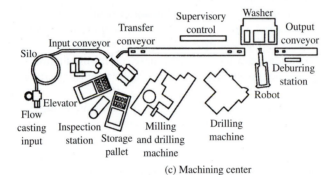

(c) Machining center

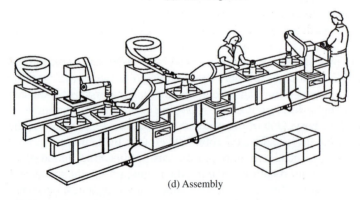

(d) Assembly

Figure 3.9   Illustration of a few common industrial robot applications.
(a) One welding robot is shown, but typically a number of robots would be used along an automotive assembly line. (b) In a die-casting application, a robot unloads die-casting machines, performs quench operations, and loads material into a press. (c) The production machining line is used for producing cam housings. (d) The assembly line uses a combination of robots, parts feeders, and human operators.

of its welding is performed by robots. In these companies, robot downtime averages less than 1 percent.

3.6 SETUP AND TOOLS

One of the most important elements of all forms of work holders, tools, and setups is economics. The amount of tooling up that proves most advantageous depends on (1) the production quantity, (2) repeat business, (3) labor, (4) delivery requirements, and (5) required capital.

The most prevalent mistake of planners and toolmakers is to tie up money in fixtures that may show a large savings when in use, but are seldom used. For example, a savings of 10 percent in direct labor cost on a job in constant use would probably justify greater expense in tools than an 80 or 90 percent savings on a small job that appears on the production schedule only a few times a year. (This is an example of Pareto analysis, from Chapter 2.) The economic advantage of lower labor costs is the controlling factor in determining the tooling; consequently, jigs and fixtures may be desirable, even when only small quantities are involved. Other considerations, such as improved interchangeability, increased accuracy, or labor trouble reduction, may provide the dominant reasons for elaborate tooling, although this is usually not the case. An example of the trade-off between fixturing and tooling costs is discussed in Chapter 9 in the section on break-even charts.

Once the needed amount of tooling has been determined (or if tooling already exists, once the ideal amount needed has been determined), specific considerations for producing the most favorable designs should be evaluated. These are outlined in the Setup and Tooling Evaluation Checklist shown in Figure 3.10.

Setup ties in very closely with tooling, because tooling invariably determines the setup and teardown time. When we speak of setup time, we usually include such items as arriving on the job; procuring instructions, drawings, tools, and material; preparing workstations so that production can begin in the prescribed manner (setting up tools; adjusting stops; setting feeds, speeds, and cut depth; and so on); tearing down the setup; and returning tools to the crib.

Setup operations are especially important in the job shop where production runs tend to be small. Even if this type of shop has modern facilities and puts forth a high effort, it may still have difficulty meeting the competition if setups are too long because of poor planning and inefficient tooling. When the ratio of setup time to production runtime is high, a methods analyst can usually develop several possibilities for setup and tool improvement. One notable option is a group technology system.

The essence of group technology is the classification of the various components of a company's products, so that parts similar in shape and processing sequence are identified numerically. Parts belonging to the same family group, such as rings, sleeves, discs, and collars, are scheduled for production over the same time interval on a general-purpose line arranged in the optimal operational

Fixtures	Yes	No
1. Can the fixture be used to produce other similar designs to advantage?	❑	❑
2. Will the fixture be similar to some other that has been used to advantage? If so, how can you improve on it?	❑	❑
3. Can any stock hardware be used for making the fixture?	❑	❑
4. Can the output be increased by placing more than one part in the fixture?	❑	❑
5. Can the chips be readily removed from the fixture?	❑	❑
6. Are the clamps on the fixture strong enough to prevent them from buckling when they are tightened down on the work?	❑	❑
7. Must any special wrenches be designed to go with the fixture?	❑	❑
8. Must special milling cutters, arbors, or collars be designed to go with the fixture?	❑	❑
9. If the fixture is of the rotary type, have you designed an accurate indexing arrangement?	❑	❑
10. Can the fixture be used on a standard rotary indexing head?	❑	❑
11. Can the fixture be made to handle more than one operation?	❑	❑
12. Have you, in designing the fixture, brought the work as close to the miller's table as possible?	❑	❑
13. Can the work be gaged in the fixture? Can a snap gage be used?	❑	❑
14. Can you use jack pins to help support the work while it is being milled?	❑	❑
15. Have you placed springs under all clamps?	❑	❑
16. Are all steel contact points, clamps, etc., hardened?	❑	❑
17. What kind or class of jigs are you going to design?	❑	❑
18. Can you use a double or triple thread on the screw that holds the work in the jig, so that it will take fewer turns to get the screw out of the way to remove the part more quickly?	❑	❑
19. Can the toolmaker make the jig?	❑	❑
20. Are the legs on the jig long enough to allow the drill, the reamer, or the pilot of the reamer to pass through the part a reasonable distance without striking the table of the drill press?	❑	❑
21. Is the jig too heavy to handle?	❑	❑
22. Is the jig identified with both a location number for storing and a part number that identifies the part or parts that the jig helps produce?	❑	❑
23. Is the work adequately supported so that the clamping force will not bend or distort it?	❑	❑

Parts	Yes	No
1. Has the part undergone any previous operations? If so, can you use any of these points or surfaces to locate or master from?	❑	❑
2. Can the part be quickly placed in the fixture?	❑	❑
3. Can the part be quickly removed from the fixture?	❑	❑
4. Is the part held firmly so that it cannot work loose, spring, or chatter while the cut is being made? (The cut should be against the solid part of the fixture and not against the clamp.)		
5. Can the part be milled in a standard vise by making up a set of special jaws, thus doing away with an expensive fixture?	❑	❑
6. If the part is to be milled at an angle, could the fixture be simplified by using a standard adjustable milling angle?	❑	❑
7. Can lugs be cast on the part to be machined to enable you to hold it?	❑	❑
8. Have you made a note on the drawing, or stamped all loose parts indicating the jig they were made for, so that lost or misplaced parts can be returned to the jig when found?	❑	❑
9. Are all necessary corners rounded?	❑	❑

Drills	Yes	No
1. What takes the thrust of the drill?	❑	❑
2. Can you use any jack pins or screws to support the work while it is being drilled?	❑	❑
3. Are your drill bushings so long that it will be necessary to make up extension drills?	❑	❑
4. Are all clamps located in such a way as to resist or help resist the pressure of the drill?	❑	❑
5. Has the drill press the necessary speeds for drilling and reaming all holes?	❑	❑
6. Must the drill press have a tapping attachment?	❑	❑
7. Drilling and reaming several small holes and only one large one in the jig is not practical since quicker results can be obtained by drilling the small holes on a small drill press, while having only one large one would require the jig to be used on a large machine:		
a. Is it cheaper to drill the large hole in another jig?	❑	❑
b. Will the result of doing so be accurate enough?	❑	❑

Others	Yes	No
1. Can a gage be designed, or hardened pins added, to help the operator set the milling cutters or check up on the work?	❑	❑
2. Is there plenty of clearance for the arbor collars to pass over the work without striking?	❑	❑

Figure 3.10 Setup and tooling evaluation checklist.

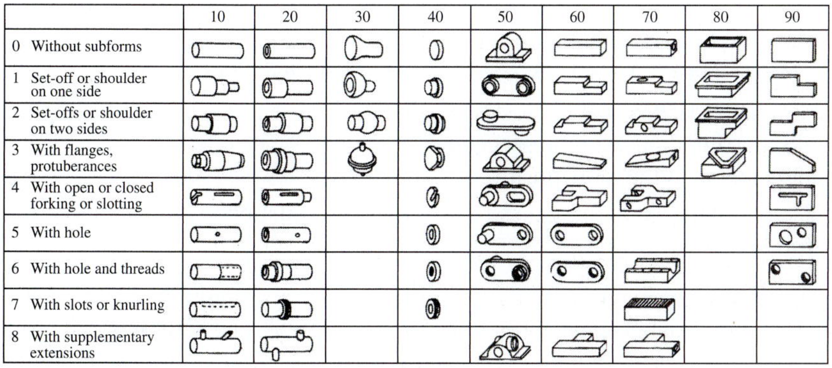

	10	20	30	40	50	60	70	80	90
0 Without subforms									
1 Set-off or shoulder on one side									
2 Set-offs or shoulder on two sides									
3 With flanges, protuberances									
4 With open or closed forking or slotting									
5 With hole									
6 With hole and threads									
7 With slots or knurling									
8 With supplementary extensions									

Figure 3.11 Subdivision of a system grouping for group technology.

sequence. Since both the size and shape of the parts in a given family vary considerably, the line is usually equipped with universal-type, quick-acting jigs and fixtures. This approach also fits in with eliminating the muda of excess inventory and the 5S pillar of standardization.

As an example, Figure 3.11 illustrates a system grouping subdivided into nine classes of parts. Note the similarity of parts within each vertical column. If we were machining a shaft with external threads and a partial bore at one end, the part would be identified as Class 206.

REDUCE SETUP TIME

Just-in-time (JIT) techniques, which have become popular in recent years, emphasize decreasing the setup times to the minimum by simplifying or eliminating them. The SMED (single minute exchange of die) System of the Toyota Production System (Shingo, 1981) is a good example of this approach. A significant portion of setup time can often be eliminated by ensuring that raw materials are within specifications, tools are sharp, and fixtures are available and in good condition. Producing in smaller lots can often prove cost-effective. Smaller lot sizes can lead to smaller inventories, with reduced carrying costs and shelf-life problems, such as contamination, corrosion, deterioration, obsolescence, and theft. The analyst must understand that decreasing the lot size will result in an increase in total setup costs for the same total production quantity over a given period. Several points should be considered in reducing setup time:

1. Work that can be done while the equipment is running should be done at that time. For example, presetting tools for numerical control (NC) equipment can be done while the machine is running.

2. Use the most efficient clamping. Usually, quick-acting clamps that employ cam action, levers, wedges, and so on are much faster, provide adequate force, and are usually a good alternative to threaded fasteners. When

threaded fasteners must be used (for clamping force), C washers or slotted holes can be used so that nuts and bolts do not have to be removed from the machine and can be reused, reducing the setup time on the next job.

3. Eliminate machine base adjustment. Redesigning part fixtures and using preset tooling may eliminate the need for spacers or guide-block adjustments to the table position.

4. Use templates or block gages to make quick adjustments to machine stops.

The time spent in requisitioning tools and materials, preparing the workstation for actual production, cleaning up the workstation, and returning the tools to the tool crib is usually included in setup time. This time is often difficult to control, and the work usually is performed least efficiently. Effective production control can often reduce this time. Making the dispatch section responsible for seeing that the tools, gages, instructions, and materials are provided at the correct time, and that the tools are returned to their respective cribs after the job has been completed, eliminates the need for the operator to leave the work area. The operator then only has to perform the actual setting up and tearing down of the machine. The clerical and routine function of providing drawings, instructions, and tools can be performed by those more familiar with this type of work. Thus, large numbers of requisitions for these requirements can be performed simultaneously, and setup time can be minimized. Here again, group technology can be advantageous.

Duplicate cutting tools should be available, rather than having the operators sharpen their tools. When the operators get new tools, the dull ones are turned in to the tool crib attendant and replaced with sharp ones. Tool sharpening becomes a separate function, and the tools can be standardized more readily.

To minimize downtime, each operator should have a constant backlog of work. The operators should always know what the next work assignment is. A technique frequently used to keep the workload apparent to the operator, supervisor, and superintendent is a board over each production facility, with three wire clips or pockets to receive work orders. The first clip contains all work orders scheduled ahead; the second clip holds the orders currently being worked on; and the last clip holds the completed orders. When issuing work orders, the dispatcher places them in the work-ahead station. At the same time, the dispatcher picks up all completed job tickets from the work-completed station and delivers them to the scheduling department for recording. This system ensures the operators of continuous loads and makes it unnecessary for them to go to the supervisor for their next work assignments.

Making a record of difficult, recurring setups can save considerable setup time when repeat business is received. Perhaps the simplest and yet most effective way to compile a record of a setup is to take a photograph of the setup once it is complete. The photograph should either be stapled to and filed with the production operation card, or placed in a plastic envelope and attached to the tooling prior to storage in the tool crib.

UTILIZE THE FULL CAPACITY OF THE MACHINE

A careful review of many jobs often reveals possibilities for utilizing a greater share of the machine's capacity. For example, a milling setup for a toggle lever was changed so that the six faces were milled simultaneously by five cutters. The old setup required that the job be done in three steps, which meant that the part had to be placed in a separate fixture three different times. The new setup reduced the total machining time and increased the accuracy of the relationship between the six machined faces.

Analysts should also consider positioning one part while another is being machined. This opportunity exists on many milling machine jobs where it is possible to conventional mill on one stroke of the table and climb mill on the return stroke. While the operator is loading a fixture at one end of the machine table, a similar fixture is holding a piece being machined by power feed. As the table of the machine returns, the operator removes the first piece from the machine and reloads the fixture. While this internal work is taking place, the machine is cutting the piece in the second fixture.

In view of the ever-increasing cost of energy, it is important to utilize the most economical equipment to do the job. Several years ago, the cost of energy was such an insignificant proportion of total cost that little attention was given to utilizing the full capacity of machines. There are literally thousands of operations where only a fraction of machine capacity is utilized, with a resulting waste of electric power. In the metal trades industry today, the cost of power is over 2.5 percent of total cost, with strong indications that the present cost of power will increase by at least 50 percent in the next decade. It is highly probable that careful planning to utilize a larger proportion of the capacity of a machine to do the work can effect a 50 percent savings in power usage in many of our plants. Typically, for most motors, if the percent of the rated full load is increased from 25 percent to 50 percent, as much as an 11 percent increase in efficiency could be realized.

INTRODUCE MORE EFFICIENT TOOLING

Just as new processing techniques are continually being developed, new and more efficient tooling should be considered. Coated cutting tools have dramatically improved the critical wear-resistance/breakage-resistance combination. For example, TiC-coated tools have provided a 50 to 100 percent increase in speed over uncoated carbide where each has the same breakage resistance. Advantages include harder surfaces, thus reducing abrasive wear; excellent adhesion to the substrates; low coefficient of friction with most workpiece materials; chemical inertness; and resistance to elevated temperatures.

Carbide tools are usually more cost-effective than high-speed steel tools on many jobs. For example, one company realized a 60 percent savings by changing the milling operation of a magnesium casting. Originally, the base was milled complete in two operations, using high-speed steel milling cutters. An analysis resulted in the employment of three carbide-tipped fly cutters mounted in a special

Figure 3.12   More efficient fixturing and tooling.
(Courtesy of Jergens, Inc.)

holder to mill parts complete. Faster feeds and speeds were possible, and surface finish was not impaired.

Savings can often be achieved by altering tool geometries. Each setup has different requirements that can be achieved only by designing an engineered system that optimizes the feed range for chip control, cutting forces, and edge strength. For example, single-sided low-force geometries may be designed to provide both good chip control and force reduction. In this case, high positive rake angles are grouped to reduce the chip thickness ratio, providing a low cutting force and cutting temperature.

While introducing more efficient tooling, the analyst should develop better methods for holding the work. The work must be held so that it can be positioned and removed quickly (see Figure 3.12). Although the loading of parts is still a manual operation, productivity, as well as equality, will be increased.

3.7   MATERIAL HANDLING

Material handling includes motion, time, place, quantity, and space constraints. First, material handling must ensure that parts, raw materials, in-process materials, finished products, and supplies are moved periodically from location to location. Second, since each operation requires materials and supplies at a particular time, material handling ensures that no production process or customer is hampered by either the early or late arrival of materials. Third, material handling must ensure that materials are delivered to the correct place. Fourth, material handling must ensure that materials are delivered at each location without damage and in the proper quantity. Finally, material handling must consider storage space, both temporary and dormant.

A study conducted by the Material Handling Institute revealed that between 30 and 85 percent of the cost of bringing a product to market is associated with material handling. Axiomatically, the best handled part is the least manually handled part. Whether the distances of the moves are large or small, these moves should be scrutinized. The following five points should be considered for reducing the time spent in handling material: (1) reduce the time spent in picking up material; (2) use mechanized or automated equipment; (3) make better use of existing handling facilities; (4) handle material with greater care; and (5) consider the application of bar coding for inventory and related applications.

A good example of the application of these five points is the evolution of warehousing; the former storage center has become an automated distribution center. Today, the automated warehouse uses computer control for material movement, as well as information flow through data processing. In this type of automated warehouse, receiving, transporting, storing, retrieving, and controlling inventory are treated as an integrated function.

REDUCE THE TIME SPENT IN PICKING UP MATERIAL

Material handling is often thought of as only transportation, neglecting consideration of positioning at the workstation, which is equally important. Since it is often overlooked, workstation positioning of material may offer even greater opportunities for savings than does transportation. Reducing the time spent in picking up material minimizes tiring, costly manual handling at the machine or the workplace. It gives the operator a chance to do the job faster with less fatigue and greater safety.

For example, consider eliminating loose piling on the floor. Perhaps the material can be stacked directly on pallets or skids after being processed at the workstation. This can result in a substantial reduction of terminal transportation time (the time that material handling equipment stands idle while loading and unloading take place). Usually some type of conveyor or mechanical fingers can bring material to the workstation, thus reducing or eliminating the time needed to pick up the material. Plants can also install gravity conveyors, in conjunction with the automatic removal of finished parts, thus minimizing material handling at the workstation. Figure 3.13 shows examples of typical handling equipment.

Interfaces between different types of handling and storage equipment should be studied to develop more efficient arrangements. For example, the sketch in Figure 3.14 shows the order picking arrangements, depicting how materials can be removed from the reserve or staging storage either by a worker-aboard order picking vehicle (left), or manually (right). A lift truck can be used to replenish pallet racks.

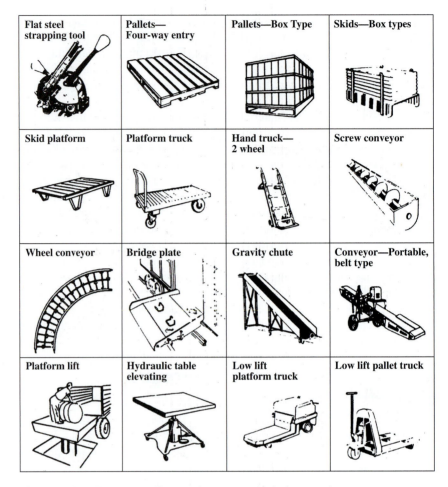

Figure 3.13 Typical handling equipment used in industry today.
(*Source:* The Material Handling Institute.)

After the required items are removed from the flow rack, they are sent by conveyor to order accumulation and packaging operations.

USE MECHANICAL EQUIPMENT

Mechanizing the handling of material usually reduces labor costs, reduces materials damage, improves safety, alleviates fatigue, and increases production. However, care must be exercised in selecting the proper equipment and methods. Equipment standardization is important because it simplifies operator training, allows equipment interchangeability, and requires fewer repair parts.

The savings possible through the mechanization of material handling equipment are typified by the following examples. In the original design of a circuit board assembly task, the operator would go to the storage crib, select the proper electronic components required for a specific board based on its "plug" list, return to the workbench, and then proceed to insert the components into the board in

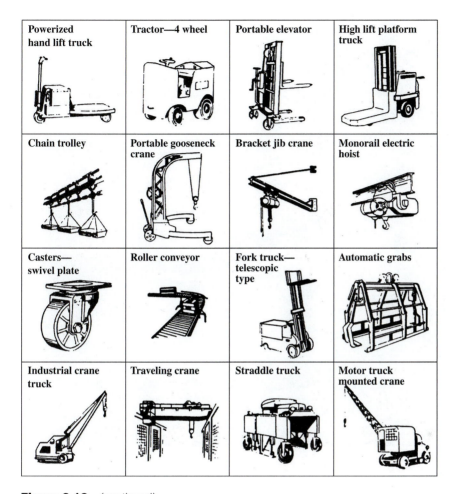

Figure 3.13 *(continued)*

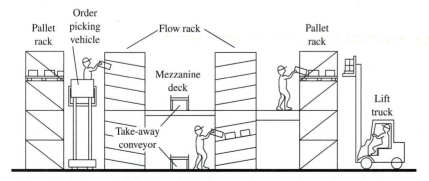

Figure 3.14 Schematic of efficient warehousing operations.

Figure 3.15 Work area of vertical storage machine used in the assembly of computer panels.

accordance with the plug list. The improved method utilizes two automated, vertical storage machines, each with 10 carriers and four pullout drawers per carrier (see Figure 3.15). The carriers move up and around in a system that is a compressed version of a Ferris wheel. With 20 possible stop positions on call, the unit always selects the closest route—either forward or backward—to bring the proper drawers to the opening in the shortest possible time. From a seated position, the operator dials the correct stop, pulls open the drawer to expose the needed components, one withdraws the proper one, and places it in the board. The improved method has reduced the required storage area by approximately 50 percent, improved workstation layout, and substantially reduced populating errors by minimizing operator handling, decision making, and fatigue.

Often, an automated guided vehicle (AGV) can replace a driver. AGVs are successfully used in a variety of applications, such as mail delivery. Typically, these vehicles are not programmed; rather, they follow a magnetic or optical guide for a planned route. Stops are made at specific locations for a predetermined

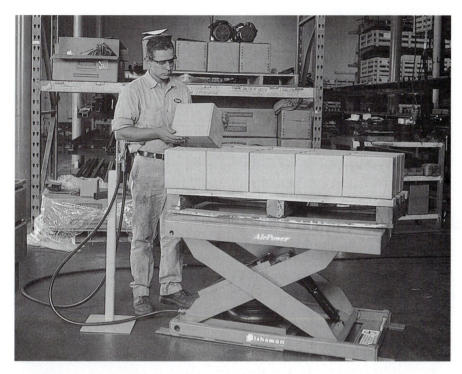

Figure 3.16 Hydraulic lift table used to minimize manual lifting. (Courtesy of Bishamon.)

period, giving an employee adequate time for unloading and loading. By pressing a "hold" button and then pressing a "start" button at the conclusion of the loading/unloading operation, the operator can lengthen the dwell period at each stop. AGVs can be programmed to go to any location over more than one path. They are equipped with sensing and control instrumentation to avoid collisions with other vehicles. Also, when such guide path equipment is used, material handling costs vary little with distance.

   Mechanization is also useful for manual materials handling, such as palletizing. There are a variety of devices under the generic label of lift tables, which eliminate most of the lifting required of the operator. Some lift tables are spring-loaded, which, when set with a proper spring stiffness, will adjust automatically to the optimal height for the operator as boxes are placed on a pallet on top of the lift. (See Chapter 4 for a discussion on the determination of optimal lifting heights.) Others are pneumatic (see Figure 3.16) and can be easily adjusted with a control, so that lifting is eliminated and material can be slid from one surface to another. Some tilt for easier access into bins, while others rotate, facilitating palletizing. In general, lift tables are probably the least expensive engineering control measure used in conjunction with the NIOSH[1] lifting guidelines (see Chapter 4).

_____

[1]NIOSH is the National Institute of Occupational Safety and Health.

EXAMPLE 3.1   Maximum Net Load That Can Safely Be Handled by Fork Trucks

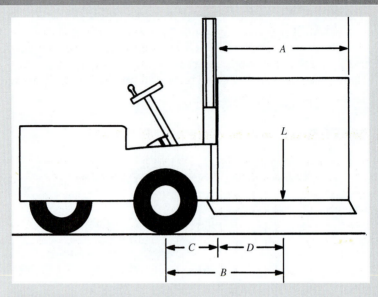

Figure 3.17   Typical forklift truck.

Start by computing the torque rating by multiplying the distance from the center of the front axle to the center of the load (see Figure 3.17):

$$Load = torque\ rating/B$$

where B is distance C + D, with D = A/2.

If the distance C from the center of the front axle to the front end of the fork truck is 18 in and the length of pallet A is 60 in, then the maximum gross weight that a 200,000. in·lb fork truck should handle would be

$$L = \frac{200,00}{18 + 60/2} = 4,167\ lb$$

By planning the pallet size to make full use of the equipment, the company can realize a greater return from the material handling equipment.

MAKE BETTER USE OF EXISTING HANDLING FACILITIES

To ensure the greatest return from material handling equipment, that equipment must be used effectively. Thus, both the methods and the equipment should be sufficiently flexible that a variety of material handling tasks can be accomplished under variable conditions. Palletizing material in temporary and permanent storage allows greater quantities to be transported faster than storing material without the use of pallets, saving up to 65 percent in labor costs. Sometimes, material can be handled in larger or more convenient units by designing special racks. When this is done, the compartments, hooks, pins, or supports for holding the work should be in multiples of 10 for ease of counting during processing and final inspection. If any material handling equipment is used only part of the time, consider the possibility of putting it to use a greater share of the time. By relocating production facilities or adapting material handling equipment to diversified areas of work, companies may achieve greater utilization.

HANDLE MATERIAL WITH GREATER CARE

Industrial surveys indicate that approximately 40 percent of plant accidents happen during material handling operations. Of these, 25 percent are caused by lifting and shifting material. By exercising greater care in handling material, and using mechanical mechanisms wherever possible for material handling, employees can reduce fatigue and accidents. Records prove that the safe factory is also an efficient factory. Safety guards at points of power transmission, safe operating practices, good lighting, and good housekeeping are essential to making material handling equipment safer. Workers should install and operate all material handling equipment in a manner compatible with existing safety codes.

Better handling also reduces product damaged. If the number of reject parts is at all significant in the handling of parts between workstations, then this area should be investigated. Usually, parts damaged during handling can be minimized if specially designed racks or trays are fabricated to hold the parts immediately after processing. For example, one manufacturer of aircraft engine parts incurred a sizable number of damaged external threads on one component that was stored in metal tote pans after the completion of each operation. When two-wheeled hand trucks moved the filled tote pans to the next workstation, the machined forgings bumped against one another and against the sides of the metal pan to such an extent that they became badly damaged. Someone investigated the cause of the rejects and suggested making wooden racks with individual compartments to support the machined forgings. This prevented the parts from bumping against one another or the metal tote pan, thereby significantly reducing the number of damaged parts. Production runs were also more easily controlled because of the faster counting of parts and rejects.

Similar considerations apply to service industries and the health care sector, not only from the standpoint of the "product," which in many cases is a person, but also with respect to the material handler. For example, patient handling in hospitals and personal care facilities is a major factor in low back and shoulder injuries of nurses. Traditionally, relatively immobile patients are moved from a bed to a wheelchair or vice versa with the use of a walking belt (see Figure 3.18a). However, these maneuvers require considerable amounts of strength and generate very high levels of low back compressive forces (see Section 4.4). Assist devices, such as the Williamson Turn Stand, require much less strength from

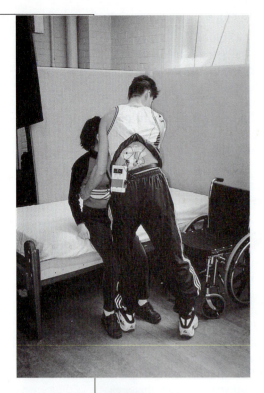

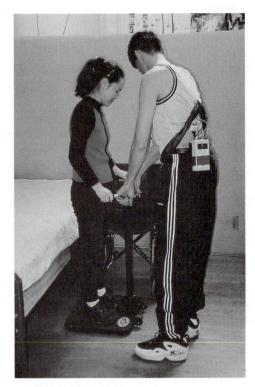

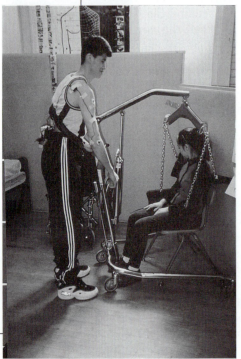

Figure 3.18 Patient handling using three different assists.
(a) Traditional walking belt requires considerable strength and generates very high low-back compressive forces. (b) The Williamson Turn Stand requires much less strength and is less stressful for the low back. (c) A Hoyer-type lift requires even less strength, but is considerably more expensive and cumbersome in small spaces.

the nurse and generate much less stress on the low back (see Figure 3.18b). However, the patient must have the leg strength to maintain body weight, with some support. Finally, a Hoyer-type lift requires even less strength, but is considerably more expensive and cumbersome in small areas (see Figure 3.18c).

CONSIDER BAR CODING FOR INVENTORY AND RELATED APPLICATIONS

The majority of technical people have some familiarity with bar coding and bar code scanning. Bar coding has shortened queues at grocery and department store checkout lines. The black bars and white spaces represent digits that uniquely identify both the item and the manufacturer. Once this Universal Product Code (UPC) is scanned by a reader at the checkout counter, the decoded data are sent to a computer that records timely information on labor productivity, inventory status, and sales. The following five reasons justify the use of bar coding for inventory and related applications:

1. *Accuracy.* Typically representative performance is less than 1 error in 3.4 million characters. This compares favorably with the 2 to 5 percent error that is characteristic of keyboard data entry.
2. *Performance.* A bar code scanner enters data three to four times faster than typical keyboard entry.
3. *Acceptance.* Most employees enjoy using the scanning wand. Inevitably, they prefer using a wand to keyboard entry.
4. *Low cost.* Since bar codes are printed on packages and containers, the cost of adding this identification is extremely low.
5. *Portability.* An operator can carry a bar code scanner into any area of the plant to determine such things as inventories and order status, etc.

Bar coding is useful for receiving, warehousing, job tracking, labor reporting, tool crib control, shipping, failure reporting, quality assurance, tracking, production control, and scheduling. For example, the typical storage bin label provides the following information: part description, size, packing quantity, department number, storage number, basic stock level, and order point. Considerable time can be saved by using a scanning wand to gather these data for inventory reordering.

Some practical applications reported by Accu-Sort Systems, Inc., include automatically controlling conveyor systems; diverting material to the location where it is needed; and providing material handlers with clear, concise instructions about where to take materials, automatically verifying that the proper material is handled. If bar coding is incorporated into programmable controllers and automatic packaging equipment, online real-time verification of packing labels with container contents can be used to avoid costly product recalls.

SUMMARY: MATERIAL HANDLING

Analysts should always be looking for ways to eliminate inefficient material handling without sacrificing safety. To assist the methods analyst in this endeavor, the Materials Handling Institute (1998) has developed 10 principles of material handling.

1. *Planning principle.* All material handling should be the result of a deliberate plan in which the needs, performance objectives, and functional specifications of the proposed methods are completely defined at the outset.

2. *Standardization principle.* Material handling methods, equipment, controls, and software should be standardized within the limits of achieving overall performance objectives and without sacrificing needed flexibility, modularity, and throughput.

3. *Work principle.* Material handling work should be minimized without sacrificing productivity or the level of service required of the operation.

4. *Ergonomic principle.* Human capabilities and limitations must be recognized and respected in the design of material handling tasks and equipment, to ensure safe and effective operations.

5. *Unit load principle.* Unit loads shall be appropriately sized and configured in a way that achieves the material flow and inventory objectives at each stage in the supply chain.

6. *Space utilization principle.* Effective and efficient use must be made of all available space.

7. *System principle.* Material movement and storage activities should be fully integrated to form a coordinated, operational system that spans receiving, inspection, storage, production, assembly, packaging, unitizing, order selection, shipping, transportation, and returns handling.

8. *Automation principle.* Material handling operations should be mechanized and/or automated where feasible, to improve operational efficiency, increase responsiveness, improve consistency and predictability, decrease operating costs, and eliminate repetitive or potentially unsafe manual labor.

9. *Environmental principle.* Environmental impacts and energy consumption are criteria to be considered when designing or selecting alternative equipment and material handling systems.

10. *Life-cycle-cost principle.* A thorough economic analysis should account for the entire life cycle of all material handling equipment and resulting systems.

To reiterate, the predominant principle is that the less a material is handled, the better it is handled, which fits in nicely with eliminating the mudas of unnecessary transportation and unnecessary motions.

3.8 PLANT LAYOUT

The principal objective of effective plant layout is to develop a production system that permits the manufacture of the desired number of products with the desired quality at the least cost. Physical layout is an important element of an entire production system that embraces operation cards, inventory control, material handling, scheduling, routing, and dispatching. All these elements must be carefully integrated to fulfill the stated objective. Poor plant layouts result in major costs. The indirect labor expense of long moves, backtracking, delays, and work stoppages due to bottlenecks in the transportation muda are characteristic of a plant with an antiquated and costly layout.

LAYOUT TYPES

Is there one type of layout that tends to be the best? The answer is no. A given layout can be best in one set of conditions and yet poor in a different set of conditions. In general, all plant layouts represent one or a combination of two basic layouts: *product* or *straight-line layouts* and *process* or *functional layouts.* In the straight-line layout, the machinery is located such that the flow from one operation to the next is minimized for any product class. In an organization that utilizes this technique, it would not be unusual to see a surface grinder located between a milling machine and a turret lathe, with an assembly bench and plating tanks in the immediate area. This type of layout is quite popular for certain mass-production manufacture, because material handling costs are lower than for process grouping.

Product layout has some distinct disadvantages. Since a broad variety of occupations are represented in a relatively small area, employee dissatisfaction can escalate. This is especially true when different opportunities carry a significant money rate differential. Because unlike facilities are grouped together, operator training can be more cumbersome, especially if an experienced employee is not available in the immediate area to train a new operator. The problem of finding competent supervisors is also exacerbated, due to the variety of facilities and jobs that must be supervised. Then, too, this type of layout invariably necessitates a larger initial investment because duplicate service lines are required, such as air, water, gas, oil, and power. Another disadvantage of product grouping is the fact that this arrangement tends to appear disorderly and chaotic. With these conditions, it is often difficult to promote good housekeeping. In general, however, the disadvantages of product grouping are more than offset by the advantages, if production requirements are substantial.

Process layout is the grouping of similar facilities. Thus, all turret lathes would be grouped in one section, department, or building. Milling machines, drill presses, and punch presses would also be grouped in their respective sections. This type of arrangement gives a general appearance of neatness and orderliness, and tends to promote good housekeeping. Another advantage of functional layout is the ease with which a new operator can be trained. Surrounded by experienced employees operating similar machines, the new worker has a greater opportunity to learn from them. The problem of finding competent supervisors is lessened, because the job demands are not as great. Since these supervisors need only be familiar with one general type or class of facilities, their backgrounds do not have to be as extensive as those of supervisors in shops using product grouping.

Also, if production quantities of similar products are limited and there are frequent "job" or special orders, a process layout is more satisfactory.

The disadvantage of process grouping is the possibility that long moves and backtracking will be needed on jobs that require a series of operations on diversified machines. For example, if the operation card of a job specifies a sequence of drill, turn, mill, ream, and grind, the movement of the material from one section to the next could prove extremely costly. Another major disadvantage of process grouping is the large volume of paperwork required to issue orders and control production between sections.

TRAVEL CHARTS

Before designing a new layout or correcting an old one, analysts must accumulate the facts that may influence that layout. *Travel* or *from-to charts* can be helpful in diagnosing problems related to the arrangement of departments and service areas, as well as the location of equipment within a given sector of the plant. The travel chart is a matrix that presents the magnitude of material handling that takes place between two facilities per time period. The unit identifying the amount of handling may be whatever seems most appropriate to the analyst. It can be pounds, tons, handling frequency, and so on. Figure 3.19 illustrates a very elementary travel chart from which the analyst can deduce that of the all machines, No. 4 W&S turret lathe and No. 2 Cincinnati Horizontal mill should be next to each other because of the high number of items (200) passing between the two machines.

MUTHER'S SYSTEMATIC LAYOUT PLANNING

A systematic approach to plant layout developed by Muther (1973) is termed systematic layout planning (SLP). The goal of SLP is to locate two areas with high frequency and logical relationships close to one another using a straightforward six-step procedure:

1. **Chart relationships.** In the first step, the relationships between different areas are established and then charted on a special form called the *relationship chart* (or rel chart for short; see Figure 3.20). A relationship is the relative degree of closeness, desired or required, among different activities, areas, departments, rooms, etc., as determined from quantitative flow information (volume, time, cost, routing) from a from-to chart, or more qualitatively from functional interactions or subjective information. For example, although painting may be the logical step between finishing and final inspection and packing, the toxic materials and hazardous or flammable conditions may require that the paint area be completely separated from the other areas. The relationship ratings range in value from 4 to –1, based on the vowels that semantically define the relationship, as shown in Table 3.3.

2. **Establish space requirements.** In the second step, space requirements are established in terms of square footage. These values can be calculated based on production requirements, extrapolated from existing areas,

From \ To	No. 4 W. & S. Turret Lathe	Delta 17" Drill Press	2-Spindle L. & G. Drill	No. 2 Cinn. Hor. Mill	No. 3B. & S. Verticle Mill	Niagara 100Ton Press	No. 2 Cinn. Centerless	No. 3 Excello Thd. Grinder
No. 4 W. & S. Turret Lathe		20	45	80	32	4	6	2
Delta 17" Drill Press			6	8	4	22	2	3
2-Spindle L. & G. Drill				22	14	18	4	4
No. 2 Cinn. Hor. Mill	120				10	5	4	2
No. 3B. & S. Verticle Mill						6	3	1
Niagara 100Ton Press		60	12	2			0	1
No. 2 Cinn. Centerless		15						15
No. 3 Excello Thd. Grinder				15	8			

Figure 3.19 The travel chart is a useful tool in solving material handling and plant layout problems related to process-type layouts. The chart enumerates the number of items (per given time period) or the volume (e.g., tons per shift) transported between the different machines.

projected for future expansion, or fixed by legal standards, such as the ADA or architectural standards. In addition to square footage, the kind and shape of the area being laid out, or the location with respect to required utilities, may be very important.

3. **Activity relationships diagram.** In the third step, a visual representation of the different activities is drawn. The analyst starts with the absolutely important relationships (A's), using four short, parallel lines to join the two areas. The analyst then proceeds to the E's, using three parallel lines approximately double the length of the A lines. The analyst continues this procedure for the I's, O's, etc., progressively increasing the length of the lines, while attempting to avoid crossing or tangling the lines. For undesirable relationships, the two areas are placed as far apart as possible, and a squiggly line (representing a spring) is drawn between them. (Some

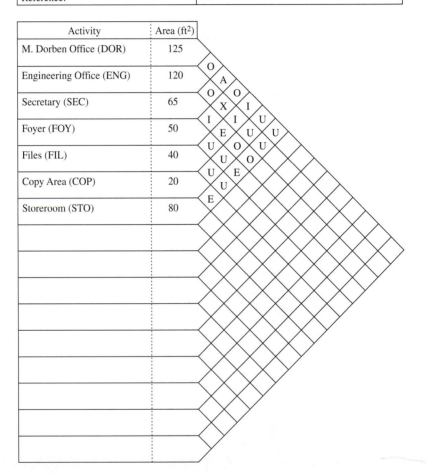

Relationship Chart   Page 1 of 1

Project: Construction of new office	Remarks:
Plant: Dorben Consulting	
Date: 6-9-97	
Charted By: AF	
Reference:	

Activity	Area (ft²)
M. Dorben Office (DOR)	125
Engineering Office (ENG)	120
Secretary (SEC)	65
Foyer (FOY)	50
Files (FIL)	40
Copy Area (COP)	20
Storeroom (STO)	80

Figure 3.20   Relationship chart for Dorben Consulting.

analysts may also define an extremely undesirable relationship with a
-2 value and a double squiggly line.)

4. **Layout space relationships.** Next, a spatial representation is created by
    scaling the areas in terms of relative size. Once the analyst is satisfied with
    the layout, the areas are compressed into a floor plan. This is typically not
    as easy as it sounds, and the analyst may want to utilize templates. In
    addition, modifications may be made to layout based on material handling

Table 3.3 SLP Relationship Ratings

Relationship	Closeness Rating	Value	Diagram Lines	Color
Absolutely necessary	A	4	≡≡≡	Red
Especially important	E	3	≡≡≡	Yellow
Important	I	2	═══	Green
Ordinary	O	1	────	Blue
Unimportant	U	0		
Not desirable	X	−1	⋀⋀⋀⋀	Brown

requirements (e.g., a shipping or receiving department would necessarily be located on an exterior wall), storage facilities (perhaps similar exterior access requirements), personnel requirements (a cafeteria or restroom located close by), building features (crane activities in a high bay area; forklift operations on the ground floor), and utilities.

5. **Evaluate alternative arrangements.** With numerous possible layouts, it would not be unusual to find that several appear to be equally likely possibilities. In that case, the analyst will need to evaluate the different alternatives to determine the best solution. First, the analyst will need to identify factors deemed important: for example, future expansion capability, flexibility, flow efficiency, material handling effectiveness, safety, supervision ease, appearance or aesthetics, etc. Second, the relative importance of these factors will need to be established through a system of weights, such as a 0-to-10 basis. Next, each alternative is rated for satisfying each factor. Muther (1973) suggests the same 4 to −1 scale: 4 is almost perfect; 3, especially good; 2, important; 1, ordinary result; 0, unimportant; and −1, not acceptable. Each rating is then multiplied by the weight. The products for each alternative are summed, with the largest value indicating the best solution.

6. **Select layout and install.** The final step is to implement the new method.

Plant Layout of Dorben Consulting Using SLP	**EXAMPLE 3.2**

The Dorben Consulting group would like to lay out a new office area. There are seven activity areas: M. Dorben's office, engineering office (occupied by two engineers), secretarial area, foyer and waiting area for visitors, file area, copy area, and storeroom. The activity relationships are subjectively assessed by M. Dorben to be as shown in the rel chart in Figure 3.20. The chart also indicates space allotments for each area, ranging from a low of 20 ft² for the copy area to 125 ft² for M. Dorben's office. For example, the relationship between M. Dorben and the secretary is deemed absolutely important (A), while the relationship between the engineering area and the foyer is deemed not desirable (X), so that the engineers are not disrupted in their work by visitors.

A relatively good first attempt at an activity relationship diagram yields Figure 3.21. Adding in the relative size of each area yields the space relationship chart in Figure 3.22. Compressing the areas yields the final floor plan in Figure 3.23.

Since Dorben's office and the engineering area are practically the same size, they could easily be interchanged, leaving two alternative layouts. These are evaluated (Figure 3.24) on the basis of personnel isolation (which is very important to M. Dorben, yielding a high weight of 8), supplies movement, visitor reception, and flexibility. The big difference in the layouts is the closeness of the engineering area to the foyer. Thus, alternative B (shown in Figure 3.23) at 68 points, compared to 60 points for alternative A, turns out to be the preferred layout.

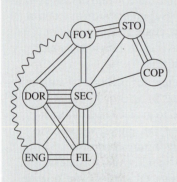

Figure 3.21 Activity relationship diagram for Dorben Consulting.

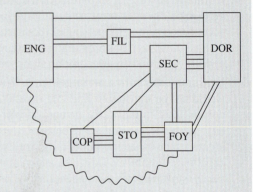

Figure 3.22 Space relationship layout for Dorben Consulting.

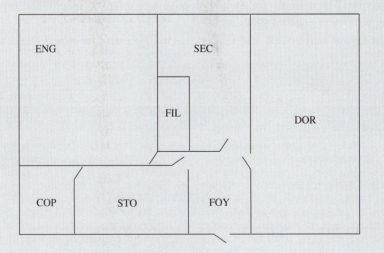

Figure 3.23 Floor plan for Dorben Consulting.

Evaluating Alternatives								Page 1 of 1
Plant: Dorben Consulting	**Alternatives**	**A**	**B**	**C**	**D**	**E**		
Project: New Office Construction		Dorben office facing west	Dorben office facing east					
Date: 6-9-97								
Analyst: AF								
Factor/Consideration	**Wt.**	**Ratings and Weighted Ratings**						**Comments**
		A		**B**		**C**		**D**		**E**		

Factor/Consideration	Wt.	A		B		C		D		E		Comments
Isolation of personnel	8	1	8	3	24							
Movement of supplies	4	3	12	3	12							
Visitor reception	4	4	16	4	16							
Flexibility	8	3	24	2	16							
Totals		60		68								

Remarks:
Alternative B, with Dorben's office facing east and engineers' office facing west, lessens disruptions of the engineers' work due to visitors.

Figure 3.24 Evaluating alternatives for Dorben Consulting.

COMPUTER-AIDED LAYOUT

Commercially available software can help analysts develop realistic layouts rapidly and inexpensively. The Computerized Relative Allocation Facilities (CRAFT) program is one that has been used extensively. An activity center could be a department or work center within a department. Any one activity center can be identified as fixed, freezing it and allowing freedom of movement in those that can be readily moved. For example, it is often desirable to freeze such activity centers as elevators, restrooms, and stairways. Input data include fixed work center numbers and locations, material handling costs, interactivity center flow, and a block layout representation. The governing heuristic algorithm asks, What change in material handling costs would result if work centers were exchanged? Once the answer is stored, the computer proceeds in an iterative manner until it converges on a good solution. CRAFT calculates the distance matrix as the rectangular distances from the department centroids.

Another layout program is CORELAP. The input requirements for CORELAP are the number of departments, the departmental areas, the departmental relationships, and the weights for these relationships. CORELAP constructs layouts by locating the departments, using rectangular areas. The objective is to provide a layout with "high-ranking" departments close together.

ALDEP, still another layout program, constructs plant layouts by randomly selecting a department and locating it in a given layout. The relationship chart is then scanned, and a department that has a high closeness rating is introduced into the layout. This process continues until the program places all departments. ALDEP then computes a score for the layout, and repeats the process a specific number of times. The program also has the ability to provide multifloor layouts.

All these plant layout programs were originally developed for large mainframe computers. With the advent of personal computers, the algorithms have been incorporated into PC programs, as have other algorithms. One such program, SPIRAL, attempts to optimize the adjacency relationship by summing the positive relationships and deducting the negative relationships for adjacent areas. This is essentially a quantified Muther's approach and is described in greater detail in Goetschalckx (1992). For example, entering the data for the Dorben Consulting example yields a slightly different layout, as shown in Figure 3.25.

Note that there is a tendency to generate long, narrow rooms, to minimize the distance between room centers. This is an especially big problem with CRAFT, ALDEP, etc. SPIRAL at least attempts to modify this tendency by adding a shape penalty. Also, there is a tendency for many of these programs (i.e., those that are *improvement* programs, such as CRAFT, that build upon an initial layout) to reach a local minimum and not attain the optimum layout. This problem can be circumvented by starting with alternate layouts. This is less a problem with *construction* programs, such as SPIRAL, which generate a solution from scratch. A more powerful and perhaps more useful set of programs are FactoryPLAN, FactoryFLOW, and FactoryCAD, which input existing AutoCAD files of floor plans and create very detailed layouts suitable for architectural planning.

3.9 WORK DESIGN

Because of the recent regulatory (i.e., OSHA) and health (i.e., rising medical and workers' compensation costs) concerns, work design techniques will be covered in detail in separate chapters. Chapter 4 addresses manual work and the principles of motion economy; Chapter 5 addresses ergonomic principles of workplace and

Figure 3.25

SPIRAL input files:
(a) DORBEN.DAT,
(b) DORBEN.DEP, and
(c) resulting layout for
the DORBEN Consulting
example.

a)

[project_name]	DORBEN
[number_of_departments]	7
[department_file_name]	DORBEN.DEP
[building_width]	25
[building_depth]	20
[seed]	12345
[tolerance]	0.00010
[time_limit]	120
[number_of_iterations]	20
[report_level]	2
[max_shape_ratio]	2.50
[shape_penalty]	500.00

b)

DOR	0	0	125	0	0	GREEN	Dorben
ENG	0	0	120	0	0	BLUE	Engineers
SEC	0	0	65	0	0	RED	Secretary
FOY	0	0	50	0	0	YELLOW	Foyer
FIL	0	0	40	0	0	BROWN	Files
COP	0	0	20	0	0	GRAY	Copy
STO	0	0	80	0	0	BLACK	Storeroom
DOR	ENG	1					
DOR	SEC	4					
DOR	FOY	1					
DOR	FIL	2					
ENG	SEC	1					
ENG	FOY	-1					
ENG	FIL	2					
SEC	FOY	2					
SEC	FIL	3					
SEC	COP	1					
SEC	STO	1					
FOY	STO	3					
COP	STO	3					
OUT	OUT	0					

c)

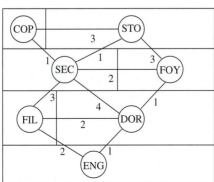

tool design; Chapter 6 covers working and environmental conditions; Chapter 7 presents cognitive work with respect to informational input from displays, information processing, and interaction with computers; and Chapter 8 addresses workplace and systems safety.

SUMMARY

The nine primary approaches to operation analysis represent a systematic approach to analyzing the facts presented on the operation and flow process charts. These principles are just as applicable to the planning of new work as to the improvement of work already in production. While decreased waste, increased output, and improved quality, consistent with lean manufacturing principles, are the primary outcomes of operation analysis, it also provides benefits to all workers with better working conditions and methods.

A systematic method for remembering and applying the nine operation analyses is offered by a checklist of pertinent questions, as shown in Figure 3.26. In the figure, the checklist demonstrates how its use resulted in a cost reduction on an electric blanket control knob shaft. Redesigning the shaft so that it could be economically produced as a die casting rather than a screw machine part reduced factory cost from $68.75 per 1,000 pieces to $17.19 per 1,000 pieces. This check sheet is also useful as an outline in providing methods training to factory foremen and superintendents. Thought-provoking questions, when intelligently used, help factory supervisors to develop constructive ideas and assist in operations analysis.

QUESTIONS

1. Explain how design simplification can be applied to the manufacturing process.
2. How is operation analysis related to methods engineering?
3. How do unnecessary operations develop in an industry?
4. Compare and contrast operations analysis with the lean manufacturing approach.
5. What are the seven mudas?
6. What are the 5S pillars?
7. What is meant by "tight" tolerances?
8. Explain why it may be desirable to "tighten up" tolerances and specifications.
9. What is meant by lot-by-lot inspection?
10. When is an elaborate quality control procedure not justified?
11. What six points should be considered when endeavoring to reduce material cost?
12. How does a changing labor and equipment situation affect the cost of purchased components?
13. Explain how rearranging operations can result in savings.
14. What process is usually considered the fastest for forming and sizing operations?
15. How should the analyst investigate the setup and tools to develop better methods?

Date 9/15	Dept. 11	Dwg. 18-4612	Sub. 2	
Mould	Die	Style	Item 2	
Pattern	Ins. Spec. C	L. Spec.	Sub.	

Part Description Blanket control knob shaft

Operation Turn, groove, drill, tap, knurl, thread, cut-off   Operator Blazer

DETERMINE AND DESCRIBE	DETAILS OF ANALYSIS
1. PURPOSE OF OPERATION To form contours of 3/8" S.A.E. 1112 rod on automatic screw machine to achieve drawing specifications.	Can purpose be accomplished better otherwise? *Yes—by die casting*

2. COMPLETE LIST OF ALL OPERATIONS PERFORMED ON PART
No. Description Work Sta. Dept.
1. Turn, groove, drill, tap, knurl, thread, cut-off B. & S. 11
2. Burr Bench 12
3. Inspect 1% Bench 18
4.
5.
6.
7.
8.
9.
10.
	Can oprn. being analyzed
	  be eliminated? *No*
	  be combined with another? *No*
	  be performed during idle period of another? *Yes, by machine coupling*
	Is sequence of oprns. best possible? *Yes*
	Should oprn. be done in another dept. to save cost or handling? *Perhaps can be purchased outside at a savings.*

3. INSPECTION REQUIREMENTS
a—Of previous oprn.
 	
b—Of this oprn. Yes. Perhaps S.Q.C. will reduce amount of inspection.
 	
c—Of next oprn.
	Are tolerance, allowance,
	finish and other requirements
	  necessary?
	  too costly?
	  suitable to purpose?

4. MATERIAL Zinc base die cast metal would be less expensive. Cutting compounds and other supply materials	Consider size, suitability, straightness, and condition. Can cheaper material be substituted?
5. MATERIAL HANDLING a—Brought by 4 wheel truck to automatics b—Removed by hand 2 wheel trucks c—Handled at work station by	Should crane, gravity conveyors, totepans, or special trucks be used? Consider layout with respect to distance moved. *Perhaps gravity to burring station.*

6. SET-UP (Accompany description with sketches if necessary)
  This is satisfactory as being done.
 	
a—Tool Equipment
  Present
 	
  Suggestions
  Redesign part to be made as zinc base die casting rather than S.A.E. 1112 screw machine part.
	How are dwgs. and tools secured?
	Can set-up be improved?
	Trial pieces.
	Machine Adjustments.
	
	Tools
	Suitable?
	Provided?
	  Ratchet Tools
	  Power Tools
	  Spl. Purpose Tools
	  Jigs, Vises
	  Special Clamps
	  Fixtures
	    Multiple
	    Duplicate

Methods Engineering Council
Form No. 101   **ANALYSIS SHEET**

Figure 3.26 Operations analysis checklist for manufacture of blanket control knob shaft.

7. CONSIDER THE FOLLOWING POSSIBILITIES.

1. Install gravity delivery chutes.
2. Use drop delivery
3. Compare methods if more than one operator is working on same job.
4. Provide correct chair for operator.
5. Improve jigs or fixtures by providing ejectors, quick-acting clamps, etc.
6. Use foot operated mechanisms.
7. Arrange for two handed operation.
8. Arrange tools and parts within normal working area.
9. Change layout to eliminate back tracking and to permit coupling of machines.
10. Utilize all improvements developed for other jobs.

RECOMMENDED ACTION

yes, to accumulate for tumbling.

8. WORKING CONDITIONS.

Generally satisfactory.

a—Other Conditions

Light *O.K.*
Heat *O.K.*
Ventilation, Fumes *O.K.*
Drinking Fountains *O.K.*
Wash Rooms *O.K.*
Safety Aspects *O.K.*
Design of Part *O.K.*
Clerical Work Required (to fill out time cards, etc.) *O.K.*
Probability of Delays *O.K.*
Probable Mfg. Quantities *O.K.*

9. METHOD (Accompany with sketches or Process Charts if necessary.)

a—Before Analysis and Motion Study.

Control knob shaft designed as screw machine part.

Arrangement of Work Area
Placement of
  Tools.
  Materials.
  Supplies.
Working Posture
Does method follow Laws of Motion Economy?
Are lowest classes of movements used?

b—After Analysis and Motion Study

Parting line

Control knob redesigned as die cast part. Threads on left-hand extension cover only 50 percent of periphery; likewise knurl on right end extension on half of periphery thus allowing piece to be easily removed from die.

See Supplementary Report Entitled *Die Cast Control shaft.*
Date

OBSERVER ___ R. Guild ___   APPROVED BY ___ R. Hussey ___

Figure 3.26 (continued)

16. Give some applications of bar coding for the improvement of productivity.
17. What are the two general types of plant layout? Explain each in detail.
18. What is the best way to test a proposed layout?
19. Which questions should the analyst ask when studying work performed at a specific workstation?
20. Explain the advantages of using a checklist.
21. In connection with automated guided vehicles, why do costs vary little with distance?
22. On what does the extent of tooling depend?
23. How can planning and production control affect setup time?
24. How can a material best be handled?
25. How is the travel chart related to Muther's SLP?
26. Why does the travel chart have greater application in process layout than in product layout?
27. Explain the fundamental purpose of group technology.
28. Explain how the conservation of welding rods can result in 20 percent material savings.
29. Identify several automobile components that have been converted from metal to plastic in recent years.
30. Where would you find application for a hydraulic elevating table?
31. What is the difference between a skid and a pallet?

PROBLEMS

1. The finish tolerance on the shaft in Figure 3.4 was changed from 0.004 in to 0.008 in. How much cost improvement resulted from this change?
2. The Dorben Company is designing a cast-iron part whose strength T is a known function of the carbon content C, where $T = 2C^2 + 3/4C - C^3 + k$. To maximize strength, what carbon content should be specified?
3. To make a given part interchangeable, it was necessary to reduce the tolerance on the outside diameter from ±0.010 to ±0.005 at a resulting cost increase of 50 percent of the turning operation. The turning operation represented 20 percent of the total cost. Making the part interchangeable meant that the volume of this part could be increased by 30 percent. The increase in volume would permit production at 90 percent of the former cost. Should the methods engineer proceed with the tolerance change? Explain.
4. The Dorben Group suite consists of five rooms, with areas and relationships as shown in Figure 3.27. Obtain an optimal layout, using Muther's SLP and SPIRAL. Compare and contrast the resulting layouts.
5. Using the from-to chart on p122 showing the number of units handled from one area to another per hour and the desired size of each area (in square feet), develop an optimal layout using Muther's SLP and SPIRAL. Note that you will need to devise a relationship scheme for the given flows. Also, * means an undesirable relationship.

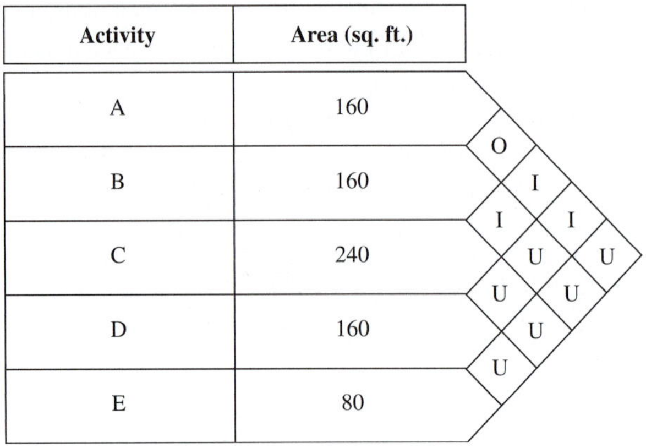

Activity	Area (sq. ft.)
A	160
B	160
C	240
D	160
E	80

O
I
I   I
U   U
U   U
U
U

Figure 3.27   Information for Problem 4.

Size	Area	A	B	C	D	E
150	A	—	1	20	8	1
50	B	0	—	30	0	8
90	C	20	5	—	40	20
90	D	0	1	2	—	*
40	E	0	0	11	0	—

REFERENCES

Bralla, James G. *Handbook of Product Design for Manufacturing.* New York: McGraw-Hill, 1986.

Buffa, Elwood S. *Modern Production Operations Management.* 6th ed. New York: John Wiley & Sons, 1980.

Chang, Tien-Chien, Richard A. Wysk, and Wang Hsu-Pin. *Computer Aided Manufacturing.* Englewood Cliffs, NJ: Prentice-Hall, 1991.

Drury, Colin G. "Inspection Performance." In *Handbook of Industrial Engineering,* 2d ed. Ed. Gavriel Salvendy. New York: John Wiley & Sons, 1992.

Francis, Richard L., and John A. White. *Facility Layout and Location: An Analytical Approach.* Englewood Cliffs, NJ: Prentice-Hall, 1974.

Goetschalckx, M. "An Interactive Layout Heuristic Based on Hexagonal Adjacency Graphs." *European Journal of Operations Research,* 63, no. 2 (December 1992), pp. 304–321.

Konz, Stephan. *Facility Design.* New York: John Wiley & Sons, 1985.

Material Handling Institute. *The Ten Principles of Material Handling.* Charlotte, NC, 1998.

Muther, R. *Systematic Layout Planning,* 2d ed. New York: Van Nostrand Reinhold, 1973.

Niebel, Benjamin W., and C. Richard Liu. "Designing for Manufacturing." In *Handbook of Industrial Engineering,* 2d ed. Ed. Gavriel Salvendy. New York: John Wiley & Sons, 1992.

Nof, Shimon Y. "Industrial Robotics." In *Handbook of Industrial Engineering,* 2d ed. Ed. Gavriel Salvendy. New York: John Wiley & Sons, 1992.

Shingo, S. *Study of Toyota Production System.* Tokyo, Japan: Japan Management Assoc. (1981), pp. 167–182.

Sims, Ralph E. "Material Handling Systems." In *Handbook of Industrial Engineering,* 2d ed. Ed. Gavriel Salvendy. New York: John Wiley & Sons, 1992.

Spur, Gunter. "Numerical Control Machines." In *Handbook of Industrial Engineering,* 2d ed. Ed. Gavriel Salvendy. New York: John Wiley & Sons, 1992.

Taguchi, Genichi. *Introduction to Quality Engineering.* Tokyo, Japan: Asian Productivity Organization, 1986.

Wemmerlov, Urban, and Nancy Lea Hyer. "Group Technology." In *Handbook of Industrial Engineering,* 2d ed. Ed. Gavriel Salvendy. New York: John Wiley & Sons, 1992.

Wick, Charles, and Raymond F. Veilleux. *Quality Control and Assembly, 4.* Detroit, MI: Society of Manufacturing Engineers, 1987.

SELECTED SOFTWARE

ALDEP, IBM Corporation, program order no. 360D-15.0.004.

CORRELAP, Engineering Management Associates, Boston, MA.

CRAFT, IBM share library No. SDA 3391.

Design Tools (available from the McGraw-Hill text website at www.mhhe.com/niebel-freivalds. New York: McGraw-Hill, 2002.

FactoryPLAN, FactoryFLOW, and FactoryCAD, (vol. 3) EDS PLM Solutions, 2321 North Loop Dr. ISU Research Park, Ames, IA, 50010, 2001. (http://www.eds.com/)

SPIRAL, *User's Manual,* 4031 Bradbury Dr., Marietta, GA, 30062, 1994.

SELECTED VIDEOTAPES/DVDS

Design for Manufacture and Assembly. DV05PUB2. Dearborn, MI: Society of Manufacturing Engineers, 2005.

Flexible Material Handling. DV03PUB104. Dearborn, MI: Society of Manufacturing Engineers, 2003.

Flexible Small Lot Production for Just-In-Time. DV03PUB107. Dearborn, MI: Society of Manufacturing Engineers, 2003.

Introduction to Lean Manufacturing. DV03PUB46. Dearborn, MI: Society of Manufacturing Engineers, 2003.

Layout Improvements for Just-In-Time. Manufacturing Insights Videotape Series. 1/29 VHS VT393–1368 & 3/49 U-Matic VT393U-1368. Dearborn, MI: Society of Manufacturing Engineers,1990.

Quick Changeover for Lean Manufacturing. DV03PUB33. Dearborn, MI: Society of Manufacturing Engineers, 2003.

Total Quality Management. Manufacturing Insights Videotape Series. VT91PUB1. Dearborn, MI: Society of Manufacturing Engineers, 1991.

Manual Work Design

- Design work according to human capabilities and limitations.

  - For manipulative tasks:
  - Use dynamic motions rather than static holds.
  - Keep the strength requirement below 15 percent of maximum.
  - Avoid extreme ranges of motion.
  - Use the smallest muscles for speed and precision.
  - Use the largest muscles for strength.
- For lifting and other heavy manual work:
  - Keep workloads below one-third of the maximum work capacity.
  - Minimize horizontal load distances.
  - Avoid twisting.
  - Use frequent, short work/rest cycles.

The design of manual work was introduced by the Gilbreths through motion study and the principles of motion economy, and later scientifically developed by human factor specialists for military applications. The principles have traditionally been broken down into three basic subdivisions: (1) the use of the human body, (2) the arrangement and conditions of the workplace, and (3) the design of tools and equipment. More important, although developed empirically, the principles are in fact based on established anatomical, biomechanical, and physiological principles of the human body. They form the scientific basis for ergonomics and work design. Accordingly, some theoretical background will be presented so that the principles of motion economy can be understood better rather than merely being accepted as memorized rules. Furthermore, the traditional principles of motion economy have been considerably expanded and are now called the principles of and guidelines for work design. This chapter presents the principles related to the human body and the guidelines for the design of work as related to physical activity. Chapter 5 covers those principles related to the design of workstations, tools, and equipment. Chapter 6 presents guidelines for the design of the work environment. Chapter 7 presents cognitive work design. Although not traditionally included as part of methods engineering, it is becoming an increasingly important aspect of work design. Chapter 8 covers workplace and systems safety.

4.1   THE MUSCULOSKELETAL SYSTEM

The human body is able to produce movements because of a complex system of muscles and bones, termed the *musculoskeletal system.* The muscles are attached to the bones on either side of a joint (see Figure 4.1), so that one or several muscles, termed *agonists*, act as the prime activators of motion. Other muscles, termed *antagonists*, counteract the agonists and oppose the motion. For elbow *flexion*, which is a decrease in the internal joint angle, the biceps, the brachioradialis, and the brachialis form the agonists, while the triceps forms the antagonist. However, on elbow *extension*, which is an increase in the joint angle, the triceps becomes the agonist, while the other three muscles become the antagonists.

There are three types of muscles in the human body: skeletal or striated muscles, attached to the bones; cardiac muscle, found in the heart; and smooth muscle, found in the internal organs and the walls of the blood vessels. Only the skeletal muscles (of which there are approximately 500 in the body) will be discussed here, because of their relevance to motion.

Each muscle is made up of a large number of muscle fibers, approximately 0.004 in (0.1 mm) in diameter and ranging in length from 0.2 to 5.5 in (5 to 140 mm), depending on the size of the muscle. These fibers are typically bound together in bundles by connective tissue, which extends to the end of the muscle and assists in

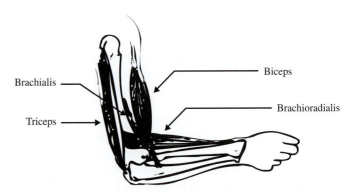

Figure 4.1   The musculoskeletal system of the arm.

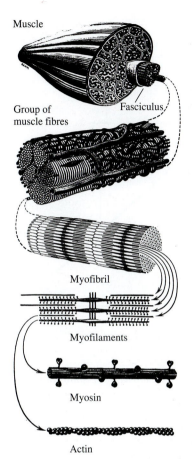

Muscle

Group of
muscle fibres

Fasciculus

Myofibril

Myofilaments

Myosin

Actin

Figure 4.2 The structure of muscle.
(*From: Gray's Anatomy,* 1973, by permission of
W. B. Saunders Co., London.)

firmly attaching the muscle and muscle fibers to the bone (see Figure 4.2). These
bundles are penetrated by tiny blood vessels that carry oxygen and nutrients to the
muscle fibers, as well as by small nerve endings that carry electrical impulses from
the spinal cord and brain.

Each muscle fiber is further subdivided into smaller *myofibrils* and ultimately into the protein filaments that provide the contractile mechanism. There
are two types of filaments: *thick filaments*, comprised of long proteins with molecular heads, called myosin; and *thin filaments*, comprised of globular proteins,
called actin. The two types of filaments are interlaced, giving rise to the striated
appearance and alternate name, as shown in Figure 4.3. This allows the muscle
to contract as the filaments slide over one another, which occurs as molecular
bridges or bonds are formed, broken, and reformed between the myosin heads
and actin globules. This *sliding filament theory* explains how the muscle length
can change from approximately 50 percent of its *resting length* (the neutral

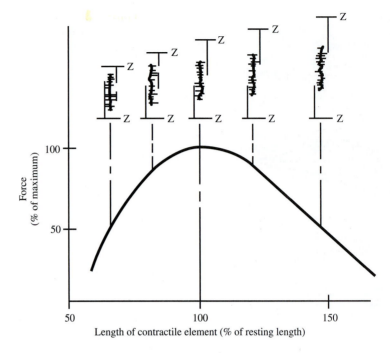

Figure 4.3 Force–length relationship of skeletal muscle.
(*From:* Winter, 1979, p. 114. Reprinted by permission of John Wiley & Sons, Inc.)

uncontracted length at approximately the midpoint in the normal range of motion) at complete contraction to 180 percent of its resting length at complete extension (see Figure 4.3).

4.2   PRINCIPLES OF WORK DESIGN: HUMAN CAPABILITIES AND MOTION ECONOMY

ACHIEVE THE MAXIMUM MUSCLE STRENGTH AT THE MIDRANGE OF MOTION

The first principle of human capability derives from the inverted-U-shaped property of muscle contraction shown in Figure 4.3. At the resting length, optimal bonding occurs between the thick and thin filaments. In the stretched state, there is minimal overlap or bonding between the thick and thin filaments, resulting in considerably decreased (almost zero) muscle force. Similarly, in the completely contracted state, interference occurs between the opposing thin filaments, again preventing optimum bonding and decreasing muscle force. This muscle property is typically termed the *force–length relationship*. Therefore, a task requiring considerable muscle force should be performed at the optimum position. For

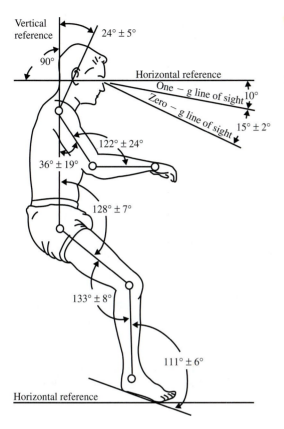

Figure 4.4 Typical relaxed posture assumed by people in weightless conditions. (*From:* Thornton, 1978, Fig. 16.)

example, the neutral or straight position will provide the strongest grip strength for wrist motions. For elbow flexion, the strongest position would be with the elbow bent somewhat beyond the 90° position. For plantar flexion (i.e., depressing a pedal), again the optimum position is slightly beyond 90°. A rough rule of thumb for finding the midrange of motion is to consider the posture assumed by an astronaut in weightless conditions when both agonist and antagonist muscles surrounding the joint are most relaxed and the limb attains a neutral position (see Figure 4.4).

ACHIEVE THE MAXIMUM MUSCLE STRENGTH WITH SLOW MOVEMENTS

The second principle of human capability is based on another property of the sliding filament theory and muscle contraction. The faster the molecular bonds are formed, broken, and reformed, the less effective is the bonding and the less muscular force is produced. This is a pronounced nonlinear effect (see Figure 4.5) with maximum muscle force being produced with no externally measurable shortening

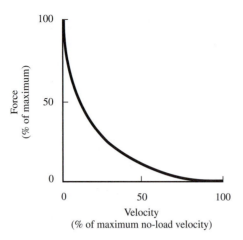

Figure 4.5
Force–velocity relationship of skeletal muscle.

(i.e., zero velocity or a static contraction), and minimal muscle force being produced at the maximum velocity of muscle shortening. The force is only sufficient to move the mass of that body segment. This muscle property is known as the *force–velocity relationship* and is especially important with respect to heavy manual work.

USE MOMENTUM TO ASSIST WORKERS WHEREVER POSSIBLE; MINIMIZE IT IF IT IS COUNTERACTED BY MUSCULAR EFFORT

There is a trade-off between the second and third principles. Faster movements produce higher momentum and higher impact forces in the case of blows. Downward motions are more effective than upward motions, because of the assistance from gravity. To make full use of the momentum built up, workstations should allow operators to release a finished part into a delivery area while their hands are on their way to get component parts or tools to begin the next work cycle.

DESIGN TASKS TO OPTIMIZE HUMAN STRENGTH CAPABILITY

Human strength capability depends on three major task factors: (1) the type of strength, (2) the muscle or joint motion being utilized, and (3) posture. There are three types of muscle exertions, defined primarily by the way the strength of the exertion is measured. Muscular exertions resulting in body motions result from dynamic strength. These are sometimes termed *isotonic* contractions, because the load and body segments lifted nominally maintain a constant external force on the muscle. (However, the internal force produced by the muscle varies due to the geometry of the effective moment arms.) Because of the many variables involved

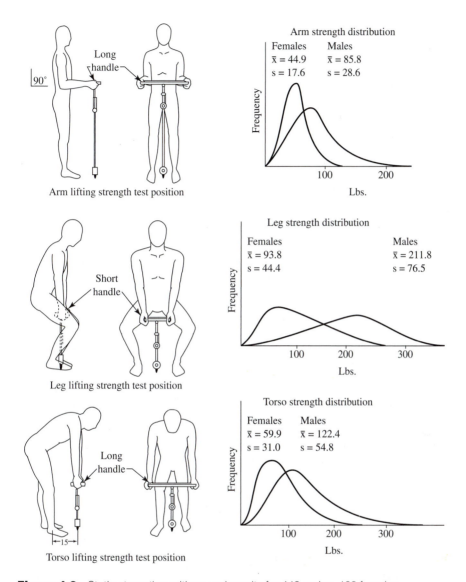

Figure 4.6 Static strength positions and results for 443 males, 108 females. (Chaffin et al., 1977.)

in such contractions, some variables necessarily need to be constrained to obtain a measurable strength. Thus, dynamic strength measurements have typically been made using constant-velocity (*isokinetic*) dynamometers, such as the Cybex or the Mini-Gym (Freivalds and Fotouhi, 1987). In the case where the body motion is restrained, an *isometric* or static strength is obtained. As seen in Figure 4.5, an isometric strength is necessarily greater than a dynamic strength because of the

Table 4.1 A. Static Muscle Strength Moment Data (ft·lb) for 25 Men and 22 Women Employed in Manual Jobs in Industry

Muscle Function	Joint Angles	Male (%ile)			Female (%ile)		
		5	**50**	**95**	**5**	**50**	**95**
Elbow flexion	90° Included to arm (arm at side)	31	57	82	12	30	41
Elbow extension	70° Included to arm (arm at side)	23	34	49	7	20	28
Medial humeral (shoulder) rotation	90° Vertical shoulder (abducted)	21	38	61	7	15	24
Lateral humeral (shoulder) rotation	5° Vertical shoulder (at side)	17	24	38	10	14	21
Shoulder horizontal flexion	90° Vertical shoulder (abducted)	32	68	89	9	30	44
Shoulder horizontal extension	90° Vertical shoulder (abducted)	32	49	76	14	24	42
Shoulder vertical adduction	90° Vertical shoulder (abducted)	26	49	85	10	22	40
Shoulder vertical abduction	90° Vertical shoulder (abducted)	32	52	75	11	27	42
Ankle extension (plantar flexion)	90° Included to shank	51	93	175	29	60	97
Knee extension	120° Included to thigh (seated)	62	124	235	38	78	162
Knee flexion	135° Included to thigh (seated)	43	74	116	16	46	77
Hip extension	100° Included to torso (seated)	69	140	309	28	72	133
Hip flexion	110° Included to torso (seated)	87	137	252	42	93	131
Torso extension	100° Included to thigh (seated)	121	173	371	52	136	257
Torso flexion	100° Included to thigh (seated)	66	106	159	36	55	119
Torso lateral flexion	Sitting erect	70	117	193	37	69	120

B. Static Muscle Strength Moment Data (N·m) for 25 Men and 22 Women Employed in Manual Jobs in Industry

Muscle Function	Joint Angles	Male (%ile)			Female (%ile)		
		5	**50**	**95**	**5**	**50**	**95**
Elbow flexion	90° Included to arm (arm at side)	42	77	111	16	41	55
Elbow extension	70° Included to arm (arm at side)	31	46	67	9	27	39
Medial humeral (shoulder) rotation	90° Vertical shoulder (abducted)	28	52	83	9	21	33
Lateral humeral (shoulder) rotation	5° Vertical shoulder (at side)	23	33	51	13	19	28
Shoulder horizontal flexion	90° Vertical shoulder (abducted)	44	92	119	12	40	60
Shoulder horizontal extension	90° Vertical shoulder (abducted)	43	67	103	19	33	57
Shoulder vertical adduction	90° Vertical shoulder (abducted)	35	67	115	13	30	54
Shoulder vertical abduction	90° Vertical shoulder (abducted)	43	71	101	15	37	57
Ankle extension (plantar flexion)	90° Included to shank	69	126	237	31	81	131
Knee extension	120° Included to thigh (seated)	84	168	318	52	106	219
Knee flexion	135° Included to thigh (seated)	58	100	157	22	62	104
Hip extension	100° Included to torso (seated)	94	190	419	38	97	180
Hip flexion	110° Included to torso (seated)	118	185	342	57	126	177
Torso extension	100° Included to thigh (seated)	164	234	503	71	184	348
Torso flexion	100° Included to thigh (seated)	89	143	216	49	75	161
Torso lateral flexion	Sitting erect	95	159	261	50	94	162

Source: Chaffin and Anderson, 1991. Reprinted by permission of John Wiley & Sons, Inc.

Consider the free-body diagram of the upper limb with the elbow at 90° in Figure 4.7. There are three muscles involved in elbow flexion: biceps brachii, brachioradialis, and brachialis (see Figure 4.1). However, the biceps is the primary flexor, and for the purpose of this example, it is the only muscle depicted. It can be also considered to be an equivalent muscle combining the characteristics of all three muscles. (Note that a solution of all three muscles independently is not possible because of a condition termed *static indeterminacy*.) The equivalent muscle inserts approximately 2 in forward of the elbow point of rotation. The lower arm weighs approximately 3 lb for an average male, and the weight can be considered to act at the lower arm center of gravity, approximately 4 in (0.33 ft) forward of the elbow. The hand bolds an unknown load L at a distance of 11 in (0.92 ft) from the elbow. The maximum load that can be held is determined by maximum voluntary elbow flexion torque, which for a 50th percentile male is 57 ft·lb (see Table 4.1). In the static equilibrium position as shown in Figure 4.7, the 57 ft·lb counterclockwise torque is balanced by two clockwise torques, one for the weight of the lower limb and the other for the load:

$$57 = 0.33 \times 3 + 0.92 \times L$$

Solving the equation yields $L = 60.9$ lb. Therefore the maximum load that an average male could lift through elbow flexion is approximately 61 lb.

It might be of interest to calculate how much force must be exerted by the equivalent muscle to lift this load. The maximum voluntary torque is produced by an unknown muscle force F_{biceps} acting through a 2-in (0.167-ft) moment arm.

$$57 = 0.167 \times F_{biceps}$$

Then F_{biceps} equals 57/0.167, or 342, lb which means that the muscle must exert close to 6 (342/61) times as much force as the load lifted. One can conclude from this that the human body is built not for strength, but for range of movement.

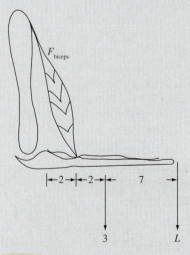

Figure 4.7

Table 4.2 Maximum Weights (in lb and kg) Acceptable to Average Males and Females Lifting Compact Boxes [14 in (34 cm) wide] with Handles

Task	1 lift per 0.5 min				1 lift per 1 min				1 lift per 30 min			
	Males		Females		Males		Females		Males		Females	
	lb	kg	lb	kg	lb	kg	lb	kg	lb	kg	lb	kg
Floor to knuckle height	42	19	26	12	66	30	31	14	84	38	37	17
Knuckle to shoulder height	42	19	20	9	55	25	29	13	64	29	33	15
Shoulder to arm reach	37	17	18	8	51	23	24	11	59	27	29	13

Note: For lowering, increase values by 6%. For boxes without handles, decrease values by 15%. Increasing the box size (away from body) to 30 in (75 cm) decreases values by 16%.
Source: Adapted from Snook and Ciriello, 1991.

more efficient bonding in the slower sliding muscle filaments. Some representative isometric muscle strengths for various postures are given in Table 4.1, and representative lifting strengths for 551 industrial workers in different postures are shown in Figure 4.6.

Most industrial tasks typically involve some movement; therefore, completely isometric contractions are relatively rare. Most typically, the movement range is somewhat limited, and the dynamic contraction is not a true isokinetic contraction, but a set of quasi-static contractions. Thus, dynamic strengths are very much task- and condition-dependent, and little is published regarding dynamic strength data.

Finally, a third type of muscle strength capability, *psychophysical* strength, has been defined for those situations in which the strength demands are required for an extended time. A static strength capability is not necessarily representative of what would be repetitively possible over an 8-h shift. Typically, the maximum acceptable load (determined by adjusting the load lifted or force exerted until the subject feels that the load or force would be acceptable on a repetitive basis for the given time period) is 40 to 50 percent less than a one-time static exertion. Extensive tables have been compiled for psychophysical strengths of various frequencies and postures (Snook and Ciriello, 1991). A summary of these values is provided in Tables 4.2, 4.3, and 4.4.

USE LARGE MUSCLES FOR TASKS REQUIRING STRENGTH

Muscle strength is directly proportional to the size of the muscle, as defined by the cross-sectional area [specifically, 87 psi (60 N/cm^2) for both males and females] (Ikai and Fukunaga, 1968). For example, leg and trunk muscles should

Table 4.3   Push Forces (in lbs and kg) at Waist Height Acceptable to Average Males and Females
(I = Initial, S = Sustained)

Distance Pushed, ft (m)	1 lift/min								1 lift per 30 min							
	Males				Females				Males				Females			
	I		S		I		S		I		S		I		S	
	lb	kg	lb	kg	lb	kg	lb	kg	lb	kg	lb	kg	lb	kg	lb	kg
150 (45)	51	23	26	12	40	18	22	10	66	30	42	19	51	23	26	12
50 (15)	77	35	42	19	44	20	29	13	84	38	51	23	53	24	33	15
7 (2)	95	43	62	28	55	25	40	18	99	45	75	34	66	30	46	21

Note: For push forces at shoulder heights or knuckle/knee heights, decrease values by 11%.
Source: Adapted from Snook and Ciriello, 1991.

Table 4.4   Pull Forces (in lbs and kg) at Waist Height Acceptable to Average Males and Females
(I = Initial, S = Sustained)

Distance Pulled, ft (m)	1 pull/min								1 pull per 30 min							
	Males				Females				Males				Females			
	I		S		I		S		I		S		I		S	
	lb	kg	lb	kg	lb	kg	lb	kg	lb	kg	lb	kg	lb	kg	lb	kg
150 (45)	37	17	26	12	40	18	24	11	48	22	42	19	48	22	26	12
50 (15)	57	26	42	19	42	19	26	12	62	28	51	23	51	23	33	15
7 (2)	68	31	57	26	55	25	35	16	73	33	70	32	66	30	44	20

Note: For pull forces at knuckle/knee heights, increases values by 75%. For pull forces at shoulder heights, decreases values by 15%.
Source: Adapted from Snook and Ciriello, 1991

be used in heavy load lifting, rather than weaker arm muscles. The posture factor, although somewhat confounded by geometric changes in the muscle moment or lever arm, is related to the resting length of the muscle fibers being roughly mid-range of motion for most joints as stated in the first principle of motion economy.

STAY BELOW 15 PERCENT OF MAXIMUM VOLUNTARY FORCE

Muscle fatigue is a very important but little utilized criterion in designing tasks appropriately for the human operator. The human body and muscle tissue rely primarily on two types of energy sources, *aerobic* and *anaerobic* (see later section on Manual Work). Since the anaerobic metabolism can supply energy for only a very small time, the oxygen supplied to the muscle fibers via peripheral blood flow becomes critical in determining how long the muscle contractions will last. Unfortunately, the harder the muscle fibers contract, the more the interlaced arterioles

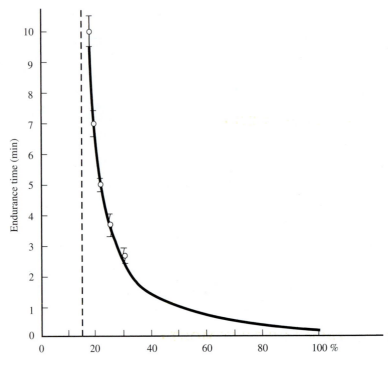

Figure 4.8   Static muscle endurance–exertion level relationship with ±1 SD ranges depicted.
(*From*: Chaffin and Anderson, 1991). Reprinted by permission of John Wiley & Sons, Inc.

and capillaries are compressed (see Figure 4.2), and the more the blood flow and oxygen supplies are restricted, the faster the muscle fatigues. The result is the endurance curve in Figure 4.8. The relationship is very nonlinear, ranging from a very short endurance time of approximately 6 s at a maximal contraction, at which point the muscle force rapidly drops off, to a rather indefinite endurance time at approximately 15 percent of a maximal contraction.

This relationship can be modeled by

$$T = 1.2/(f - 0.15)^{0.618} - 1.21$$

where T = endurance time, min

f = required force, expressed as a fraction of maximum isometric strength

For example, a worker would be able to sustain a force level of 50 percent of maximum strength for only about 1 min:

$$T = 1.2/(0.5 - 0.15)^{0.618} - 1.21 = 1.09 \text{ min}$$

The indefinite asymptote is due to early researchers stopping their experimentation without reaching complete muscle fatigue. Later researchers suggested reducing this level of acceptable static force levels from 15 percent to below 10 percent, perhaps even 5 percent (Jonsson, 1978). The amount of rest needed to recover from a static hold will be presented as a set of relaxation allowances that depend on the force exerted and the holding time (see Chapter 11).

USE SHORT, FREQUENT, INTERMITTENT WORK/REST CYCLES

Whether performing repeated static contractions (such as holding a load in elbow flexion) or a series of dynamic work elements (such as cranking with the arms or legs), work and recovery should be apportioned in short, frequent cycles. This is due primarily to a fast initial recovery period, which then tends to level off with increasing time. Thus, most of the benefit is gained in a relatively short period. A much higher percentage of maximum strength can be maintained if the strength is exerted as a series of repetitive contractions rather than one sustained static contraction (see Figure 4.9). However, if the person is driven to complete muscle (or whole body) fatigue, full recovery will take a fairly long time, perhaps several hours.

DESIGN TASKS SO THAT MOST WORKERS CAN DO THEM

As can be seen in Figure 4.6, for a given muscle group, there is a considerable range of strength in the normal, healthy adult population, with the strongest being five to eight times stronger than the weakest. These large ranges are due to

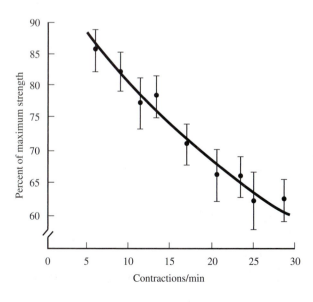

Figure 4.9 Percentage of maximum isometric strength that can be maintained in a steady state during rhythmic contractions.
Points are averages for finger muscles, hand muscles, arm muscles, and leg muscles, combined. Vertical lines denote ± standard error. (*From:* Åstrand and Rodahl, 1986.)

individual factors that affect strength performance: gender, age, handedness, and fitness/training. Gender accounts for the largest variation in muscle strength, with average female strength ranging from 35 to 85 percent of average male strength, with an average effect of 66 percent (see Figure 4.10). The difference is greatest for upper extremity strengths and smallest for lower extremity strengths. However, this effect is primarily due to average body size (i.e., total muscle mass) and not strictly to gender; the average female is considerably smaller and lighter than the average male. Furthermore, with the wide distribution for a given muscle strength, there are many females who are stronger than many males.

In terms of age, muscle strength appears to peak in the mid-20s and then decreases linearly by 20 to 25 percent by the mid-60s (see Figure 4.10). The decrease in strength is due to reduced muscle mass and a loss of muscle fibers. However, whether this loss is due to physiological changes of aging or just a gradual reduction in activity levels is not well known. It has definitely been shown that by starting a strength training program, a person can increase strength by 30 percent in the first several weeks, with maximum increases approaching 100 percent (Åstrand and Rodahl, 1986). In terms of handedness, the nondominant hand typically produces about 90 percent of the dominant hand's grip strength, with the effect being less pronounced in lefties, probably because they have been forced to adapt to a right-handed world (Miller and Freivalds, 1987). In any case, it is best to design tools and machines such that they can be used by either hand, to avoid placing any individual at a strength disadvantage.

USE LOW FORCE FOR PRECISE MOVEMENTS OR FINE MOTOR CONTROL

Muscle contractions are initiated by neural innervation from the brain and spinal cord, which together comprise the central nervous system. A typical motor

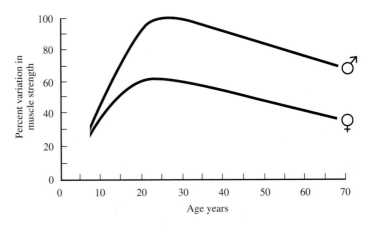

Figure 4.10 Changes in maximal isometric strength with age in women and men. (*From:* Åstrand and Rodahl, 1986.)

neuron, or nerve cell leading to the muscle from the central nervous system, may innervate or have connections with several hundred muscle fibers. The innervation ratio of the number of fibers per neuron ranges from less than 10 in the small muscles of the eye to over 1,000 for the large calf muscles, and can vary considerably even within the same muscle. Such functional arrangement is called a *motor unit* and has important implications in movement control. Once a neuron is stimulated, the electrical potential is transferred simultaneously to all the muscle fibers innervated by that neuron, and the motor unit acts as one contractile or motor control unit. Also, the central nervous system tends to recruit these motor units selectively by increasing size as higher muscle forces are needed (Figure 4.11). The initial motor units recruited are small in size, with few muscle fibers and low produced forces. However, since these are small and low in tension, the change in force production from one to two or more motor units recruited is very gradual, and very fine precision in motor control can be produced. Near the end of motor recruitment, the total muscle force is high, and each additional motor unit recruited becomes a large increment in force, with little sensitivity in terms of precision or control. This muscle property is sometimes termed the *size principle.*

The electrical activity of muscles, termed *electromyograms* (EMGs), is a useful measure of local muscle activity. Such activity is measured by placing recording electrodes on the skin surface over the muscles of interest, then modifying and processing to the amplitude and frequency of the signal. For amplitude analysis, the signal is typically rectified and smoothed (with a resistor-capacitor circuit). The result has a reasonably linear relationship to the muscle force exerted (Bouisset, 1973). The frequency approach involves digitizing the signal and performing a fast Fourier transform analysis to yield a frequency spectrum. As the muscle begins to fatigue, muscle activity shifts from high frequency (>60 Hz) to lower frequencies (<60 Hz) (Chaffin, 1969). Also, the EMG amplitude tends to increase with fatigue, for a given level of exertion.

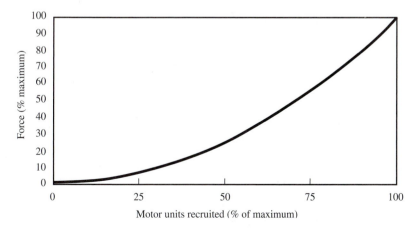

Figure 4.11 Muscle recruitment demonstrating size principle.

DO NOT ATTEMPT PRECISE MOVEMENTS OR FINE CONTROL IMMEDIATELY AFTER HEAVY WORK

This is a corollary to the previous principle of human capability. The small motor units tend to be used continually during normal motion, and although more resistant to fatigue than the large motor units, they can still experience fatigue. A typical example where this principle is violated occurs when operators load their workstations before their shift or replenish parts during a shift. Lifting heavy parts containers requires the recruitment of the small motor units, as well as the later larger motor units, to generate the necessary muscle forces. During lifting and restocking, some of the motor units will fatigue and others will be recruited to compensate for the fatigued ones. Once the operator has restocked the bins and returned to more precise assembly work, some of the motor units, including the smaller precision ones, will not be available for use. The larger motor units recruited to replace the fatigued ones will provide larger increments in force and less-precise motor control. After several minutes, the motor units will have recovered and will be available, but in the meantime, the quality and speed of the assembly work may suffer. One solution would be to provide less-skilled laborers to restock the bins on a regular basis.

USE BALLISTIC MOVEMENTS FOR SPEED

Through spinal reflexes, cross innervation of agonists and antagonists always occurs. This minimizes any unnecessary conflict between the muscles as well as the consequent excess energy expenditure. Typically, in a short (less than 200 ms), gross, voluntary motion, the agonist is activated and the antagonist is inhibited (termed *reciprocal inhibition*), to reduce counterproductive muscle contractions. On the other hand, for precise movements, feedback control from both sets of muscles is utilized, increasing motion time. This is sometimes referred to as the *speed–accuracy trade-off*.

BEGIN AND END MOTIONS WITH BOTH HANDS SIMULTANEOUSLY

When the right hand is working in the normal area to the right of the body and the left hand is working in the normal area to the left of the body, a feeling of balance tends to induce a rhythm in the operator's performance, which leads to maximum productivity. The left hand, in right-handed people, can be just as effective as the right hand, and it should be used. A right-handed boxer learns to jab just as effectively with the left hand as with the right hand. A speed typist is just as proficient with one hand as the other. In a large number of instances, workstations can be designed to do "two at a time." Using dual fixtures to hold two components, both hands can work at the same time, making symmetric moves in opposite directions. A corollary to this principle is that both hands should not be idle at the same time, except during rest periods. (This principle was the one followed by Frank Gilbreth in shaving with both hands simultaneously.)

MOVE THE HANDS SYMMETRICALLY AND SIMULTANEOUSLY TO AND FROM THE CENTER OF THE BODY

It is natural for the hands to move in symmetric patterns. Deviations from symmetry in a two-handed workstation result in slow, awkward movements of the operator. The difficulty of patting the stomach with the left hand while rubbing the top of the head with the right hand is familiar to many. Another experiment that can readily illustrate the difficulty of performing nonsymmetric operations is to try to draw a circle with the left hand while trying to draw a square with the right hand. Figure 4.12 illustrates an ideal workstation that allows the operator to assemble a product by going through a series of symmetric, simultaneous motions away from and toward the center of the body.

USE THE NATURAL RHYTHMS OF THE BODY

The spinal reflexes that excite or inhibit muscles also lead to natural rhythms in the motion of body segments. These can be logically compared to second-order mass–spring–dashpot systems, with the body segments providing mass and the muscle having internal resistance and damping. The natural frequency of the system will depend on all three parameters, but the segment mass will have the greatest effect. This natural frequency is essential to the smooth and automatic performance of a task. Drillis (1963) has studied a variety of very common manual tasks and has suggested optimum work tempos, as follows:

Filing metal   60–78 strokes per minute
Chiseling   60 strokes per minute
Arm cranking   35 rpm
Leg cranking   60–72 rpm
Shoveling   14–17 tosses per minute

Figure 4.12 An ideal workstation that permits the operator to assemble a product by going through a series of symmetric motions made simultaneously away from and toward the center of the body.

USE CONTINUOUS CURVED MOTIONS

Because of the nature of body segment linkages (typically approximating pin joints), it is easier for the human to produce curved motions, that is, to pivot around a joint. Straight-line motions involving sudden and sharp changes in direction require more time and are less accurate. This law is very easily demonstrated by moving either hand in a rectangular pattern and then moving it in a circular pattern of about the same magnitude. The greater amount of time required to make the abrupt 90° directional changes is quite apparent. To make a directional change, the hand must decelerate, change direction, and accelerate until it is time to decelerate again for the next directional change. Continuous curved motions do not require deceleration and are consequently performed faster per unit of distance. This is demonstrated very nicely in Figure 4.13, with subjects who made positioning movements with the right hand in eight directions in a horizontal plane from a center starting point. Motion from the lower left to the upper right (pivoting about the elbow) required 20 percent less time than the perpendicular motion from the lower right to upper left (additional awkward shoulder and arm line movements).

USE THE LOWEST PRACTICAL CLASSIFICATION OF MOVEMENT

Understanding the classifications of motions plays a major role in using this fundamental law of motion economy appropriately in methods studies. The classifications are as follows:

1. Finger motions are made by moving the finger or fingers while the
    remainder of the arm is kept stationary. They are first-class motions and

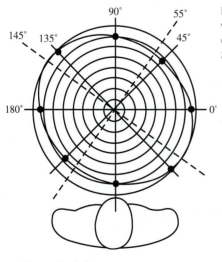

Figure 4.13 Forearm motion is best while pivoting on elbow.
(*Source:* Adapted from Schmidtke and Stier 1960.)

Concentric circles represent
equal time intervals.

the fastest of the five motion classes. Typical finger motions are running a nut down on a stud, depressing the keys of a typewriter, or grasping a small part. There is usually a significant difference in the time required to perform finger motions with the various fingers with the index finger being the fastest. Because repetitive finger motions can result in cumulative trauma disorders (see Chapter 5), finger forces should be kept low by using bar switches in place of trigger switches.

2. Finger and wrist motions are made while the forearm and upper arm are stationary and are referred to as second-class motions. In the majority of cases, finger and wrist motions consume more time than strictly finger motions. Typical finger and wrist motions occur when a part is positioned in a jig or fixture, or when two mating parts are assembled.

3. Finger, wrist, and lower arm motions are commonly referred to as forearm or third-class motions and include those movements made by the arm below the elbow while the upper arm is stationary. Since the forearm includes relatively strong and nonfatiguing muscles, workstations should be designed to utilize these third-class motions, rather than fourth-class motions. However, repetitive work involving force with the arms extended can induce injury and the workstation should be designed so that the elbows can be kept at 90° while work is being done.

4. Finger, wrist, lower arm, and upper arm motions, commonly known as fourth-class or shoulder motions, require considerably more time for a given distance than the three classes previously described. Fourth-class motions are required to perform transport motions for parts that cannot be reached without extending the arm. To reduce static loading of shoulder motions, tools should be designed so that the elbow is not elevated while the work is being performed.

5. Fifth-class motions include body motions such as of the trunk, which are the most time-consuming and should generally be avoided.

First-class motions require the least amount of effort and time, while fifth-class motions are considered the least efficient. Therefore, always utilize the lowest practicable motion classification to perform the work properly. This will involve careful consideration of the location of tools and materials, so that ideal motion patterns can be arranged.

This classification of movement was shown experimentally by Langolf et al. (1976), in a series of positioning movements to and from targets, known as Fitts' tapping task (Fitts, 1954), discussed in greater detail in Chapter 7. The movement time increases with the difficulty of the task (see Figure 4.14), but also increases with higher levels of classification; that is, the slope for the arm (105 ms) is steeper than for the wrist (45 ms), which in turn is steeper than for the finger (26 ms). The effect is due simply to the added time required for the central nervous system to process additional joints, motor units, and receptors.

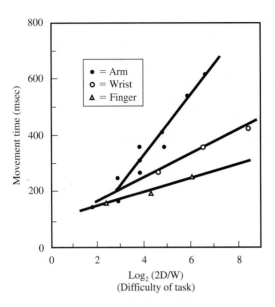

Figure 4.14 Classifications of movements
(*Source:* Data from Langolf et al., 1976. Reproduced with permission of the McGraw-Hill Companies.)

WORK WITH BOTH HANDS AND FEET SIMULTANEOUSLY

Since the major part of work cycles is performed by the hands, it is economical to relieve the hands of work that can be done by the feet, but only if this work is performed while the hands are occupied. Since the hands are more skillful than the feet, it would be foolish to have the feet perform elements while the hands are idle. Foot pedal devices that allow clamping, parts ejection, or feeding can often be arranged, freeing the hands for other, more useful work and consequently reducing the cycle time (see Figure 4.15). When the hands are moving, the feet should not be moving, since the simultaneous movement of the hands and feet is difficult. However, the feet can be applying pressure to something, such as a foot pedal. Also, the operator should be seated, as it is difficult to operate a foot pedal while standing, which would mean maintaining the full body weight on the other foot.

MINIMIZE EYE FIXATIONS

Although eye fixations or eye movements cannot be eliminated for most work, the location of the primary visual targets should be optimized with respect to the human operator. The normal line of sight is roughly about 15° below the horizontal (see Figure 5.5), and the primary visual field is roughly defined as a cone ±15° in arc centered on the line of sight. The implication is that within this area, no head movements are needed and eye fatigue is minimized.

SUMMARY

The principles of human capabilities and motion economy are based on an elementary understanding of human physiology and should be very useful in

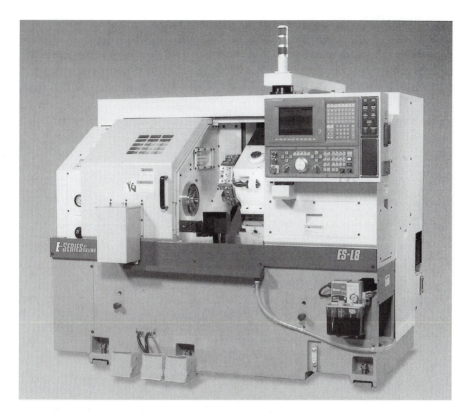

Figure 4.15 Foot-operated machine tool.
(Courtesy of Okuma.)

applying methods analysis with the human operator in mind. However, the analyst need not be an expert in human anatomy and physiology to be able to apply these principles. In fact, for most task analysis purposes, it may be sufficient to use the Motion Economy Checklist, which summarizes most of the principles in a questionnaire format (see Figure 4.16).

4.3 MOTION STUDY

Motion study is the careful analysis of body motions employed in doing a job. The purpose of motion study is to eliminate or reduce ineffective movements, and facilitate and speed effective movements. Through motion study, in conjunction with the principles of motion economy, the job is redesigned to be more effective and to produce a higher rate of output. The Gilbreths pioneered the study of manual motion and developed basic laws of motion economy that are still considered fundamental. They were also responsible for the development of detailed motion picture studies, known as *micromotion studies*, which have proved invaluable in studying highly repetitive manual operations. Motion study, in the

Suboperations	Yes	No
1. Can a suboperation be eliminated?	❑	❑
a. As unnecessary?	❑	❑
b. By a change in the order of the work?	❑	❑
c. By a change of tools or equipment?	❑	❑
d. By a change in layout of the workplace?	❑	❑
e. By combining tools?	❑	❑
f. By a slight change of material?	❑	❑
g. By a slight change in product?	❑	❑
h. By a quick-acting clamp on the jigs or fixtures?	❑	❑
2. Can a suboperation be made easier?	❑	❑
a. By better tools?	❑	❑
b. By changing leverages?	❑	❑
c. By changing positions of controls or tools?	❑	❑
d. By better material containers?	❑	❑
e. By using inertia where possible?	❑	❑
f. By lessening visual requirements?	❑	❑
g. By better workplace heights?	❑	❑

Movements	Yes	No
1. Can a movement be eliminated?	❑	❑
a. As unnecessary?	❑	❑
b. By a change in the order of work?	❑	❑
c. By combining tools?	❑	❑
d. By a change in tools or equipment?	❑	❑
e. By a drop disposal of finished material?	❑	❑
2. Can a movement be made easier?	❑	❑
a. By a change in layout, shortening distances?	❑	❑
b. By changing the direction of movements?	❑	❑
c. By using different muscles?	❑	❑
Use the first muscle group that is strong enough for the task:		
(1) Finger?	❑	❑
(2) Wrist?	❑	❑
(3) Forearm?	❑	❑
(4) Upper arm?	❑	❑
(5) Trunk?	❑	❑
d. By making movements continuous rather than jerky?	❑	❑

Holds	Yes	No
1. Can a hold be eliminated? (Holding is extremely fatiguing.)	❑	❑
a. As unnecessary?	❑	❑
b. By a simple holding device or fixture?	❑	❑
2. Can a hold be made easier?	❑	❑
a. By shortening its duration?	❑	❑
b. By using stronger muscle groups, such as the legs with foot-operated vises?	❑	❑

Delays	Yes	No
1. Can a delay be eliminated or shortened?	❑	❑
a. As unnecessary?	❑	❑
b. By a change in the work each body member does?	❑	❑
c. By balancing the work between the body members?	❑	❑
d. By working simultaneously on two items?	❑	❑
e. By alternating the work, each hand doing the same job, but out of phase?	❑	❑

Cycles	Yes	No
1. Can the cycle be rearranged so that more of the handwork is done during running time?	❑	❑
a. By automatic feed?	❑	❑
b. By automatic supply of material?	❑	❑
c. By change of man and machine phase relationship?	❑	❑
d. By automatic power cutoff at completion of cut or in case of tool or material failure?	❑	❑

Machine Time	Yes	No
1. Can the machine time be shortened?	❑	❑
a. By better tools?	❑	❑
b. By combined tools?	❑	❑
c. By higher feeds or speeds?	❑	❑

Figure 4.16 Motion Economy Checklist.

broad sense, covers both studies that are performed as a simple visual analysis and studies that utilize more expensive equipment. Traditionally, motion picture film cameras were utilized, but today the videotape camera is used exclusively, because of the easy ability to rewind and replay sections, the freeze-frame capability in four-head videotape cassette recorders (VCRs), and the elimination of the need for film development. In view of its much higher cost, micromotion is usually practical only on extremely active jobs with high repetitiveness.

The two types of studies may be compared to viewing a part under a magnifying glass versus viewing it under a microscope. The added detail revealed by the microscope is needed only on the most productive jobs. Traditionally, micromotion studies were recorded on a *simultaneous motion (simo) chart,* while motion studies are recorded on a *two-hand process chart.* A true simo chart is hardly used today, but the term is sometimes applied to a two-hand process chart.

BASIC MOTIONS

As part of motion analysis, the Gilbreths concluded that all work, whether productive or nonproductive, is done by using combinations of 17 basic motions that they called *therbligs* (Gilbreth spelled backward). The therbligs can be either effective or ineffective. Effective therbligs directly advance the progress of the work. They can frequently be shortened, but typically cannot be completely eliminated. Ineffective therbligs do not advance the progress of the work and should be eliminated by applying the principles of motion economy. The 17 therbligs, along with their symbols and definitions, are shown in Table 4.5.

THE TWO-HAND PROCESS CHART

The *two-hand process chart,* sometimes referred to as an operator process chart, is a motion study tool. This chart shows all movements and delays made by the right and left hands, and the relationships between them. The purpose of the two-hand process chart is to identify inefficient motion patterns and observe violations of the principles of motion economy. This chart facilitates changing a method so that a balanced two-handed operation can be achieved and a smoother, more rhythmic cycle that keeps both delays and operator fatigue to a minimum.

As usual, the analyst heads the chart Two-Hand Process Chart and adds all necessary identifying information, including the part number, drawing number, operation or process description, present or proposed method, date, and name of the person doing the charting. Immediately below the identifying information, the analyst sketches the workstation, drawn to scale. The sketch materially aids in presenting the method under study. Figure 4.17 shows a typical two-hand process chart for a cable-clamp assembly, with the times for each therblig obtained from stopwatch timing.

Next the analyst begins constructing the two-hand process chart by observing the duration of each element and determining the amount of time to be represented on the chart drawn to scale. For example, in Figure 4.17, the first element, Get U-bolt, has a time of 1.00 min and one large or five small vertical

Table 4.5 Gilbreths' Therbligs

Effective Therbligs (Directly advance progress of work. May be shortened but difficult to eliminate completely.)		
Therblig	**Symbol**	**Description**
Reach	RE	Motion of empty hand to or from object; time depends on distance moved; usually preceded by Release and followed by Grasp.
Move	M	Movement of loaded hand; time depends on distance, weight, and type of move; usually preceded by Grasp and followed by Release or Position.
Grasp	G	Closing fingers around an object; begins as the fingers contact the object and ends when control has been gained; depends on type of grasp; usually preceded by Reach and followed by Move.
Release	RL	Relinquishing control of object, typically the shortest of the therbligs.
Preposition	PP	Positioning object in predetermined location for later use; usually occurs in conjunction with Move, as in orienting a pen for writing.
Use	U	Manipulating tool for intended use; easily detected, as it advances the progress of work.
Assemble	A	Bringing two mating parts together; usually preceded by Position or Move; followed by Release.
Disassemble	DA	Opposite of Assemble, separating mating parts; usually preceded by Grasp and followed by Move or Release.

Ineffective Therbligs (Do not advance progress of work. Should be eliminated if possible.)		
Therblig	**Symbol**	**Description**
Search	S	Eyes or hands groping for object; begins as the eyes move in to locate an object.
Select	SE	Choosing one item from several; usually follows Search.
Position	P	Orienting object during work, usually preceded by Move and followed by Release (as opposed to *during* for Preposition).
Inspect	I	Comparing object with standard, typically with sight, but could also be with the other senses.
Plan	PL	Pausing to determine next action; usually detected as a hesitation preceding Motion.
Unavoidable Delay	UD	Beyond the operator's control due to the nature of the operation, e.g., left hand waiting while right hand completes a longer Reach.
Avoidable Delay	AD	Operator solely responsible for idle time, e.g., coughing.
Rest to Overcome Fatigue	R	Appears periodically, not every cycle, depends on the physical workload.
Hold	H	One hand supports object while other does useful work.

Two-hand process chart   Page 1 of 1

Operation: Assemble cable clamps	Part: SK-112	Summary	Left hand	Right hand
Operator name and no.: J.B. #1157		Effective time:	2.7	11.6
Analyst: G. Thuering	Date: 6-11-98	Ineffective time:	11.6	2.7
Method (circle choice) Present Proposed		Cycle time = 14.30 sec.		

Sketch:

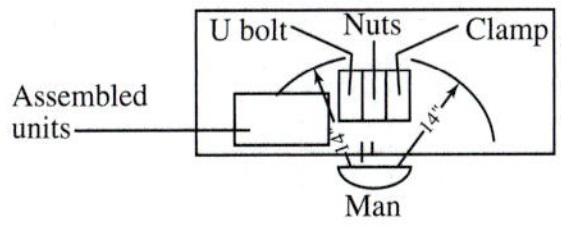

U bolt Nuts Clamp   Note: Gravity feed chutes for assembly parts

Assembled units

Man

Left hand description	Sym-bol	Time		Time	Sym-bol	Right hand description
Get U-bolt (10")	RE G	1.00		1.00	RE G	Get cable clamp (10")
Place U-bolt (10")	M P	1.20		1.20	M P RL	Place cable clamp (10")
				1.00	RE G	Get first nut (9")
				1.20	M P	Place first nut (9")
				3.40	U	Run down first nut
					RL	
Hold U-bolt	H	11.00		1.00	RE G	Get second nut (9")
				1.20	M P	Place second nut (9")
				3.40	U	Run down second nut
					RL	
Dispose of assembly	M RL	1.10		0.90	UD	Wait

Figure 4.17
Two-hand process chart for assembly of cable clamps.

spaces are marked. Under the "Symbols" column RE (for reach) is written, indicating that an effective motion has been accomplished. Note also that a grasp (G) is involved, but is not measured separately, since it is not possible in most instances to time individual therbligs. Next, the analyst charts "Place U-bolt" and continuing on with the left hand. Usually it is less confusing to chart the activities of one hand completely before examining the other hand.

After the activities of both the right and the left hand have been charted, the analyst creates a summary at the bottom of the sheet, indicating the cycle time, pieces per cycle, and time per piece. Once the two-hand process chart has been completed for an existing method, the analyst can determine what improvements can be introduced. Several important corollaries to the principles of motion economy should be applied at this point:

1. Establish the best sequences of therbligs.
2. Investigate any substantial variation in the time required for a given therblig and determine the cause.
3. Examine and analyze hesitations, to determine and then eliminate their causes.
4. As a goal, aim for cycles and portions of cycles completed in the least amount of time. Study deviations from these minimum times to determine the causes.

In the example, the "delays" and "holds" are good places to begin. For example, in Figure 4.17, the left hand acted as a holding device for almost the entire cycle. This would suggest the development of a fixture to hold the U-bolt. Further considerations to achieve balanced motions of both hands would suggest that when the fixture holds the U-bolts, the left hand and the right hand can be used simultaneously so that each completely assembles a cable clamp. Additional study of this chart might result in the introduction of an automatic ejector and gravity chute, to eliminate the final cycle element "dispose of assembly." The use of the Therblig Analysis Checklist (see Figure 4.18) may also be helpful in this analysis.

4.4   MANUAL WORK AND DESIGN GUIDELINES

Although automation has significantly reduced the demands for human power in the modern industrial environment, muscular strength still remains an essential part of many occupations, particularly those involving manual materials handling (MMH) or manual work. In these activities, overexertion from moving heavy loads can highly stress the musculoskeletal system, resulting in nearly one-third of all occupational injuries. The low back alone accounts for almost one-quarter of these injuries and one-quarter of the annual workers' compensation costs (National Safety Council, 2003). Back injuries are especially detrimental because they often result in permanent disorders, with considerable discomfort and limitations for the employee as well as a large expense for the employer (an average case involving surgery may exceed $60,000 in direct costs).

ENERGY EXPENDITURE AND WORKLOAD GUIDELINES

Energy is required for the muscle contraction process. The molecule called *ATP* (adenosine triphosphate) is the immediate energy source, which physically interacts with the protein cross bridging as one of ATP's high-energy phosphate bonds

Reach and Move   Yes No

1. Can either of these therbligs be eliminated?   ☐ ☐
2. Can distances be shortened to advantage?   ☐ ☐
3. Are the best means (conveyors, tongs, tweezers) being used?   ☐ ☐
4. Is the correct body member (fingers, wrist, forearm, shoulder) being used?   ☐ ☐
5. Can a gravity chute be employed?   ☐ ☐
6. Can transports be effected through mechanization and foot-operated devices?   ☐ ☐
7. Will time be reduced by transporting in larger units?   ☐ ☐
8. Is time increased because of the nature of the material being moved or because of a subsequent
    delicate positioning?   ☐ ☐
9. Can abrupt changes in direction be eliminated?   ☐ ☐

Grasp   Yes No

1. Would it be advisable for the operator to grasp more than one part or object at a time?   ☐ ☐
2. Can a contact grasp be used rather than a pickup grasp?   ☐ ☐
3. In other words, can objects be slid instead of carried?   ☐ ☐
4. Will a lip on the front of bins simplify grasping small parts?   ☐ ☐
5. Can tools or parts be prepositioned for easy grasp?   ☐ ☐
6. Can a vacuum, magnet, rubber fingertip, or other device be used to advantage?   ☐ ☐
7. Can a conveyor be used?   ☐ ☐
8. Has the jig been designed so that operators may easily grasp the part when removing it?   ☐ ☐
9. Can the previous operator preposition the tool or the work, simplifying grasp for the next operator? ☐ ☐
10. Can tools be prepositioned on a swinging bracket?   ☐ ☐
11. Can the work table surface be covered with a layer of sponge material so that the fingers can more
    easily enclose small parts?   ☐ ☐

Release   Yes No

1. Can the release be made in transit?   ☐ ☐
2. Can a mechanical ejector be used?   ☐ ☐
3. Are the bins that contain the part after release of the proper size and design?   ☐ ☐
4. At the end of the therblig release, are the hands in the most advantageous position for the next
    therblig?   ☐ ☐
5. Can multiple units be released?   ☐ ☐

Preposition   Yes No

1. Can a holding device at the workstation keep tools in proper positions and the handles in upright
    positions?   ☐ ☐
2. Can tools be suspended?   ☐ ☐
3. Can a guide be used?   ☐ ☐
4. Can a magazine feed be used?   ☐ ☐
5. Can a stacking device be used?   ☐ ☐
6. Can a rotating fixture be used?   ☐ ☐

Use   Yes No

1. Can a jig or fixture be used?   ☐ ☐
2. Does the activity justify mechanized or automated equipment?   ☐ ☐
3. Would it be practical to make the assembly in multiple units?   ☐ ☐
4. Can a more efficient tool be used?   ☐ ☐
5. Can stops be used?   ☐ ☐
6. Is the tool being operated at the most efficient feeds and speeds?   ☐ ☐
7. Should a power tool be employed?   ☐ ☐

Figure 4.18   Therblig Analysis Checklist.

is broken. This source is very limited, lasting only several seconds, and the ATP must immediately be replenished from another molecule termed *CP* (creatine phosphate). The CP source is also limited, less than 1 min of duration (see Figure 4.19), and must ultimately be regenerated from the metabolism of the basic foods we eat: carbohydrates, fats, and proteins. This metabolism can occur

Search   Yes   No

1. Are articles properly identified? ☐ ☐
2. Perhaps labels or color could be utilized? ☐ ☐
3. Can transparent containers be used? ☐ ☐
4. Will a better layout of the workstation eliminate searching? ☐ ☐
5. Is proper lighting being used? ☐ ☐
6. Can tools and parts be prepositioned? ☐ ☐

Select   Yes   No

1. Are common parts interchangeable? ☐ ☐
2. Can tools be standardized? ☐ ☐
3. Are parts and materials stored in the same bin? ☐ ☐
4. Can parts be prepositioned in a rack or tray? ☐ ☐

Position   Yes   No

1. Can such devices as a guide, funnel, bushing, stop, swinging bracket, locating pin, recess, key, pilot, or chamfer be used? ☐ ☐
2. Can tolerances be changed? ☐ ☐
3. Can the hole be counterbored or countersunk? ☐ ☐
4. Can a template be used? ☐ ☐
5. Are burrs increasing the problem of positioning? ☐ ☐
6. Can the article be pointed to act as a pilot? ☐ ☐

Inspect   Yes   No

1. Can inspection be eliminated or combined with another operation or therblig? ☐ ☐
2. Can multiple gages or tests be used? ☐ ☐
3. Will inspection time be reduced by increasing the illumination? ☐ ☐
4. Are the articles being inspected at the correct distance from the worker's eyes? ☐ ☐
5. Will a shadowgraph facilitate inspection? ☐ ☐
6. Does an electric eye have application? ☐ ☐
7. Does the volume justify automatic electronic inspection? ☐ ☐
8. Would a magnifying glass facilitate the inspection of small parts? ☐ ☐
9. Is the best inspection method being used? ☐ ☐
10. Has consideration been given to polarized light, template gages, sound tests, performance tests, and so on? ☐ ☐

Rest to Overcome Fatigue   Yes   No

1. Is the best order-of-muscles classification being used? ☐ ☐
2. Are temperature, humidity, ventilation, noise, light, and other working conditions satisfactory? ☐ ☐
3. Are benches of the proper height? ☐ ☐
4. Can the operator alternately sit and stand while performing work? ☐ ☐
5. Does the operator have a comfortable chair of the right height? ☐ ☐
6. Are mechanical means being used for heavy loads? ☐ ☐
7. Is the operator aware of his or her average intake requirements in calories per day? ☐ ☐

Hold   Yes   No

1. Can a mechanical jig, such as a vise, pin, hook, rack, clip, or vacuum, be used? ☐ ☐
2. Can friction be used? ☐ ☐
3. Can a magnetic device be used? ☐ ☐
4. Should a twin holding fixture be used? ☐ ☐

Figure 4.18   (*continued*)

in two different modes: *aerobic,* requiring oxygen, and *anaerobic,* not using oxygen. Aerobic metabolism is much more efficient, generating 38 ATPs for each *glucose* molecule (basic unit of carbohydrates), but it is relatively slow. Anaerobic metabolism is very inefficient, producing only 2 ATPs for each glucose molecule, but it is much quicker. Also, the glucose molecule is only partially broken down into two lactate molecules, which in the watery environment of the body

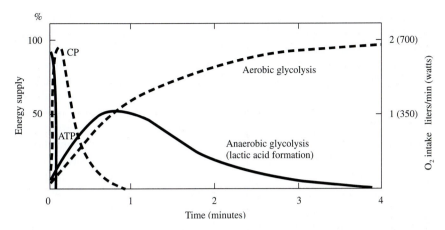

Figure 4.19 Sources of energy during the first few minutes of moderately heavy work. High-energy phosphate stores (ATP and CP) provide most of the energy during the first seconds of work. Anaerobic glycolysis supplies less and less of the energy required as the duration of work increases, and aerobic metabolism takes over.
(*Source:* Jones, Moran-Campbell, Edwards and Robertson, 1975.)

forms *lactic acid*, a direct correlate of fatigue. Thus, during the first few minutes of heavy work, the ATP and CP energy sources are used up very quickly, and anaerobic metabolism must be utilized to regenerate the ATP stores. Eventually, as the worker reaches steady state, the aerobic metabolism catches up and maintains the energy output, as the anaerobic metabolism slows down. By warming up and starting heavy work slowly, the worker can minimize the amount of anaerobic metabolism and the concurrent buildup of lactic acid associated with feelings of fatigue. This delay of full aerobic metabolism is termed *oxygen deficit* and must eventually be repaid by *the oxygen debt* of a cooling down period, which is always larger than the oxygen deficit.

The energy expended on a task can be estimated by assuming that most of the energy is produced through aerobic metabolism and measuring the amount of oxygen consumed by the worker. The amount of inspired air is measured with a flowmeter and assumed to contain 21 percent oxygen. However, not all this oxygen is utilized by the body; therefore, the expired oxygen must also be measured. Typically, the volume of air inspired and expired is the same, and only the percentage of expired oxygen must be found, using an oxygen meter. A conversion factor is included for a typical diet in which 4.9 kcal (19.6 Btu) of energy is produced for each liter of oxygen used in metabolism.

$$E \text{ (kcal/min)} = 4.9 * \dot{V}(0.21 - E_{O_2})$$

where
E = energy expenditure, kcal/min
$\dot{V}$ = volume of air inspired, L/min
E_{O_2} = fraction of oxygen (O_2) in expired air (roughly 0.17)

The energy expended on a task varies by the type of task being performed, the posture maintained during the task, and the type of load carriage. Energy expenditure data on several hundred different types of tasks have been collected, with the most common summarized in Figure 4.20. Alternatively, one may also estimate energy expenditure by using Garg's (1978) metabolic prediction model. For manual materials handling, the manner in which the load is carried is most critical, with lowest energy costs for balanced loads held closest to the center of gravity of the body, which has the largest muscle groups. For example, a backpack supported by the trunk muscles is less demanding than holding an equal weight in two suitcases, one in each arm. Although balanced, the latter situation places the load far from the center of gravity and on the smaller arm muscles. Posture also plays an important role, with the least amount of energy expenditure for supported postures. Thus, a posture with the trunk bent over, with no arm support, will expend 20% more energy than a standing posture.

A 5.33 kcal/min (21.3 Btu/min) limit for acceptable energy expenditure for an 8-h workday has been proposed by Bink (1962). This number corresponds to one-third the maximum energy expenditure of the average U.S. male [for females, it would be $1/3 \times 12 = 4$ kcal/min (16 Btu/min)]. If the overall workload is exceedingly high (i.e., exceeds the recommended limits), aerobic metabolism may

Figure 4.20 Examples of energy costs of various types of human activity. Energy costs are given in kilocalories per minute.
(*Source:* Passmore and Dumin, 1955, as adapted and presented by Gordon, 1957.)

not be sufficient to provide all the energy requirements, and the worker may rely on greater amounts of anaerobic metabolism, resulting in fatigue and the buildup of lactic acid. Sufficient recovery must then be provided to allow the body to recover from fatigue and recycle the lactic acid. One guideline for rest allocation was developed by Murrell (1965):

$$R = (W - 5.33)/(W - 1.33)$$

where R = time required for rest, as percent of total time

  W = average energy expenditure during work, kcal/min

The value of 1.33 kcal/min (5.3 Btu/min) is the energy expenditure during rest. Consider a strenuous task of shoveling coal into a hopper, which has an energy expenditure of 9.33 kcal/min (37.3 Btu/min). Entering $W = 9.33$ into the equation yields $R = 0.5$. Therefore, to provide adequate time for recovery from fatigue, the worker would need to spend roughly one-half of an 8-h shift, or 4-h, resting.

The manner in which rest is also allocated is important. It serves no purpose to have the laborer work for 4-h straight at a rate of 9.33 kcal/min (37.3 Btu/min), suffer from extreme fatigue, and then rest for 4 h. In general, the duration of the work cycle is the primary determinant of fatigue buildup. With heavy work, blood flow tends to be occluded, further accelerating the use of anaerobic pathways. In addition, the recovery process tends to be exponential, with later times providing minimal incremental benefits. Therefore, short bursts (approximately 1/2 to 1 min) of heavy work interspersed with short rest periods provide maximum benefit. During the 1/2 to 1 min periods, the immediate energy sources of ATP and CP get used up, but they can also be quickly replenished. Once lactic acid builds up during longer work periods, it becomes more difficult to remove. Micropauses of 1 to 3s are also useful for flushing any occluded blood vessels, and active breaks, during which the worker alternates hands or uses other muscles, serve to relieve the fatigued muscles. Also, it is best for workers to decide when to take the breaks, whenever they feel the need for rest (self-paced), as opposed to prescribed (or machine-paced) breaks. In summary, the use of frequent, short work/rest cycles is highly recommended.

HEART RATE GUIDELINES

Unfortunately, the measurement of oxygen consumption and the computation of energy expenditure are both costly and cumbersome in an industrial work situation. The equipment costs several thousand dollars and interferes with the worker performing the job. An alternative indirect measure of energy expenditure is the heart rate level. Since the heart pumps the blood carrying oxygen to the working muscles, the higher the required energy expenditure, the higher the corresponding heart rate (Figure 4.21). The instrumentation needed to measure heart rate is inexpensive (less than $100 for a visual readout, several hundred dollars for a PC interface) and relatively nonintrusive (worn regularly by athletes to monitor performance). On the other hand, the analyst must be careful, since heart rate

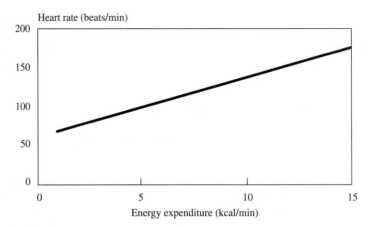

Figure 4.21   Linear increase in heart rate with physical workload, as measured by energy expenditure.

measurement is most appropriate for dynamic work involving the large muscles of the body at fairly high levels (40 percent of maximum) and can vary considerably between individuals, depending on their fitness levels and age. In addition, heart rate can be confounded by other stressors including heat, humidity, emotional levels, and mental stress. Limiting these external influences will result in a better estimate of physical workload. However, if the desired goal is to obtain the overall stress on the worker on the job, this may not be necessary.

A methodology for interpreting heart rate has been proposed by various German researchers (cited in Grandjean, 1988). The average working heart rate is compared to the resting prework heart rate, with 40 beats/min being proposed as an acceptable increase. This increase corresponds very nicely with the recommended working energy expenditure limits. The average increase in heart rate per increase in energy expenditure for dynamic work (i.e., the slope in Figure 4.21) is 10 beats/min per 1 kcal/min. Thus, a 5.33 kcal/min workload (4 kcal/min above the resting level of 1.33 kcal/min) produces a 40 beat/min increase in heart rate, which is the limit for an acceptable workload. This value also corresponds closely to the heart rate recovery index presented by Brouha (1967).

The average heart rate is measured in two time periods (cross hatched areas) during recovery after the cessation of work (see Figure 4.22): (1) between 1/2 and 1 min after cessation and (2) between 21/2 and 3 min after cessation. Acceptable heart rate recovery (and therefore acceptable workload) occurs if the first reading does not exceed 110 beats/min and the difference between the two readings is at least 20 beats. Given a typical resting heart rate of 72 beats/min, the addition of the acceptable increment of 40 beats/min yields a working heart rate of 112 beats/min, which corresponds closely to Brouha's first criterion.

As a final note on heart rate, it is very important to observe the course of heart rate during the working hours. An increase in heart rate during steady-state work (see upper curve in Figure 4.22), termed *heart rate creep,* indicates an increasing

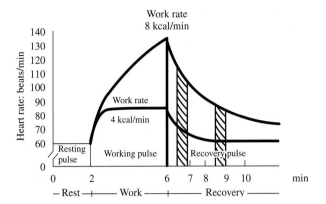

Figure 4.22 Heart rate for two different workloads.
The work rate of 8 kcal/min exhibits heart rate creep. The two marked time periods are used in Brouha's criteria.

buildup of fatigue and insufficient recovery during rest pauses (Brouha, 1967). This fatigue most likely results from the physical workload, but could also result from heat and mental stress, and a greater proportion of static rather than dynamic work. In any case, heart rate creep should be avoided by providing additional rest.

SUBJECTIVE RATINGS OF PERCEIVED EXERTION

An even simpler approach to estimating workload and the stress on the worker is the use of *subjective ratings of perceived exertion*. These can replace the expensive and relatively cumbersome equipment required for physiological measurements with the simplicity of verbal ratings. Borg (1967) developed the most popular scale for assessing the perceived exertion during dynamic whole-body activities—the *Borg Rating of Perceived Exertion (RPE) scale.* The scale is constructed such that the ratings 6 through 20 correspond directly to the heart rate (divided by 10) expected for that level of exertion (Table 4.6). Verbal anchors are provided to assist the worker in performing the ratings. Therefore, to ensure an acceptable heart rate recovery, based on the previous heart rate guidelines, the Borg scale should probably not exceed a rating of 11.

Note that the ratings, being subjective, can be affected by previous experience and the individual's level of motivation. Therefore, the ratings should be used with caution and should perhaps be normalized to each individual's maximum rating.

LOW BACK COMPRESSIVE FORCES

The adult human spine, or vertebral column, is an S-shaped assembly of 25 separate bones (*vertebrae*) divided into four major regions: 7 *cervical* vertebrae in the neck, 12 *thoracic* vertebrae in the upper back, 5 *lumbar* vertebrae in the low back, and the sacrum in the pelvic area (Figure 4.23). The bones have a roughly cylindrical body, with several bony processes emanating from the rear, which

Table 4.6 Borg's (1967) RPE Scale with Verbal Anchors

Rating	Verbal Anchor
6	No exertion at all
7	Extremely light
8	
9	Very light
10	
11	Light
12	
13	Somewhat hard
14	
15	Hard
16	
17	Very hard
18	
19	Extremely hard
20	Maximal exertion

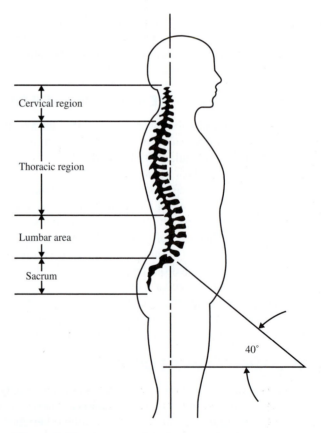

Cervical region

Thoracic region

Lumbar area

Sacrum

40°

Figure 4.23 Anatomy of the human spine.
(*From:* Rowe, 1983.)

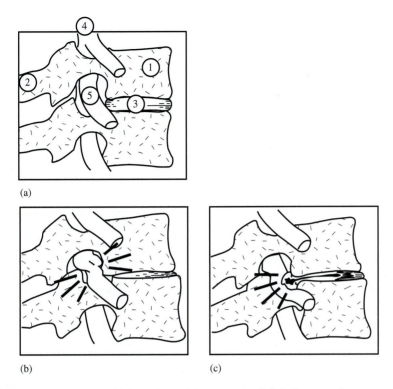

(a)

(b)   (c)

Figure 4.24 Anatomy of a vertebra and the process of disk degeneration.
(a) Normal state: (1) body of vertebra; (2) spinous process, serves as muscle attachment
point; (3) intervertebral disk; (4) spinal cord; (5) nerve root. (b) Narrowing of the disk
space, allowing the nerve root to be pinched. (c) Herniated disk, allowing the gel
material to extrude and impinge upon the nerve root. (Adapted from Rowe, 1983.)

serve as attachments for the back muscles, the *erector spinae*. Through the center of each vertebra is an opening that contains and protects the spinal cord as it
travels from the brain to the end of the vertebral column (Figure 4.24). At various points along the way, spinal nerve roots separate from the spinal cord and
pass between the vertebral bones out to the extremities, heart, organs, and other
parts of the body.

The vertebral bones are separated by softer tissue, the intervertebral *disks*.
These serve as joints, allowing a large range of motion in the spine, although
most trunk flexion occurs in the two lowest joints, the one on the border between
the lowest lumbar vertebra and the sacrum (termed the L_5/S_1 disk, where the
numbering of vertebrae is top down by region), and the next one up (L_4/L_5 disk).
The disks also act as cushions between the vertebral bones, and along with the
S-shaped spine, they help to protect the head and brain from the jarring impacts
of walking, running, and jumping. The disks are composed of a gellike center
surrounded by onionlike layers of fibers, separated from the bone by a cartilage
end plate. Considerable movement of fluid occurs between the gel center and the

surrounding tissue, depending on the pressure on the disk. Consequently, the length of the vertebral column (measured by change in overall stature) can change by as much as 1/2 to 1 in (1.3 to 2.5 cm) over the course of a workday and is sometimes used as an independent measure of an individual's physical workload. (Interestingly, astronauts in space, removed from the effects of gravity, can be as much as 2 in taller.)

Unfortunately, due to the combined effects of aging and heavy manual work exposure (these effects are hard to separate), the disks can weaken with time. Some of the enclosing fiber can become frayed, or the cartilage end plate can suffer microfractures, releasing some of the gelatinous material, reducing inner pressures, and allowing the center to start drying up. Correspondingly, the disk space narrows, allowing the vertebral bones to come closer together and eventually even touch, causing irritation and pain. Even worse, the nerve roots are impinged upon, leading to pain and sensory and motor impairments. As the fibers lose integrity, the vertebral bones may shift, causing uneven pressure on the disks and even more pain. In more catastrophic cases, termed *disk herniation,* or more commonly, a *slipped disk,* the fiber casings can actually rupture, allowing large amounts of the gel substance to extrude and impinge upon the nerve roots even more (Figure 4.24c).

The causes for low-back problems are not always easy to identify. As with most occupational diseases, both job and individual factors are at play. The latter may include a genetic predisposition toward weaker connective tissues, disks, and ligaments, and personal lifestyle conditions, such as smoking and obesity, over which the industrial engineer has very little control. Changes can only be made with the job factors. Although epidemiological data are easily confounded with survivor population effects or individual compensatory mechanisms, it can be shown statistically that heavy work leads to an increase in low back problems. Heavy work includes more than just the frequent lifting of large loads; it also encompasses the static maintenance of forward-bending trunk postures for long periods. Long periods of immobility, even in sitting postures, and whole-body vibration are also contributing factors. Therefore, scientists have associated the buildup of high disk pressures with eventual disk failures and have resorted to biomechanical calculations or estimations of disk compressive forces from intraabdominal or direct intradisk pressure measurements, neither of which are practical for industry.

A crude but useful analogy (Figure 4.25) considers a free-body diagram of the L_5/S_1 disk (where most of trunk flexion and disk herniation occurs) and models the components as a first-class lever, with the center of the disk acting as the fulcrum. The load acting through a moment arm determined by the distance from the center of the hands to the center of the disk creates a clockwise moment, while erector spinae muscle is modeled as a force acting downward through a very small moment arm [approximately 2 in (5 cm)], creating a counterclockwise moment barely sufficient to maintain equilibrium. Thus, the two moments must be equal, allowing calculation of the internal force of the erector spinae muscle:

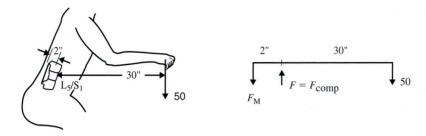

Figure 4.25 Back compressive forces modeled as a first-class lever.

$$2 \times F_M = 30 \times 50$$

where F_M = muscle force. Then F_M = 1,500/2, or 750, lb (341 kg). Solving for the total compressive force F_{comp} exerted on the disk yields

$$F_{comp} = F_M + 50 = 800$$

This disk compressive force of 800 lb (364 kg) is a considerable load, which may cause injury in certain individuals.

Note that this simple analogy neglects the offset alignment of the disks, weights of the body segments, multiple action points of erector spinae components, and other factors, and probably underpredicts the extremely high compressive forces typically obtained in the low back area. More accurate values for various loads and horizontal distances are presented in Figure 4.26. Due to the considerable individual variation in force levels resulting in disk failures, Waters (1994) recommended that a compressive force of 770 lb (350 kg) be considered the danger threshold.

The hand calculation of such compressive forces through biomechanical modeling is exceedingly time-consuming and has led to the development of various computerized biomechanical models, the best known of which is the 3D Static Strength Prediction Program.

Note that although disk herniation may be the most severe of low back injuries, there are other problems, such as soft tissue injuries involving ligaments, muscles, and tendons. These are probably more common, resulting in the backache that most people associate with manual work. Such pain, although uncomfortable, will probably recede over the course of several days with moderate rest. Physicians are currently recommending moderate daily activity to accelerate recovery, rather than the traditional complete bed rest. In addition, researchers are incorporating the soft tissue components in ever more complex back models.

NIOSH LIFTING GUIDELINES

Recognizing and attempting to control the growing problem of work-related back injuries, the National Institute for Occupational Safety and Health (NIOSH) issued what is commonly referred to as the *NIOSH lifting guidelines* (Waters et al., 1994).

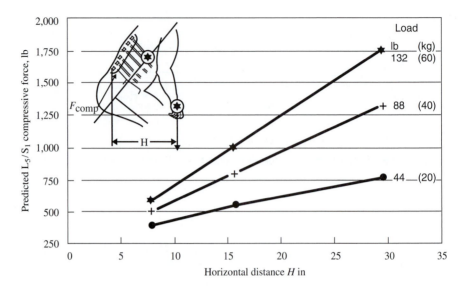

Figure 4.26
Effect of weight of load and horizontal distance between the load center of gravity and
the L_5/S_1 disk on the predicted compressive force on the L_5/S_1 disk.
(*Source:* Adapted from NIOSH, 1981, Figs 3.4 and 3.5.)

Although these are only guidelines, OSHA uses them extensively in its workplace
inspections and will issue citations based on these through the General Duty Clause.

The key output is the *recommended weight limit* (RWL), which is based on
the concept of an optimum weight, with adjustments for various factors related
to task variables. The RWL is meant to be a load that can be handled by most
workers:

1. The 770-lb (350-kg) compression force on the L_5/S_1 disk, created by the
    RWL, can be tolerated by most young, healthy workers.
2. Over 75 percent of women and over 99 percent of men have the strength
    capability to lift a load described by the RWL.
3. Maximum resulting energy expenditures of 4.7 kcal/min (18.8 Btu/min)
    will not exceed recommended limits.

Once the RWL is exceeded, musculoskeletal injury incidences and severity
rates increase considerably. The formulation for RWL is based on a maximum
load that can be handled in an optimum posture. As the posture deviates from the
optimum, adjustments for various task factors, in the form of multipliers, decrease the acceptable load.

$$RWL = LC * HM * VM * DM * AM * FM * CM$$

where   LC = load constant = 51 lb
          HM = horizontal multiplier = $10/H$
          VM = vertical multiplier = $1 - 0.0075|V - 30|$

$$DM = \text{distance multiplier} = 0.82 + 1.8/D$$

$$AM = \text{asymmetry multiplier} = 1 - 0.0032*A$$

$$FM = \text{frequency multiplier from Table 4.7}$$

$$CM = \text{coupling multiplier from Table 4.8}$$

H = horizontal location of the load cg forward of the midpoint between the ankles, $10 \le H \le 25$ in

V = vertical location of the load cg, $0 \le V \le 70$ in

D = vertical travel distance between origin and destination of lift, $10 \le D \le 70$ in

A = angle of asymmetry between the hands and feet (degrees), $0° \le A \le 135°$

Table 4.7 Frequency Multiplier (FM) Table

Frequency Lifts/min (F)[‡]	Work Duration					
	≤ 1 h		>1 but ≤ 2 h		>2 but ≥ 8 h	
	$V < 30$[†]	$V \ge 30$	$V < 30$	$V \ge 30$	$V < 30$	$V \ge 30$
≤ 0.2	1.00	1.00	0.95	0.95	0.85	0.85
0.5	0.97	0.97	0.92	0.92	0.81	0.81
1	0.94	0.94	0.88	0.88	0.75	0.75
2	0.91	0.91	0.84	0.84	0.65	0.65
3	0.88	0.88	0.79	0.79	0.55	0.55
4	0.84	0.84	0.72	0.72	0.45	0.45
5	0.80	0.80	0.60	0.60	0.35	0.35
6	0.75	0.75	0.50	0.50	0.27	0.27
7	0.70	0.70	0.42	0.42	0.22	0.22
8	0.60	0.60	0.35	0.35	0.18	0.18
9	0.52	0.52	0.30	0.30	0.00	0.15
10	0.45	0.45	0.26	0.26	0.00	0.13
11	0.41	0.41	0.00	0.23	0.00	0.00
12	0.37	0.37	0.00	0.21	0.00	0.00
13	0.00	0.34	0.00	0.00	0.00	0.00
14	0.00	0.31	0.00	0.00	0.00	0.00
15	0.00	0.28	0.00	0.00	0.00	0.00
>15	0.00	0.00	0.00	0.00	0.00	0.00

[†] Values of V are in inches.
[‡] For lifting less frequently than once per 5 min, set $F = 0.2$ lift/min.

Table 4.8 Coupling Multiplier

Coupling Type	Coupling Multiplier	
	$V < 30$ in (75 cm)	$V \ge 30$ in (75 cm)
Good	1.00	1.00
Fair	0.95	1.00
Poor	0.90	0.90

More simply,

$$\text{RWL (lb)} = 51(10/H)(1 - 0.0075|V - 30|)(0.82 + 1.8/D)$$
$$(1 - 0.0032A) \times \text{FM} \times \text{CM}$$

Note that these multipliers range from a minimum value of 0 for extreme postures to a maximum value of 1 for an optimal posture or condition. Table 4.7 provides frequency multipliers for three different work durations and for frequencies varying from 0.2/min to 15/min. Work duration is divided into three categories:

1. *Short duration.* One hour or less followed by a recovery time equal to 1.2 times the work time. (Thus, even though an individual works for three 1-h periods, as long as these work periods are interspersed with recovery times of 1.2 h, the overall work will still be considered of short duration.)
2. *Moderate duration.* Between 1 and 2 h of work, followed by a recovery period of at least 0.3 times the work time.
3. *Long duration.* Anything longer than 2 h but less than 8 h.

The coupling multiplier depends on the nature of the hand-to-object interface. In general, a good interface or grip will reduce the maximum grasp forces required and increase the acceptable weight for lifting. On the other hand, a poor interface will require large grasp forces and decrease the acceptable weight. For the revised NIOSH guidelines, three classes of couplings are used: good, fair, and poor.

A good coupling is obtained if the container is of optimal design, such as boxes and crates with well-defined handles or hand-hold cutouts. An optimal container has a smooth, nonslip texture, is no greater than 16 in (40 cm) in the horizontal direction, and is no greater than 12 in (30 cm) high. An optimal handle is cylindrical, with a smooth, nonslip surface, 0.75 to 1.5 in (1.9 to 3.8 cm) in diameter, more than 4.5 in (11.3 cm) long, and with 2-in (5-cm) clearance. For loose parts or irregular objects that are not found in containers, a good coupling would consist of a comfortable grip in which the hand can comfortably wrap around the object without any large wrist deviations (typically, small parts in a *power grip*).

A fair coupling results from less than optimal interfaces due to less than optimal handles or hand-hold cutouts. For containers of optimal design but with no handles or cutouts, or for loose parts, a fair coupling results if the hand cannot wrap all the way around but is flexed to only 90°. This would typically apply to most industrial packaging boxes.

A poor coupling results from containers of less than optimal design with no handles or hand-hold cutouts, or from loose parts that are bulky or hard to handle. Any container with rough or slippery surfaces, sharp edges, an asymmetric center of gravity, or unstable contents, or one that requires gloves would result in a poor coupling, by definition. To assist in the classification of couplings, the decision tree shown in Figure 4.27 may be useful.

The multipliers for each variable act as simple design tools for fairly straightforward job redesign. For example, if HM = 0.4, 60 percent of the potential lifting capability is lost due to a large horizontal distance. Therefore, the horizontal distance should be reduced as much as possible.

Object Lifted

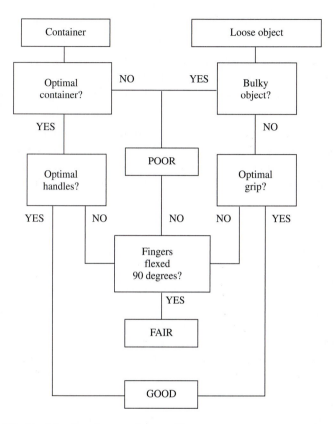

Figure 4.27 Decision tree for coupling quality

NIOSH also devised a *lifting index* (LI) to provide a simple estimate of the hazard level of lifting a given load, with values exceeding 1.0 deemed to be hazardous. Also, the LI is useful in prioritizing jobs for ergonomic redesign.

$$LI = \text{load weight/RWL}$$

In terms of controlling the hazard, NIOSH recommends engineering controls, physical changes, or a job and workplace redesign rather than administrative controls consisting of specialized selection and training of workers. Most common changes include avoiding high and low locations, using lift and tilt tables, using handles or specialized containers for handling loads, and reducing the horizontal distance by cutting out work surfaces and bringing loads closer to the body.

MULTITASK LIFTING GUIDELINES

For jobs with a variety of lifting tasks, the overall physical/metabolic load is increased compared to the single lifting task. This is reflected in a decreased RWL and an increased LI, and there is a special procedure to handle such situations. The concept is a *composite lifting index* (CLI), which represents the collective demands of the job. The CLI equals the largest *single-task lifting index* (STLI) and increases incrementally for each subsequent task. The multitask procedure is as follows:

1. Compute a *single-task RWL* (STRWL) for each task.
2. Compute a *frequency-independent RWL* (FIRWL) for each task by setting FM = 1.
3. Compute a *single-task LI* (STLI) by dividing the load by STRWL.
4. Compute a *frequency-independent LI* (FILI) by dividing the load by FIRWL.
5. Compute the CLI for the overall job by rank-ordering the tasks according to decreasing physical stress, that is, the STLI for each task. The CLI is then

$$CLI = STLI_1 + \Sigma\Delta LI$$

where $\Sigma\Delta LI = FILI_2 (1/FM_{1,2} - 1/FM_1) + FILI_3 (1/FM_{1,2,3} - 1/FM_{1,2}) + \cdots$ Consider the three-task lifting job shown in Table 4.9. The multitask lifting analysis is as follows:

1. The task with the greatest lifting index is new task 1 (old task 2) with STLI = 1.6.
2. The sum of the frequencies for new tasks 1 and 2 is 1 + 2 = 3.
3. The sum of the frequencies for new tasks 1, 2, and 3 is 1 + 2 + 4 = 7.
4. From Table 4.7, the new frequency multipliers are $FM_1 = 0.94$, $FM_{1,2} = 0.88$, and $FM_{1,2,3} = 0.70$.
5. The combined lifting index is therefore

$$CLI = 1.6 + 1 (1/0.88 - 1/0.94) + 0.67 (1/0.7 - 1/0.88)$$
$$= 1.60 + 0.07 + 0.20 = 1.90$$

| EXAMPLE 4.2 | NIOSH Analysis of Lifting a Box into the Trunk of a car |

Figure 4.28
Posture for trunk loading example

Before recent automotive design changes, it was not unusual to have to lean forward and extend the arms while placing an object into the trunk of a car (Figure 4.28). Assume the occupant lifts a 30-lb box from the ground into a trunk. Being lazy, the occupant simply twists 90° to pick up the box from the ground level ($V = 0$) at a short horizontal distance ($H \sim 10$ in). The vertical travel distance is the difference between the vertical location of the box at the destination (assume the bottom of the trunk is 25 in from the ground) and the vertical location of the box at the origin ($V = 0$), yielding $D = 25$. Assume that this is a one-time lift; therefore, FM = 1. Also assume that the box is fairly small and compact, but has no handles. Thus, coupling is fair with CM = 0.95. This yields the following calculation for the origin:

$$RWL_{ORG} = 51(10/10) (1 - 0.0075|0 - 30|) (0.82 + 1.8/25)$$
$$\times (1 - 0.0032 * 90) (1) (0.95)$$

$$= 51(1)(0.775)(0.892)(0.712)(1)(0.95) = 23.8$$

Assuming a larger reach ($H = 25$ in) into the trunk because of the bumper and high trunk lip, no twisting, the distance traveled remaining the same, and the coupling remaining fair, the calculation at the destination is

$$RWL_{DEST} = 51(10/25)(1 - 0.0075|25 - 30|)(0.82 + 1.8/25)$$
$$\times (1 - 0.0032 * 0)(1)(0.95)$$
$$= 51(0.4)(0.963)(0.892)(1)(1)(0.95) = 16.6$$

and

$$LI = 30/16.6 = 1.8$$

Thus, in the worst-case approach, only 16.6 lb could be lifted safely by most individuals, and the 30-lb box would create a hazard almost twice the acceptable level. The biggest reduction in capability is the horizontal distance at the destination, due to the trunk design. Decreasing the horizontal distance to 10 in would increase the H factor to $10/10 = 1$ and increase RWL to 41.5 lb. For most newer cars, this has been accomplished by the auto manufacturers by opening the front part of the trunk such that once the load is lifted to the lower lip, minimum horizontal lifting is needed; the load can be simply pushed forward. However, the limiting case is now the origin, which can be improved by moving the feet and eliminating the twist, increasing RWL to 33.4 lb. Note that a two-step analysis is necessary if the occupant lifts the load from ground level to the lip of the trunk and then lowers the box into the trunk. This lift is also improved in newer-model cars because of a decrease in the vertical height of the lip, which also decreases the distance lifted.

Table 4.9 A Sample Three-Task Lifting Job's Characteristics

Task Number	1	2	3
Load weight L	20	30	10
Task frequency F	2	1	4
FIRWL	20	20	15
FM	0.91	0.94	0.84
STRWL	18.2	18.8	12.6
FILI	1.0	1.5	0.67
STLI	1.1	1.6	0.8
New task number	2	1	3

This procedure is facilitated by the NIOSH Multitask Job Analysis Worksheet (see Figure 4.29). However, once the number of tasks exceeds three or four, it becomes very time-consuming to calculate CLI by hand. A variety of software programs and websites are now available to assist the user in this effort, including Design Tools. Of course, the best solution overall is to avoid manual materials handling and use mechanical assist devices or completely automated material handling systems (see Chapter 3).

Multitask job analysis worksheet

Department   _____   Job description

Job title   _____

Analyst's name   _____   _____

Date   _____   _____

Step 1. Measure and record task variable data

Task no.	Object weight (lbs)		Hand location (in)				Vertical distance (in)	Asymmetry angle (degs)		Frequency rate Lifts/min	Duration Hrs	Coupling
			Origin		Dest.			Origin	Dest.			
	L (avg.)	L (max.)	H	V	H	V	D	A	A	F		C
_____												
_____												
_____												
_____												
_____												

Step 2. Compute multipliers and FIRWL, STRWL, FILI, and STLI for each task

Task no.	LC x HM x VM x DM x AM x CM	FIRWL x FM	STRWL	FILI = L/FIRWL	STLI = L/STRWL	New task no.	F
51							
51							
51							
51							
51							

Step 3. Compute the composite lifting index for the job (after renumbering tasks)

$\text{CLI} = \text{STLI}_1, + \Delta \text{FILI}_2 + \Delta \text{FILI}_3 + \Delta \text{FILI}_4 + \Delta \text{FILI}_5$

	$\text{FILI}_2(1/\text{FM}_{1,2} - 1/\text{FM}_1)$	$\text{FILI}_3(1/\text{FM}_{1,2,3} - 1/\text{FM}_{1,2})$	$\text{FILI}_4(1/\text{FM}_{1,2,3,4} - 1/\text{FM}_{1,2,3})$	$\text{FILI}_5(1/\text{FM}_{1,2,3,4,5} - 1/\text{FM}_{1,2,3,4})$	
CLI =					

Figure 4.29   Multitask Job Analysis Worksheet.

GENERAL GUIDELINES: MANUAL LIFTING

Although no one optimal lifting technique is suitable for all individuals or task conditions, several guidelines are generally appropriate overall (see Figure 4.30). First, plan the lift by evaluating the size and shape of the load, determining whether assistance is needed, and ascertaining what worksite conditions may interfere with the lift. Second, determine the best lifting technique. In general, a squat lift, keeping the back relatively straight and lifting with the knees bent, is the safest in terms of lowback compressive forces. However, bulky loads may interfere with the knees, and a stoop lift, in which the individual bends over and then extends the back, may be required. Third, spread the feet apart, both side ways and fore–aft, to maintain a good balance and stable posture. Fourth, secure a good grip on the load. These last two guidelines are especially important in avoiding sudden twisting and jerking movements, both of which are extremely detrimental to the low back. Fifth, hold the load close to the body to minimize the horizontal moment arm created by the load and the resulting moment on the low back.

Avoiding twisting and jerky motions is critical. The first produces an asymmetric orientation of the disks, leading to increased disk pressures, while the

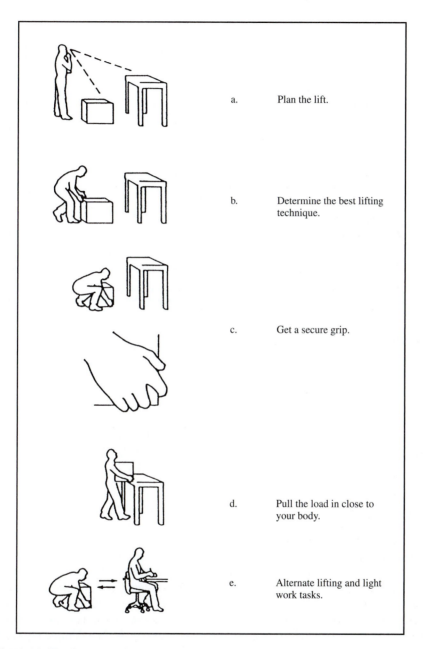

a.   Plan the lift.

b.   Determine the best lifting technique.

c.   Get a secure grip.

d.   Pull the load in close to your body.

e.   Alternate lifting and light work tasks.

Figure 4.30   Safe lifting procedure.
(Available through S. H. Rodgers, Ph.D. P.O. Box 23446, Rochester, NY 14692.)

General Posture Evaluation	Yes	No
1. Are the joints maintained in a neutral position (most are straight, elbow is at 90°)?	❑	❑
2. Is the work or load held close to the body?	❑	❑
3. Are forward bending postures avoided?	❑	❑
4. Are twisting postures of the trunk avoided?	❑	❑
5. Are sudden movements or jerks avoided?	❑	❑
6. Are static postures avoided? I.e., are there changes in posture?	❑	❑
7. Are excessive reaches avoided?	❑	❑
8. Are the hands utilized in front of the body?	❑	❑

Task Evaluation	Yes	No
1. Are static muscle exertions avoided?	❑	❑
a. Are repetitive static exertions limited to < 15% maximum strength?	❑	❑
b. Are durations of static exertions limited to several seconds?	❑	❑
2. Are pinch grips only used for low-force precision tasks?	❑	❑
3. Are large muscle groups and power grips utilized for tasks requiring force?	❑	❑
4. Is momentum utilized to assist the operator?	❑	❑
5. Are curved motions pivoting around the lowest order joints utilized?	❑	❑
6. Are materials and tools placed within the normal working area?	❑	❑
7. Are gravity bins and drop deliveries utilized?	❑	❑
8. Are tasks carried out below shoulder level and above knuckle height?	❑	❑
9. Are lifts performed slowly with knees bent?	❑	❑
10. Are mechanical assists or additional help utilized for loads exceeding 50 lbs?	❑	❑
11. Is the workload low enough such that the heart rate is steady and below 110 beats/min?	❑	❑
12. Are frequent short rest breaks provided?	❑	❑

Figure 4.31   General Posture and Task Evaluation Checklist.

second generates additional accelerative forces on the back. One nonintuitive method for discouraging twisting in workers is actually to increase the travel distance between the origin and the destination. This will force the worker to take a step and, in so doing, turn the whole body, rather than twisting the trunk. Carrying uneven loads in both arms or an entire load in only one arm generates similar asymmetric disk orientations and should also be avoided.

The General Posture and Task Evaluation Checklist (see Figure 4.31) can be very useful in reminding the analyst of the basic principles of good work design.

BACK BELTS

A cautionary note should be given regarding back belts. Although commonly found on many workers and automatically prescribed in some companies, back belts are not the ultimate panacea and must be regarded with caution. Back belts originated from early studies of weightlifting, showing that for extreme loads, belts relieved 15 to 30 percent of low back compressive forces, as estimated from back electromyograms (Morris et al., 1961). However, these studies were performed on trained weightlifters, lifting much larger loads, in a completely sagittal plane. Industrial workers lift much lighter loads, producing a much lower effect. Twisting because of misaligned muscles probably reduces this effect even further. There are also anecdotal data of the "superman" effect—industrial workers with back belts selecting heavier loads than those without back belts—and some workers having coronary incidents from the 10 to 15 mmHg increase in blood pressure due to the abdominal compression.

Finally, a longitudinal study of airline baggage handlers (Ridell et al., 1992) concluded there was no significant difference in back injuries between workers using back belts and "control" workers without back belts. Surprisingly, a smaller group of workers, who for one reason or another (e.g., discomfort, heat) quit wearing the belts but continued in the study, had significantly higher injuries. This may be attributed to atrophy of the abdominal muscles, which should naturally provide an internal back belt but were weakened due to decreased stress. A positive approach may be to encourage workers to strengthen these abdominal muscles through abdominal crunches (modified sit-ups), regular exercise, and body weight reduction. Back belts should only be used with proper training and only after engineering controls have been attempted.

SUMMARY

Chapter 4 introduces some of the theoretical concepts of the human musculoskeletal and physiological systems as a means of providing a framework for a better understanding of the principles of motion economy and work design. These principles are presented as a set of rules to be utilized in redesigning manual assembly work as part of the motion study. Hopefully, with a better understanding of the functioning of the human body, the analyst will view these rules as less arbitrary. These same concepts will be elaborated on in Chapter 5, for the discussions on the design of the workplace, tools, and equipment.

QUESTIONS

1. What structural components are found in muscles? What do these components have to do with muscle performance?
2. Explain the elements of static and dynamic muscle performance with the sliding filament theory.
3. Describe the different types of muscle fibers and relate their properties to muscle performance.
4. Why does a change in the number of active motor units not result in a proportional change in muscle tension?
5. What does EMG measure? How is EMG interpreted?
6. Explain why workstation designers should endeavor to have operators perform work elements without lifting their elbows.
7. What viewing distance would you recommend for a seated operator working at a computer terminal?
8. Define and give examples of the 17 fundamental motions, or therbligs.
9. How may the basic motion Search be eliminated from the work cycle?
10. What basic motion generally precedes Reach?
11. What three variables affect the time for the basic motion Move?
12. How does the analyst determine when the operator is performing the element Inspect?
13. Explain the difference between avoidable and unavoidable delays.

14. Which of the 17 therbligs are classed as effective and usually cannot be removed from the work cycle?

15. Why should fixed locations be provided at the workstation for all tools and materials?

16. Which of the five classes of motions is preferred for industrial workers? Why?

17. Why is it desirable to have the feet working only when the hands are occupied?

18. In a motion study, why is it inadvisable to analyze both hands simultaneously?

19. What task factors increase the index of difficulty in a Fitts' tapping task?

20. What factors affect back compressive forces during lifting?

21. What factors influence the measurement of isometric muscle strength?

22. Why do psychophysical, dynamic, and static strength capabilities differ?

23. What methods can be used to estimate the energy requirements of a job?

24. What factors change the energy expended for a given job?

25. How does work capability vary with gender and age?

26. What limits endurance in a whole-body manual task?

PROBLEMS

1. What is the maximum load that can be raised by an outstretched stiff arm by a 50th-percentile female? (For estimating anthropometry use Table 5.1.)

2. In the packing department, a worker stands sideways between the end of a conveyor and a pallet. The surface of the conveyor is 40 in from the floor, and the top of the pallet is 6 in from the floor. As a box moves to the end of the conveyor, the worker twists 90° to pick up the box, then twists 180° in the opposite direction and sets the box down on the pallet. Each box is 12 in on a side and weighs 25 lb. Assume the worker moves five boxes per minute for an 8-h shift and a horizontal distance of 12 in. Using the NIOSH lifting equation, calculate RWL and LI. Redesign the task to improve it. What are the RWL and LI now?

3. For Problem 2 calculate the low back compressive forces incurred in the performance of this job, using the University of Michigan 3D Static Strength Prediction Program.

4. A 95th percentile male is holding a 20-lb load in his outstretched arm in 90° abduction. What is the voluntary torque required at the shoulder to be able to hold this load?

5. A worker is shoveling sand at a rate of 8 kcal/min. How much rest does he need during an 8-h shift? How should the rest be allocated?

6. A current problem in the U.S. Army is the neck/shoulder fatigue experienced by helicopter pilots. To be able to fly missions at night, the pilots wear night vision goggles, which are attached to the front of the helmet. Unfortunately, these are fairly heavy, causing a large downward torque of the head. This torque must be counteracted by the neck muscles, which then fatigue. To alleviate this problem, many pilots have started attaching random lead weights to the back of the helmet. Find the appropriate weight that would best balance the head and minimize neck fatigue. Assumptions: (a) cg of goggles is 8 in in front of neck pivot point; (b) goggles weigh 2 lb; (c) maximum volitional neck torque is 480 in·lb; (d) cg of

lead weight is 5 in behind neck pivot point; (e) bare helmet weighs 4 lb; and (f) cg of helmet is 0.5 in in front of neck pivot.

7. The laborer on a palletizing operation has been complaining about fatigue and the lack of rest. You measure his heart rate and find it to be 130 beats/min and slowly increasing during work. When he sits down, his heart rates dropped to 125 beats/min by the end of the first minute of rest and 120 beats/min by the end of the third minute. What do you conclude?

8. A grievance has been filed by the union at Dorben Co. regarding the final inspection station, in which the operator slightly lifts a 20-lb assembly, examines all sides, and, if acceptable, sets it back down on the conveyor that takes the assembly to the packing station. On average, the inspector examines five assemblies per minute, at an energy expenditure level of 6 kcal/min. The conveyor is 40 in off the floor, and the assembly is roughly 20 in from the inspector while being inspected. Evaluate the job with respect to the NIOSH lifting guidelines and metabolic energy expenditure considerations. Indicate whether the job exceeds allowable limits. If it does, calculate how many assemblies the inspector may inspect per minute without exceeding acceptable guidelines.

9. A relatively unfit worker with resting heart rate of 80 beat/min starts his job of palletizing boxes. At the morning break, an industrial engineer quickly measures the worker's heart rate and finds a peak value of 110 beats/min, a value of 105 beats/min 1 min after stopping work, and a value of 95 beat/min 3 min after stopping work. What can you conclude about the workload for this worker?

REFERENCES

Åstrand, P. O., and K. Rodahl. *Textbook of Work Physiology.* 3d ed. New York: McGraw-Hill, 1986.

Bink, B. "The Physical Working Capacity in Relation to Working Time and Age." *Ergonomics,* 5, no.1 (January 1962), pp. 25–28.

Borg, G., and H. Linderholm. "Perceived Exertion and Pulse Rate During Graded Exercise in Various Age Groups." *Acta Medica Scandinavica,* Suppl. 472 (1967), pp. 194–206.

Bouisset, S. "EMG and Muscle Force in Normal Motor Activities." In *New Developments in EMG and Clinical Neurophysiology.* Ed. J. E. Desmedt. Basel, Switzerland: S. Karger, 1973.

Brouha, L. *Physiology in Industry.* New York: Pergamon Press, 1967.

Chaffin, D. B. "Electromyography—A Method of Measuring Local Muscle Fatigue." *The Journal of Methods-Time Measurement,* 14 (1969), pp. 29–36.

Chaffin, D. B., and G. B. J. Andersson. *Occupational Biomechanics.* New York: John Wiley & Sons, 1991.

Chaffin, D. B., G. D. Herrin, W. M. Keyserling, and J. A. Foulke. *Preemployment Strength Testing.* NIOSH Publication 77-163. Cincinnati, OH: National Institute for Occupational Safety and Health, 1977.

Drillis, R. "Folk Norms and Biomechanics." *Human Factors,* 5 (October 1963), pp. 427–441.

Dul, J., and B. Weerdmeester. *Ergonomics for Beginners.* London: Taylor & Francis, 1993.

Eastman Kodak Co., Human Factors Section. *Ergonomic Design for People at Work.* New York: Van Nostrand Reinhold, 1983.

Fitts, P. "The Information Capacity of the Human Motor System in Controlling the Amplitude of Movement." *Journal of Experimental Psychology,* 47, no. 6 (June 1954), pp. 381–391.

Freivalds, A., and D. M. Fotouhi. "Comparison of Dynamic Strength as Measured by the Cybex and Mini-Gym Isokinetic Dynamometers." *International Journal of Industrial Ergonomics,* 1, no. 3 (May 1987), pp. 189–208.

Garg, A. "Prediction of Metabolic Rates for Manual Materials Handling Jobs," *American Industrial Hygiene Association Journal*, 39, pp.661–674.

Gordon, E. (1957). The use of energy costs in regulating physical activity in chronic disease. *A.M.A. Archives of Industrial Health*, 16, 437–441.

Grandjean, E. *Fitting the Task to the Man.* New York: Taylor & Francis, 1988.

Gray, H. *Gray's Anatomy.* 35th ed. Eds. R. Warrick and P. Williams. Philadelphia: W.B. Saunders, 1973.

Ikai, M., and T. Fukunaga. "Calculation of Muscle Strength per Unit Cross-Sectional Area of Human Muscle by Means of Ultrasonic Measurement." *Internationale Zeitschrift für angewandte Physiologie einschließlich Arbeitsphysiologie,* 26 (1968), pp. 26–32.

Jonsson, B. "Kinesiology." In *Contemporary Clinical Neurophysiology (EEG Sup. 34).* New York: Elsevier-North-Holland, 1978.

Langolf, G., D. G. Chaffin, and J. A. Foulke. "An Investigation of Fitt's Law Using a Wide Range of Movement Amplitudes." *Journal of Motor Behavior,* 8, no.2 (June 1976), pp. 113–128.

Miller, G. D., and A. Freivalds. "Gender and Handedness in Grip Strength—A Double Whammy for Females." *Proceedings of the Human Factors Society,* 31 (1987), pp. 906–910.

Morris, J. M., D. B. Lucas, and B. Bressler. "Role of the Trunk in Stability of the Spine." *Journal of Bone and Joint Surgery,* 43-A, no. 3 (April 1961), pp. 327–351.

Mundel, M. E., and D. L. Danner. *Motion and Time Study.* 7th ed. Englewood Cliffs, NJ: Prentice-Hall, 1994.

Murrell, K. F. H. *Human Performance in Industry.* New York: Reinhold Publishing, 1965.

National Safety Council. *Accident Facts.* Chicago: National Safety Council, 2003.

Passmore, R., and J. Durnin, (1955). Human energy expenditure. *Physiological Reviews,* 35, 801–875.

Ridell, C. R., J. J. Congleton, R. D. Huchingson, and J. T. Montgomery. "An Evaluation of a Weightlifting Belt and Back Injury Prevention Training Class for Airline Baggage Handlers." *Applied Ergonomics,* 23, no. 5 (October 1992), pp. 319–329.

Rodgers, S. H. *Working with Backache.* Fairport, NY: Perinton Press, 1983.

Rowe, M. L. *Backache at Work.* Fairport, NY: Perinton Press, 1983.

Sanders, M. S., and E. J. McCormick. *Human Factors in Engineering and Design.* New York: McGraw-Hill, 1993.

Snook, S. H., and V. M. Ciriello. "The Design of Manual Handling Tasks: Revised Tables of Maximum Acceptable Weights and Forces." *Ergonomics,* 34, no. 9 (September 1991), pp. 1197–1213.

Thorton, W. "Anthropometric Changes in Weightlessness." In *Anthropometric Source Book,* 1, ed. Anthropology Research Project, Webb Associates. NASA RP1024. Houston, TX: National Aeronautics and Space Administration, 1978.

Waters, T. R., V. Putz-Anderson, and A. Garg. *Revised NIOSH Lifting Equation,* Pub. No. 94-110, Cincinnati, OH: National Institute for Occupational Safety and Health, 1994.

Winter, D. A. *Biomechanics of Human Movement.* New York: John Wiley & Sons, 1979.

SELECTED SOFTWARE

3D Static Strength Prediction Program. University of Michigan Software, 475 E. Jefferson, Room 2354, Ann Arbor, MI 48109. (http://www.umichergo.org)

Design Tools (available from the McGraw-Hill website at www.mhhe.com/neibel-freivalds), New York: McGraw-Hill, 2002.

Energy Expenditure Prediction Program. University of Michigan Software, 475 E. Jefferson, Room 2354, Ann Arbor, MI 48109. (http://www.umichergo.org)

Ergointelligence (Manual Material Handling). Nexgen Ergonomics, 3400 de Maisonneuve Blvd. West, Suite 1430, Montreal, Quebec, Canada H3Z 3B8. (http://www.nexgenergo.com/)

ErgoTRACK (NIOSH Lifting Equation). ErgoTrack.com, P.O. Box 787, Carrboro, NC 27510.

WEBSITES

NIOSH Homepage–http://www.cdc.gov/niosh/homepage.html
NIOSH Lifting Guidelines–http://www.cdc.gov/niosh/94-110.html
NIOSH Lifting Calculator–http://www.industrialhygiene.com/calc/lift.html
NIOSH Lifting Calculator–http://tis.eh.doe.gov/others/ergoeaser/download.html

Workplace, Equipment, and Tool Design

Designing the workplace, tools, equipment, and work environment to fit the human operator is called *ergonomics*. Rather than devoting a lot of space to the underlying theory of the physiology, capabilities, and limitations of the human, this chapter presents the principles of work design and appropriate checklists to facilitate the use of these design principles. With each design principle, a brief explanation of its origin or relationship to the human is provided. This approach will better assist the methods analyst in designing the workplace, equipment, and tools to meet the simultaneous goals of (1) increased production and efficiency of the operation and (2) decreased injury rates for the human operator.

5.1 ANTHROPOMETRY AND DESIGN

The primary guideline is to design the workplace to accommodate most individuals with regard to structural size of the human body. The science of measuring the human body is termed *anthropometry* and typically utilizes a variety of caliperlike devices to measure structural dimensions, for example, stature and forearm length. Practically speaking, however, few ergonomists or engineers collect their own data, because of the wealth of data that has already been collected and tabulated. Close to 1,000 different body dimensions, for close to 100 mostly military population types, are available in the somewhat dated *Anthropometric Source Book* (Webb Associates, 1978). More recently the CAESAR (Civilian American and European Surface

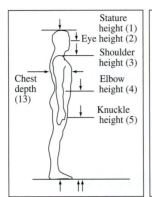
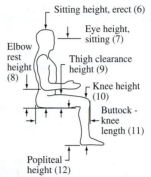
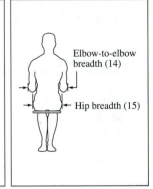

Table 5.1   Selected Body Dimensions and Weights of U.S. Adult Civilians

Body dimension	Sex	Dimension (in)			Dimension (cm)		
		5th	**50th**	**95th**	**5th**	**50th**	**95th**
1. Stature (height)	Male	63.7	68.3	72.6	161.8	173.6	184.4
	Female	58.9	63.2	67.4	149.5	160.5	171.3
2. Eye height	Male	59.5	63.9	68.0	151.1	162.4	172.7
	Female	54.4	58.6	62.7	138.3	148.9	159.3
3. Shoulder height	Male	52.1	56.2	60.0	132.3	142.8	152.4
	Female	47.7	51.6	55.9	121.1	131.1	141.9
4. Elbow height	Male	39.4	43.3	46.9	100.0	109.9	119.0
	Female	36.9	39.8	42.8	93.6	101.2	108.8
5. Knuckle height	Male	27.5	29.7	31.7	69.8	75.4	80.4
	Female	25.3	27.6	29.9	64.3	70.2	75.9
6. Height, sitting	Male	33.1	35.7	38.1	84.2	90.6	96.7
	Female	30.9	33.5	35.7	78.6	85.0	90.7
7. Eye height, sitting	Male	28.6	30.9	33.2	72.6	78.6	84.4
	Female	26.6	28.9	30.9	67.5	73.3	78.5
8. Elbow rest height, sitting	Male	7.5	9.6	11.6	19.0	24.3	29.4
	Female	7.1	9.2	11.1	18.1	23.3	28.1
9. Thigh clearance height	Male	4.5	5.7	7.0	11.4	14.4	17.7
	Female	4.2	5.4	6.9	10.6	13.7	17.5
10. Knee height, sitting	Male	19.4	21.4	23.3	49.3	54.3	59.3
	Female	17.8	19.6	21.5	45.2	49.8	54.5
11. Buttock-knee distance, sitting	Male	21.3	23.4	25.3	54.0	59.4	64.2
	Female	20.4	22.4	24.6	51.8	56.9	62.5
12. Popliteal height, sitting	Male	15.4	17.4	19.2	39.2	44.2	48.8
	Female	14.0	15.7	17.4	35.5	39.8	44.3
13. Chest depth	Male	8.4	9.5	10.9	21.4	24.2	27.6
	Female	8.4	9.5	11.7	21.4	24.2	29.7
14. Elbow-elbow breadth	Male	13.8	16.4	19.9	35.0	41.7	50.6
	Female	12.4	15.1	19.3	31.5	38.4	49.1
15. Hip breadth, sitting	Male	12.1	13.9	16.0	30.8	35.4	40.6
	Female	12.3	14.3	17.2	31.2	36.4	43.7
X. Weight (lb and kg)	Male	123.6	162.8	213.6	56.2	74.0	97.1
	Female	101.6	134.4	197.8	46.2	61.1	89.9

Source: Kroemer, 1989.

Probability Distributions and Percentiles   **EXAMPLE 5.1**

A *kth percentile* is defined as a value such that *k* percent of the data values (plotted in ascending order) are at or below this value and $100 - k$ percent of the data values are at or above this value. A histogram plot of U.S. adult male statures shows a bell-shaped curve, termed a *normal distribution,* with a median value of 68.3 in (see Figure 5.1). This is also the 50th percentile value; for example, one-half of all males are shorter than 68.3 in, while one-half are taller. The 5th percentile male is only 63.7 in tall, while a 95th percentile male is 72.6 in tall. The proof is as follows.

Typically, in a statistical approach, the approximately bell-shaped curve is normalized by the transformation

$$z = (x - \mu)/\sigma$$

where $\mu =$ mean
          $\sigma =$ standard deviation (measure of dispersion)

to form a standard normal distribution (also termed a *z* distribution; see Figure 5.2). Once normalized, any approximately bell-shaped population distribution will have the same statistical properties. This allows easy calculation of any percentile value desired, using the appropriate *k* and *z* values, as follows:

κth percentile	10 or 90	5 or 95	2.5 or 97.5	1 or 99
z value	±1.28	±1.645	±1.96	±2.33

$$k\text{th percentile} = \mu \pm z\sigma$$

Given that the mean stature for males in the United States is 68.3 in (173.6 cm), while the standard deviation is 2.71 in (6.9 cm) (Webb Associates, 1978), the 95th percentile male stature is calculated as

$$68.3 + 1.645(2.71) = 72.76 \text{ in}$$

while the 5th percentile male stature is

$$68.3 - 1.645(2.71) = 63.84 \text{ in}$$

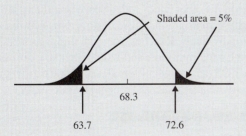

Figure 5.1 Normal distribution of U.S. adult male statures.

Note that the calculated values of 72.76 and 63.84 in are not exactly equal to the actual values of 72.6 and 63.7 in. This is so because the U.S. male height distribution is not a completely normal distribution.

(continued)

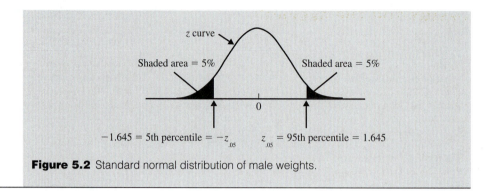

Figure 5.2 Standard normal distribution of male weights.

Anthropometry Resource) project collected over 100 dimensions on 5,000 civilians using three-dimensional body scans. A summary of useful dimensions that apply to the particular postures needed for workplace design for U.S. males and females is given in Table 5.1. Much of this anthropometric data is included in computerized human models such as COMBIMAN, Jack, MannequinPro, and Safeworks that provide easy size adjustments and limitations in ranges of motion or visibility as part of the computer-aided design process.

DESIGN FOR EXTREMES

Designing for most individuals is an approach that involves the use of one of three different specific design principles, as determined by the type of design problem. *Design for extremes* implies that a specific design feature is a limiting factor in determining either the maximum or minimum value of a population variable that will be accommodated. For example, clearances, such as a doorway or an entry opening into a storage tank, should be designed for the maximum individual, that is, a 95th percentile male stature or shoulder width. Then 95 percent of all males and almost all females will be able to enter the opening. Obviously, for doorways, space is usually not at a premium, and the opening can be designed to accommodate even larger individuals. On the other hand, added space in military aircraft or submarines is expensive, and these areas are therefore designed to accommodate only a certain (smaller) range of individuals. Reaches, for such things as a brake pedal or control knob, are designed for the minimum individual, that is, a 5th percentile female leg or arm length. Then 95 percent of all females and practically all males will have a longer reach and will be able to activate the pedal or control.

DESIGN FOR ADJUSTABILITY

Design for adjustability is typically used for equipment or facilities that can be adjusted to fit a wider range of individuals. Chairs, tables, desks, vehicle seats, steering columns, and tool supports are devices that are typically adjusted to accommodate the worker population ranging from 5th percentile females to 95th percentile males. Obviously, designing for adjustability is the preferred method of design, but there is a trade-off with the cost of implementation. (Specific adjustment ranges for seat design are given later in Table 5.2)

DESIGN FOR THE AVERAGE

Design for the average is the cheapest but least preferred approach. Even though there is no individual with all average dimensions, there are certain situations where it would be impractical or too costly to include adjustability for all features. For example, most industrial machine tools are too large and too heavy to include height adjustability for the operator. Designing the operating height at the 50th percentile of the elbow height for the combined female and male populations (roughly the average of the male and female 50th percentile values) means that most individuals will not be unduly inconvenienced. However, the exceptionally tall male or very short female may experience some postural discomfort.

PRACTICAL CONSIDERATIONS

Finally, the industrial designer should also consider the legal ramifications of design work. Due to the passage of the Americans with Disabilities Act of 1990 (see Section 9.6), reasonable effort must be made to accommodate individuals with all abilities. Special accessibility guidelines (U.S. Department of Justice, 1991) have been issued regarding parking lots, entryways into buildings, assembly areas, hallways, ramps, elevators, doors, water fountains, lavatories, restaurant or cafeteria facilities, alarms, and telephones.

It is also very useful, if practical and cost-effective, to build a full-scale mock-up of the equipment or facility being designed and then have the users evaluate the mock-up. Anthropometric measurements are typically made in standardized postures. In real life, people slouch or have relaxed postures, changing the effective dimensions and the ultimate design. Many costly errors have occurred during production, because of the lack of mock-up evaluations. In Example 5.2, the final design actually accommodates more than 95 percent of the population, yielding a rise height larger than necessary. The true design should have used the body dimensions for a combined male and female population. However, such combined data are rarely available. The data can be created through statistical techniques, but the general design approach is sufficient for most industrial applications.

5.2 PRINCIPLES OF WORK DESIGN: THE WORKPLACE

DETERMINE WORK SURFACE HEIGHT BY ELBOW HEIGHT

The work surface height (whether the worker is seated or standing) should be determined by a comfortable working posture for the operator. Typically, this means that the upper arms are hanging down naturally and the elbows are flexed at 90° so that the forearms are parallel to the ground (see Figure 5.4). The elbow height becomes the proper operation or work surface height. If the work surface is too high, the upper arms are abducted, leading to shoulder fatigue. If the work surface is too low, the neck or back is flexed forward, leading to back fatigue.

ADJUST THE WORK SURFACE HEIGHT BASED ON THE TASK BEING PERFORMED

There are modifications to the first principle. For rough assembly involving the lifting of heavy parts, it is more advantageous to lower the work surface by as much as 8 in (20 cm) to take advantage of the stronger trunk muscles (see Figure 5.5). For fine assembly involving minute visual details, it is more advantageous to raise the work surface by up to 8 in (20 cm) to bring the details closer to the optimum line of sight of 15°(principle from Chapter 4). Another, perhaps better, alternative is to slant the work surface approximately 15° then both principles can be satisfied. However, rounded parts then have a tendency to roll off the surface.

These principles also apply to a seated workstation. A majority of tasks, such as writing or light assembly, are best performed at the resting-elbow height. If the job requires the perception of fine detail, it may be necessary to raise the work to bring it closer to the eyes. Seated workstations should be provided with adjustable chairs and adjustable footrests (see Figure 5.6). Ideally, after the operator is comfortably seated with both feet on the floor, the work surface is positioned at the appropriate elbow height to accommodate the operation. Thus, the workstation also needs to be adjustable. Short operators whose feet do not reach the floor, even after adjusting the chair, should utilize a footrest to provide support for the feet.

PROVIDE A COMFORTABLE CHAIR FOR THE SEATED OPERATOR

The seated posture is important from the standpoint of reducing both the stress on the feet and the overall energy expenditure. Because comfort is a very individual response, strict principles for good seating are somewhat difficult to define. Furthermore, few chairs will comfortably adapt to the many possible seating postures (see Figure 5.7). However, several general principles hold true for all seats. When a person is standing erect, the lumbar portion of the spine (the small of the back, approximately at the belt level) curves naturally inward, which is termed *lordosis*. However, as a person sits down, the pelvis rotates backward, flattening the lordotic curve and increasing the pressure on the disks in the vertebral column (see Figure 5.8). Therefore, it is very important to provide *lumbar support* in the form of an outward bulge in the seat back, or even a simple lumbar pad placed at the belt level.

Another approach to preventing flattening of the lordotic curve is to reduce the pelvic rotation by maintaining a large angle between the torso and thighs, via a forward-tilting seat (kneeling posture in Figure 5.7). The theory is that this is a shape maintained by astronauts in the weightless environment of space (see Figure 4.4). The disadvantage of this type of seat is that it may put additional

Designing Seating in a Large Training Room	**EXAMPLE 5.2**

This example will show the step-by-step procedures utilized in a typical design problem—arranging seating in an industrial training room such that most individuals will have an unobstructed view of the speaker and screen (see Figure 5.3).

1. Determine the body dimensions critical to the design—sitting height, erect; and eye height, sitting.

2. Define the population being served—U.S. adult males and females.

3. Select a design principle and the percentage of the population to be accommodated—Designing for extremes and accommodating 95 percent of the population. The key principle is to allow a 5th percentile female sitting behind a 95th percentile male to have an unimpeded line of sight.

4. Find appropriate anthropometric values from Table 5.1. The 5th percentile female seated eye height is 26.6 in (67.5 cm), while the 95th percentile male erect sitting height is 38.1 in (96.7 cm). Thus, for the small female to see over the large male, a rise height of 11.5 in (29.2 cm) is necessary between the two rows. This would be a very large rise height, which would create a very steep slope. Typically, therefore, the seats are staggered, so that the individual in the back is looking over the head of an individual two rows in front, decreasing the rise height by one-half.

5. Add allowances and test. Many anthropometric measurements have been made on nude human bodies. Therefore, allowances for heavy clothing, hats, or shoes may be necessary. For example, if all the trainees will be wearing hard hats, an additional 2 to 3 in might be needed for the rise height. It would be much more practical to remove the hard hats in the training room.

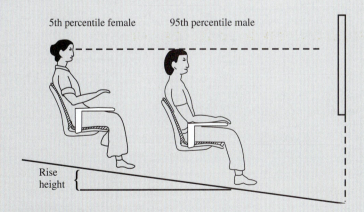

Figure 5.3 Seating design in a large training room.

stress on the knees. The addition of a pommel to the forward-sloping seat, forming a saddlelike seat, may be a better overall approach, as it eliminates the need for knee supports and still allows for back support (see Figure 5.9).

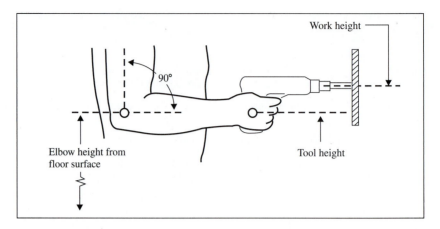

Figure 5.4   Graphic aid for determining correct work surface height.
(*From:* Putz-Anderson, 1988.)

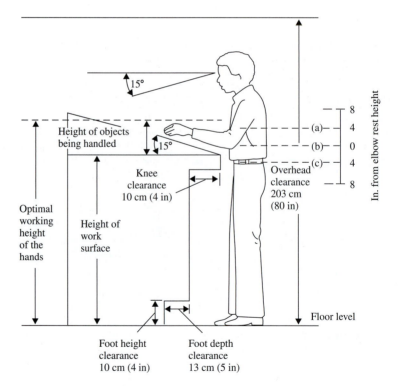

Figure 5.5   Recommended standing workplace dimensions.
(a) For precision work with armrest, (b) for light assembly, (c) for heavy work.

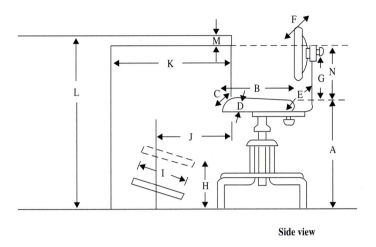

Side view

Figure 5.6 Adjustable chair (specific seat parameter values found in Table 5.2).

Table 5.2 Recommended Seat Adjustment Ranges

Seat parameter	Design Value [in (cm) unless specified]	Comments
A–Seat height	16–20.5 (40–52)	Too high—compresses thighs; too low—disk pressure increases
B–Seat depth	15–17 (38–43)	Too long—cuts popliteal region, use waterfall contour
C–Seat width	≥18.2 (≥46.2)	Wider seats recommended for heavy individuals
D–Seat pan angle	−10°– +10°	Downward tilting requires greater friction in the fabric
E–Seat back to pan angle	>90°	>105°preferred, but requires workstation modifications
F–Seat back width	>12 (>30.5)	Measured in the lumbar region
G–Lumbar support	6–9 (15–23)	Vertical height from seat pan to center of lumbar support
H–Footrest height	1–9 (2.5–23)	
I–Footrest depth	12 (30.5)	
J–Footrest distance	16.5 (42)	
K–Leg clearance	26 (66)	
L–Work surface height	~32 (~81)	Determined by elbow rest height
M–Work surface thickness	<2 (<5)	Maximum value
N–Thigh clearance	>8 (>20)	Minimum value

Note: A–G from ANSI (1988); H–M from Eastman Kodak (1983).

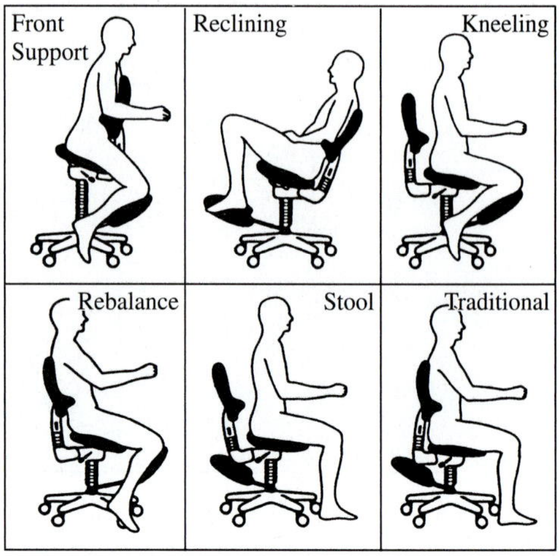

Figure 5.7 Six basic seating postures.
(*From:* Serber, 1990. Reprinted with permission of the Human Factors and Ergonomics Society. All rights reserved.)

Figure 5.8 Posture of the spine when standing and sitting.
Lumbar portion of spine is lordotic when standing (a) and kyphotic when sitting (b). The shaded vertebrae are the lumbar portion of the spine.
(*Source:* Grandjean 1988, Fig.47.)

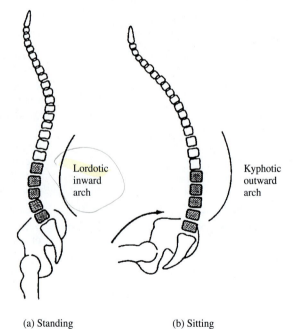

Figure 5.9 Saddle chair.
(Courtesy: This version of the Nottingham Chair, called the checkmate, is made by the
Osmond Group: Photograph by Nigel Corlett. For further details of the Nottingham
chair see Nottinghamchair.com; for the Checkmate see ergonomics.co.uk and search for
the Checkmate chair.)

PROVIDE ADJUSTABILITY IN THE SEAT

A second consideration is the reduction of disk pressure, which can increase considerably with a forward tilt of the trunk. Reclining the backrest from the vertical also has a dramatic effect in reducing disk pressures (Andersson et al., 1974). Unfortunately, there is a trade-off. With increasing angles, it becomes more difficult to look down and perform productive work.

Another factor is the need to provide easy adjustability for specific seat parameters. Seat height is most critical, with ideal height being determined by the person's popliteal height, which is defined in the figure accompanying Table 5.1. A seat that is too high will uncomfortably compress the underside of the thighs. A seat that is too low will raise the knees uncomfortably high and decrease trunk angle, again increasing disk pressure. Specific recommendations for seat height and other seat parameters (shown in Figure 5.6) are given in Table 5.2.

In addition, armrests for shoulder and arm support and footrests for shorter individuals are recommended. Casters assist in movement and ingress/egress from workstations. However, there may be situations where a stationary chair is desired. In general, the chair should be slightly contoured, slightly cushioned, and covered in a breathable fabric to prevent moisture buildup. Overly soft cushioning restricts posture and may restrict circulation in the legs. An overall optimum working posture and workstation is shown in Figure 5.10.

ENCOURAGE POSTURAL FLEXIBILITY

The workstation height should be adjustable so that the work can be performed efficiently either standing or sitting. The human body is not designed for long periods of sitting. The disks between the vertebrae do not have a separate blood supply, and they rely on pressure changes resulting from movement to receive nutrients and remove wastes. Postural rigidity also reduces blood flow to the muscles and induces muscle fatigue and cramping. An alternate compromise is to provide a sit/stand stool so that the operator can change postures easily. Two key features for a sit/stand stool are height adjustability and a large base of support so that the stool does not tip, preferably long enough that the feet can rest on and counterbalance it (see Figure 5.11).

PROVIDE ANTIFATIGUE MATS FOR A STANDING OPERATOR

Standing for extended periods on a cement floor is fatiguing. The operators should be provided with resilient antifatigue mats. The mats allow small muscle contractions in the legs, forcing the blood to move and keeping it from tending to pool in the lower extremities.

LOCATE ALL TOOLS AND MATERIALS WITHIN THE NORMAL WORKING AREA

In every motion, a distance is involved. The greater the distance, the larger the muscular effort, control, and time. It is therefore important to minimize distances.

Arms: When operator's hands are on keyboard, upper arm and forearm should form right angle; hands should be lined up with forearm; if hands are angled up from the wrist, try lowering or downward tilting the keyboard; optional arm rests should be adjustable.

Telephone: Cradling telephone receiver between head and shoulder can cause muscle strain; headset allows head, neck to remain straight while keeping hands free.

Document holder: Same height and distance from user as the screen, so eyes can remain focused as they look from one to

Screen: Positioned so that midscreen is 15⁰ down from eye level.

Backrest: Adjustable for occasional variations; shape should match contour of lower back, providing even pressure and support.

Keyboard: Positioned to allow hands, forearms to remain straight, level.

Posture: Sit all the way back into chair for proper back support; back, neck should be comfortably erect; knees should be slightly lower than hips; do not cross legs or shift weight one side; give joints, muscles a chance to relax; periodically, get up and walk around.

Seat: Adjustable height, angle; firm cushion; "waterfall" front helps circulation to legs.

Feet: Entire sole should rest comfortably on floor or foot rest.

Desk: Thin work surface to allow leg room and posture adjustments; adjustable surface height preferable; table should be large enough for books, files, telephone while permitting different positions of screen, keyboard, mouse pad.

Avoiding eye strain:
1. Get glasses that improve focus on screen; measure distance before visiting eye doctor.
2. Try to position screen or lamps so that lighting is indirect; do not have light shining directly at screen or into eyes.

3. Use a glare-reducing screen.
4. Periodically rest eyes by looking into the distance.

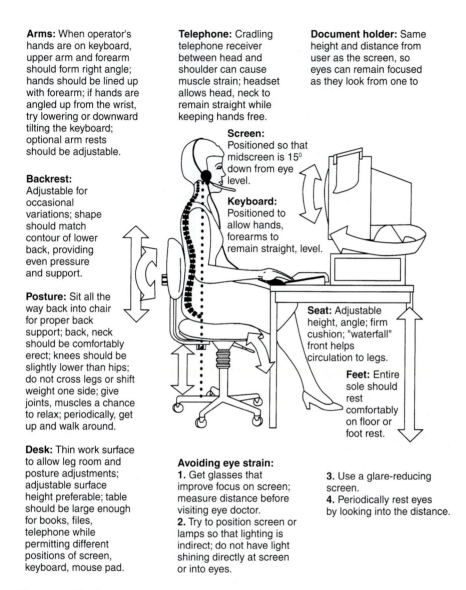

Figure 5.10 Properly adjusted workstation.

The normal working area in the horizontal plane of the right hand includes the area circumscribed by the arm below the elbow when it is moved in an arc pivoted at the elbow (see Figure 5.12). This area represents the most convenient zone within which motions may be made by that hand with a normal expenditure of energy. The normal area of the left hand may be similarly established. Since movements are made in the third dimension, as well as in the horizontal plane, the normal working area also applies to the vertical plane. The normal area relative to

Figure 5.11 Industrial sit/stand stools.
(*Courtesy:* BioFit, Waterville, OH)

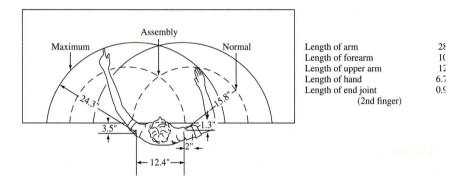

Figure 5.12 Normal and maximum working areas in the horizontal plane for women
(for men, multiply by 1.09).

height for the right hand includes the area circumscribed by the lower arm in an upright position hinged at the elbow moving in an arc. There is a similar normal area in the vertical plane (see Figure 5.13).

FIX LOCATIONS FOR ALL TOOLS AND MATERIALS TO PERMIT THE BEST SEQUENCE

In driving an automobile, we are all familiar with the short time required to apply the foot brake. The reason is obvious: since the brake pedal is in a fixed cation, no time is required to decide where the brake is located. The body responds instinctively and applies pressure to the area where the driver knows the foot pedal is located. If the location of the brake foot pedal varied, the driver would need considerably more time to brake the car. Similarly, providing fixed locations for all tools and materials at the workstation eliminates, or at least

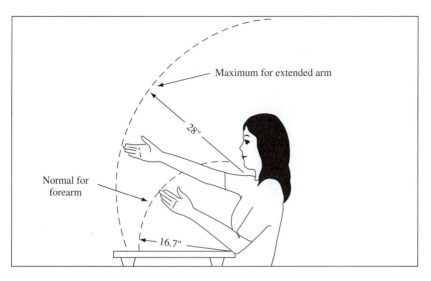

Figure 5.13 Normal and maximum working areas in the vertical plane for women (for men, multiply by 1.09).

minimizes, the short hesitations required to search for and select the objects needed to do the work. These are the ineffective Search and Select therbligs discussed in Chapter 4 (see Figure 5.14).

USE GRAVITY BINS AND DROP DELIVERY TO REDUCE REACH AND MOVE TIMES

The time required to perform both of the transport therbligs Reach and Move is directly proportional to the distance that the hands must move in performing these therbligs. Utilizing gravity bins, components can be continuously brought to the normal work area, thus eliminating long reaches to get these supplies (see Figure 5.15). Likewise, gravity chutes allow the disposal of completed parts within the normal area, eliminating the necessity for long moves to do so. Sometimes, ejectors can remove finished products automatically. Gravity chutes make a clean work area possible, as finished material is carried away from the work area, rather than stacked up all around it. A bin raised off the work surface (so that the hand can partially slide underneath) will also decrease the time required to perform this task by approximately 10 to 15 percent.

ARRANGE TOOLS, CONTROLS, AND OTHER COMPONENTS OPTIMALLY TO MINIMIZE MOTIONS

The optimum arrangement depends on many characteristics, both human (strength, reach, sensory) and task (loads, repetition, orientation). Obviously, not all factors can be optimized. The designer must set priorities and make trade-offs in the layout

Figure 5.14 Tool balancers provide fixed locations for tools. (Courtesy of Packers Kromer.)

of the workplace. However, certain basic principles should be followed. First, the designer must consider the general location of components relative to other components, using the *importance* and *frequency-of-use* principles. The most important, as determined by overall goals or objectives, or most frequently used components, should be placed in the most convenient locations. For example, an emergency stop button should be placed in a readily visible, reachable, or convenient position. Similarly, a regularly used activation button, or the most often used fasteners, should be within easy reach of the operator.

Once the general location has been determined for a group of components, that is, the most frequently used parts for assembly, the principles of *functionality* and *sequence of use* must be considered. Functionality refers to the grouping of components by similar function, for example, all fasteners in one area, all

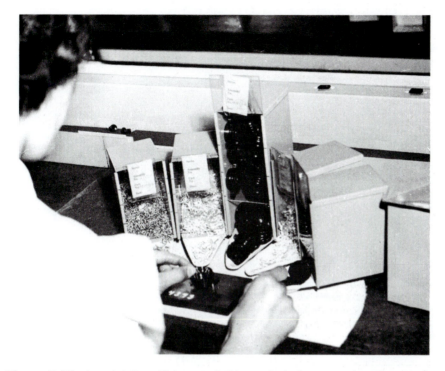

Figure 5.15 A workstation utilizing gravity bins and a belt conveyor to reduce reach and move times.
The conveyor in the background carries other parts past this particular workstation.
The operator is feeding the conveyor from under the platform by merely dropping assembled parts onto the feeder belt. (Source: Alden Systems Co.)

gaskets and rubber components in another area. Since many products are assembled in a strict sequence, cycle after cycle, it is very important to place the components or subassemblies in the order that they are assembled, since this will have a very large effect on reducing wasteful motions. The designer should also consider using Muther's Systematic Layout Planning (see Chapter 3) or other types of adjacency layout diagramming techniques, to develop a quantitative or relative comparison of the various layouts of components on a work surface. The relationships between components can be modified from traditional data on the flow from one area to another, and should include visual links (eye movements), auditory links (voice communications or signals), and tactile and control motions.

These principles of work design for workstations are summarized in the Workstation Evaluation Checklist (see Figure 5.16). The analyst may find this useful in evaluating existing workstations or implementing new workstations.

Sitting Workstation	Yes	No
1. Is the chair easily adjustable according to the following features:	☐	☐
a. Is the seat height adjustable from 15 to 22 inches?	☐	☐
b. Is the seat width a minimum of 18 inches?	☐	☐
c. Is the seat depth 15 to 16 inches?	☐	☐
d. Can the seat be sloped $\pm$ 10° from horizontal?	☐	☐
e. Is a back rest with lumbar support provided?	☐	☐
f. Is the back rest a minimum of 8 x 12 inches in size?	☐	☐
g. Can the back rest be moved 7 to 10 inches above the seat?	☐	☐
h. Can the back rest be moved 12 to 17 inches from the front of the seat?	☐	☐
i. Does the chair have five legs for support?	☐	☐
j. Are casters and swivel capability provided for mobile tasks?	☐	☐
k. Is the chair covering breathable?	☐	☐
l. Is a footrest (large, stable, and adjustable in height and slope) provided?	☐	☐
2. Has the chair been adjusted properly?	☐	☐
a. Is the seat height adjusted to the popliteal height with the feet flat on the floor?	☐	☐
b. Is there approximately a 90° angle between the trunk and thigh?	☐	☐
c. Is the lumbar area of the back support in the small of the back (~ belt line)?	☐	☐
d. Is there sufficient legroom (i.e., to the back of the workstation)?	☐	☐
3. Is the workstation surface adjustable?	☐	☐
a. Is the workstation surface roughly at elbow rest height?	☐	☐
b. Is the surface lowered 2 to 4 inches for heavy assembly?	☐	☐
c. Is the surface raised 2 to 4 inches (or tilted) for detailed assembly or visually intensive tasks?	☐	☐
d. Is there sufficient thigh room (i.e., from the bottom of the worksurface)?	☐	☐
4. Is sitting alternated with standing or walking?	☐	☐

Computer Workstation	Yes	No
1. Has the chair been adjusted first, then keyboard and mouse, finally the monitor?	☐	☐
2. Is the keyboard as low as possible (without hitting the legs)?	☐	☐
a. Are the shoulders relaxed, upper arms hanging down comfortably, and forearms below horizontal (i.e., elbow angle >90°)?	☐	☐
b. Is a keyboard shelf utilized (i.e., lower than a normal 28-inch writing surface)?	☐	☐
c. Is the keyboard sloped downward so as to maintain a neutral wrist position?	☐	☐
d. Is the mouse positioned next to the keyboard at the same height?	☐	☐
e. Are armrests (adjustable in height at least 5 inches) provided?	☐	☐
f. If no armrest, are wrist rests provided?	☐	☐
3. Is the monitor positioned 16 to 30 inches (roughly arm's length) from the eyes?	☐	☐
a. Is the top of the screen slightly below eye level?	☐	☐
b. Is the bottom of the screen roughly 30° down from horizontal eye level?	☐	☐
c. Is the monitor positioned at a 90° angle to windows to minimize glare?	☐	☐
d. Can the windows be covered with curtains or blinds to reduce bright light?	☐	☐
e. Is the monitor tilted to minimize ceiling light reflections?	☐	☐
f. If glare still exists, is an antiglare filter utilized?	☐	☐
g. Is a document holder utilized for data transfer from papers?	☐	☐
h. Is the main visual task (monitor or documents) placed directly in front?	☐	☐

Standing Workstation	Yes	No
1. Is the workstation surface adjustable?	☐	☐
a. Is the workstation surface roughly at elbow rest height?	☐	☐
b. Is the surface lowered 4 to 8 inches for heavy assembly?	☐	☐
c. Is the surface raised 4 to 8 inches (or tilted) for detailed assembly or visually intensive tasks?	☐	☐
2. Is there sufficient legroom?	☐	☐
3. Is a sit/stand stool (adjustable in height) provided?	☐	☐
4. Is standing alternated with sitting?	☐	☐

Figure 5.16 Workstation Evaluation Checklist.

5.3 PRINCIPLES OF WORK DESIGN: MACHINES AND EQUIPMENT

TAKE MULTIPLE CUTS WHENEVER POSSIBLE BY COMBINING TWO OR MORE TOOLS IN ONE, OR BY ARRANGING SIMULTANEOUS CUTS FROM BOTH FEEDING DEVICES

Advanced production planning for the most efficient manufacture includes taking multiple cuts with combination tools and simultaneous cuts with different tools. Of course, the type of work to be processed and the number of parts to be produced determine the desirability of combining cuts, such as cuts from both the square turret and the hexagonal turret.

USE A FIXTURE INSTEAD OF THE HAND AS A HOLDING DEVICE

If either hand is used as a holding device during the processing of a part, then the hand is not performing useful work. Invariably, a fixture can be designed to hold the work satisfactorily, thus allowing both hands to do useful work. Fixtures not only save time in processing parts, but also permit better quality in that the work can be held more accurately and firmly. Many times, foot-operated mechanisms allow both hands to perform productive work.

An example will help clarify the principle of using a fixture, as opposed to the hands, for holding work. A company that produced specialty windows needed to remove a 0.75-in-wide strip of protective paper from around all four edges of both sides of Lexan panels. An operator would pick up a single sheet of Lexan and bring it to the work area. The operator would then pick up a pencil and square and mark the four corners of the Lexan panel. The pencil and square would be laid aside and a template would be picked up and located on the pencil marks. The operator would then strip the protective paper from around the periphery of the Lexan panels. The standard time developed by MTM-1 was 1.063 min per piece.

A simple wood fixture was developed to hold three Lexan panels while each was stripped of the 0.75-in-wide protection paper around the periphery. With the fixture method, the worker picked up three Lexan sheets and located them in the fixture (see Figure 5.17). The protective paper was stripped, the sheets were turned 180°, and the protective paper was removed from the remaining two sides. The improved method resulted in a standard of 0.46 min per panel, or a savings of 0.603 min of direct labor per panel.

LOCATE ALL CONTROL DEVICES FOR BEST OPERATOR ACCESSIBILITY AND STRENGTH CAPABILITY

Many of our machine tools and other devices are mechanically perfect, yet incapable of effective operation, because the facility designer overlooked various

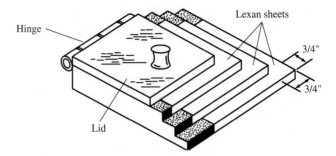

Figure 5.17   Fixture for stripping ¾ in-wide protection paper around periphery of Lexan sheets.

human factors. Handwheels, cranks, and levers should be of such a size and placed in such positions that operators can manipulate them with maximum proficiency and minimum fatigue. Frequently used controls should be positioned between elbow and shoulder height. Seated operators can apply maximum force to levers located at elbow level; standing operators, to levers located at shoulder height. Handwheel and crank diameters depend on the torque to be expended and the mounting position. The maximum diameters of handgrips depend on the forces to be exerted. For example, for a 10- to 15-lb (4.5- to 6-kg) force, the diameter should be no less than 0.25 in (0.6 cm) and preferably larger; for 15 to 25 lb (6.8 to 11.4 kg) , a minimum of 0.5 in (1.3 cm) should be used; and for 25 lb or more (11.4 kg), a minimum of 0.75 in (1.9 cm). However, diameters should not exceed 1.5 in (3.8 cm), and the grip length should be at least 4 in (10 cm), to accommodate the breadth of the hand.

Guidelines for crank and handwheel radii are as follows: light loads, radii of 3 to 5 in (7.6 to 12.7 cm); medium to heavy loads, radii of 4 to 7 in (10.2 to 17.8 cm); very heavy loads, radii of more than 8 in (20 cm) but not in excess of 20 in (51 cm). Knob diameters of 0.5 to 2 in (1.3 to 5.1 cm) are usually satisfactory. The diameters of knobs should be increased as greater torques are needed.

USE SHAPE, TEXTURE, AND SIZE CODING FOR CONTROLS

Shape coding, using two- or three-dimensional geometric configurations, permits both tactual and visual identification. It is especially useful under low-light conditions or in situations where redundant or double-quality identification is desirable, thus helping to minimize errors. Shape coding permits a relatively large number of discriminable shapes. An especially useful set of known shapes that are seldom confused is shown in Figure 5.18. Multiple rotation knobs are used for continuous controls in which the adjustment range is more than one full turn. Fractional rotation knobs are used for continuous controls with a range less than a full turn, while detent positioning knobs are for discrete settings. In addition to shape, the surface texture can provide cues for discrimination by touch. Typically,

TURN MORE THAN ONE O

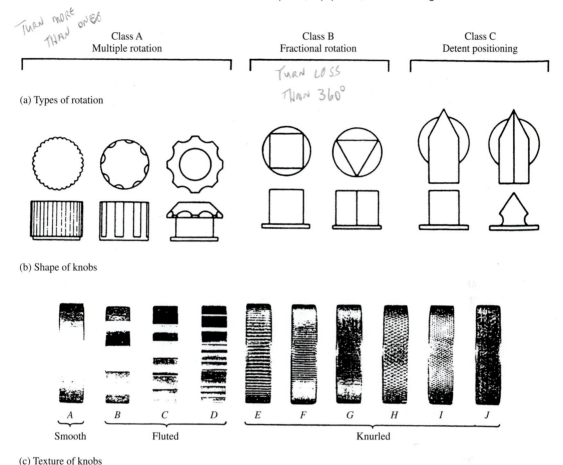

Class A
Multiple rotation

Class B
Fractional rotation
TURN LESS THAN 360°

Class C
Detent positioning

(a) Types of rotation

(b) Shape of knobs

| A | B | C | D | E | F | G | H | I | J |

Smooth | Fluted | Knurled

(c) Texture of knobs

Figure 5.18 Examples of knob designs for three classes of use that are seldom confused by touch. The diameter or length of these controls should be between 0.5 and 4.0 in (1.3 and 10 cm), except for class C, where 0.75 in (1.9 cm) is the minimum suggested. The height should be between 0.5 and 1 in (1.3 and 2.5 cm). (a&b) Adapted from Hunt, 1953 (c) *Source*: Bradley, 1967.

three textures are rarely confused: smooth, fluted, and knurled. However, as the number of shapes and textures increases, discrimination can be difficult and slow if the operator must identify controls without vision. If the operator is obliged to wear gloves, then shape coding is only desirable for visual discrimination, or for the tactual discrimination of only two to four shapes.

Size coding, analogous to shape coding, permits both tactual and visual identification of controls. Size coding is used principally where the controls cannot be seen by the operators. Of course, as is the case with shape coding, size coding permits redundant coding, since controls can be discriminated both tactually and visually. In general, try to limit the size categories to three or four, with at least a 0.5-in size difference between controls. Operational coding requiring a unique movement (e.g., putting the gearshift into reverse) is especially useful for critical controls that shouldn't be activated inadvertently.

USE PROPER CONTROL SIZE, DISPLACEMENT, AND RESISTANCE

In their work assignments, workers continually use various types and designs of controls. The three parameters that have a major impact on performance are control size, control-response ratio, and control resistance when engaged. A control that is either too small or too large cannot be activated efficiently. Tables 5.3, 5.4, and 5.5 provide helpful design information about minimum and maximum dimensions for various control mechanisms.

The *control-response (C/R) ratio* is defined as the amount of movement in a control divided by the amount of movement in the response (see Figure 5.19).

Table 5.3 Control Size Criteria

Control	Dimension		Control size	
			Minimum (mm)	Maximum (mm)
Pushbutton	Fingertip	Diameter	13	*
	Thumb/palm	Diameter	19	*
	Foot	Diameter	8	*
Toggle switch		Tip diameter	3	25
		Lever arm length	13	50
Rotary selector		Length	25	*
		Width	*	25
		Depth	16	*
Continuous adjustment knob	Finger/thumb	Depth	13	25
		Diameter	10	100
	Hand/palm	Depth	19	*
		Diameter	38	75
Cranks	For rate	Radius	13	113
	For force	Radius	13	500
Handwheel		Diameter	175	525
		Rim thickness	19	50
Thumbwheel		Diameter	38	*
		Width	*	*
		Protrusion from surface	3	*
Lever handle	Finger	Diameter	13	75
	Hand	Diameter	38	75
Crank handle		Grasp area	75	*
Pedal		Length	88	†
		Width	25	†
Valve handle		Diameter 75 in per inch of valve size		

* No limit set by operator performance.
† Dependent on space available.

Table 5.4 Control Displacement Criteria

		Displacement	
Control	**Condition**	**Minimum**	**Maximum**
Pushbutton	Thumb/fingertip operation	3 mm	25 mm
	Foot normal	13 mm	—
	Heavy boot	25 mm	—
	Ankle flexion only	—	63 mm
	Leg movement	—	100 mm
Toggle switch	Between adjacent positions	30°	—
	Total	—	120°
Rotary selector	Between adjacent detents: Visual	15°	—
	Nonvisual	30°	—
	For facilitating performance	—	40°
	When special engineering is required	—	90°
Continuous adjustment knob	Determined by desired control/display ratio (mm of control movement for each mm of display movement)		
Crank	Determined by desired control/display ratio		
Handwheel	Determined by desired control/display ratio 90°–120°†		
Thumbwheel	Determined by number of positions		
Lever handle	Fore-aft movement	*	350 mm
	Lateral movement	*	950 mm
Pedals	Normal	13 mm	—
	Heavy boot	25 mm	—
	Ankle flexion (raising)	—	63 mm
	Leg movement	—	175 mm

*None established.
†Provided optimum control/display ratio is not hindered.

A low C/R ratio indicates high sensitivity, such as in the coarse adjustment of a micrometer. A high C/R ratio means low sensitivity, such as the fine adjustment on a micrometer. Overall control movement depends on the combination of the primary travel time to reach the approximate target setting and the secondary adjust time to reach the exact target setting accurately. The optimum C/R ratio that minimizes this total movement time depends on the type of control and task conditions (see Figure 5.20). Note that there is also a *range effect*—the tendency to overshoot short distances and undershoot long distances.

Control resistance is important in terms of providing feedback to the operator. Ideally, it can be of two types: pure displacement with no resistance or pure force with no displacement. The first has the advantage of being less fatiguing, while the second is a *deadman's control,* that is, the control returns to zero upon release. Real-life controls are typically spring-loaded, incorporating the features of both. Faulty

Table 5.5   Control Resistance Criteria

Control	Condition	Resistance Minimum (kg)	Resistance Maximum (kg)
Pushbutton	Fingertip	0.17	1.14
	Foot: Normally off control	1.82	9.10
	Rested on control	4.55	9.10
Toggle switch	Finger operation	0.17	1.14
Rotary selector	Torque	1 cm·kg	7 cm·kg
Continuous	Torque: Fingertip <1-in dia	*	0.3 cm·kg
adjustment knob	Fingertip > 1-in dia	*	0.4 cm·kg
Crank	Rapid, steady turning: <3-in radius	0.91	2.28
	5–8 in radius	2.28	4.55
	Precise settings	1.14	3.64
Handwheel†	Precision operation: <3-in radius	*	*
	5–8 in radius	1.14	3.64
	Resistance at rim: One-hand	2.28	13.64
	Two-hand	2.28	22.73
Thumbwheel	Torque	1 cm·kg	3 cm·kg
	Finger grasp	0.34	1.14
	Hand grasp: One-hand	0.91	—
	Two-hand	1.82	—
	Fore-aft: Along median plane:		
Lever handle	One-hand—10 in forward SRP§	—	13.64
	—16–24 in forward SRP	—	22.73
	Two-hand—10–19 in forward SRP	—	45
	Lateral:		
	One-hand—10–19 in forward SRP	—	9.09
	Two-hand—10–19 in forward SRP	—	22.73
Pedal	Foot: Normally off control	1.82	—
	Rested on control	4.55	—
	Ankle flexion only	—	4.55
	Leg movement	—	80

* Not established.
† For valve handles/wheels: 25 ± cm·kg of torque/cm of valve size (8 cm·kg of torque/cm of handle diameter).
§ SRP = Seat reference point.

control aspects include high initial static friction, excessive viscous damping, and *deadspace,* that is, control movement with no response. All three impair tracking and use performance. However, the first two are sometimes incorporated purposely to prevent inadvertent activation of the control (Sanders and McCormick, 1993).

ENSURE PROPER COMPATIBILITY BETWEEN CONTROLS AND DISPLAYS

Compatibility is defined as the relationship between controls and displays that is consistent with human expectations. Basic principles include *affordance,* the

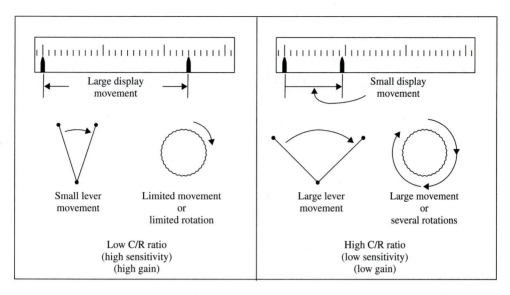

Figure 5.19 Generalized illustrations of low and high control–response ratios (C/R ratios) for lever and rotary controls. The C/R ratio is a function of the linkage between the control and the display.

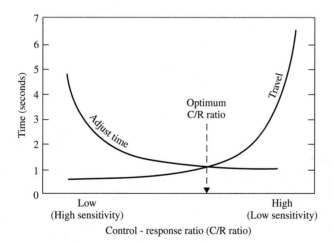

Figure 5.20 Relationship between C/R ratio and movement time (travel time and adjust time). The specific C/R ratios are not meaningful out of their original context, so are omitted here. These data, however, very typically depict the nature of the relationships, especially for knob controls.
(*Source:* Jenkins and Connor, 1949.)

perceived property results in the desired action; *mapping,* the clear relationship between controls and responses; and *feedback,* so that the operator knows that the function has been accomplished. For example, good affordance is a door with a handle that pulls open or a door with a plate that pushes open. Spatial mapping is provided on well-designed stoves. Movement compatibility is provided by direct drive action, scale readings that increase from left to right, and clockwise movements that increase settings. For circular displays, the best compatibility is accomplished with a fixed scale and moving pointer display (see Section 7.4). For vertical or horizontal displays, *Warrick's principle,* which says that points closest on the display and control move in the same direction, provides the best compatibility (see Figure 5.19). For controls and displays in different planes, a clockwise movement for increases and the right-hand screw rule (the display advances in the direction of motion of a right-handed screw or control) are most compatible. For stick controls of a direct drive, the best approach is up results in up movement (Sanders and McCormick, 1993).

The principles of work design for machines and equipment are summarized in the Machine Evaluation Checklist (Figure 5.21). The analyst may find this useful in evaluating and designing machines or other equipment.

5.4 CUMULATIVE TRAUMA DISORDERS

The cost of work-related musculoskeletal disorders such as cumulative trauma disorders (CTDs) in U.S. industry, although not all due to improper work design, is quite high. Data from the National Safety Council (2003) suggest that 15 to 20 percent of workers in key industries (meatpacking, poultry processing, auto assembly, and garment manufacturing) are at potential risk for CTD, and 61 percent of all occupational illnesses are associated with repetitive motions. The worst industry is manufacturing, while the worst occupational title is butchering, with 222 CTD claims per 100,000 workers. With such high rates, and with average medical costs of $30,000 per case, NIOSH and OSHA have focused on the reduction of incidence rates for work-related musculoskeletal disorders as one of their main objectives.

Cumulative trauma disorders (sometimes called *repetitive motion injuries,* or *work-related musculoskeletal disorders*) are injuries to the musculoskeletal system that develop gradually as a result of repeated microtrauma due to poor design and the excessive use of hand tools and other equipment. Because of the slow onset and relatively mild nature of the trauma, the condition is often ignored until the symptoms become chronic and more severe injury occurs. These problems are a collection of a variety of problems, including repetitive motion disorders, carpal tunnel syndrome, tendinitis, ganglionitis, tenosynovitis, and bursitis, with these terms sometimes being used interchangeably.

Four major work-related factors seem to lead to the development of CTD: (1) excessive force, (2) awkward or extreme joint motions, (3) high repetition, and (4) duration of work. The most common symptoms associated with CTD include: pain, joint movement restriction, and soft tissue swelling. In the early stages, there may be few visible signs; however, if the nerves are affected, sensory

Machine Efficiency and Safety	**Yes**	**No**
1. Are multiple or simultaneous cuts possible?	❏	❏
2. Are handles, wheels, and levers readily accessible?	❏	❏
3. Are handles, wheels, and levers designed for best mechanical advantage?	❏	❏
a. Are knobs at least 0.5–2 inches in diameter, with larger sizes for greater torque?	❏	❏
b. Are cranks and handwheels a minimum of 3–5 inches in diameter for low loads?	❏	❏
c. Are cranks and handwheels more than 8 inches in diameter for heavy loads?	❏	❏
4. Are fixtures used to avoid holding with the hand?	❏	❏
5. Are guards or interlocks used to prevent unintended entry?	❏	❏

Design of General Controls	**Yes**	**No**
1. Are different colors used for different controls?	❏	❏
2. Are controls clearly labeled?	❏	❏
3. Are shape and texture coding used for tactual identification?	❏	❏
a. Are no more than seven unique codes being utilized?	❏	❏
4. Is size coding used for tactual identification?	❏	❏
a. Are no more than three unique codes being utilized?	❏	❏
b. Are size differences greater than 0.5 inch?	❏	❏

Design of Emergency Controls	**Yes**	**No**
1. Are power-on controls designed to prevent accidental activation?	❏	❏
2. Do activation controls require a unique or dual-action motion?	❏	❏
3. Are power-on buttons recessed?	❏	❏
4. Are activation controls colored green?	❏	❏
5. Are deadman controls utilized for continually activated controls?	❏	❏
6. Are emergency controls designed for quick activation?	❏	❏
7. Are stop buttons protruding?	❏	❏
8. Are emergency controls large and easy to activate?	❏	❏
9. Are emergency controls easily reachable?	❏	❏
10. Are emergency controls visible and colored red?	❏	❏
11. Are emergency controls placed away from other normally used controls?	❏	❏

Control Placement	**Yes**	**No**
1. Are primary controls placed in front of the operator at elbow height?	❏	❏
a. Are frequency-of-use and importance principles used to identify primary controls?	❏	❏
2. Are secondary controls placed next to primary controls, but still within reach?	❏	❏
3. Is twisting avoided in reaching for controls?	❏	❏
4. Are controls located in the proper sequence of operation?	❏	❏
5. Are mutually related controls grouped together?	❏	❏
6. Are hand-operated controls separated by at least 2 inches?	❏	❏
7. Are three or less foot pedals utilized?	❏	❏
8. Are foot pedals located at floor level to avoid raising the leg?	❏	❏
9. Is a sit/stand stool provided for extended foot pedal operation?	❏	❏

Display Design	**Yes**	**No**
1. Are displays located on the visual cone of sight (horizontal to 30° down)?	❏	❏
2. Are indicator lights used to attract the operator's attention?	❏	❏
3. Are acoustic signals used for critical warnings?	❏	❏
4. Are movable pointers used to indicate trends?	❏	❏
5. Are counters provided for accurate readings?	❏	❏
6. Are displays grouped so as to accentuate an abnormal display?	❏	❏
7. Are mutually related displays grouped together?	❏	❏

Figure 5.21 Machine Evaluation Checklist   *(continued)*

Control-Display Compatibility

	Yes	No
1. Is affordance (perceived property results in desired action) used?	❏	❏
2. Is feedback utilized to indicate completion of action?	❏	❏
3. Does the control and display have a direct-drive relationship?	❏	❏
4. Does the display reading increase from left to right?	❏	❏
5. Do clockwise motions increase settings?	❏	❏
6. Do clockwise motions close valves?	❏	❏
7. For stick controls, does upward or backward motion produce upward motion?	❏	❏
8. For controls out of plane, does the right-hand rule apply?	❏	❏

Label Design

	Yes	No
1. Is clear and concise wording used?	❏	❏
2. Do the letters subtend at least 12 arcminutes of visual angle?	❏	❏
3. Are dark letters used on a white background?	❏	❏
4. Are uppercase letters used for only a few words?	❏	❏
5. Are symbols (preferably simple) used only if clearly understood?	❏	❏

Figure 5.21   Machine Evaluation Checklist

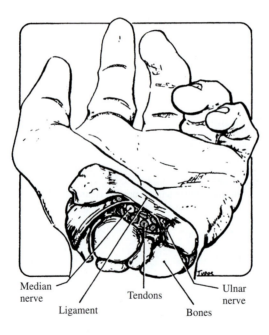

Figure 5.22   A pictorial view of the carpal tunnel (*From:* Putz-Anderson, 1988.)

Median nerve   Tendons   Ulnar nerve

Ligament   Bones

responses and motor control may be impaired. If left untreated, CTD can result in permanent disability.

The human hand is a complex structure of bones, arteries, nerves, ligaments, and tendons. The fingers are controlled by the extensor carpi and flexor carpi muscles in the forearm. The muscles are connected to the fingers by tendons, which pass through a channel in the wrist, formed by the bones of the back of the hand on one side and the transverse carpal ligament on the other. Through this channel, called the carpal tunnel, also pass various arteries and nerves (see Figure 5.22). The bones of the wrist connect to two long bones in the forearm, the ulna and the radius. The radius connects to the thumb side of the wrist, and the ulna connects to the little-finger side of the wrist. The orientation of the wrist joint allows movement in two planes, at 90°to each other (see Figure 5.23). The first permits *flexion* and *extension.* The second movement plane permits *ulnar* and *radial deviation.* Also, rotation of the forearm can result in *pronation* with the palm down or *supination* with the palm up.

Tenosynovitis, one of the more common CTDs, is the inflammation of the tendon sheaths due to overuse or unaccustomed use of improperly designed

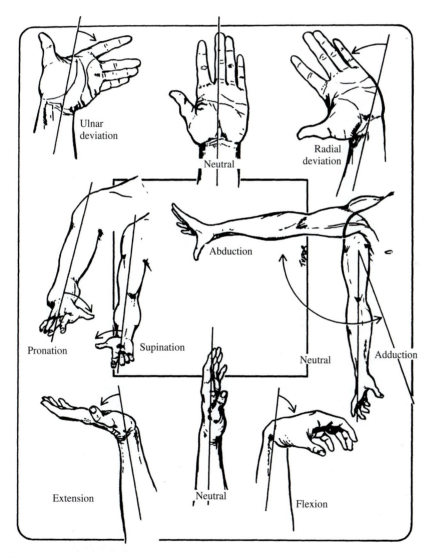

Figure 5.23 Positions of the hand and arm.
(*From:* Putz-Anderson, 1988.)

tools. If the inflammation spreads to the tendons, it becomes *tendinitis.* It is often experienced by trainees exposed to large ulnar deviations, coupled with supination of the wrist. Repetitive motions and impact shocks may further aggravate the condition. *Carpal tunnel syndrome* is a disorder of the hand caused by injury of the median nerve inside the wrist. Repetitive flexion and extension of the wrist under stress may cause inflammation of the tendon sheaths. The sheaths, sensing increased friction, secrete more fluid to lubricate the sheaths and facilitate tendon movement. The resulting buildup of fluid in the carpal tunnel increases pressure, which in turn compresses the median nerve. Symptoms include impaired or lost nervous function in the first $3^1/_2$ fingers, manifesting as numbness, tingling, pain, and loss of dexterity. Again, proper tool design is very important for avoiding these extreme wrist positions. Extreme radial deviations of the wrist result in pressure between the head of the radius and the adjoining part of the humerus, resulting in tennis elbow, a form of tendinitis. Similarly, simultaneous extension of the wrist, concurrent with full pronation, is equally stressful on the elbow.

Trigger finger is a form of tendinitis resulting from a work situation in which the distal phalanx of the index finger must be bent and flexed against resistance before more proximal phalanges are flexed. Excessive isometric forces impress a groove on the bone, or the tendon enlarges due to inflammation. When the tendon moves within the sheath, it may jerk or snap with an audible click. *White finger* results from excessive vibration from power tools, inducing the constriction of arterioles within the digits. The resulting lack of blood flow appears as a blanching of the skin, with a corresponding loss of motor control. A similar effect can occur as a result of exposure to cold and is termed *Raynaud's syndrome.* A very good introduction to these and other CTDs can be found in Putz-Anderson (1988).

Not all incidences are traumatic. Short-term fatigue and discomfort have also been shown to result from poor handle and work orientation in hammering, and improper tool shape and work height in work with screwdrivers. Typically, a poor tool grip design leads to the exertion of higher grip forces and to extreme wrist deviations, resulting in greater fatigue (Freivalds, 1996).

To evaluate the level of CTD problems in a plant, the methods analyst or ergonomist typically starts out by surveying the workers to determine their health and discomfort at work. One common tool utilized for this purpose is the *body discomfort chart* (Corlett and Bishop, 1976; see Figure 5.24) in which the worker rates the level of pain or body discomfort for various parts of the body, on a scale from 0 (nothing at all) to 10 (almost maximum). The rating scale is based on Borg's (1990) *category ratio scale (CR-10)* with verbal anchors shown in Figure 5.24.

A more quantitative approach is the novel CTD risk analysis procedure that sums risk values for all three major causative factors into one risk score (see Figure 5.25; Seth, et al., 1999). A frequency factor is determined by the number of damaging wrist motions and then scaled by a threshold value of 10,000. A posture factor is determined from the degree of deviation from the neutral posture for major upper extremity motions. A force factor is determined from the relative percentage of maximum muscle exertion required for the task, and then scaled by

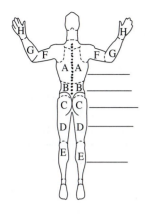

0	Nothing at all
.5	Extremely weak (just noticeable)
1	Very weak
2	Weak (light)
3	Moderate
4	
5	Strong (heavy)
6	
7	Very strong
8	
9	
10	Extremely strong (almost max)
•	Maximal

Figure 5.24 Body discomfort chart.
(Adapted from Corlett and Bishop, 1976.)

15 percent, the maximum allowed for extended static contractions (see Chapter 4). A final miscellaneous factor incorporates a variety of conditions that may have a role in CTD causation, such as vibration and temperature. They are weighted appropriately and then summed to yield a final CTD risk index. For relatively safe conditions, the index should be less than 1 (similar to the NIOSH lifting index, Chapter 4).

One example (see Figure 5.25) analyzes the CTD stress incurred on a highly repetitive cutoff operation described in greater detail in Example 8-1. Both the frequency factor of 1.55 and the force factor of 2.00 exceed the safety threshold of 1.0, leading to a total risk value of 1.34, which also exceeds 1.0. Thus, the most cost-effective approach is to decrease the frequency by eliminating or combining unnecessary motions (which may or may not be possible) and decrease the force component by modifying the grasp utilized (the basis for methods change in Example 8.1).

The CTD index has proved to be quite successful at identifying injurious jobs, but it works much better on a relative basis, rather than an absolute basis, for example, rank-ordering critical jobs. Note that the CTD risk index also serves as both a useful checklist for identifying poor postures and a design tool for selecting key conditions to redesign.

5.5 PRINCIPLES OF WORK DESIGN: TOOLS

USE A POWER GRIP FOR TASKS REQUIRING FORCE AND PINCH GRIPS FOR TASKS REQUIRING PRECISION

Prehension of the hand can basically be defined as variations of grip between two extremes: a power grip and a pinch grip. In a *power grip*, the cylindrical handle

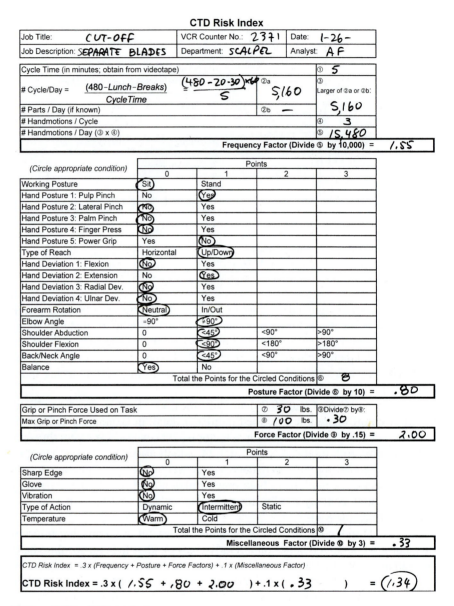

Figure 5.25 CTD risk index.

of the tool, whose axis is more or less perpendicular to the forearm, is held in a clamp formed by the partly flexed fingers and the palm. Opposing pressure is applied by the thumb, which slightly overlaps the middle finger (see Figure 5.26). The line of action of the force can vary with (1) the force parallel to the forearm,

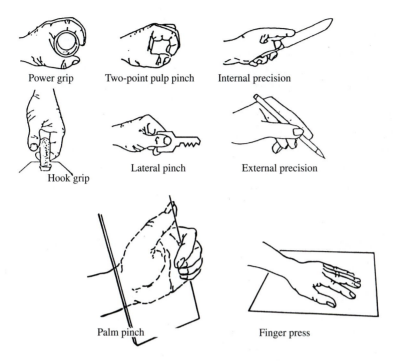

Power grip

Two-point pulp pinch

Internal precision

Hook grip

Lateral pinch

External precision

Palm pinch

Finger press

Figure 5.26 Types of grip.

as in sawing; (2) the force at an angle to the forearm, as in hammering; and (3) the force acting on a moment arm, creating torque about the forearm, as in using a screwdriver. As the name implies, the power grip is used for power or for holding heavy objects. However, the more the fingers or the thumb deviate from the cylindrical grip, the less force is produced and the greater the precision that can be provided. For example, in holding a light hammer as in tacking, the thumb may deviate from opposing the fingers to aligning with the handle. If the index finger also deviates to the tool axis, as in holding a knife for a precise cut, then a pinch grip is approached, with the blade being pinched between the thumb and index finger. This grip is sometimes called an internal precision grip (Konz and Johnson, 2000). A hook grip, used for holding a box or a handle, is an incomplete power grip in which the thumb counterforce is not applied, thereby considerably reducing the available grip force.

The *pinch grip* is used for control or precision. In a pinch grip, the item is held between the distal ends of one or more fingers and the opposing thumb (the thumb is sometimes omitted). The relative position of the thumb and fingers determines how much force can be applied and provides a sensory surface for receiving the feedback necessary to give the precision needed. There are four basic types of pinch grips, with many variations (see Figure 5.26): (1) lateral pinch, thumb opposes the side of the index finger; (2) two- and three-point tip (or pulp) pinches, in which the tip (or palmar pad) of the thumb opposes to the tips (or palmar pads) of one or more fingers (for a relatively small cylindrical object, the three digits act as a chuck, resulting in a chuck grip); (3) palm pinch, the fingers oppose the palm of the hand without the thumb participating, as in glass windshield handling; and (4) finger press, the thumbs and fingers press against a surface, as in garment workers pushing cloth into a sewing machine. One specialized grip is an external precision or writing grip, which is a combination of a lateral pinch with the middle finger and a two-point pinch to hold the writing implement (Konz and Johnson, 2000).

Complete gradation and naming of grips can be found in Kroemer (1986). Note the significantly decreased strength capability of the various pinch grips as compared to the power grip (see Table 5.6). Large forces should never be applied with pinch grips.

AVOID PROLONGED STATIC MUSCLE LOADING

When tools are used in situations in which the arms must be elevated or the tools must be held for extended periods, muscles of the shoulders, arms, and hands may be statically loaded, resulting in fatigue, reduced work capacity, and soreness. Abduction of the shoulder, with corresponding elevation of the elbow, will occur if work must be done with a pistol-grip tool on a horizontal workplace. An in-line or straight tool reduces the need to raise the arm and also permits a neutral wrist posture. Prolonged work with arms extended, as in assembly tasks done with force, can produce soreness in the forearm. Rearranging the workplace so as to keep the elbows at 90° eliminates most of the problem (see Figure 5.4). Similarly, continuously holding an activation switch can result in fatigue of the fingers and reduced flexibility.

PERFORM TWISTING MOTIONS WITH THE ELBOWS BENT

When the elbow is extended, tendons and muscles in the arm stretch out and provide low force capability. When the elbow is bent 90° or less, the biceps brachii has a good mechanical advantage and can contribute to forearm rotation.

Table 5.6   Relative Strengths for Different Types of Grips

Grip	Male		Female		Mean % of Power Grip
	lb	kg	lb	kg	
Power	89.9	40.9	51.2	23.3	100
Tip pinch	14.6	6.6	10.1	4.6	17.5
Pulp pinch	13.7	6.2	9.7	4.4	16.6
Lateral pinch	24.5	11.1	17.1	7.8	29.5

Source: Adapted from An et al., 1986.

MAINTAIN A STRAIGHT WRIST

As the wrist is moved from its neutral position, a loss of grip strength occurs. Starting from a neutral wrist position, pronation decreases grip strength by 12 percent, flexion/extension by 25 percent, and radial/ulnar deviation by 15 percent (see Figure 5.27). Furthermore, awkward hand positions may result in soreness of the wrist, loss of grip, and, if sustained for extended periods, the occurrence of carpal tunnel syndrome. To reduce this problem, the workplace or tools should be redesigned to allow for a straight wrist; for example, lower work surface and edges of containers, and tilt jigs toward the user. Similarly, the tool handle should reflect the axis of grasp, which is about 78° from horizontal, and should be oriented such that the eventual tool axis is in line with the index finger; examples are bent plier handles and a pistol-grip knife (see Figure 5.28).

AVOID TISSUE COMPRESSION

Often, in the operation of hand tools, considerable force is applied by the hand. Such actions can concentrate considerable compressive force on the palm of the hand or the fingers, resulting in ischemia, which is the obstruction of blood flow to the tissues and eventual numbness and tingling of the fingers. Handles should

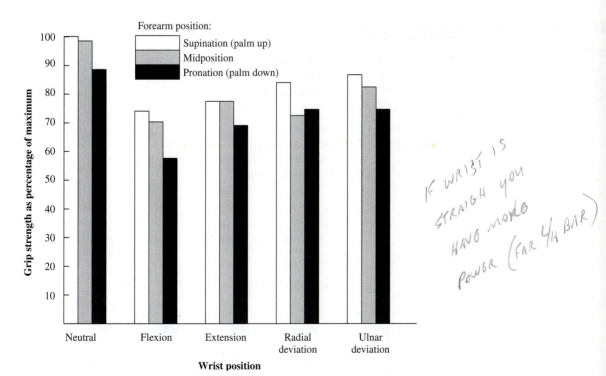

Figure 5.27 Grip strength as a function of wrist and forearm position.
(*Source:* Based on data form Terrell and Purswell, 1976, Table 1.)

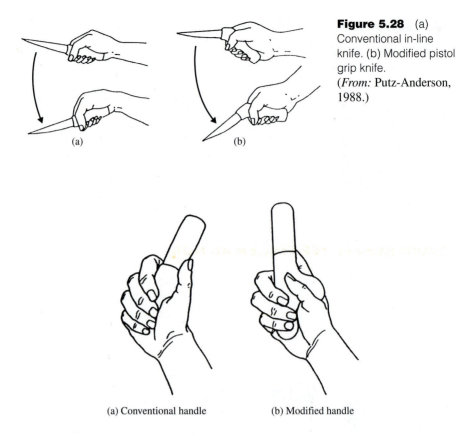

Figure 5.28 (a) Conventional in-line knife. (b) Modified pistol grip knife. (*From:* Putz-Anderson, 1988.)

(a)   (b)

(a) Conventional handle   (b) Modified handle

Figure 5.29 Handle design.
Shown here is a conventional paint scraper (a) that presses on the ulnar artery and a modified handle (b) that rests on the tough tissues between thumb and index finger and prevents pressure on the critical areas of the hand. Note that the handle extends beyond the base of the palm. (*Source:* Tichauer, 1967.)

be designed with large contact surfaces, to distribute the force over a larger area (see Figure 5.29) or to direct it to less sensitive areas, such as the tissue between the thumb and index finger. Similarly, finger grooves or recesses in tool handles should be avoided. Since hands vary considerably in size, such grooves would accommodate only a fraction of the population.

DESIGN TOOLS SO THAT THEY CAN BE USED BY EITHER HAND AND BY MOST INDIVIDUALS

Alternating hands allows the reduction of local muscle fatigue. However, in many situations, this is not possible, as the tool use is one-handed. Furthermore, if the tool is designated for the user's preferred hand, which for 90 percent of the population is the right hand, then 10 percent are left out. Good examples of right-handed tools that cannot be used by a left-handed person include a power drill with side handle on the left side only, a circular saw, and a serrated knife leveled on one side only. Typically, right-handed males show a 12 percent strength decrement in the left hand, while right-handed females show a 7 percent strength decrement. Surprisingly, both left-handed males and females had nearly equal strengths in both hands. One conclusion is that left-handed subjects are forced to adapt to a right-handed world (Miller and Freivalds, 1987).

Female grip strength typically ranges from 50 to 67 percent of male strength (Pheasant and Scriven, 1983); for example, the average male can be expected to exert approximately 110 lb (50 kg), while the average female can be expected to exert approximately 60 lb (27.3 kg). Females have a twofold disadvantage: an average lower grip strength and an average smaller grip span. The best solution is to provide a variety of tool sizes.

AVOID REPETITIVE FINGER ACTION

If the index finger is used excessively for operating triggers, symptoms of trigger finger develop. Trigger forces should be kept low, preferably below 2 lb (0.9 kg) (Eastman Kodak, 1983), to reduce the load on the index finger. Two or three finger-operated controls are preferable (see Figure 5.30); finger strip controls or a power grip bar is even better, because they require the use of more and stronger fingers. Absolute finger flexion strengths and their relative contributions to grip are shown in Table 5.7.

For a two-handled tool, a spring-loaded return saves the fingers from having to return the tool to its starting position. In addition, the high number of repetitions must be reduced. Although critical levels of repetitions are not known, NIOSH (1989) found high rates of muscle–tendon disorders in workers exceeding 10,000 motions per day.

USE THE STRONGEST WORKING FINGERS: THE MIDDLE FINGER AND THE THUMB

Although the index finger is usually the finger that is capable of moving the fastest, it is not the strongest finger (see Table 5.7). Where a relatively heavy load is involved, it is usually more efficient to use the middle finger, or a combination of the middle finger and the index finger.

DESIGN 1.5-IN HANDLE DIAMETERS FOR POWER GRIPS

Power grips around a cylindrical object should entirely surround the circumference of the cylinder, with the fingers and thumb barely touching. For most individuals, this would entail a handle diameter of approximately 1.5 in (3.8 cm), resulting in minimum EMG activity, minimum grip endurance deterioration, and maximum thrust forces. In general, the upper end of the range is best for maximum torque, and the lower end is best for dexterity and speed. The handle diameter for precision grips should be approximately 0.5 in (1.3 cm) (Freivalds, 1996).

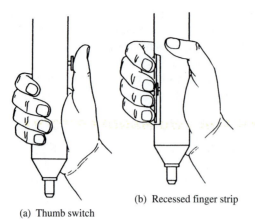

Figure 5.30 Thumb-operated and finger-strip-operated pneumatic tool.
Thumb operation (a) results in overextension of the thumb. Finger-strip control (b,c) allows all the fingers to share the load and the thumb to grip and guide the tool.

(a) Thumb switch

(b) Recessed finger strip

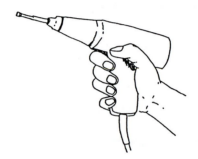

(c) Three-finger trigger for power tools

Table 5.7 Maximal Static Finger Flexion Forces

Digit	Max Force		% Force (of thumb)	% Contribution to Power Grip
	lb	**kg**		
Thumb	16	7.3	100	—
Index	13	5.9	81	29
Middle	14	6.4	88	31
Ring	11	5.0	69	24
Little	7	3.2	44	16

Source: Adapted from Hertzberg, 1973.

DESIGN HANDLE LENGTHS TO BE A MINIMUM OF 4 IN

For both handles and cutouts, there should be enough space to allow for all four fingers. Hand breadth across the metacarpals ranges from 2.8 in (7.1 cm) for a 5th percentile female to 3.8 in (9.7 cm) for a 95th percentile male (Garrett, 1971).

Thus, 4 in (10 cm) may be a reasonable minimum, but 5 in (12.5 cm) may be more comfortable. If the grip is enclosed, or if gloves are used, even larger openings are recommended. For an external precision grip, the tool shaft must be long enough to be supported at the base of the first finger or thumb. For an internal precision grip, the tool should extend past the palm, but not far enough to hit the wrist (Konz and Johnson, 2000).

DESIGN A 3-IN GRIP SPAN FOR TWO-HANDLED TOOLS

Grip strength and the resulting stress on finger flexor tendons vary with the size of the object being grasped. On a dynamometer with handles angled inward, a maximum grip strength is achieved at about 3 to 3.2 in (7.68.1 cm) (Chaffin and Andersson, 1991). At distances different from the optimum, the percent grip strength decreases (see Figure 5.31), as defined by

$$\%\text{Grip strength} = 100 - 0.28*S - 65.8*S^2$$

where S is the given grip span minus the optimum grip span (3 in for females and 3.2 in for males). For dynamometers with parallel sides, this optimum span decreases to 1.8 to 2 in (4.5 to 5 cm) (Pheasant and Scriven, 1983). Because of the large variation in individual strength capacities, and the need to accommodate most of the working population (i.e., the 5th percentile female), maximal grip requirements should be limited to less than 20 lb. A similar effect is found for pinch

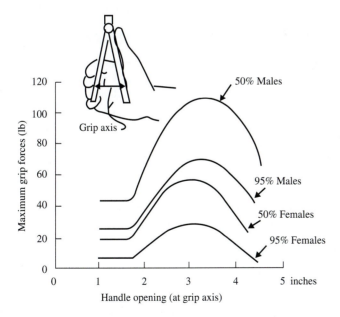

Figure 5.31 Grip strength capability for various population distributions as a function of grip span.
(*From*: Greenberg and Chaffin, 1976)

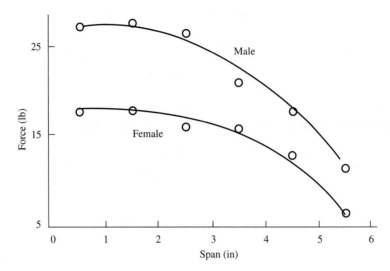

Figure 5.32 Pulp pinch strength capability for various spans. (From Heffernan and Freivalds, 2000.)

strength (see Figure 5.32). However, the overall pinch force is at a much more re-duced force level (approximately 20 percent of power grip) and the optimum pinch span (for a 4-point pulp pinch) ranges from 0.5 to 2 in (1.3 to 5.1 cm) and then drops sharply for larger spans (Heffernan and Freivalds, 2000).

DESIGN APPROPRIATELY SHAPED HANDLES

For a power grip, design for maximum surface contact to minimize unit pressure of the hand. Typically, a tool with a circular cross section is thought to give the largest torque. However, the shape may be dependent on the type of task and the motions involved (Cochran and Riley, 1986). For example, the maximum pull force and the best thrusting actions are actually obtained with a triangular cross section. For a rolling type of manipulation, the triangular shape is slowest. A rec-tangular shape (with corners rounded) with width to height ratios from 1:1.25 to 1:1.5 appears to be a good compromise. A further advantage of a rectangular cross section is that the tool does not roll when placed on a table. Also, the han-dles should not have the shape of a true cylinder, except for a hook grip. For screwdriver-type tools, the handle end should be rounded to prevent undue pres-sure at the palm; for hammer-type tools, the handle may have some flattening curving, to indicate the end of the handle.

In a departure from the circular, cylindrically shaped handles, Bullinger and Solf (1979) proposed a more radical design using a hexagonal cross section, shaped as two truncated cones joined at the largest ends. Such a shape fits the contours of the palm and thumb best, in both precision and power grips, and it yielded the highest torques in comparison with more conventional handles. A similar dual-truncated conical shape was also developed for a file handle. In this case, the heavily rounded square-shaped cross section was found to be markedly superior to more conventional shapes.

A final note on shape is that T-handles yield a much higher torque (up to 50 percent more) than straight screwdriver handles. The slanting of the T-handle generates even larger torques by allowing the wrist to remain straight (Saran, 1973).

DESIGN GRIP SURFACE TO BE COMPRESSIBLE AND NONCONDUCTIVE

For centuries, wood was the material of choice for tool handles. Wood is readily available and easily worked. It has good resistance to shock and thermal and electrical conductivity, and it has good frictional qualities, even when wet. Since wooden handles can break and stain with grease and oil, there has recently been a shift to plastic and even metal. However, metal should be covered with rubber or leather to reduce shock and electrical conductivity and increase friction (Fraser, 1980). Such compressible materials also dampen vibration and allow a better distribution of pressure, reducing fatigue and hand tenderness (Fellows and Freivalds, 1991). However, the grip material should not be too soft; otherwise sharp objects, such as metal chips, will get embedded in the grip and make it difficult to use. The grip surface area should be maximized to ensure pressure distribution over as large an area as possible. Excessive localized pressure may cause pain sufficient to interrupt the work.

The frictional characteristics of the tool surface vary with the pressure exerted by the hand, the smoothness and porosity of the surface, and the type of contamination (Bobjer et al., 1993). Sweat increases the coefficient of friction, while oil and fat reduce it. Adhesive tape and suede provide good friction when moisture is present. The type of surface pattern, as defined by the ratio of ridge area to groove area, shows some interesting characteristics. When the hand is clean or sweaty, the maximum frictions are obtained with high ratios (maximizing the hand–surface contact area); when the hand is contaminated, maximum frictions are obtained with low ratios (maximizing the capacity to channel contaminants away).

KEEP THE WEIGHT OF THE TOOL BELOW 5 LB

The weight of the hand tool will determine how long it can be held or used and how precisely it can be manipulated. For tools held in one hand with the elbow at 90° for extended periods, Greenberg and Chaffin (1976) recommend loads of no more than 5 lb (2.3 kg). In addition, the tool should be well balanced, with the center of gravity as close as possible to the center of gravity of the hand (unless the purpose of the tool is to transfer force, as in a hammer). Thus, the hand or arm muscles do not need to oppose any torque development by an unbalanced tool. Heavy tools used to absorb impact or vibration should be mounted on telescoping arms or tool balancers to reduce the effort required by the operator. For precision operations, tool weights greater than 1 lb are not recommended, unless a counterbalanced system is used.

USE GLOVES JUDICIOUSLY

Gloves are often used with hand tools for safety and comfort. Safety gloves are seldom bulky, but gloves worn in subfreezing climates can be very heavy and can interfere with grasping ability. Wearing woolen or leather gloves may add 0.2 in (0.5 cm) to the hand thickness and 0.3 in (0.8 cm) to the hand breadth at the thumb, while heavy mittens add 1 in (2.5 cm) and 1.6 in (4.0 cm), respectively (Damon et al., 1966). More important, gloves reduce grip strength 10 to 20 percent (Hertzberg, 1973), torque production, and manual dexterity performance times. Neoprene gloves slow performance times by 12.5 percent over bare-handed performance, terry cloth by 36 percent, leather by 45 percent, and PVC by 64 percent (Weidman, 1970). A trade-off between increased safety and decreased performance with gloves must be considered.

USE POWER TOOLS SUCH AS NUT AND SCREWDRIVERS INSTEAD OF MANUAL TOOLS

Power hand tools not only perform work faster than manual tools, but also do the work with considerably less operator fatigue. Greater uniformity of product can be expected when power hand tools are used. For example, a power nut driver can drive nuts consistently to a predetermined tightness in inch-pounds, while a manual nut driver cannot be expected to maintain a constant driving pressure due to operator fatigue.

There is, however another trade-off. Powered hand tools produce vibration, which can induce white finger syndrome, the primary symptom of which is a reduction in blood flow to the fingers and hand due to *vasoconstriction* of the blood vessels. As a result, there is a sensory feedback loss and decreased performance, and the condition may contribute to the development of carpal tunnel syndrome, especially in jobs with a combination of forceful and repetitive exertions. It is generally recommended that vibrations in the critical range of 40 to 130 Hz or a slightly larger (but safer) range of 2 to 200 Hz (Lundstrom and Johansson, 1986) be avoided. The exposure to vibration can be minimized through a reduction in the driving force, the use of specially designed vibration damping handles (Andersson, 1990) or vibration-absorbing gloves, and better maintenance to decrease misalignments or unbalanced shafts.

USE THE PROPER CONFIGURATION AND ORIENTATION OF POWER TOOLS

In a power drill or other power tools, the major function of the operator is to hold, stabilize, and monitor the tool against a workpiece, while the tools perform the main effort of the job. Although the operator may at times need to shift or orient the tool, the main function for the operator is effectively to grasp and hold the tool. A hand drill is comprised of a head, body, and handle, with all three ideally being in line. The line of action is the line from the extended index finger, which means that in the ideal drill, the head is off-center with respect to the central axis of the body.

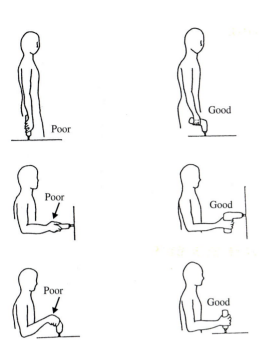

Figure 5.33 Proper orientation of power tools in the workplace. (From: Armstrong, 1983)

Handle configuration is also important, with the choices being pistol grip, in-line, or right-angle. As a rule of thumb, in-line and right-angle are best for tightening downward on a horizontal surface, while pistol grips are best for tightening on a vertical surface, with the aim being to obtain a standing posture with a straight back, upper arms hanging down, and the wrist straight (see Figure 5.33). For the pistol grip, this results in the handle being at an angle of approximately 78° with the horizontal (Fraser, 1980).

Another important factor is the center of gravity. If it is too far forward in the body of the tool, a turning moment is created, which must be overcome by the muscles of the hand and forearm. This requires muscular effort additional to that required for holding, positioning, and pushing the drill into the workpiece. The primary handle is placed directly under the center of gravity, so that the body juts out behind the handle, as well as in front. For heavy drills, a secondary supportive handle may be needed, either to the side or preferably below the tool, so that the supporting arm can be tucked in against the body, rather than being abducted.

CHOOSE A POWER TOOL WITH THE PROPER CHARACTERISTICS

Power tools, such as nut-runners used to tighten nuts, are commercially available in a variety of handle configurations, spindle diameters, speeds, weights, shutoff mechanisms, and torque outputs. The torque is transferred from the motor to the spindle through a variety of mechanisms, so that the power (often compressed air) can be shut off quickly once the nut or other fastener is tight. The simplest and cheapest mechanism is a direct drive, which is under the operator's control, but because of the long time to release the trigger once the nut is tightened, direct drive transfers a very large reaction torque to the operator's arm. Mechanical friction clutches will allow the spindle to slip, reducing some of this reaction torque. A better mechanism for reducing the reaction torque is the airflow shutoff, which automatically senses when to cut off the air supply as the nut is tightened. A still faster mechanism is an automatic mechanical clutch shutoff. Recent mechanisms include the hydraulic pulse system, in which the rotational energy from the motor is transferred over a pulse unit containing an oil cushion (filtering off the high-frequency pulses, as well as noise), and a similar electrical pulse system, both of which reduce the reaction torque to a large extent (Freivalds and Eklund, 1993).

Variations of torque delivered to the nut depend on several conditions including: properties of the tool; the operator; properties of the joint, for example, the combination of the fastener and the material being fastened (ranging from soft, in which the materials have elastic properties, such as body panels, to hard, in which there are two stiff surfaces, such as pulleys on a crankshaft); and stability of the air supply. The torque experienced by the user (the reaction torque) depends on these factors plus the torque shutoff system. In general, using electrical tools at lower than normal rpm levels, or underpowering pneumatic tools, results in larger reaction torques and more stressful ratings. Pulse-type tools produce the lowest reaction torques, perhaps because the short pulses "chop up" the reaction torque. Other potential problems include noise from the pneumatic mechanism reaching levels as high as 95 dB(A), vibration levels exceeding 132 dB(V), and dust or oil fumes emanating from the exhaust (Freivalds and Eklund, 1993).

USE REACTION BARS AND TOOL BALANCERS FOR POWER TOOLS

Reaction torque bars should be provided if the torque exceeds 53 in·lb (6 N·m) for in-line tools used for a downward action, 106 in·lb (12 N·m) for pistol grip tools used in a horizontal mode, and 444 in·lb (50 N·m) for right-angled tools used in a downward or upward motion (Mital and Kilbom, 1992).

This information is summarized in an evaluative checklist for tools (see Figure 5.34). If the tool does not conform to the recommendations and desired features, it should be redesigned or replaced.

SUMMARY

Many factors significantly impact both the productivity and the well-being of the operator at the workstation. Sound ergonomics technology applies to both the equipment being used and the general conditions surrounding the work area. For the equipment point and the workstation environment, adequate flexibility should

Basic Principles	**Yes**	**No**
1. Does the tool perform the desired function effectively?	❏	❏
2. Does the tool match the size and strength of the operator?	❏	❏
3. Can the tool be used without undue fatigue?	❏	❏
4. Does the tool provide sensory feedback?	❏	❏
5. Are the tool capital and maintenance costs reasonable?	❏	❏

Anatomical Concerns	**Yes**	**No**
1. If force is required, can the tool be grasped in a power grip (i.e., handshake)?	❏	❏
2. Can the tool be used without shoulder abduction?	❏	❏
3. Can the tool be used with a 90° elbow angle (i.e., forearms horizontal)?	❏	❏
4. Can the tool be used with the wrist straight?	❏	❏
5. Does the tool handle have large contact surfaces to distribute forces?	❏	❏
6. Can the tool be used comfortably by a 5th percentile female operator?	❏	❏
7. Can the tool be used in either hand?	❏	❏

Handles and Grips	**Yes**	**No**
1. For power uses, is the tool grip 1.5–2 inches in diameter?	❏	❏
a. Can the handle be grasped with the thumb and fingers slightly overlapped?	❏	❏
2. For precision tasks, is the tool grip $\frac{5}{16} - \frac{5}{8}$ inches in diameter?	❏	❏
3. Is the grip cross section circular?	❏	❏
4. Is the grip length at least 4 inches (5 inches if gloves are worn)?	❏	❏
5. Is the grip surface finely textured and slightly compressible?	❏	❏
6. Is the handle nonconductive and stain free?	❏	❏
7. For power uses, does the tool have a pistol grip angled at 78°?	❏	❏
8. Can a two-handled tool be operated with less than 20 pounds grip force?	❏	❏
9. Is the span of the tool handles between $2\frac{3}{4} - 3\frac{1}{4}$ inches?	❏	❏

Power Tool Considerations	**Yes**	**No**
1. Are trigger activation forces less than 1 pound?	❏	❏
2. For repetitive use, is a finger strip trigger present?	❏	❏
3. Are less than 10,000 triggering actions required per shift?	❏	❏
4. Is a reaction bar provided for torques exceeding:	❏	❏
a. 50 inch-pounds for in-line tools?	❏	❏
b. 100 inch-pounds for pistol-grip tools?	❏	❏
c. 400 inch-pounds for right-angled tools?	❏	❏
5. Does the tool create less than 85 dBA for a full day of noise exposure?	❏	❏
6. Does the tool vibrate?	❏	❏
a. Are the vibrations outside the 2–200 Hz range?	❏	❏

Miscellaneous and General Considerations	**Yes**	**No**
1. For general use, is the weight of the tool less than 5 pounds?	❏	❏
2. For precision tasks, is the weight of the tool less than 1 pound?	❏	❏
3. For extended use, is the tool suspended?	❏	❏
4. Is the tool balanced (i.e., center of gravity on the grip axis)?	❏	❏
5. Can the tool be used without gloves?	❏	❏
6. Does the tool have stops to limit closure and prevent pinching?	❏	❏
7. Does the tool have smooth and rounded edges?	❏	❏

Figure 5.34 Tool Evaluation Checklist

be provided so that variations in employee height, reach, strength, reflex time, and so on can be accommodated. A workbench that is 32 in (81 cm) high may be just right for a 75-in (191-cm) tall worker, but would definitely be too high for a 66-in (167.6-cm) tall employee. Adjustable-height workstations and chairs are desirable to accommodate the full range of workers, based on plus or minus two standard deviations from the norm. The better able we are to provide a flexible work center to accommodate the total range of the workforce, the better will be the productivity results and worker satisfaction.

Just as there are significant variations in height and size in the workforce, there are equal or greater variations in visual capacity, hearing ability, feeling ability, and manual dexterity. The vast majority of workstations can be improved. Applying ergonomic considerations along with methods engineering will lead to more efficient competitive work environments that will improve the well-being of the workers, the quality of the product, the labor turnover of the business, and the prestige of the organization.

QUESTIONS

1. What seat width would accommodate 90 percent of adults?
2. Compare and contrast the three different design strategies.
3. Explain how a proper work surface height would be determined.
4. What are the most critical features in a good ergonomic chair? Which should be adjustable?
5. What is the principle behind the design of a saddle seat?
6. What is lordosis and how does it relate to a lumbar pad?
7. What is the principle behind antifatigue mats?
8. What is the principle behind the proper layout of bins, parts, and tools on a work surface?
9. Why is a fixture so important in workplace? List as many reasons as possible.
10. What does Warrick's principle refer to in designing controls and displays?
11. What is the optimum line of sight?
12. List three principles for arranging components on a panel.
13. What is the range effect?
14. List the three principles for effective control–display compatibility.
15. What is operational coding?
16. What is the main disadvantage of tactile controls?
17. What is "control movement without system response" known as?
18. If the C/R ratio is increased from 1.0 to 4.0, what happens to travel time, adjust time, and total time?
19. What are the three most important task factors leading to cumulative trauma disorders?
20. What is the most important factor leading to white finger?
21. What is trigger finger?

22. Describe the progression of the disease state for carpal tunnel syndrome.

23. Design an ergonomic handle, indicating all the principles used in the design.

24. What are the key concerns in the design of a power tool?

PROBLEMS

1. Because of the Challenger disaster, NASA has decided to include a personal escape capability (i.e., a launch compartment) for each space shuttle astronaut. Because space is at a premium, proper anthropometric design is crucial. Also, because of budget restrictions, the design is to be nonadjustable; for example, the same design must fit all present and future astronauts, both males and females. For each launch compartment feature, indicate the body feature used in the design, the design principle used, and the actual value (in inches) to be used in its construction.

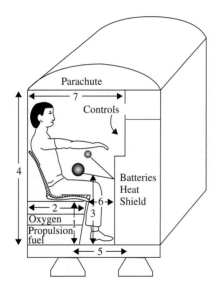

Launch Compartment Feature	Body Feature	Design Principle	Actual Value
1. Height of seat			
2. Seat depth			
3. Height of joystick			
4. Height of compartment			
5. Depth of foot area			
6. Depth of leg area			
7. Depth of chamber			
8. Width of compartment			
9. Weight limit			

2. You are asked to design a control/display panel for the NASA escape launch. After the initial escape, propulsion is to be used to decelerate against the earth's

gravitational field. The parachute can be released only within a given, narrow altitude range. Arrange the seven displays/controls, using the same-sized dials as shown on the following control panel. Explain the logic for your arrangement.

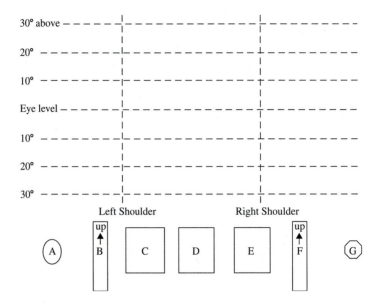

Control/Display	% of Viewing Time	Importance	# Times Used
A Launch release	1	Critical	1
B Propulsion fuel level	20	Very important	10
C Air speed indicator	15	Important	5
D Oxygen pressure	1	Unimportant	2
E Electric power level	2	Important	3
F Altitude indicator	60	Critical	50
G Parachute release	1	Critical	1

3. The Dorben Foundry uses an overhead crane with a magnetic head to load scrap iron into the blast furnace. The crane operator uses various levers to control the three degrees of freedom needed for the crane and its magnetic head. A control is used to activate/deactivate the magnetic pull of the head. The operator is above the operation and looking down much of the time. Operators frequently complain of

Lever	Throw Distance (in)	Crane Movement (ft)	Time to Target(s)
A	20	20	1.2
B	20	10	2.2
C	20	80	1.8
D	20	40	1.2

back pain. Information on various commercially available lever controls is as follows:

a.   Design an appropriate control system for the crane operator. Indicate the number of controls needed, their location (especially in reference to the operator's line of sight), their direction of movement, and their type of feedback.

b.   Indicate an appropriate control-response ratio for these controls.

c.   What other factors may be important in designing these controls?

4.   The following are data on two different control-response setups. Based on these data, what is the optimum C/R ratio for each setup? Which setup would you recommend as best?

Setup	C/R Ratio	Travel Time (s)	Adjust Time (s)
	1	0.1	5
	5	1	2
A	10	5	0.5
	20	10	0.1
	2	1	6
	10	3	5
B	15	5	4
	20	10	3

5.   In a small manufacturing plant, the soldering iron shown here is used to solder connections on a large vertical panel. Several musculoskeletal injuries have been reported on this job over the last year in addition to many operator complaints. In general, it seems that

a.   It is difficult to see the point of application when using this tool.

b.   The operators are gripping the tool unnecessarily tightly.

c.   The power cord tends to get entangled.

d.   Operators complain about wrist pain.

Redesign the soldering iron to eliminate the problems outlined above. Point out the ergonomic or other special features that have been incorporated into the design.

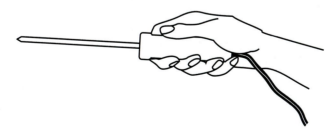

6.   Use the CTD risk index to calculate the potential risk of injury for the right hand on the following jobs shown on the Web page:

a.   Stamping extrusions—assume a grip force of 30% MVC.

b.   Stamping end couplings—assume a grip force of 15% MVC.

c.   Flashlight assembly—assume a grip force of 15% MVC.

d.   Union assembly—assume a grip force of 15% MVC.

e.   Hospital bed rail assembly—assume a grip force of 30% MVC.

f.   Stitching (garments)—assume a grip force of 30% MVC.

g.   Labeling (garments)—assume a grip force of 15% MVC.

h.   Cut and tack (garments)—assume a grip force of 30% MVC.

7.   The worker shown below is driving 4 screws into a panel with a powered driver. His production quota is 2,300 panes per 8-h shift. What specific ergonomic problems are found on this job? For each problem: (a) specify an ergonomics improvement that would correct the problem and (b) provide a specific work design principle to support this methods change.

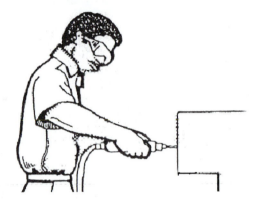

REFERENCES

An, K., L. Askew, and E. Chao. "Biomechanics and Functional Assessment of Upper Extremities." In *Trends in Ergonomics/Human Factors III*. Ed. W. Karwowski. Amsterdam: Elsevier, 1986, pp. 573–580.

Andersson, E. R. "Design and Testing of a Vibration Attenuating Handle." *International Journal of Industrial Ergonomics,* 6, no. 2 (September 1990), pp. 119–125.

Andersson, G. B. J., R. Örtengren, A. Nachemson, and G. Elfström. "Lumbar Disc Pressure and Myoelectric Back Muscle Activity During Sitting." I, Studies on an Experimental Chair. *Scandinavian Journal of Rehabilitation Medicine,* 6 (1974), pp. 104–114.

Armstrong, T. J. *Ergonomics Guide to Carpal Tunnel Syndrome.* Fairfax, VA: American Industrial Hygiene Association, 1983.

Bobjer, O., S. E. Johansson, and S. Piguet. "Friction Between Hand and Handle. Effects of Oil and Lard on Textured and Non-textured Surfaces; Perception of Discomfort." *Applied Ergonomics,* 24, no.3 (June 1993), pp. 190–202.

Borg, G. "Psychophysical Scaling with Applications in Physical Work and the Perception of Exertion." *Scandinavian Journal of Work Environment and Health,* 16, Supplement 1 (1990), pp. 55–58.

Bradley, J. V. (1967). Tactual coding of cylindrical knobs. *Human Factors,* 9(5), 483–496.

Bullinger, H. J., and J. J. Solf. *Ergononomische Arbeitsmittel-gestaltung, II - Handgeführte Werkzeuge - Fallstudien.* Dortmund, Germany: Bundesanstalt fr Arbeitsschutz und Unfallforschung, 1979.

Chaffin, D. B., and G. Andersson. *Occupational Biomechanics.* New York: John Wiley & Sons, 1991, pp. 355–368.

Cochran, D. J., and M. W. Riley. "An Evaluation of Knife Handle Guarding." *Human Factors,* 28, no. 3 (June 1986), pp. 295–301.

Congleton, J. J. *The Design and Evaluation of the Neutral Posture Chair.* Doctoral dissertation. Lubbock, TX: Texas Tech University, 1983.

Corlett, E. N., and R. A. Bishop. "A Technique for Assessing Postural Discomfort." *Ergonomics,* 19, no. 2 (March 1976), pp. 175–182.

Damon, A., H. W. Stoudt, and R. A. McFarland. *The Human Body in Equipment Design.* Cambridge, MA: Harvard University Press, 1966.

Eastman Kodak Co. *Ergonomic Design for People at Work.* Belmont, CA: Lifetime Learning Pub., 1983.

Fellows, G. L., and A. Freivalds. "Ergonomics Evaluation of a Foam Rubber Grip for Tool Handles." *Applied Ergonomics,* 22, no. 4 (August 1991), pp. 225–230.

Fraser, T. M. *Ergonomic Principles in the Design of Hand Tools.* Geneva, Switzerland: International Labor Office, 1980.

Freivalds, A. "Tool Evaluation and Design." In *Occupational Ergonomics.* Eds. A. Bhattacharya and J. D. McGlothlin. New York: Marcel Dekker, 1996, pp. 303–327.

Freivalds, A., and J. Eklund. "Reaction Torques and Operator Stress While Using Powered Nutrunners." *Applied Ergonomics,* 24, no. 3 (June 1993), pp. 158–164.

Garrett, J. "The Adult Human Hand: Some Anthropometric and Biomechanical Considerations." *Human Factors,* 13, no. 2 (April 1971), pp. 117–131.

Grandjean, E. (1998). *Fitting the task to the man* (4th ed.). London: Taylor & Francis.

Greenberg, L., and D. B. Chaffin. *Workers and Their Tools.* Midland, MI: Pendell Press, 1976.

Heffernan, C., and A. Freivalds. "Optimum Pinch Grips in the Handling of Dies." *Applied Ergonomics,* 31 (2000), pp. 409–414.

Hertzberg, H. "Engineering Anthropometry." In *Human Engineering Guide to Equipment Design.* Eds. H. Van Cott, and R. Kincaid. Washington, DC: U.S. Government Printing Office, 1973, pp. 467–584.

Hunt, D. P. (1953). *The coding of aircraft controls* (Tech. Rept. 53–221). U. S. Air Force, Wright Air Development Center.

Jenkins, W., and Conner, M. B. (1949). Some design factors in making settings on a linear scale. *Journal of Applied Psychology,* 33, 395–409.

Konz, S., and S. Johnson. *Work Design,* 5th ed. Scottsdale, AZ: Holcomb Hathaway Publishers, 2000.

Kroemer, K. H. E. "Coupling the Hand with the Handle: An Improved Notation of Touch, Grip and Grasp." *Human Factors,* 28, no. 3 (June 1986), pp. 337–339.

Kroemer, K. (1989). Engineering anthropometry. *Ergonomics,* 32(7), 767–784.

Lundstrom, R., and R. S. Johansson. "Acute Impairment of the Sensitivity of Skin Mechanoreceptive Units Caused by Vibration Exposure of the Hand." *Ergonomics,* 29, no. 5 (May 1986), pp. 687–698.

Miller, G., and A. Freivalds. "Gender and Handedness in Grip Strength." *Proceedings of the Human Factors Society 31st Annual Meeting.* Santa Monica, CA, 1987, pp. 906–909.

Mital, A., and Å. Kilbom. "Design, Selection and Use of Hand Tools to Alleviate Trauma of the Upper Extremities." *International Journal of Industrial Ergonomics,* 10, no. 1 (January 1992), pp. 1–21.

National Safety Council. Accident Facts. Chicago: National Safety Council, 2003.

NIOSH, *Health Hazard Evaluation-Eagle Convex Glass, Co.* HETA-89-137-2005. Cincinnati, OH: National Institute for Occupational Safety and Health, 1989.

Pheasant, S. T., and S. J. Scriven. "Sex Differences in Strength, Some Implications for the Design of Handtools." In *Proceedings of the Ergonomics Society.* Ed. K. Coombes. London: Taylor & Francis, 1983, pp. 9–13.

Putz-Anderson, V. *Cumulative Trauma Disorders.* London: Taylor & Francis, 1988.

Sanders, M. S., and E. J. McCormick. *Human Factors in Engineering and Design.* New York: McGraw-Hill, 1993.

Saran, C. "Biomechanical Evaluation of T-handles for a Pronation Supination Task." *Journal of Occupational Medicine,* 15, no. 9 (September 1973), pp. 712–716.

Serber, H. "New Developments in the Science of Seating." *Human Factors Bulletin,* 33, no. 2 (February 1990), pp. 1–3.

Seth, V., R. Weston, and A. Freivalds. "Development of a Cumulative Trauma Disorder Risk Assessment Model." *International Journal of Industrial Ergonomics,* 23, no. 4 (March 1999), pp. 281–291.

Terrell, R., and J. Purswell. "The Influence of Forearm and Wrist Orientation on Static Grip Strength as a Design Criterion for Hand Tools." *Proceedings of the Human Factors Society 20th Annual Meeting.* Santa Monica, CA, 1976, pp. 28–32.

Tichauer, E. (1967). Ergonomics: The state of the Art. *American Industrial Hygiene Association Journal,* 28, 105–116.

U.S. Department of Justice. *Americans with Disabilities Act Handbook.* EEOC-BK-19. Washington, DC: U.S. Government Printing Office, 1991.

Webb Associates. *Anthropometric Source Book.* II, Pub. 1024. Washington, DC: National Aeronautics and Space Administration, 1978.

Weidman, B. *Effect of Safety Gloves on Simulated Work Tasks.* AD 738981. Springfield, VA: National Technical Information Service, 1970.

SELECTED SOFTWARE

COMBIMAN, *User's Guide for COMBIMAN, CSERIAC.* Dayton, OH: Wright-Patterson AFB. (http://dtica.dtic.mil/hsi/srch/hsi5.html)

Design Tools (available from the McGraw-Hill text website at www.mhhe.com/niebel-freivalds), New York: McGraw-Hill, 2002.

Ergointelligence (Upper Extremity Analysis). 3400 de Maisonneuve Blvd. West, Suite 1430, Montreal, Quebec, Canada H3Z 3B8.

Jack®. Engineering Animation, Inc., 2321 North Loop Dr., Ames, IA 50010. (http://www.eai.com/)

Job Evaluator ToolBox™. ErgoWeb, Inc., P.O. Box 1089, 93 Main St., Midway, UT 84032.

ManneQuinPRO. Nexgen Ergonomics, 3400 de Maisonneuve Blvd. West, Suite 1430, Montreal, Quebec, Canada H3Z 3B8. (http://www.nexgenergo.com/)

Multimedia Video Task Analysis. Nexgen Ergonomics, 3400 de Maisonneuve Blvd. West, Suite 1430, Montreal, Quebec, Canada H3Z 3B8.

Safework. Safework (2000) Inc., 3400 de Maisonneuve Blvd. West, Suite 1430, Montreal, Quebec, Canada H3Z 3B8.

WEBSITES

CTD News— http://ctdnews.com/
CTD Resource Network—http://www.ctdrn.org/
ErgoWeb—http://www.ergoweb.com/
Examples of bad ergonomic design—http://www.baddesigns.com/
CAESAR—http://store.sae.org/caesar

Work Environment Design

KEY POINTS

- Provide both general and task lighting—avoid glare.
- Control noise at the source.
- Control heat stress with radiation shielding and ventilation.
- Provide both overall air movement and local ventilation for hot areas.
- Dampen tool handles and seats to reduce vibration exposure.
- Use rapid, forward-rotating shifts, if shiftwork can't be avoided.

Methods analysts should provide good, safe, comfortable working conditions for the operator. Experience has conclusively proved that plants with good working conditions outproduce those with poor conditions. The economic return from investment in an improved working environment is usually significant. In addition to increasing production, ideal working conditions improve the safety record; reduce absenteeism, tardiness, and labor turnover; raise employee morale; and improve public relations. The acceptable levels for working conditions and the recommended control measures for problem areas are presented in greater detail in this chapter.

6.1 ILLUMINATION

THEORY

Light is captured by the human eye (see Figure 6.1) and processed into an image by the brain. It is a fairly complicated process with the light rays passing through the *pupil*, an opening in the eye, and through the *cornea* and *lens*, which focus the light rays on the retina at the back of the eyeball. The *retina* is composed of photosensitive receptors, the *rods*, which are sensitive to black and white, especially at night, but have poor visual acuity, and the *cones*, which are sensitive to colors in daylight and have good visual acuity. The cones are concentrated in the *fovea*, while the rods are spread out over the retina. Electrical signals from the photoreceptors are

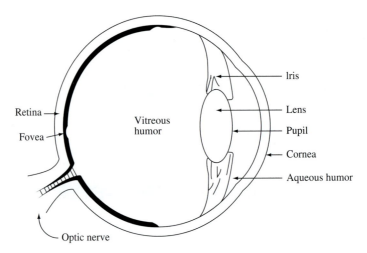

Figure 6.1   The human eye.

collected and passed by the optic nerve to the brain where the light from external illumination is processed and interpreted.

The basic theory of illumination applies to a point source of light (such as a candle) of a given *luminous intensity*, measured in *candelas* (cd) (see Figure 6.2). Light emanates spherically in all directions from the source with 1-cd sources emitting 12.57 lumens (lm) (as determined from the surface area of a sphere, $4pr^2$). The amount of light striking a surface, or a section of this sphere, is termed *illumination or illuminance* and is measured in *footcandles* (fc). The amount of illumination striking a surface drops off as the square of the distance d in feet from the source to the surface:

$$\text{Illuminance} = \text{intensity}/d^2$$

Some of that light is absorbed and some of it is reflected (for translucent materials, some is also transmitted), which allows humans to "see" that object and provides a perception of brightness. The amount reflected is termed *luminance* and is measured in *foot-lamberts* (fL). It is determined by the reflective properties of the surface, known as *reflectance:*

$$\text{Luminance} = \text{illuminance} \times \text{reflectance}$$

Reflectance is a unitless proportion and ranges from 0 to 100 percent. High-quality white paper has a reflectance of about 90 percent, newsprint and concrete around 55 percent, cardboard 30 percent, and matte black paint 5 percent. The reflectances for various color paints or finishes are presented in Table 6.1.

VISIBILITY

The clarity with which the human sees something is usually referred to as *visibility*. The three critical factors of visibility are *visual angle, contrast*, and most

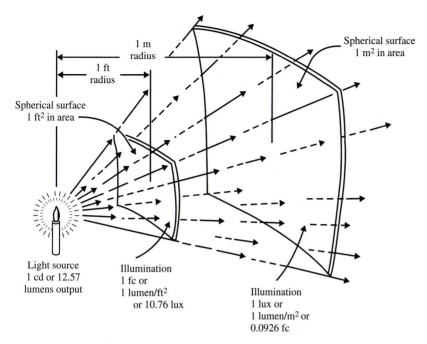

1 m
radius

1 ft
radius

Spherical surface
1 m² in area

Spherical surface
1 ft² in area

Light source
1 cd or 12.57
lumens output

Illumination
1 fc or
1 lumen/ft²
or 10.76 lux

Illumination
1 lux or
1 lumen/m² or
0.0926 fc

Figure 6.2 Illustration of the distribution of light from a light source following the inverse-square law.
(*Source*: General Electric Company, 1965, p. 5.)

Table 6.1 Reflectances of Typical Paint and Wood Finishes

Color or finish	Percent of reflected light	Color or finish	Percent of reflected light
White	85	Medium blue	35
Light cream	75	Dark gray	30
Light gray	75	Dark red	13
Light yellow	75	Dark brown	10
Light buff	70	Dark blue	8
Light green	65	Dark green	7
Light blue	55	Maple	42
Medium yellow	65	Satinwood	34
Medium buff	63	Walnut	16
Medium gray	55	Mahogany	12
Medium green	52		

important, *illuminance*. Visual angle is the angle subtended at the eye by the target, and contrast is the difference in luminance between a visual target and its background. Visual angle is usually defined in arc minutes (1/60 of a degree) for small targtets by

$$\text{Visual angle (arc min)} = 3{,}438 \times h/d$$

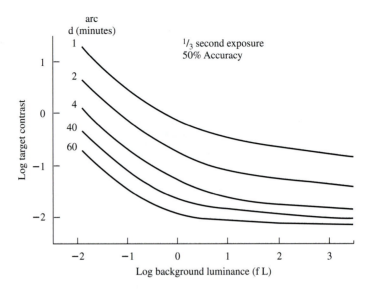

Figure 6.3 Smoothed threshold contrast curves for disks of diameter *d*. (*Adapted from:* Blackwell, 1959)

where h is the height of the target or critical detail (or stroke width for printed matter) and d is the distance from the target to the eye (in the same units as h).

Contrast can be defined in several ways. A typical one is

$$\text{Contrast} = (L_{max} - L_{min})/L_{max}$$

where L is luminance. Contrast, then, is related to the difference in maximum and minimum luminances of the target and background. Note that contrast is unitless.

Other less important factors for visibility are exposure time, target motion, age, known location, and training, which will not be included here.

The relationship between these three critical factors was quantified by Blackwell (1959) in a series of experiments that led to the development of the Illuminating Engineering Society of North America (IESNA, 1995) standards for illumination. Although the Blackwell curves (see Figure 6.3) as such are not often used today, they show the trade-off between the size of the object, the amount of illumination (in this case, measured as luminance reflected from the target), and the contrast between the target and background. Thus, although increasing the amount of illumination is the simplest approach to improving task visibility, it can also be improved by increasing the contrast or increasing the size of the target.

ILLUMINANCE

Recognizing the complexity of extending the point source theory to real light sources (which can be anything but a point source) and some of the uncertainties and constraints of Blackwell's (1959) laboratory setting, the IESNA adopted a

Table 6.2 Recommended Illumination Levels for Use in Interior Lighting Design

Category	Range of illuminance (fc)	Type of activity	Reference area
A	2-3-5	Public areas with dark surroundings.	
B	5-7.5-10	Simple orientation for short temporary visits.	General lighting throughout room or area.
C	10-15-20	Working spaces where visual tasks are performed only occasionally.	
D	20-30-50	Performance of visual tasks of high contrast or large size, e.g., reading printed material, typed originals, handwriting in ink and xerography; rough bench and machine work; ordinary inspection; rough assembly.	
E	50-75-100	Performance of visual tasks of medium contrast or small size, e.g., reading medium-pencil handwriting, poorly printed or reproduced material; medium bench and machine work; difficult inspection; medium assembly.	Illuminance on task.
F	100-150-200	Performance of visual tasks of low contrast or very small size, e.g., reading handwriting in hard pencil on poor-quality paper and very poorly reproduced material; highly difficult inspection, difficult assembly.	
G	200-300-500	Performance of visual tasks of low contrast and very small size over a prolonged period, e.g., fine assembly; very difficult inspection; fine bench and machine work; extra fine assembly.	
H	500-750-1,000	Performance of very prolonged and exacting visual tasks, e.g., the most difficult inspection; extra fine bench and machine work; extra fine assembly.	Illuminance on task via a combination of general and supplementary local lighting.
I	1,000-1,500-2,000	Performance of very special visual tasks of extremely low contrast and small size, e.g., surgical procedures.	

Source: Adapted from IESNA, 1995.

much simpler approach for determining minimum levels of illumination (IESNA, 1995). The first step is to identify the general type of activity to be performed and classify it into one of nine categories, shown in Table 6.2. A more extensive list of specific tasks for this process can be found in IESNA (1995). Note that categories A, B, and C do not involve specific visual tasks. For each category, there is a range of illuminances (low, middle, high). The appropriate value is selected by calculating a weighting factor (-1, 0, $+1$) based on three task and worker characteristics, shown in Table 6.3. These weights are then summed to obtain the total weighing factor. Note that since categories A, B, and C do not involve visual

Table 6.3 Weighting Factors to Be Considered in Selecting Specific Illumination Levels Within Each Category of Table 6.2

Task and worker characteristics	Weight		
	−1	0	+1
Age	<40	40–55	>55
Reflectance of task/surface background	>70%	30–70%	<30%
Speed and accuracy (only for categories D – I)	Not important	Important	Critical

(Adapted from IESNA, 1995)

tasks, the speed/accuracy characteristic is not utilized for these categories, and overall room surfaces are utilized in place of task background. If the total sum of the two or three weighting factors is −2 or −3, the low value of the three illuminances is used; if −1, 0, or +1, the middle value is used; and if +2 or +3, the high value is used.

In practice, illumination is typically measured with a light meter (similar to one found on cameras, but in different units), while luminance is measured with a photometer (typically, a separate attachment to the light meter). Reflectance is usually calculated as the ratio between the luminance of the target surface and the luminance of a standard surface of known reflectance (e.g., a Kodak neutral test card of reflectance = 0.9) placed at the same position on the target surface. The reflectance of the target is then

$$\text{Reflectance} = 0.9 \times L_{\text{target}} / L_{\text{standard}}$$

LIGHT SOURCES AND DISTRIBUTION

After determining the illumination requirements for the area under study, analysts select appropriate artificial light sources. Two important parameters related to artificial lighting are *efficiency* [light output per unit energy; typically, lumens per watt (lm/W)] and *color rendering*. Efficiency is particularly important, since it is related to cost; efficient light sources reduce energy consumption. Color rendering relates to the closeness with which the perceived colors of the object being observed match the perceived colors of the same object when illuminated by standard light sources. The more efficient light sources (high- and low-pressure sodium) have only fair to poor color rendering characteristics and consequently may not be suitable for certain inspection operations where color discrimination is necessary. Table 6.4 provides efficiency and color rendering information for the principal types of artificial light. Typical industrial lighting sources, that is, luminaires, are shown in Figure 6.4.

Luminaires for general lighting are classified in accordance with the percentage of total light output emitted above and below the horizontal (see Figure 6.5). *Indirect lighting* illuminates the ceiling, which in turn reflects light downward. Thus, the ceilings should be the brightest surface in the room

Table 6.4 Artificial Light Sources

Type	Efficiency (lm/W)	Color rendering	Comments
Incandescent	17–23 *lowest*	Good	A commonly used light source, but the least efficient. Lamp cost is low. Lamp life is typically less than 1 year.
Fluorescent	50–80	Fair to good	Efficiency and color rendering vary considerably with type of lamp: cool white, warm white, deluxe cool white. Significant energy cost reductions are possible with new energy-saving lamps and ballasts. Lamp life is typically 5–8 years.
Mercury	50–55	Very poor to fair	A very long lamp life (9–12 years), but efficiency drops off substantially with age.
Metal halide	80–90	Fair to moderate	Color rendering is adequate for many applications. Lamp life is typically 1–3 years.
High-pressure sodium	85–125	Fair	Very efficient light source. Lamp life is 3–6 years at average burning rates, up to 12 h/day.
Low-pressure sodium	100–180	Poor	The most efficient light source. Lamp life is 4–5 years at average burning rate of 12 h/day. Mainly used for roadways and warehouse lighting.

Source: Courtesy Human Factors Section, Eastman Kodak Co.
Note: The efficiency (column 2), in lumens per watt (lm/W), and color rendering (column 3) of six frequently used light sources (column 1) are indicated. Lamp life and other features are given in column 4. Color rendering is a measure of how colors appear under any of these artificial light sources compared with their color under a standard light source. Higher values for efficiency indicate better energy conservation.

(see Figure 6.6), with reflectances above 80 percent. The other areas of the room should reflect lower and lower percentages of the light as one moves downward from the ceiling until the floor is reached, which should reflect no more than 20 to 40 percent of the light, to avoid glare. To avoid excessive luminance, the luminaires should be evenly distributed across the ceiling.

Calculation of Required Illumination	EXAMPLE 6.1

Consider workers of all ages performing an important, medium-difficulty assembly on a dingy metal workstation with a reflectance of 35 percent. The appropriate weights would be age = +1, reflectance = 0, and accuracy = 0. The total weight of +1 implies that the middle value of category E is utilized with a required illumination of 75 fc.

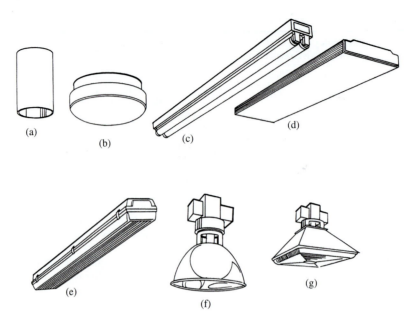

Figure 6.4
Types of industrial ceiling-mounted luminaires: (a,c) downlighting, (b,d) diffuse, (e) damp location, (f) high bay, (g) low bay. (*From:* IESNA, 1995)

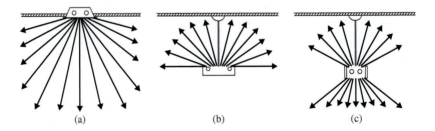

Figure 6.5
Luminaires for general lighting are classified in accordance with the percentage of total light output emitted above and below the horizontal. Three of the classifications are (a) direct lighting, (b) indirect lighting, and (c) direct–indirect lighting. (*From:* IESNA, 1995)

Direct lighting deemphasizes the ceiling surface and places more of the light on the work surfaces and the floor. Direct–indirect lighting is a combination of both. This distribution of lighting is important, as IESNA (1995) recommends that the ratio of luminances of any adjacent areas in the visual field not exceed 3/1. The purpose of this is to avoid glare and problems in adaptation.

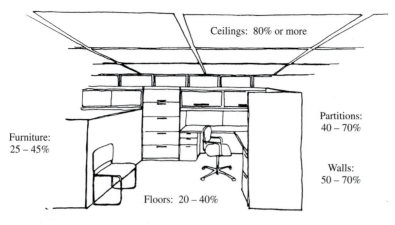

Figure 6.6 Reflectances recommended for room and furniture surfaces in offices. (*From:* IESNA, 1995)

GLARE

Glare is the excessive brightness in the field of vision. This excessive light is scattered in the cornea, lens, and even corrective lenses (Freivalds, Harpster, and Heckman, 1983), decreasing visibility so that additional time is required for the eyes to adapt from light to darker conditions. Also, unfortunately, the eyes tend to be drawn directly to the brightest light source, which is known as *phototropism*. Glare can be either direct, as caused by light sources directly in the field of view, or indirect, as reflected from a surface in the field of view. Direct glare can be reduced by using more luminaires with lower intensities, using baffles or diffusers on luminaires, placing the work surface perpendicular to the light source, and increasing overall background lighting so as to decrease the contrast.

Reflected glare can be reduced by using nonglossy or matte surfaces and reorienting the work surface or task, in addition to the modifications recommended for direct glare. Also, polarizing filters can be used at the light source as part of glasses worn by the operator. A special problem is the stroboscopic effect caused by the reflections from moving parts or machinery. Avoiding polished mirrorlike surfaces is important here. For example, the mirrorlike qualities of the glass screen on computer monitors is a problem in office areas. Repositioning the monitor or using a screen filter is helpful. Typically, most jobs will require supplementary task lighting. This can be provided in a variety of forms, depending on the nature of the task (see Figure 6.7).

COLOR

Both color and texture have psychological effects on people. For example, yellow is the accepted color of butter; therefore, margarine must be made yellow to appeal to the appetite. Steak is another example. Cooked in 45 s on an electronic grill, it does not appeal to customers because it lacks a seared, brown, "appetizing"

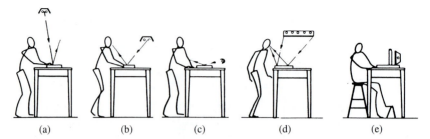

Figure 6.7 Examples of placement of supplementary luminaires.
(a) Luminaire located to prevent veiling reflections and reflected glare; reflected light
does not coincide with angle of view. (b) Reflected light coincides with angle of view.
(c) Low-angle (grazing) lighting to emphasize surface irregularities. (d) Large-area
surface source and pattern are reflected toward the eye. (e) Transillumination from
diffuse source. (*From:* IESNA, 1995)

surface. A special attachment had to be designed to sear the steak. In a third example, employees in an air-conditioned Midwestern plant complained of feeling cold,
although the temperature was maintained at 72°F (22.2°C). When the white walls
of the plant were repainted in a warm coral color, complaints ceased.

Perhaps the most important use of color is to improve the environmental conditions of the workers by providing more visual comfort. Analysts use colors to
reduce sharp contrasts, increase reflectance, highlight hazards, and call attention
to features of the work environment.

Sales are also affected or conditioned by colors. People recognize a company's products instantly by the pattern of colors used on packages, trademarks,
letterheads, trucks, and buildings. Some research has indicated that color prefer-
ences are influenced by nationality, location, and climate. Sales of a product formerly made in one color increased when several colors suited to the differences
in customer demands were supplied. Table 6.5 illustrates the typical emotional
effects and psychological significances of the principal colors.

6.2 NOISE

THEORY

From the analyst's point of view, noise is any unwanted sound. Sound waves
originate from the vibration of some object, which in turns sets up a succession
of compression and expansion waves through the transporting medium (air, water, and so on). Thus, sound can be transmitted not only through air and liquids,
but also through solids, such as machine tool structures. We know that the velocity of sound waves in air is approximately 1,100 ft/s (340 m/s). In viscoelastic
materials, such as lead and putty, sound energy is dissipated rapidly as viscous
friction.

Table 6.5 The Emotional and Psychological Significance of the Principal Colors

Color	Characteristics
Yellow	Has the highest visibility of any color under practically all lighting conditions. It tends to instill a feeling of freshness and dryness. It can give the sensation of wealth and glory, yet can also suggest cowardice and sickness.
Orange	Tends to combine the high visibility of yellow and the vitality and intensity characteristic of red. It attracts more attention than any other color in the spectrum. It gives a feeling of warmth and frequently has a stimulating or cheering effect.
Red	A high-visibility color having intensity and vitality. It is the physical color associated with blood. It suggests heat, stimulation, and action.
Blue	A low-visibility color. It tends to lead the mind to thoughtfulness and deliberation. It tends to be a soothing color, although it can promote a depressed mood.
Green	A low-visibility color. It imparts a feeling of restfulness, coolness, and stability.
Purple and violet	Low-visibility colors. They are associated with pain, passion, suffering, heroism, and so on. They tend to bring a feeling of fragility, limpness, and dullness.

Sound can be defined in terms of the frequencies that determine its tone and quality, along with the amplitudes that determine its intensity. Frequencies audible to the human ear range from approximately 20 to 20,000 cycles per second, commonly called hertz and abbreviated Hz. The fundamental equation of wave propagation is

$$c = f\lambda$$

where c = sound velocity (1,100 ft/s)
   f = frequency, (Hz)
   λ = wavelength (ft)

Note that as the wavelength increases, the frequency decreases.

The sound pressure waves are captured by the human ear (see Figure 6.8) through a complex process. The outer ear funnels the pressure waves onto the *eardrum* or *tympanic membrane*, which starts vibrating. The membrane is attached to three little bones (malleus, incus, and stapes), which transmit the vibrations to the oval window of the *cochlea*. The cochlea is a coiled, fluid-filled structure split lengthwise by the *basilar membrane* containing hair cells with nerve endings. The vibrations from the bones set the fluid into a wavelike motion, which then causes the hair cells to vibrate, activating these-nerve endings, which transmit the impulses via the auditory nerve to the brain for further processing. Note the series of energy transformations: the original pneumatic pressure waves are converted to mechanical vibrations, then to hydraulic waves, back to mechanical vibrations, and finally to electrical impulses.

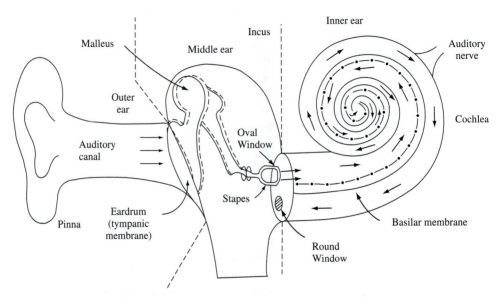

Figure 6.8   The human ear.

MEASUREMENT

Because of the very large range of sound intensities encountered in the normal human environment, the *decibel* (dB) scale has been chosen. In effect, it is the logarithmic ratio of the actual sound intensity to the sound intensity at the threshold of hearing of a young person. Thus, the sound pressure level L in decibels is given by

$$L = 20 \log_{10} P_{\text{rms}}/P_{\text{ref}}$$

where P_{rms} = root-mean-square sound pressure [microbars (μbar)]
   P_{ref} = sound pressure at the threshold of hearing of a young person at 1,000 Hz (0.0002 μbar)

Since sound pressure levels are logarithmic quantities, the effect of the coexistence of two or more sound sources in one location requires that a logarithmic addition be performed as follows:

$$L_{\text{TOT}} = 10 \log_{10}(10^{L_1/10} + 10^{L_2/10} + \cdots)$$

where L_{TOT} = total noise
   L_1 and L_2 = two noise sources

The A-weighted sound level used in Figure 6.9 is the most widely accepted measure of environmental noise. The A weighting recognizes that from both the psychological and physiological points of view, the low frequencies (50 to 500 Hz) are far less annoying and harmful than sounds in the critical frequency range of

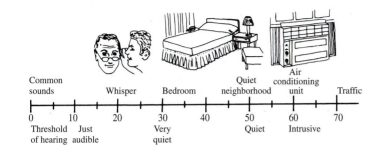

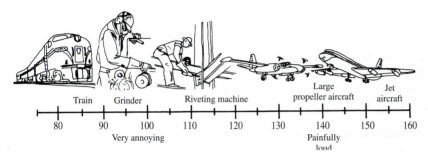

Figure 6.9 Decibel values of typical sounds (dBA).

1,000 to 4,000 Hz. Above frequencies of 10,000 Hz, hearing acuity (and therefore noise effects) again drops off (see Figure 6.10). The appropriate electronic network is built into sound level meters to attenuate low and high frequencies, so that the sound level meter can read in dBA units directly, to correspond to the effect on the average human ear.

HEARING LOSS

The chances of damage to the ear, resulting in "nerve" deafness, increase as the frequency approaches the 2,400- to 4,800-Hz range. This loss of hearing is a result of a loss of receptors in the inner ear, which then fail to transmit the sound waves further to the brain. Also, as the exposure time increases, especially where higher intensities are involved, there will eventually be an impairment in hearing. Nerve deafness is due most commonly to excessive exposure to occupational noise. Individuals vary widely in their susceptibility to noise-induced deafness.

In general, noise is classified as either broadband noise or meaningful noise. Broadband noise is made up of frequencies covering a significant part of the sound spectrum. This type of noise can be either continuous or intermittent. Meaningful noise represents distracting information that impacts the worker's efficiency. In long-term situations, broadband noise can result in deafness; in day-to-day operations, it can result in reduced worker efficiency and ineffective communication.

Continuous broadband noise is typical of such industries as the textile industry and an automatic screw machine shop, where the noise level does not

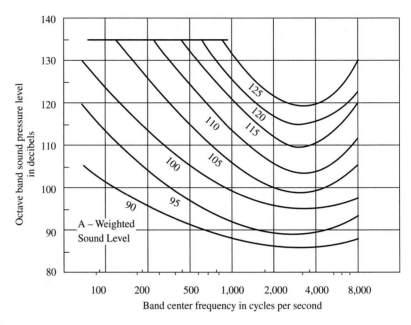

Figure 6.10 Equivalent sound level contours.

deviate significantly during the entire working day. Intermittent broadband noise is characteristic of a drop forge plant and a lumber mill. When a person is exposed to noise that exceeds the damage level, the initial effect is likely to be a temporary hearing loss from which there is complete recovery within a few hours after leaving the work environment. If repeated exposure continues over a long period, irreversible hearing damage can result. The effects of excessive noise depend on the total energy that the ear has received during the work period. Thus, reducing the time of exposure to excessive noise during the work shift reduces the probability of permanent hearing impairment.

Both broadband and meaningful noise have proved to be sufficiently distracting and annoying to result in decreased productivity and increased employee fatigue. However, federal legislation was enacted primarily because of the possibility of permanent hearing damage due to occupational noise exposure. The OSHA (1997) limits for permissible noise exposure are contained in Table 6.6.

When noise levels are determined by *octave-band analysis* (a special filter attachment to the sound level meter that decomposes the noise into component frequencies), the equivalent A-weighted sound level may be determined as follows: Plot the octave-band sound pressure levels on the graph in Figure 6.10, and note the A-weighted sound level corresponding to the point of highest penetration into the sound level contours. This is the dBA value to be used in further calculations.

NOISE DOSE

OSHA uses the concept of *noise dose*, with the exposure to any sound level above 80 dBA causing the listener to incur a partial dose. If the total daily exposure

Table 6.6 Permissible Noise Exposures

Duration per day (h)	Sound level (dBA)
8	90
6	92
4	95
3	97
2	100
1.5	102
1	105
0.5	110
0.25 or less	115

Note: When the daily noise exposure is composed of two or more periods of noise exposure of different levels, their combined effect should be considered rather than the individual effects of each. If the sum of the following fractions $C_1/T_1 + C_2/T_2 + \cdots + C_n/T_n$ exceeds unity, then the mixed exposure should be considered to exceed the limit value. C_n indicates the total time of exposure at a specified noise level, and T_n equals the total time of exposure permitted during the workday.

Exposure to impulsive or impact noise should not exceed 140-dB peak sound pressure level.

consists of several partial exposures to different noise levels, then the several partial doses are added to obtain a combined exposure:

$$D = 100 \times (C_1/T_1 + C_2/T_2 + \cdots + C_n/T_n) \leq 100$$

where D = noise dose
  C = time spent at specified noise level (h)
  T = time permitted at specified noise level (h) (see Table 6.6)

The total exposure to various noise levels cannot exceed a 100-percent dose.

Calculation of OSHA Noise Dose	EXAMPLE 6.2

A worker is exposed to 95 dBA for 3 h and to 90 dBA for 5 h. Although each partial dose is separately permissible, the combined dose is not:

$$D = 100 \times \left(\frac{3}{4} + \frac{5}{8}\right) = 137.5 > 100$$

Thus, 90 dBA is the maximum permissible level for an 8-h day, and any sound level above 90 dBA will require some noise abatement. All sound levels between 80 and 130 dBA must be included in the noise dose computations (although continuous levels above 115 dBA are not allowed at all). Since Table 6.6 provides only certain key times, a computational formula can be used for intermediate noise levels:

$$T = 8/2^{(L - 90)/5}$$

where L = noise level (dBA).

The noise dose can also be converted to an *8-h time-weighted average* (TWA) sound level. This is the sound level that would produce a given noise dose if a worker were exposed to that sound level continuously over an 8-h workday. The TWA is defined by

$$\text{TWA} = 16.61 \times \log_{10}(D/100) + 90$$

Thus, in the last example, a 139.3 percent dose would yield a TWA of

$$\text{TWA} = 16.61 \times \log_{10}(139.3/100) + 90 = 92.39 \text{ dB}$$

Today, OSHA also requires a mandatory hearing conservation program, including exposure monitoring, audiometric testing, and training, for all employees who have occupational noise exposures equal to or exceeding TWA of 85 dB. Although noise levels below 85 dB may not cause hearing loss, they contribute to distraction and annoyance, resulting in poor worker performance. For example, typical office noises, although not loud, can make it difficult to concentrate, resulting in low productivity in design and other creative work. Also, the effectiveness of telephone and face-to-face communications can be considerably distracted by noise levels less than 85 dB.

PERFORMANCE EFFECTS

Generally, performance decrements are most often observed in difficult tasks that place high demands on perceptual, information processing, and short-term memory capacities. Surprisingly, noise may have no effect, or may even improve performance, on simple routine tasks. Without the noise source, the person's attention may wander due to boredom.

EXAMPLE 6.3	Calculation of OSHA Noise Dose with Additional Exposures

A worker is exposed to 1 h at 80 dBA, 4 h at 90 dBA, and 3 h at 96 dBA. The worker is permitted 32 h for the first exposure, 8 h for the second exposure, and

$$T = 8/2^{(96-90)/5} = 3.48$$

hours for the third exposure. The total noise dose becomes

$$D = 100 \times (1/32 + 4/8 + 3/3.48) = 139.3$$

Thus, for this worker, the 8-h noise exposure dosage exceeds OSHA requirements, and either the noise must be abated or the worker must be provided with a rest allowance (see Chapter 11) to comply with OSHA requirements.

Annoyance is even more complicated and is fraught with emotional issues. Acoustic factors, such as intensity, frequency, duration, fluctuations in level, and spectral composition, play a major role, as do nonacoustic factors, such as past noise experience, activity, personality, noise occurrence predictability, time of day and year, and type of locale. There are approximately a dozen different methods for evaluating annoyance aspects (Sanders and McCormick, 1993). However, most of these measures involve community-type issues with noise levels in the 60- to 70-dBA range, much lower than could reasonably be applied in an industrial situation.

NOISE CONTROL

Management can control the noise level in three ways. The best, and usually the most difficult, is to reduce the noise level at its source. However, it would be very difficult to redesign such equipment as pneumatic hammers, steam forging presses, board drop hammers, and woodworking planers and joiners so that the efficiency of the equipment would be maintained while the noise level was being brought into a tolerable range. In some instances, however, more quietly operating facilities may be substituted for those operating at a high noise level. For example, a hydraulic riveter may be substituted for a pneumatic riveter, an electrically operated apparatus for a steam-operated apparatus, and an elastomer-lined tumble barrel for an unlined barrel. Low-frequency noise at the source is effectively controlled at the source by using rubber mounts and better alignment and maintenance of the equipment.

If the noise cannot be controlled at its source, then analysts should investigate the opportunity to isolate the equipment responsible for the noise; that is, control the noise that emanates from a machine by housing all or a substantial portion of the facility in an insulating enclosure. This has frequently been done in connection with power presses having automatic feeds. Ambient noise can frequently be reduced by isolating the noise source from the remainder of the structure, thus preventing a sounding board effect. This can be done by mounting the facility on a shear-type elastomer, thus damping the telegraphing of noise.

In situations where enclosing the facility would not interfere with operation and accessibility, the following steps can ensure the most satisfactory enclosure design:

1. Clearly establish design goals and the acoustical performance required of the enclosure.
2. Measure the octave-band noise levels of the equipment to be enclosed, at 3 ft (1 m) from the major machine surfaces.
3. Determine the spectral attenuation of each enclosure. This is the difference between the design criteria determined in step 1 and the noise level determined in step 2.
4. Select the materials from Table 6.7 that are popular for relatively small enclosures and will provide the protection needed. A viscoelastic damping material should be applied if any of these materials (with the exception of lead) is used. This can provide an additional attenuation of 3 to 5 dB.

Figure 6.11 illustrates the amount of noise reduction typically possible through various acoustical treatments and enclosures.

Table 6.7 Octave-Band Noise Reduction of Single-Layer Materials Commonly Used for Enclosures

Octave-band center frequency	125	250	500	1,000	2,000	4,000
16-gage steel	15	23	31	31	35	41
7-mm steel	25	38	41	45	41	48
7-mm plywood 0.32 kg/0.1 m^2	11	15	20	24	29	30
3/4-in plywood 0.9 kg/0.1 m^2	19	24	27	30	33	35
14-mm gypsum board 1 kg/0.1 m^2	14	20	30	35	38	37
7-mm fiberglass 0.23 kg/0.1 m^2	5	15	23	24	32	33
0.2-mm lead 0.45 kg/0.1 m^2	19	19	24	28	33	38
0.4-mm lead 0.9 kg/0.1 m^2	23	24	29	33	40	43

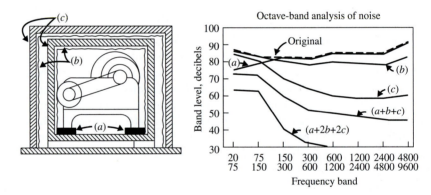

Figure 6.11 Illustrations of the possible effects of some noise control measures. The lines on the graph show the possible reductions in noise (from the original level) that might be expected by vibration insulation, *a;* an enclosure of acoustic absorbing material, *b;* a rigid, sealed enclosure, *c;* a single combined enclosure plus vibration insulation, *a + b + c;* and a double combined enclosure plus vibration insulation, *a + 2b + 2c.* (*Adapted from*: Peterson and Gross, 1978)

Note that some sounds are desirable in a work environment. For example, background music has been used in factories for many years to improve the work environment, especially where voice communications aren't critical. The majority of production and indirect workers (maintenance, shipping, receiving, etc.) enjoy listening to music while they work. However, first consult the employees on the type of music to be played.

The third level of noise control is with hearing protection, though in most cases OSHA accepts this as only a temporary solution. Personal protective equipment can include various types of earplugs, some of which are able to attenuate noises in all frequencies up to sound pressure levels of 110 dB or more. Also available are earmuffs that attenuate noises to 125 dB above 600 Hz and up to 115 dB below this frequency. Earplug effectiveness is measured quantitatively by a *noise reduction rating* (NRR), which is marked on the packaging. The equivalent noise exposure for the listener is equal to the TWA plus 7 minus the NRR (NIOSH, 1998). In general, insert-type (e.g., expandable foam) devices provide better protection than muff-type devices. A combination of an insert device and a muff device can yield NRR values as high as 30. Note that this is a laboratory value obtained under ideal conditions. Typically, in a real-world setting, with hair, beards, eyeglasses, and improper fit, the NRR value is going to be considerably lower, perhaps by as much as 10 (Sanders and McCormick, 1993).

6.3 TEMPERATURE

Most workers are exposed to excessive heat at one time or another. In many situations, artificially hot climates are created by the demands of the particular industry. Miners are subjected to hot working conditions due to the increase of temperature with depth, as well as a lack of ventilation. Textile workers are subjected to the hot, humid conditions needed for weaving cloth. Steel, coke, and aluminum workers are subject to intense radiative loads from open-hearth furnaces and refractory ovens. Such conditions, while present for only a limited part of the day, may exceed the climatic stress found in the most extreme, naturally occurring climates.

THEORY

The human is typically modeled as a cylinder with a shell, corresponding to the skin, surface tissues, and limbs, and with a core, corresponding to the deeper tissues of the trunk and head. Core temperatures exhibit a narrow range around a normal value of 98.6°F (37°C). At values between 100 and 102°F (37.8 and 38.9°C), physiological performance drops sharply. At temperatures above 105°F (40.6°C), the sweating mechanism may fail, resulting in a rapid rise in core temperature and eventual death. The shell tissues of the body, on the other hand, can vary over a much wider range of temperatures without serious loss of efficiency, and can act as a buffer to protect core temperatures. Clothing, if worn, acts as a second shell to insulate the core temperature further.

The heat exchanges between the body and its environment can be represented by the following heat balance equation:

$$S = M \pm C \pm R - E$$

where M = heat gain of metabolism
C = heat gained (or lost) due to convection
R = heat gained (or lost) due to radiation
E = heat lost through evaporation of sweat
S = heat storage (or loss) of body

For thermal neutrality, S must be zero. If the summation of the various heat exchanges across the body results in a heat gain, the resulting heat will be stored in the tissues of the body, with a concomitant increase in core temperature and a potential heat stress problem.

A thermal comfort zone, for areas where 8 h of sedentary or light work is done, has been defined as the range of temperatures of from 66 to 79°F (18.9 to 26.1°C), with a relative humidity ranging from 20 to 80 percent (see Figure 6.12). Of course, the workload, clothing, and radiant heat load all affect the individual's sense of comfort within the comfort zone.

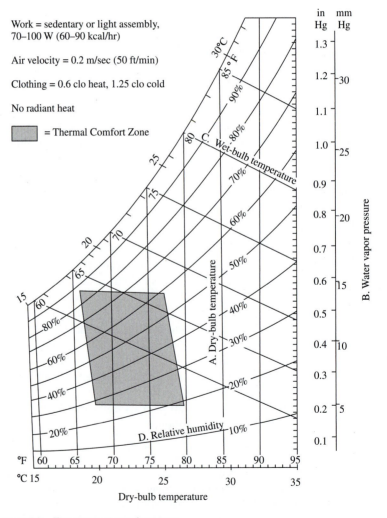

Figure 6.12 The thermal comfort zone.
(*Courtesy:* Eastman Kodak Co.)

HEAT STRESS: WBGT

Many attempts have been made to combine into one index the physiological manifestations of these heat exchanges with environmental measurements. Such attempts have centered on designing instruments intended to simulate the human body, or devising formulas and models based on theoretical or empirical data to estimate the environmental stresses or the resulting physiological strains. In the simplest form, an index consists of the dominant factor, such as the dry-bulb temperature, which is used by most people in temperate zones.

Probably the most commonly used index in industry today establishes heat exposure limits and work/rest cycles based on the *wet-bulb globe temperature*, or WBGT (Yaglou and Minard, 1957), and the metabolic load. In slightly different forms it is recommended by ACGIH (1985), NIOSH (1986), and ASHRAE (1991). For outdoors with a solar load, the WBGT is defined as

$$\text{WBGT} = 0.7\,\text{NWB} + 0.2\,\text{GT} + 0.1\,\text{DB}$$

and indoors or outdoors with no solar load, the WBGT is

$$\text{WBGT} = 0.7\,\text{NWB} + 0.3\,\text{GT}$$

where   NWB = *natural wet-bulb temperature* (measure of evaporative cooling, using a thermometer with a wet wick and natural air movement)

GT = *globe temperature* (measure of radiative load, using a thermometer in a 6-in-diameter black copper sphere)

DB = *Dry-bulb temperature* (basic ambient temperature; thermometer shielded from radiation)

Note that NWB is different from a psychometric wet bulb, which uses maximum air velocity and is used in conjunction with DB to establish relative humidity and thermal comfort zones.

Once the WBGT is measured (commercially available instruments provide instantaneous weighted readings), it is used along with the metabolic load of the workers to establish the amount of time an unacclimatized worker and an acclimatized worker are allowed to work under the given conditions (see Figure 6.13). These limits are based on the individual's core temperature having increased by approximately 1.8°F (1°C) as calculated by the heat balance equation. The 1.8°F increase has been established by NIOSH (1986) as the upper acceptable limit for heat storage in the body. The appropriate amount of rest is assumed to be under the same conditions. Obviously, if the worker rests in a more comfortable area, less rest time will be needed.

CONTROL METHODS

Heat stress can be reduced by implementing either engineering controls, that is, modifying the environment, or administrative controls. Modifying the environment follows directly from the heat balance equation. If the metabolic load is a significant contributor to heat storage, the workload should be reduced by mechanization of the operation. Working more slowly will also decrease the workload, but will

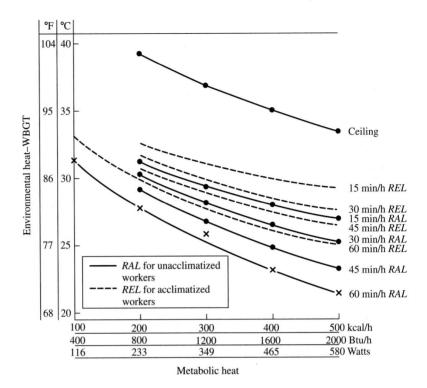

Figure 6.13 Recommended heat stress levels based on metabolic heat (1-h time-weighted average), acclimatization, and workrest cycle.
A rough approximation for metabolic heat from Figure 4.21 is $W = (HR - 50) \times 6$. Temperature limits are 1-h time-weighted average WGBT. RAL = recommended alert limit for unacclimatized workers. REL = recommended exposure limit for acclimatized workers. (*From:* NIOSH, 1986, Figs. 1 and 2)

have the negative effect of decreasing productivity. The radiative load can be decreased by controlling the heat at the source: insulating hot equipment, providing drains for hot water, maintaining tight joints where steam may escape, and using local exhaust ventilation to disperse heated air rising from a hot process. Radiation can also be intercepted before it reaches the operator, via radiation shielding: sheets of reflective material, such as aluminum or foil-covered plasterboard, or metal chain curtains, wire mesh screens, or tempered glass, if visibility is required. Reflective garments, protective clothing, or even long-sleeved clothing will also help in reducing the radiative load.

EXAMPLE 6.4   Calculation of WBGT and Heat Stress Level

Consider an unacclimatized worker palletizing a skid at 400 kcal/h (1,600 BU/h) with a thermal load of WBGT = 77°F (25°C). This individual would be able to work for 45 min and would then need to rest. At this point, the worker must rest for at least 15 min in the same environment or a shorter time in a less stressful environment.

Convective heat loss from the operator can be increased by increasing air movement through ventilation, as long as the dry-bulb temperature is less than skin temperature, which is typically around 95°F (35°C) in such environments. Convection is more effective over bare skin; however, bare skin also absorbs more radiation. Thus, there is a trade-off between convection and radiation. Evaporative heat loss from the operator can be improved by again increasing air movement and decreasing the ambient water vapor pressure, using dehumidifiers or air-conditioning. Unfortunately, the latter approach, although creating a very pleasant environment, is quite costly and is often not practical for the typical production facility.

Administrative measures, though less effective, include modifying work schedules to decrease the metabolic load, using work/rest schedules per Figure 6.13, acclimatizing workers (this may take close to two weeks, and the effect is lost over a similar time period), rotating workers into and out of the hot environment, and using cooling vests. The cheapest vests utilize ice frozen in small, plastic packets placed into numerous pockets in the vest (Kamon et al., 1986).

COLD STRESS

The most commonly used cold stress index is the *wind chill index*. It describes the rate of heat loss by radiation and convection as a function of ambient temperature and wind velocity. Typically, the wind chill index is not used directly, but is converted to an *equivalent wind chill temperature*. This is the ambient temperature that, in calm conditions, would produce the same wind chill index as the actual combination of air temperature and wind velocity (Table 6.8). For the operator to maintain thermal balance under such low-temperature conditions, there must be a close relationship between the worker's physical activity (heat production) and the insulation provided by protective clothing (see Figure 6.14). Here, *clo* represents the insulation needed to maintain comfort for a person sitting where the relative humidity is 50 percent, the air movement is 20 ft/min, and the dry-bulb temperature is 70°F (21.1°C). A light business suit is approximately equivalent to 1 clo of insulation.

Table 6.8 Equivalent Wind Chill Temperatures (°F) of Cold Environments Under Calm Conditions

Wind speed (mi/h)	Actual thermometer reading (°F)							
	40	30	20	10	0	−10	−20	−30
5	36	25	13	1	−11	−22	−34	−46
10	34	21	9	−4	−16	−28	−41	−53
15	32	19	6	−7	−19	−32	−45	−58
20	30	17	4	−9	−22	−35	−48	−61
30	28	15	1	−12	−26	−39	−53	−67
40	27	13	−1	−15	−29	−43	−57	−71

Little Danger:	**Increasing Danger:**	**Great Danger:**
Exposed, dry flesh won't freeze for 5 h.	Frostbite occurs within 30 mins.	Frostbite occurs within 5 min.

Source: National Weather Service

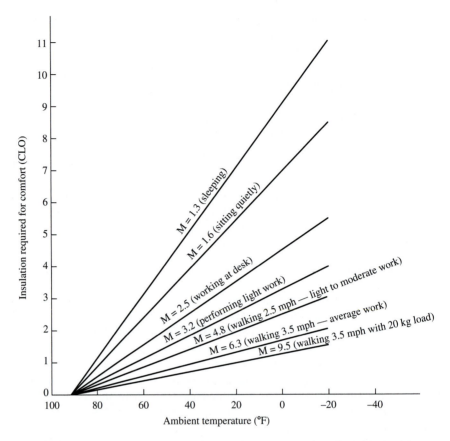

Figure 6.14 Prediction of the total insulation required as a function of the ambient temperature for a 50th percentile male (*M* = heat production in kcal/min). (*From*: Redrawn from Belding and Hatch, 1955.)

Probably the most critical effects for industrial workers exposed to outdoor conditions are decreased tactile sensitivity and manual dexterity due to vasodilation and decreased blood flow to the hands. Manual performance may decrease as much as 50 percent as the hand skin temperature drops from 65 to 45°F (18.3 to 7.2°C) (Lockhart, Kiess, and Clegg, 1975). Auxiliary heaters, hand warmers, and gloves are potential solutions to the problem. Unfortunately, as indicated in Chapter 5, gloves can impair manual performance and decrease grip strength. A compromise that protects the hands and minimally affects performance may be fingerless gloves (Riley and Cochran, 1984).

6.4 VENTILATION

If a room has people, machinery, or activities in it, the air in the room will deteriorate due to the release of odors, the release of heat, the formation of water

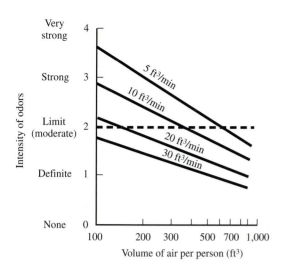

Figure 6.15 Guidelines for ventilation requirements for sedentary workers given the available volume of room air. (Flow rates are per person.) (*Adapted from:* Yaglou, Riley, and Coggins, 1936)

vapor, the production of carbon dioxide, and the production of toxic vapors. Ventilation must be provided to dilute these contaminants, exhaust the stale air, and supply fresh air. This can be done in one or more of three approaches: general, local, or spot. General or displacement ventilation is delivered at the 8- to 12-ft (2.4- to 3.6-m) level and displaces the warm air rising from the equipment, lights, and workers. Recommended guidelines for fresh air requirements, based on the room volume per person, are shown in Figure 6.15 (Yaglou, Riley, and Coggins, 1936). A rough rule of thumb is 300 ft^3 (8.5 m^3) of fresh air per person per hour.

In a building with only a few work areas, it would be impractical to ventilate the whole building. In that case, local ventilation can be provided at a lower level, or perhaps in an enclosed area, such as a ventilated control booth or crane cab. Note that fan velocity drops rapidly with increasing distance from the fan (see Figure 6.16), and directionality of airflow is very critical. Acceptable air velocities at the worker are specified in Table 6.9 (ASHRAE, 1991). A rough rule of thumb is that at a distance of 30 fan diameters, the fan velocity drops to less than 10 percent of its face velocity (Konz, 1995). Finally, in areas with localized heat sources, such as refractory ovens, spot cooling with a direct high-velocity airstream at the worker will increase convective and evaporative cooling.

6.5 VIBRATION

Vibration can cause detrimental effects on human performance. Vibrations of high amplitude and low frequency have especially undesirable effects on body organs and tissue. The parameters of vibration are frequency, amplitude, velocity,

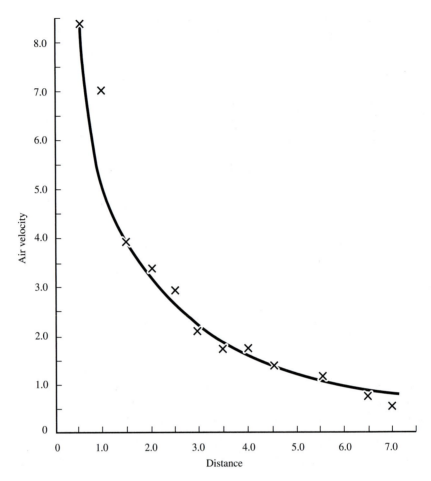

Figure 6.16　Air velocity versus distance for fan placement.
(*From:* Konz and Johnson, 2000)

Table 6.9　Acceptable Air Motion at the Worker

Exposure	Air Velocity (ft/min)
Continuous	
Air-conditioned space	50 to 75
Fixed workstation, general ventilation, or spot cooling	
Sitting	75 to 125
Standing	100 to 200
Intermittent, spot cooling, or relief stations	
Light heat loads and activity	1,000 to 2,000
Moderate heat loads and activity	2,000 to 3,000
High heat loads and activity	3,000 to 4,000

Source: Reprinted with permission from ASHRAE, 1991.

Table 6.10   Resonant Frequencies for Different Body Parts

Frequency (Hz)	Body part affected
3–4	Cervical vertebrae
4	Lumbar vertebrae (key for forklift and truck operators)
5	Shoulder girdle
20–30	Between head and shoulder
>30	Fingers, hands, and arms (key for power tool operators)
60–90	Eyeballs (key for pilots and astronauts)

acceleration, and jerk. For sinusoidal vibrations, amplitude and its derivations with respect to time are

$$\text{Amplitude } s = \text{maximum displacement from static position (in)}$$

$$\text{Maximum velocity } \frac{ds}{dt} = 2\pi(s)(f) \qquad \text{in/s}$$

$$\text{Maximum acceleration } \frac{d^2s}{dt^2} = 4\pi^2(s)(f^2) \qquad \text{in/s}^2$$

$$\text{Maximum jerk } \frac{d^3s}{dt^3} = 8\pi^3(s)(f^3) \qquad \text{in/s}^3$$

where f = frequency
 s = displacement amplitude

Displacement and maximum acceleration are the principal parameters used to characterize the intensity of vibration.

There are three classifications of vibration exposure:

1. Circumstances in which the whole or a major portion of the body surface is affected, for example, when high-intensity sound in air or water excites vibration.

2. Cases in which vibrations are transmitted to the body through a supporting area, for example, through the buttocks of a person driving a truck, or through the feet of a person standing by a shakeout facility in a foundry.

3. Instances in which vibrations are applied to a localized body area, for example, to the hand when holding and operating a power tool.

Every mechanical system can be modeled using a mass, spring, and dashpot, which, in combination, result in the system having its own *natural frequency*. The nearer the vibration comes to this frequency, the greater the effect on that system. In fact, if the forced vibrations induce larger-amplitude vibrations in the system, then the system is in *resonance*. This can have dramatic effects, for example, large winds causing the Tacoma Narrows bridge in Washington to oscillate and eventually collapse, or soldiers breaking step in crossing bridges. For a sitting person, the critical resonant frequencies are given in Table 6.10.

On the other hand, oscillations in the body, or any system, tend to be dampened. Thus, in a standing posture, the muscles of the legs heavily dampen vibrations.

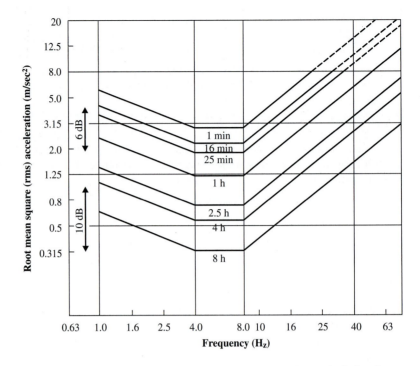

Figure 6.17 The fatigue-decreased proficiency boundary for vertical vibration contained in ISO 2631 and ANSI S3.18-1979.
To obtain the boundary for reduced comfort, subtract 10 dB (i.e., divide each value by 3.15); to obtain the boundary for safe physiological exposure, add 6 dB (i.e., multiply each value by 2.0). (*Source:* Acoustical Society of America, 1980)

Frequencies above 35 Hz are especially dampened. Amplitudes of oscillations induced in the fingers will reduce 50 percent in the hands, 66 percent in the elbows, and 90 percent in the shoulders.

The human tolerance for vibration decreases as the exposure time increases. Thus, the tolerable acceleration level increases with decreasing exposure time. The limits for whole-body vibration have been developed by both the International Standards Organization (ISO) and the American National Standards Institute (ANSI) (ASA, 1980) for transportation and industrial applications. The standards specify limits in terms of acceleration, frequency, and time duration (Figure 6.17). The plotted lines show fatigue/performance limits. For comfort limits, the acceleration values are divided by 3.15; for safety limits, the values are multiplied by 2. Unfortunately, no limits have been developed for the hands and upper extremities.

Low-frequency (0.2- to 0.7-Hz), high-amplitude vibrations are the principal cause of motion sickness in sea and air travel. Workers experience fatigue much more rapidly when they are exposed to vibrations in the range of 1 to 250 Hz. Early symptoms of vibration fatigue are headache, vision problems, loss of appetite, and loss of interest. Later problems include motor control impairments, disk degeneration, bone atrophy, and arthritis. Vibrations experienced in this range are often characteristic of the trucking industry. The vertical vibrations of many rubber-tired trucks when traveling at typical speeds over ordinary roads range from 3 to about 7 Hz, which are exactly in the critical range for resonances in the human trunk.

Power tools with frequencies of 40 to 300 Hz tend to occlude blood flow and affect nerves, resulting in the *white fingers syndrome*. The problem is exacerbated in cold conditions, with the additional problem of cold-induced occlusion of blood flow, or *Raynaud's syndrome*. Better dampened tools, the exchange of detachable handles with special vibration-absorbing handles, and the wearing of gloves, especially those padded with vibration-absorbing gel, will help reduce the problem.

Management can protect employees against vibration in several ways. The applied forces responsible for initiating the vibration may be reduced by modifying the speed, feed, or motion, and by properly maintaining the equipment, balancing and/or replacing worn parts. Analysts can place equipment on antivibration mountings (springs, shear-type elastomers, compression pads) or alter workers' body positions to lessen the disturbing vibratory forces. They can also reduce the time workers are exposed to the vibration by alternating work assignments within a group of employees. Last, they can introduce supports that cushion the body and thus dampen higher-amplitude vibrations. Seat suspension systems involving hydraulic shock absorbers, coil or leaf springs, rubber shear-type mountings, or torsion bars may be used. In standing operations, a soft, elastomer floor mat usually proves helpful.

6.6 RADIATION

Although all types of ionizing radiation can damage tissue, beta and alpha radiation are so easy to shield that most attention today is given to gamma ray, X-ray, and neutron radiation. High-energy electron beams impinging on metal in vacuum equipment can produce very penetrating X-rays that may require much more shielding than the electron beam itself.

The absorbed dose is the amount of energy imparted by ionizing radiation to a given mass of material. The unit of absorbed dose is the *rad*, which is equivalent to the absorption of 0.01 joule per kilogram (J/kg) [100 ergs per gram (erg/g)]. The dose equivalent is a way of correcting for the differences in the biological effect of various types of ionizing radiation on humans. The unit of dose equivalent is the *rem*, which produces a biological effect essentially the same as that of 1 rad of absorbed dose of X or gamma radiation. The *roentgen* (R) is a unit of exposure that measures the amount of ionization produced in air by X or gamma radiation. Tissue located at a point where the exposure is 1 R receives an absorbed dose of approximately 1 rad.

Very large doses of ionizing radiation—100 rads or more—received over a short time span by the entire body can cause radiation sickness. An absorbed dose of about 400 rads to the whole body would be fatal to approximately one-half of adults. Small doses received over a longer period may increase the probability of contracting various types of cancers or other diseases. The overall risk of a fatal cancer from a radiation dose equivalent of 1 rem is about 10^{-4}; that is, a person receiving a dose equivalent of 1 rem has about 1 chance in 10,000 of dying from a cancer produced by the radiation. The risk can also be expressed as the expectation of one fatal cancer in a group of 10,000 persons, if each person receives a dose equivalent of 1 rem.

Persons working in areas where access is controlled for the purpose of radiation protection are generally limited to a dose equivalent of 5 rem/yr. The limit in uncontrolled areas is usually the same. Working within these limits should have no significant effect on the health of the individuals involved. All persons are exposed to radiation from naturally occurring radioisotopes in the body, cosmic radiation, and radiation emitted from the earth and building materials. The dose equivalent from natural background sources is about 0.1 rem/yr (100 mrem/yr).

6.7   SHIFTWORK AND WORKING HOURS

SHIFTWORK

Shiftwork, defined as working other than daytime hours, is becoming an ever-increasing problem for industry. Traditionally, the need for continuous services from police, fire, and medical personnel, or for continuous operations in the chemical or pharmaceutical industries, has required the use of shiftwork. More recently, however, the economics of manufacturing, that is, the capitalization or payback of ever more expensive automated machinery, are also increasing the demand for shiftwork. Similarly, just-in-time production and seasonal demands for products (i.e., decreased inventory space) have also required more shiftwork.

The problem with shiftwork is the stress on *circadian rhythms*, which are the roughly 24-h variations in bodily functions in humans (as well as other organisms). The length of the cycle varies from 22 to 25 h, but is kept synchronized into a 24-h cycle by various timekeepers, such as the daily light–dark changes, social contacts, work, and clock time. The most marked cyclic changes occur in sleep, core temperature, heart rate, blood pressure, and task performance, such as critical tracking capability (see Figure 6.18). Typically, bodily functions and performance start increasing upon awakening, peak in midafternoon, then steadily decline to a low point in the middle of the night. There may also be a dip after midday, typically known as the *postlunch dip*. Thus, individuals who are asked to work on night shift will exhibit a marked degradation in performance, from truck drivers falling asleep at the wheel to gas inspectors reading meters (Grandjean, 1988).

It could be assumed that night workers would adapt to night work because of the change in work patterns. Unfortunately, the other social interactions still play a very important role, and the circadian rhythm never truly shifts (as it would for individuals traveling to the other side of the globe for extensive periods) but rather flattens, which some researchers consider to be a worse scenario. Thus, night workers

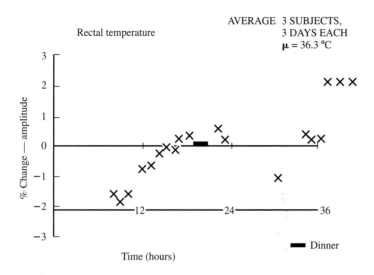

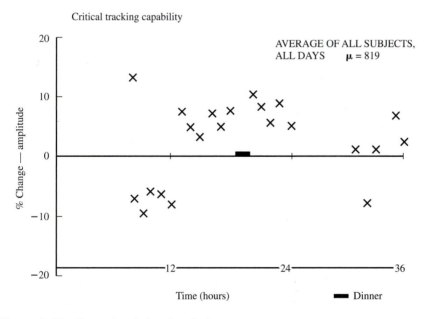

Figure 6.18 Examples of circadian rhythms.
(*From:* Freivalds, Chaffin, and Langolf, 1983)

also experience health problems, such as appetite loss, digestive problems, ulcers, and increased sickness rates. The problems become even worse as the worker ages.

There are many ways to organize shiftwork. Typically, a three-shift system has an early (E) shift from 8 A.M. to 4 P.M., a late afternoon shift (L) from 4 P.M.

to 12 P.M. (midnight), and a night (N) shift from 12 P.M. to 8 A.M.. In the simplest case, because of short-term increased production demands, a company may go from just an early shift to both an early and a late afternoon shift. Usually, because of seniority, the early shift is requested by older, established workers, while new hires start on the late afternoon shift. The rotation of the two shifts on a weekly basis does not cause major physiological problems, since the sleep pattern is not disrupted. However, the social patterns can be considerably disrupted.

Progressing to a third, night shift becomes more problematic. Since there is difficulty adjusting to a new circadian rhythm, even over the course of several weeks, most researchers advocate a *rapid rotation*, with shift changes every two or three days. This maintains the quality of sleep as well as possible and does not disrupt family life and social contacts for extended periods. The weekly rotation typically found in the United States is perhaps the worst scenario, because the workers never truly adjust to any one shift.

One rapid-rotation shiftwork system for a 5-day production system (i.e., weekends off) is given in Table 6.11. However, in many companies, the night shift is mainly a maintenance shift with limited production. In that case, a full crew is not needed, and it may be simpler to rotate only the early and late afternoon shifts and to operate a smaller, fixed night shift, which can be staffed primarily by volunteers who can better adapt to that shift.

For continuous round-the-clock operations, a rapid-rotation seven-day shift system is needed. Two plans commonly used in Europe are the 2-2-2 system, with no more than two days on any one shift (see Table 6.12), and the 2-2-3 system, with no more than three days on any one shift (see Table 6.13). There are trade-offs with each of these systems. The 2-2-2 system provides a free weekend

Table 6.11   Eight-Hour Shift Rotation (Weekends Off)

Week	M	T	W	Th	F	S	Su
1	E	E	L	L	L	—	—
2	N	N	E	E	E	—	—
3	L	L	N	N	N	—	—

Note: E = early, L = late afternoon, N = night.

Table 6.12   The 2-2-2 Shift Rotation (8-h Continuous)

Week	M	T	W	Th	F	S	Su
1	E	E	L	L	N	N	—
2	—	E	E	L	L	N	N
3	—	—	E	E	L	L	N
4	N	—	—	E	E	L	L
5	N	N	—	—	E	E	L
6	L	N	N	—	—	E	E
7	L	L	N	N	—	—	E
8	E	L	L	N	N	—	—

Note: E = early, L = late afternoon, N = night.

Table 6.13 The 2-2-3 Shift Rotation (8-h Continuous)

Week	M	T	W	Th	F	S	Su
1	E	E	L	L	N	N	N
2	—	—	E	E	L	L	L
3	N	N	—	—	E	E	E
4	L	L	N	N	—	—	—

Note: E = early, L = late afternoon, N = night.

Table 6.14 A 12-h Shift Rotation (3 Days On, 3 Days Off)

Week	M	T	W	Th	F	S	Su
1	D	D	D	—	—	—	N
2	N	N	—	—	—	D	D
3	D	—	—	—	N	N	N
4	—	—	—	D	D	D	—
5	—	—	N	N	N	—	—
6	—	D	D	D	—	—	—

Note: D = day, N = night.

Table 6.15 A 12-h Shift Rotation (Every Other Weekend Off)

Week	M	T	W	Th	F	S	Su
1	D	—	—	N	N	—	—
2	—	D	D	—	—	N	N
3	N	—	—	D	D	—	—
4	—	N	N	—	—	D	D

Note: D = day, N = night.

only once in eight weeks. The 2-2-3 system provides a free three-day weekend once in four weeks, but requires workers to work seven days straight, which is not appealing. A basic problem in both systems is that with 8-h shifts, a total of 42 h/week is worked. Alternative systems with more crews and shorter hours may be required (Eastman Kodak, 1986).

Another possible approach is to schedule 12-h shifts. Under these systems, workers either work 12-h day (D) shifts or 12-h night (N) shifts, on either a regular three days on and three days off schedule (see Table 6.14), or a more complicated two or three days on or off, with every other weekend free (see Table 6.15). There are several advantages in that there are longer rest periods between workdays, and at least one-half of the rest days coincide with a weekend. Of course, the obvious disadvantage is having to work extended days or essentially regular overtime (see next section).

More complicated systems exist with reduced hours (40 or less) per week. These can be studied in further detail in Eastman Kodak (1986) or Schwarzenau et al. (1986).

In summary, definite health and accident risks are associated with shiftwork. However, if shiftwork is unavoidable, due to manufacturing process considerations, the following recommendations should be considered:

1. Avoid shiftwork for workers older than 50.
2. Use rapid rotations as opposed to weekly or monthly cycles.
3. Schedule as few night shifts (three or less) in succession as possible.
4. Use forward rotation of shifts if possible (e.g., E-L-N or D-N).
5. Limit the total number of working shifts in succession to seven or less.
6. Include some free weekends, with at least two successive full days off.
7. Schedule rest days after night shifts.
8. Keep the plans simple, predictable, and equitable for all workers.

OVERTIME

Many studies have shown that changes in the length of the workday or workweek have a direct effect on work output. Unfortunately, the result is not typically the direct proportionality expected. Note that in Figure 6.19, the theoretical daily performance is linear (line 1), but, in practice, is more S-shaped (curve 2). For example, there is an initial setup or preparatory period with little productivity (area A), a gradual warming up, a steeper section with greater than the theoretical productivity (area B), and graduate leveling off as the end of the shift approaches. In an 8-h shift, the two areas, subpar (area A) and excess productivity (area B), are equal, whereas in longer than 8-h shifts for heavy manual work (curve 3), the subpar productivity is greater than the excess productivity,

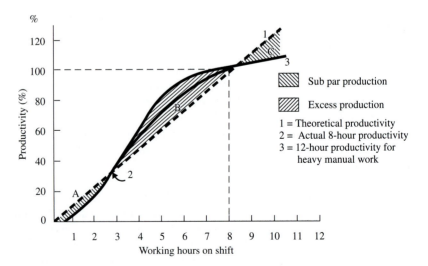

Figure 6.19 Productivity as a function of working hours.
(*Adapted from:* Lehmann, 1953)

especially with additional subpar performance (area C) in the last few hours (Lehmann, 1953).

The results of an old British survey (cited in Grandjean, 1988) found that *shortening* the working day resulted in a *higher* hourly output, with fewer rest pauses taken. This change in the working performance required at least several days (sometimes longer) before steady state was reached. Conversely, making the working day longer, that is, assigning overtime, causes productivity to fall, sometimes to the point that the total output over the course of the shift actually drops, even though the total hours worked are longer (see curve 3 in Figure 6.19). Therefore, any expected benefit from increased hours is typically offset by decreased productivity. This effect depends on the level of the physical workload: the more strenuous the work, the greater the decrease in productivity, with the worker using more rest to pace himself or herself.

More recent data (cited in Eastman Kodak, 1986) indicate that the expected increase in output is approximately 10 percent for each 25 percent increase in hours worked. This definitely does not justify the time-and-a-half pay expended for overtime work. This discussion presumes a day work pay scale (see Chapter 17). With an incentive scheme throughout the extended hours, the fall in productivity may not be so great. Similarly, if the work is machine-paced, productivity is tied to that machine-pace. However, the operator may reach unacceptable fatigue levels, and additional rest per appropriate allowances (see Chapter 11) may be needed. A secondary effect of overtime is that excessive or continuous overtime is accompanied by increased accident rates and sick leave (Grandjean, 1988).

Scheduling overtime on a regular basis cannot be recommended. However, overtime may be necessary for transient short periods, to maintain production or alleviate temporary labor shortages. In such cases, the following guidelines should be followed:

1. Avoid overtime for heavy manual work.
2. Reevaluate machine-paced work for appropriate rest periods or lowered rates.
3. For continuous or long periods of overtime, rotate the work among several workers, or examine alternate shift systems.
4. In choosing between extending a series of workdays by 1 or 2 h versus extending the workweek by 1 day, most workers will opt for the former, to avoid losing a weekend day with the family (Eastman Kodak, 1986).

COMPRESSED WORKWEEK

A compressed workweek implies that 40 h is performed in fewer than 5 days. Typically, this occurs in the form of four 10-h days, three 12-h days, or four 9-h days with a half-day on Friday. From the management perspective, this concept offers several advantages: reduced absenteeism, relatively less time spent on coffee or lunch breaks, and reduced start-up and shutdown costs (relative to operating time). For example, heat treating, forging, and melting facilities require a significant amount of time, up to 15 percent of the 8-h workday or more, to bring the facility and material up to the required temperature before production can begin. By going to a 10-h day, the operation can gain an additional 2 h of production time with no additional setup time. Here, the economic savings from the longer workday can be significant. Workers also gain from increased leisure time, reduced commuter time (relative to working time), and lower commuting costs.

On the other hand, based on the discussions on overtime, a compressed workweek essentially amounts to continuous overtime. Although the total hours worked are less, the hours worked on a given day are more. Therefore, many of the same disadvantages of overtime would apply to a compressed workweek (Eastman Kodak, 1986). Other objections to the 10-h day, 4-day week stem from members of management who state that they are obliged to be on the job not only 10 h for 4 days, but at least 8 h on the fifth day.

ALTERNATIVE WORK SCHEDULES

With the greater influx of women, especially mothers with school-age children, single parents, older workers, and dual-career-family workers into the workforce, and with the increased concerns for the cost and time of commuting and the value of quality of life, alternate work schedules are needed. One such schedule is *flextime*, where the starting and stopping times are established by the workers, within limits set up by management. Various plans of this nature currently exist. Some require employees to work at least 8 h/day, others require a specified number of hours in a week or a month, while still others require all workers to be on site four or five middle hours of the shift.

There are many advantages to these plans for both employees and management. Employees can work the morning or evening hours most conducive to their circadian rhythms, they can better handle family needs or emergencies, and they can take care of personal business during business hours without requiring special leave from work. Management gains from reduced tardiness and sick leave. Even the surrounding community gains from decreased traffic congestion and better use of recreational and service facilities. On the other hand, flextime may have limited use in manufacturing, machine-paced, and continuous-process operations, because of problems in scheduling and coordinating the labor force. However, in situations where work groups (see Chapter 18) are utilized, flextime may still be possible (Eastman Kodak, 1986).

Part-time employment and job sharing may be especially useful to single parents with children or to retirees seeking to supplement their retirement incomes. Both groups can provide considerable talent and services to a company, but may be limited by circumstances from performing traditional 8-h shifts. While there may be problems regarding benefits or other fixed employee costs, these may be handled on a prorated or other creative basis.

SUMMARY

A proper working environment is important not only from the standpoint of increasing productivity and improving the physical health and safety of the workers, but also for promoting worker morale and consequent reductions in worker absenteeism and labor turnover. Although many of these factors may seem intangible or of marginal effect, controlled scientific studies have shown the positive benefits of improved illumination, decreased noise and heat stress, and better ventilation.

Visibility is directly dependent on the illumination provided, but is also affected by the visual angle of the target viewed and the contrast of the target with the background. Consequently, improvement in task visibility can be accomplished through various means and does not always depend on increasing the light source.

Extended exposures to loud noise, although not directly affecting productivity, can cause hearing loss and are definitely annoying. The control of noise (and vibration) is simplest at the source and typically becomes more costly farther away. Although using hearing protection may seem the simplest approach, it requires the expense of continuous motivation and enforcement.

Similarly, the effect of climate on productivity is quite variable, depending on individual motivation. A comfortable climate is a function of the amount and velocity of air exchange, the temperature, and the humidity. For hot areas, the climate is controlled most easily through adequate ventilation to remove pollutants and improve the evaporation of sweat. (Air-conditioning is more effective, but is also more expensive.) For cold climates, adequate clothing is the primary control. Shiftwork should utilize short, rapid, forward-rotating schedules with limited overtime.

To assist the methods analyst in utilizing the various factors discussed in this chapter, they have been summarized in the Work Environment Checklist shown in Figure 6.20.

QUESTIONS

1. What factors affect the quantity of light needed to perform a task satisfactorily?
2. Explain the color rendering effect of low-pressure sodium lamps.
3. What is the relationship between contrast and visibility?
4. What footcandle intensity would you recommend 30 in above the floor in the company washroom?
5. Explain how sales may be influenced by colors.
6. What color has the highest visibility?
7. How is sound energy dissipated in viscoelastic materials?
8. A frequency of 2,000 Hz would have approximately what wavelength in meters?
9. What would be the approximate decibel value of a grinder being used to grind a high-carbon steel?
10. Distinguish between broadband noise and meaningful noise.
11. Would you advocate background music at the workstation? What results would you anticipate?
12. According to the present OSHA law, how many continuous hours per day of a 100-dBA sound level would be permissible?

Illumination	**Yes**	**No**
1. Is the illumination sufficient for the task, per IESNA recommendations?	❏	❏
a. To increase illumination, are more luminaires provided, rather than increasing the wattage of existing ones?	❏	❏
2. Is there general lighting as well as supplementary lighting?	❏	❏
3. Are the workplace and lighting arranged so as to avoid glare?	❏	❏
a. Are direct luminaires placed away from the field of vision?	❏	❏
b. Do the luminaires have baffles or diffusers?	❏	❏
c. Are work surfaces laid out perpendicular to the luminaires?	❏	❏
d. Are surfaces matted or nonglossy?	❏	❏
4. If necessary, are screen filters available for computer monitors?	❏	❏

Thermal conditions—Heat	**Yes**	**No**
1. Is the worker within the thermal comfort zone?	❏	❏
a. If not within the thermal comfort zone, has the WBGT of the working environment been measured?	❏	❏
2. Are the thermal conditions within ASHRAE guidelines?	❏	❏
a. If not within guidelines, is sufficient recovery time provided?	❏	❏
3. Are procedures in place for control of potential heat stress conditions?	❏	❏
a. Is the escape of heat controlled at the source?	❏	❏
b. Are radiation shields in place?	❏	❏
c. Is ventilation provided?	❏	❏
d. Is the air dehumidified?	❏	❏
e. Is air-conditioning provided?	❏	❏

Thermal conditions—Cold	**Yes**	**No**
1. Is the worker adequately clothed for the equivalent wind chill temperature?	❏	❏
2. Are auxiliary heaters provided?	❏	❏
3. Are gloves provided?	❏	❏

Ventilation	**Yes**	**No**
1. Are ventilation levels acceptable per guidelines?	❏	❏
a. Is a minimum of 300 ft³/h/per person provided?	❏	❏
2. If necessary, are local fans provided for workers?	❏	❏
a. Are these tans within a distance of 30 times the fan diameter?	❏	❏
3. For local heat sources, is spot cooling provided?	❏	❏

Noise Levels	**Yes**	**No**
1. Are noise levels below 90 dBA?	❏	❏
a. If the noise levels exceed 90 dBA, is there sufficient rest such that the 8-h dose is less than 100%?	❏	❏
2. Are noise control measures in place?	❏	❏
a. Is the noise controlled at the source with better maintenance, mufflers, and rubber mounts?	❏	❏
b. Is the noise source isolated?	❏	❏
c. Are acoustical treatments being utilized?	❏	❏
d. As a last resort, are earplugs (or earmuffs) being used properly?	❏	❏

Vibration	**Yes**	**No**
1. Are vibration levels within acceptable ANSI standards?	❏	❏
2. If there is vibration, can the vibration-causing sources be eliminated?	❏	❏
3. Have specially dampened seats been installed on vehicles?	❏	❏
4. Have vibration-absorbing handles been attached to power tools?	❏	❏
5. Have resilient, fatigue-resistant mats been supplied to standing operators?	❏	❏

Figure 6.20 Work Environment Checklist.

13. What three classifications have been identified from the standpoint of exposure to vibration?
14. In what ways can workers be protected from vibration?
15. What is meant by the *environmental temperature*?
16. Explain what is meant by the *thermal comfort zone*.
17. What is the maximum rise in body temperature that analysts should allow?
18. How would you go about estimating the maximum length of time that a worker should be exposed to a particular heat environment?
19. What is WBGT?
20. What is the WBGT with a dry-bulb temperature of 80°F, a wet-bulb temperature of 70°F, and a globe temperature of 100°F?
21. Which type of radiation is given the most attention by the safety engineer?
22. What is meant by *absorbed dose of radiation*? What is the unit of absorbed dose?
23. What is meant by *rem*?
24. What steps would you take to increase the amount of light in the following assembly department by about 15 percent? The department currently uses fluorescent fixtures, and the walls and ceiling are painted a medium green. The assembly benches are a dark brown.
25. What color combination would you use to attract attention to a new product being displayed?
26. When would you advocate that the company purchase aluminized clothing?
27. Are possible health hazards associated with electron beam machining? With laser beam machining? Explain.
28. Explain the impact of noise levels below 85 dBA on office work.
29. What environmental factors affect heat stress? How can each be measured?
30. How would you determine if a job places an excessive heat load on the worker?

PROBLEMS

1. A work area has a reflectivity of 60 percent, based on the color combinations of the workstations and the immediate environment. The seeing task of the assembly work could be classified as difficult. What would be your recommended illumination?
2. What is the combined noise level of two sounds of 86 and 96 dB?
3. In the Dorben Company, an industrial engineer designed a workstation where the seeing task was difficult because of the size of the components going into the assembly. The desired brightness was 100 fL, and the workstation was painted a medium green with a reflectance of 50 percent. What illumination in footcandles would be required at this workstation to provide the desired brightness? Estimate the required illumination if you repainted the workstation with a light cream paint.
4. In the Dorben Company, an industrial engineer (IE) was assigned to alter the work methods in the press department to meet OSHA standards relative to permissible noise exposures. The IE found a time-weighted average sound level of 100 dBA. The 20 operators in this department wore earplugs provided by Dorben with an NRR value of 20 dB. What improvement resulted? Do you feel that this department is now in compliance with the law? Explain.

5. In the Dorben Company, an all-day study revealed the following noise sources: 0.5 h, 100 dBA; 1 h, less than 80 dBA; 3.5 h, 90 dBA; 3 h, 92 dBA. Is this company in compliance? What is the dose exposure? What is the TWA noise level?

6. In Problem 5, consider that the last exposure is in the press room, which currently has five presses operating. Assuming that Dorben Company can eliminate some of the presses and transfer production to the remaining presses, how many presses should Dorben eliminate so as not to exceed 100 percent dose exposure for the workers?

7. What is the illumination on a surface 6 in from a 2-cd source?

8. What is the luminance of a surface having a 50 percent reflectance and 4-fc illumination?

9. What is the contrast created by black text (reflectance = 10 percent) on white paper (reflectance = 90 percent)?

10. How much louder is an 80-dB noise than a 60-dB noise?

11. What is the increase in decibels of a noise that doubles in intensity?

12. A supervisor is sitting at her desk illuminated by a 180-cd source 3 ft above it. She is writing with green ink (reflectance = 30 percent) on a yellow notepad (reflectance = 60 percent). What is the illumination on the notepad? Is that sufficient? If not, what amount of illumination is needed? What is the contrast of the writing task? What is the luminance of the notepad?

13. How much ventilation would be recommended for a classroom of area 1,000 ft^2 with 12-ft ceilings? Assume that the class size may reach 40 students.

REFERENCES

ACGIH. *Threshold Limit Values for Chemical Substances and Physical Agents in the Work Environment*, Cincinnati, OH: American Conference of Government Industrial Hygienists, 1985.

ASA. *American National Standard: Guide for the Evaluation of Human Exposure to Whole Body Vibration (ANSI S3.18-1979)*. New York: Acoustical Society of America, 1980.

ASHRAE. *Handbook, Fundamentals*. Chapter 8. Atlanta, GA: American Society of Heating, Refrigeration and Air Conditioning Engineers, 1993.

ASHRAE. *Handbook, Heating, Ventilation and Air Conditioning Applications*. Chapter 25. Atlanta, GA: American Society of Heating, Refrigeration and Air Conditioning Engineers, 1991.

Belding, H. S., and T. F. Hatch. "Index for Evaluating Heat Stress in Terms of Physiological Strains." *Heating, Piping and Air Conditioning*, 27(August 1955), pp. 129–136.

Blackwell, H. R. "Development and Use of a Quantitative Method for Specification of Interior Illumination Levels on the Basis of Performance Data." *Illuminating Engineer*, 54 (June 1959), pp. 317–353.

Eastman Kodak Co. *Ergonomic Design for People at Work*, vol. 1. New York: Van Nostrand Reinhold, 1983.

Eastman Kodak Co. *Ergonomic Design for People at Work*, vol. 2. New York: Van Nostrand Reinhold, 1986.

Freivalds, A., D. B. Chaffin, and G. D. Langolf. "Quantification of Human Performance Circadian Rhythms." *Journal of the American Industrial Hygiene Association*, 44, no. 9 (September 1983), pp. 643–648.

Freivalds, A., J. L. Harpster, and L. S. Heckman. "Glare and Night Vision Impairment in Corrective Lens Wearers," *Proceedings of the Human Factor Society*, (27th Annual Meeting, 1983), pp. 324–328.

General Electric Company. *Light Measurement and Control* (TP-118). Nela Park, Cleveland, OH: Large Lamp Department, G.E., March 1965.

Grandjean, E. *Fitting the Task to Man*. 4th ed. London: Taylor & Francis, 1988.

IESNA. *Lighting Handbook*, 8th ed. Ed. M. S. Rea. New York: Illuminating Engineering Society of North America, 1995, pp. 459–478.

Kamon, E., W. L. Kenney, N. S. Deno, K. J. Soto, and A. J. Carpenter. "Readdressing Personal Cooling with Ice." *Journal of the American Industrial Hygiene Association*, 47, no. 5 (May 1986), pp. 293–298.

Konz, S., and S. Johnson. *Work Design*, 5th ed. Scottsdale, AZ: Holcomb Hathaway Publishers, 2000.

Lehmann, G. *Praktische Arbeitsphysiologie*. Stuttgart: G. Thieme, 1953.

Lockhart, J. M., H. O. Kiess, and T. J. Clegg. "Effect of Rate and Level of Lowered Finger-Surface Temperature on Manual Performance." *Journal of Applied Psychology*, 60, no. 1 (February 1975), pp. 106–113.

NIOSH. *Criteria for a Recommended Standard … Occupational Exposure to Hot Environments, Revised Criteria*. Washington, DC: National Institute for Occupational Safety and Health, Superintendent of Documents, 1986.

NIOSH. *Occupational Noise Exposure, Revised Criteria 1998*. DHHS Publication No. 98-126. Cincinnati, OH: National Institute for Occupational Safety and Health, 1998.

OSHA. *Code of Federal Regulations—Labor. (29 CFR 1910)*. Washington, DC: Office of the Federal Register, 1997.

OSHA. *Ergonomics Program Management Guidelines for Meatpacking Plants*. OSHA 3123. Washington, DC: The Bureau of National Affairs, Inc., 1990.

Peterson, A., and E. Gross, Jr. *Handbook of Noise Measurement*, 8th ed. New Concord, MA: General Radio Co., 1978.

Riley, M. W., and D. J. Cochran. "Partial Gloves and Reduced Temperature." In *Proceedings of the Human Factors Society 28th Annual Meeting*. Santa Monica, CA: Human Factors and Ergonomics Society, 1984, pp. 179–182.

Sanders, M. S., and E. J. McCormick. *Human Factors in Engineering and Design*, 7th ed. New York: McGraw-Hill, 1993.

Schwarzenau, P., P. Knauth, E. Kiessvetter, W. Brockmann, and J. Rutenfranz. "Algorithms for the Computerized Construction of Shift Systems Which Meet Ergonomic Criteria." *Applied Ergonomics*, 17, no. 3 (September 1986), pp. 169–176.

Yaglou, C. P., and D. Minard. "Control of Heat Casualties at Military Training Centers." *AMA Archives of Industrial Health*, 16 (1957), pp. 302–316.

Yaglou, C. P., E. C. Riley, and D. I. Coggins. "Ventilation Requirements." *American Society of Heating, Refrigeration and Air Conditioning Engineers Transactions*, 42 (1936), pp. 133–158.

SELECTED SOFTWARE

DesignTools (available from the McGraw-Hill text website at www.mhhe.com/niebel-freivalds), New York: McGraw-Hill, 2002.

WEBSITES

American Society for Safety Engineers—http://www.ASSE.org/
CalOSHA Standard—http://www.ergoweb.com/Pub/Info/Std/calstd.html
National Safety Council— http://www.nsc.org/
NIOSH homepage—http://www.cdc.gov/niosh/homepage.html
OSHA homepage—http://www.osha.gov/
OSHA Proposed Ergonomics Standard—http://www.osha-slc.gov/FedReg_osha_data/FED20001114.html

Design of Cognitive Work

KEY POINTS

- Minimize informational workload.
- Limit absolute judgments to 7 ± 2 items.
- Use visual displays for long, complex messages in noise areas.
- Use auditory displays for warnings and short, simple messages.
- Use color, symbols, and alphanumerics in visual displays.
- Use color and flashing lights to get attention.

The design of cognitive work has not been traditionally included as part of methods engineering. However, with ongoing changes in jobs and the working environment, it is becoming increasingly important to study not only the manual components of work but also the cognitive aspects of work. Machines and equipment are becoming increasingly complex and semiautomated, if not fully automated. The operator must be able to perceive and interpret large amounts of information, make critical decisions, and control these machines quickly and accurately. Furthermore, there has been a gradual shift of jobs from manufacturing to the service sector. In either case, there typically will be less emphasis on gross physical activity and a greater emphasis on information processing and decision making, especially via computers and associated modern technology. Thus, this chapter explains information theory, presents a basic conceptual model of the human as an information processor, and details how best to code and display information for maximum effectiveness, especially with auditory and visual displays. Also, a final section outlines both software and hardware considerations of the human interacting with computers.

7.1 INFORMATION THEORY

Information, in the everyday sense of the word, is knowledge received regarding a particular fact. In the technical sense, information is the reduction of uncertainty about that fact. For example, the fact that the engine (oil) light comes on when a car is started provides very little information (other than the fact that the lightbulb is functioning) because it is expected. On the other hand, when that same light comes on when you are driving down a road, it conveys considerable information about the status of the engine because it is unexpected and a very unlikely event. Thus, there is a relationship between the likelihood of an event and the amount of information it conveys, which can be quantified through the mathematical definition of information. Note that this concept is irrespective of the importance of the information; that is, the status of the engine is quite a bit more important than whether the windshield washer container is empty.

Information theory measures information in bits, where a *bit* is the amount of information required to decide between two equally likely alternatives. The term *bit* came from the first and last part of the words *bi*nary dig*it* used in computer and communication theory to express the on/off state of a chip or the polarized/reverse-polarized position of small pieces of ferromagnetic core used in archaic computer memory. Mathematically this can be expressed as

$$H = \log_2 n$$

where H = the amount of information
n = the number of equally likely alternatives

With only two alternatives, such as the on/off state of a chip or the toss of an unweighted coin, there is 1 bit of information presented. With 10 equally likely alternatives, such as the numbers from 0 to 9, 3.322 bits of information can be conveyed ($\log_2 10 = 3.322$). An easy way of calculating $\log_2$ is to use the following formula:

$$\log_2 n = 1.4427 \times \ln n$$

When the alternatives are not equally likely, the information conveyed is determined by

$$H = \Sigma p_i \times \log_2(1/p_i)$$

where p_i = probability of ith event
i = alternatives from 1 to n

As an example, consider a coin weighted so that heads comes up 90 percent of the time and tails only 10 percent of time. The amount of information conveyed in a coin toss becomes

$$H = 0.9 \times \log_2(1/0.9) + 0.1 \times \log_2(1/0.1) = 0.9 \times 0.152 + 0.1 \times 3.32$$
$$= 0.469 \text{ bit}$$

Note that the amount of information (0.469) conveyed by a weighted coin is less than the amount of information conveyed by an unweighted coin (1.0). The maximum amount of information is always obtained when the probabilities are equally likely. This is so because the more likely an alternative becomes, the less information is being conveyed (i.e., consider the engine light upon starting a car). This leads to the concept of *redundancy* and the reduction of information from the maximum possible due to unequal probabilities of occurrence. Redundancy can be expressed as

$$\% \text{ redundancy} = (1 - H/H_{\max}) \times 100$$

For the case of the weighted coin, the redundancy is

$$\% \text{ redundancy} = (1 - 0.469/1) \times 100 = 53.1\%$$

An interesting example relates to the use of the English language. There are 26 letters in the alphabet (A through Z) with a theoretical informational content for a randomly chosen letter of 4.7 bits ($\log_2 26 = 4.7$). Obviously, with the combinations of letters into words, considerably more information can be presented. However, there is a considerable reduction in the amount of information that can be actually presented due to the unequal probabilities of occurrence. For example, letters s, t, and e are much more common than q, x, and z. It has been estimated that the redundancy in the English language amounts to 68 percent (Sanders and McCormick, 1993). On the other hand, redundancy has some important advantages that will be discussed later with respect to designing displays and presenting information to users.

One final related concept is the *bandwidth* or *channel capacity*, the maximum information processing speed of a given communication channel. In terms of the human operator, the bandwidth for motor-processing tasks could be as low as 6 to 7 bits/s or as high as 50 bits/s for speech communication. For purely sensory storage of the ear (i.e., information not reaching the decision-making stage), the bandwidth approaches 10,000 bits/s (Sanders and McCormick, 1993). The latter value is much higher than the actual amount of information that is processed by the brain in that time because most of the information received by our senses is filtered out before it reaches the brain.

7.2 HUMAN INFORMATION PROCESSING MODEL

Numerous models have been put forward to explain how people process information. Most of these models consist of black boxes (because of relatively incomplete information) representing various processing stages. Figure 7.1 presents one such generic model consisting of four major stages or components: perception, decision and response selection, response execution, memory, and attentional resources distributed over various stages. The decision-making component, when combined with working memory and long-term memory, can be considered as the central processing unit while the sensory store is a very transient memory, located at the input stage (Wickens, Gordon, and Liu, 1997).

PERCEPTION AND SIGNAL DETECTION THEORY

Perception is the comparison of incoming stimulus information with stored knowledge to categorize the information. The most basic form of perception is

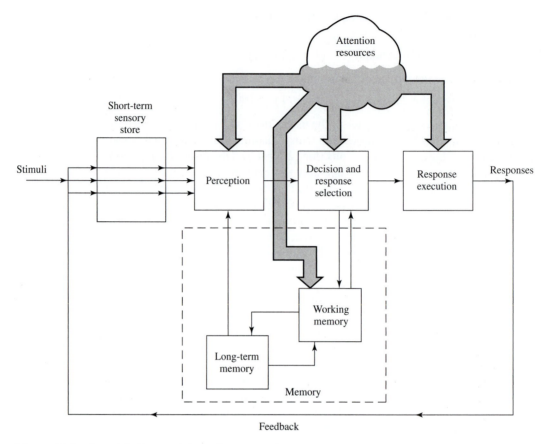

Figure 7.1   A model of human information processing.
(*Source*: Wickens, 1984, Fig. 1.1. Reprinted by permission of the publisher.)

simple *detection*, that is, determining whether the stimulus is actually present. It becomes more complicated if the person is asked to indicate the type of stimulus or the stimulus class to which it belongs and then gets into the realm of identification and recognition with the use of prior experiences and learned associations. The consequent linkage between long-term memory and perceptual encoding is shown in Figure 7.1. This latter more complex perception can be explained in terms of feature analysis, breaking down objects into component geometric shapes or text into words and character strings, and, simultaneously, of top-down or bottom-up processing to reduce the amount of information entering central processing. Top-down processing is conceptually driven using high-level concepts to process low-level perceptual features, while bottom-up processing is data-driven and guided by sensory features.

The detection part of perceptual encoding can be modeled or, in fairly simple tasks, even quantified through *signal detection theory* (SDT). The basic concept of SDT is that in any situation, an observer needs to identify a signal (i.e., whether it is present or absent) from confounding noise. For example, a quality inspector in an electronics operation must identify and remove defective chip capacitors from the good capacitors being used in the assembly of printed-circuit boards. The defective chip capacitor is the signal, which could be identified by excessive solder on the capacitor that shorts out the capacitor. The good capacitors, in this case, would be considered noise. Note that one could just as easily reverse the decision process and consider good capacitors as the signal and defective capacitors as noise. This would probably depend on the relative proportions of each.

Given that the observer must identify whether the signal is present and that only two possible states exist (i.e., the signal is either there or not there), there are a total of four possible outcomes:

1. *Hit*—saying there is a signal when the signal is present
2. *Correction rejection*—saying there is no signal when no signal is present
3. *False alarm*—saying there is a signal when no signal is present
4. *Miss*—saying there is no signal when the signal is present

Both the signal and noise can vary over time, as is the case with most industrial processes. For example, the soldering machine may warm up and initially expel a larger drop of solder on the capacitors, or there may be simply "random" variation in the capacitors with no cause yet determined. Therefore both the signal and noise form distributions of varying solder quantity from low to high, which typically are modeled as overlapping normal distributions (Figure 7.2). Note that the distributions overlap because excessive solder on the body of the capacitor would cause it to short out, causing a defective product (in this case a signal). However, if there is excessive solder, but primarily on the leads, the capacitor may not short out and thus is still a good capacitor (in this case noise). With ever-shrinking electronic products, chip capacitors are smaller than pinheads, and the visual inspection of these is not a trivial task.

When a capacitor appears, the inspector needs to decide if the quantity of solder is excessive and whether to reject the capacitor. Either through instructions and/or sufficient practice, the inspector has made a mental standard of judgment, which is depicted as the vertical line in Figure 7.2 and termed the *response criterion*. If the detected quantity of the solder, which enters the visual system as a high level of sensory stimulation, exceeds the criterion, the inspector will say there is a signal. On the other hand, if the detected quantity is small, a smaller level of sensory stimulation is received, landing below the criterion, and the inspector will say there is no signal.

Related to the response criterion is the quantity *beta*. Numerically beta is the ratio of the height of the two curves (signal to noise) in Figure 7.2 at the given criterion point. If the criterion shifts to the left, beta decreases with an increase of hits but at the cost of a corresponding increase of false alarms. This behavior on the part of the observer is termed *risky*. If the criterion were at the point where the two curves intersect, beta would be 1.0. On the other hand, if the criterion

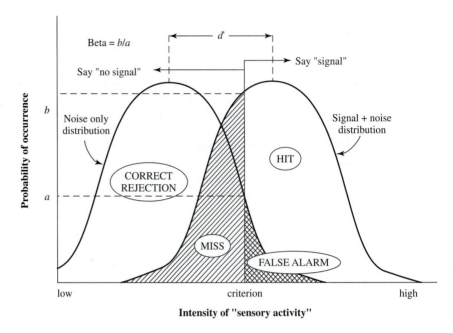

Figure 7.2 Conceptual illustration of signal detection theory.

shifts to the right, beta increases with a decrease of both hits and false alarms. This behavior on the part of the observer would be termed *conservative*.

The response criterion (and beta) can easily change depending on the mood or fatigue of the visual inspector. It would not be unexpected for the criterion to shift to the right and the miss rate to increase dramatically late Friday afternoons shortly before quitting times. Note that there will be a corresponding decrease in the hit rate because the two probabilities sum to 1. Similarly, the probabilities of a correct rejection and false alarms also sum to 1. The change in the response criterion is termed *response bias* and could also change with prior knowledge or changes in expectancy. If it were known that the soldering machine was malfunctioning, the inspector would most likely shift the criterion to the left, increasing the number of hits. The criterion could also change due to the costs or benefits associated with the four outcomes. If a particular batch of capacitors were being sent to NASA for use in the space shuttle, the costs of having a defect would be very high, and the inspector would set a very low criterion, producing many hits but also many false alarms with corresponding increased costs (e.g., losing good products). On the other hand, if the capacitors were being used in cheap give-away cell phones, the inspector might set a very high criterion, allowing many defective capacitors to pass through the checkpoint as misses.

A second important concept in SDT is that of *sensitivity* or the resolution of the sensory system. In SDT, sensitivity is measured as the separation between the two distributions shown in Figure 7.2 and labeled as d'. The greater the separation, the greater the observer's sensitivity and the more correct responses (more hits and more correct rejections) and fewer errors (fewer false alarms and misses) are made. Usually sensitivity will improve with greater training and alertness (e.g., through more frequent rest breaks) on the part of the inspector, better illumination at the workstation, and slowing the rate of signal presentation (which has the trade-off of decreasing productivity). Other factors that may help increase sensitivity involve supplying visual templates of the defective parts, providing redundant representations or clues for the defective parts, and providing knowledge of results. Note that providing incentives will help increase hit rates. However, this is typically due to a shift of the response bias (not an increase in sensitivity) with a corresponding increase in false-alarm rates. Similarly introducing "false signals" to increase alertness will again have a greater tendency to shift the response bias. More information on signal detection theory can be found in Green and Swets (1988).

Signal Detection Theory as Applied to the Inspection of Glass | EXAMPLE 7.1

A good application of signal detection theory was detailed by Drury and Addison (1973) in the visual inspection of glass. The inspection was in two stages: (1) 100 percent general inspection in which each item was either accepted or rejected and (2) a sample inspection by special examiners who reexamined the previous results and provided feedback to the general inspectors. Considering the quality of the items being inspected, a proportion was good and the rest was bad. The general inspector could make only two decisions: accept or reject. The proper responses would be the acceptance of a good item (hit) and the rejection of a bad item (correct rejection). However, some good items could be rejected (misses), and some bad items could be accepted (false alarms). Consider four different cases of varying conditions.

Case 1—Conservative Inspector   A conservative inspector sets the criterion far to the right (Figure 7.3a). In such a situation, the probability of hits (saying yes to the signal of good glass) is low (e.g., 0.30). The probability of false alarms (saying yes to the noise of bad glass) is even lower (e.g., 0.05). Beta is determined by the ratio of the ordinates of the signal curve over the noise curve at the criterion. The ordinate for a standard normal curve is

$$y = \frac{e^{-z^2/2}}{\sqrt{2\pi}}$$

For the signal curve, a probability of 0.30 yields a z of 0.524 and an ordinate of 0.348. For the noise curve, a probability of 0.05 yields a z of 1.645 and an ordinate of 0.103. Beta then becomes 3.38 (0.348/0.103). Note that the probability of hits and misses equals 1.0 (i.e., 0.30 + 0.70 = 1.0). The same is true for false alarms and correct rejections.

Case 2—Average Inspector   If the inspector is average—neither conservative nor risky—the probability of hits roughly equals the probability of correct rejections (Figure 7.3b). The curves intersect symmetrically, resulting in the same ordinate values and a value of 1.0 for beta.

Case 3—Risky Inspector   A risky inspector (Figure 7.3c) sets the criterion far to the left, increasing the probability of hits (e.g., 0.95) at the cost of a high probability of false alarms (e.g., 0.70). In this case, for the signal curve, a probability of 0.95 yields a

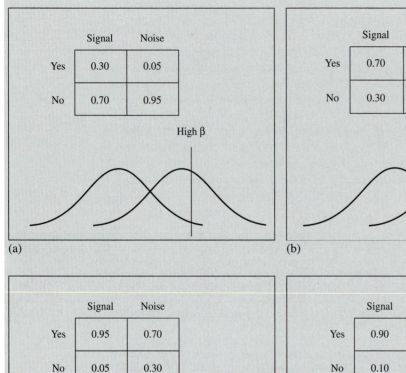

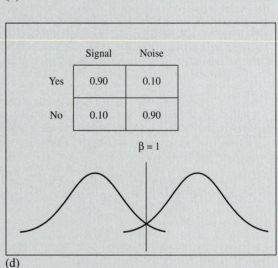

Figure 7.3 Signal detection theory as applied to inspection: (a) conservative inspector, (b) average inspector, (c) risky inspector, and (d) increased sensitivity.

z of -1.645 and an ordinate of 0.103. For the noise curve, a probability of 0.70 yields a z of -0.524 and an ordinate of 0.348. Beta then becomes 0.296 (0.103/0.348).

Case 4—Increased Sensitivity Sensitivity can be calculated as the difference of the z value for the same abscissa for both signal and noise curves (Figure 7.3):

$$d' = z(\text{false alarms}) - z(\text{hits})$$

Using the criterion of case 1,

$$d' = 1.645 - 0.524 = 1.121$$

The same is found from the criterion in case 3:

$$d' = -0.524 - (-1.645) = 1.121$$

If the signal can be better separated from the noise, the probability of hits will increase (e.g., up to 0.90), while the probability of false alarms will remain fairly low (e.g., 0.10). Using the criterion as the comparison point, the probability of hits is 0.90 with a corresponding z value of -1.283 and an ordinate of 0.175. The probability of false alarms is 0.10, with a corresponding z value of 1.283 and an ordinate of 0.175. The sensitivity then becomes

$$d' = 1.283 - (-1.283) = 2.566$$

With increased sensitivity there is better performance in identifying defective parts. Sometimes, the hit rate is plotted against the false alarm rate to yield a receiver operator characteristic curve in which the deviation of the curve from the 45 degree slope indicates sensitivity.

In the Drury and Addison (1973) case study, weekly data on glass inspection were collected from which the value d' was calculated. A change in inspection policy of providing more rapid feedback to the general inspector resulted in d' increasing from a mean value of 2.5 to a mean value of 3.16, a 26 percent increase in sensitivity over the course of 10 weeks (see Figure 7.4). This represented a 60 percent increase in the signal-to-noise ratio (i.e., beta) and a 50 percent decrease in the probability of missing a defect.

MEMORY

Once the stimulus has been perceptually encoded, it goes into *working memory*, one of the three components of the human memory system. The other two are sensory store and long-term memory. Each sensory channel has a temporary storage mechanism that prolongs the stimulus for it to be properly encoded. This storage is very short, on the order of 1 or 2 s depending on the sensory channel, before the stimulus representation disappears. It is also fairly automatic, in that it doesn't require much attention to maintain it. On the other hand, there is very little that can be done to maintain this storage or increase the length of the time period. Note also from Figure 7.1 that although there may be a vast amount of stimuli, which could be represented on the order of millions of bits of information entering the sensory storage, only a very small portion of that information is actually encoded and sent on to working memory.

As opposed to long-term memory, working memory is a means of temporarily storing information or keeping it active while it is being processed for a response. Thus, it sometimes is termed *short-term memory*. Looking up a phone number and retaining it until the number has been dialed and finding a processing code from a list and entering it into the control pad of a machine are good examples of working memory. Working memory is limited in both the amount of information and the length of time that information can be maintained.

The upper limit for the capacity of working memory is approximately 7 ± 2 items, sometimes known as *Miller's rule* after the psychologist who defined it (Miller, 1956). For example, the 11 digits 12125551212 would be very difficult,

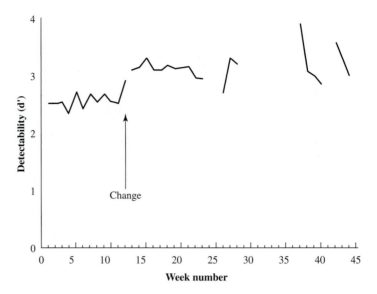

Figure 7.4 Change in sensitivity with time for glass inspection.
(*From:* Drury and Addison, 1973; Taylor & Francis, Philadelphia, PA.)

if not impossible, to recall. Recall can be improved by the use of *chunking*, or the grouping of similar items. The above numbers, when properly grouped as 1-212-555-1212, are much more easily remembered as three chunks (the 1 is the obvious standard for long distance). Similarly, *rehearsal*, or mentally repeating the numbers, which shifts additional attentional resources (see Figure 7.1) to working memory can improve recall.

Working memory also decays quickly, in spite of rehearsing or serially cycling through the items being actively maintained. The more items in working memory, the longer it takes to cycle through and the greater likelihood of one or more of the items being lost. It is estimated that the half-life for a memory store of three items is 7 s. This can be easily demonstrated by presenting a subject three random numbers (e.g., 5 3 6). After 7 s counting backward to prevent rehearsal, most individuals would have forgotten at least one number, if not two numbers.

Some recommendations for minimizing errors on tasks requiring the use of working memory are to (Wickens, Gordon, and Liu, 1997)

■ Minimize the memory load, in terms of both capacity and time to maintain recall

■ Utilize chunking, especially in terms of meaningful sequences and use of letters over numbers (e.g., the use of words or acronyms instead of numbers for toll-free telephone numbers, such as 1-800-CTD-HELP)

■ Keep chunks small, no more than three or four items of arbitrary nature

- Keep numbers separate from letters (e.g., chunks should contain similar entities)
- Minimize confusion of similar-sounding items (e.g., letters D, P, and T are easily confused, as opposed to letters J, F, and R)

Information from working memory may be transferred to *long-term memory* if it will be needed for later use. This could be information related to general knowledge in semantic memory or information on specific events in one's life in the form of event memory. This transfer must be done in an orderly manner so as to be able to easily retrieve the data at a later time in a process we know as learning. The process of retrieving the information is the weak link and can be facilitated by frequent activation of that memory trace (e.g., a social security or phone number used every day) and the use of *associations* with previous knowledge. These associations should be concrete rather than abstract and meaningful to the user, utilizing the user's expectations and stereotypes. For example, the name *John Brown* can be associated with the image of a brown house.

If there is the lack of clear or organized associations, the process can be done artificially in the form of mnemonics—an acronym or phrase—the letters of which represent a series of items. For example, the resistor color code (black, brown, red, orange, yellow, green, blue, violet, gray, white) can be remembered from the first letters of each word in the expression "big brown rabbits often yield great big vocal groans when gingerly slapped." Standardization of procedures or use of memory aids (signs or notes) for complicated procedures also helps by decreasing the load on long-term memory. Unfortunately, long-term memory decays exponentially, with the most rapid decline in the first few days. Because of this, the effectiveness of training programs should not be evaluated immediately after the program.

DECISION MAKING AND RESPONSE SELECTION

Decision making is really the core of information processing, in which people evaluate alternatives and select an appropriate response. This is a relatively long-term process and should be distinguished from short-term processing as in choice reaction time. Unfortunately people are not optimal decision makers and frequently do not make rational decisions based on objective numbers or hard information. The rational approach in classical decision theory would be to calculate an expected value based on the sum of products of each outcome multiplied by its expected probability:

$$E = \Sigma p_i v_i$$

where E = expected value
p_i = probability of *i*th outcome
v_i = value of the *i*th outcome

Unfortunately, people typically use a variety of heuristics to make decisions, in which case a variety of biases may influence the way they seek information, attach values to outcomes, and make overall decisions. A short list of such biases is derived from Wickens, Gordon, and Liu (1997):

- A limited number of *cues* or pieces of information are used.
- Undue weight is given to early cues.
- Inattention is given to later cues.
- Prominent cues are given greater weight.
- All information is weighted equally regardless of true weight.
- A limited number of hypotheses are generated.
- Once a hypothesis has been selected, later cues are omitted.
- Only confirming information is sought for the chosen hypothesis.
- Only a small number of responses are chosen.
- Potential losses are weighted more heavily than comparable potential gains.

By understanding these biases, the industrial engineer may be able to better present information and better set up the overall process to improve the quality of decision making and minimize errors.

In addition, current theories on decision making center around *situational awareness*, which is an evaluation of all the cues received from the surrounding environment. It requires the integration of cues or information into mental representations, ranging from simple schemata to complex mental models. To improve situational awareness, operators need to be trained to recognize and consider appropriate cues, to check the situation for consistency within the cues, and to analyze and resolve any conflicts in the cues or the situation. Decision aides, such as simple decision tables (discussed in Chapter 9) or more complex, expert systems, may assist in the decision-making process. Also, the display of key cues, the filtering out of undesirable cues, and the use of spatial techniques and display integration can assist in this process. Some of these will be discussed further in the section below on display modality.

The speed and difficulty of decision making and response selection, as discussed above, are influenced by many factors. One attempt at quantifying this process is typically performed through a *choice–reaction time* experiment, in which the operator will respond to several stimuli with several appropriate responses (see Figure 7.5a). This can be considered as simple decision making, and based on the human information processing system, the response time should increase as the number of alternative stimuli increases. The response is nonlinear (see Figure 7.5b), but when decision complexity is quantified in terms of the amount of information conveyed in bits, then the response becomes linear and is referred to as the *Hick-Hyman law* (Hick, 1952; Hyman, 1953; see Figure 7.5c)

$$\mathrm{RT} = a + bH$$

where   RT = response time (s)
H = amount of information (bits)
a = intercept
b = slope, sometimes referred to as information processing rate

Note that when there is only one choice (e.g., when the light appears, press the button), $H = 0$ and the response time is equal to the intercept. This is known as *simple reaction time*. It can vary depending on the type of stimulus (auditory reaction times are about 40 ms faster than visual reaction times), the intensity of the stimulus, and preparedness for the signal.

Overall choice reaction times also vary considerably due to a variety of factors. The greater the compatibility (see also Section 5.3) between the stimulus and the response, the faster the response. The more practice, the faster the response. However, the faster the operator tries to respond, the greater the number of errors. Similarly, if there is a requirement for very high accuracy (e.g., air-traffic control), the response time will become slower. This inverse relationship is known as the *speed-accuracy trade-off*.

The use of multiple dimensions, in another form of redundancy, can also decrease response time in decision making; or, conversely, if there is conflicting information, response time will be slowed. A classic example is the *Stroop Color-Word Task* (Stroop, 1935), in which the subject is asked to read a series of words expressing colors as rapidly as possible. In the control redundant case, having red ink spell out the word *red* will result in a fast response. In the conflict case, red ink letters spelling out *blue* will slow response time due to the semantic and visual conflicts.

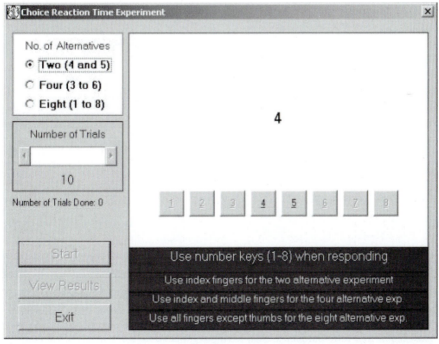

(a)

Figure 7.5
Hick-Hyman law illustrated in a choice–reaction time experiment: (a) choice–reaction time experiment from Design Tools, (b) raw data, (c) information expressed in bits.

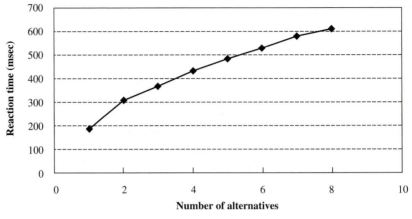

(b)

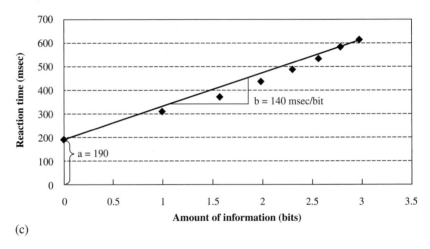

(c)

Figure 7.5   (continued)

EXAMPLE 7.2
Human Information Processing in a Wiring Task

A good example of quantifying the amount of information processed in an industrial task was presented by Bishu and Drury (1988). In a simulated wiring task, industrial operators moved a stylus to the proper terminal or location on a control panel, which consisted of four different plates, each having eight possible components. Each component area was divided into 128 terminals in an array of eight columns and 16 rows. The most complex task involved all four plates ($\log_2 4 = 2$ bits of information), all eight components (3 bits), eight columns (3 bits), and 16 rows (4 bits) for a total complexity of 12 bits (sum of 2, 3, 3, and 4). From this control panel, others of less complexity could be produced by reducing the number of plates, components, columns, and rows. A low-complexity task involved only two plates (1 bit), four components (2 bits), four columns (2 bits), and 8 rows (3 bits) for a total complexity of 8 bits (sum of 1, 2, 2, and 3). Other intermediate-complexity tasks were also utilized.

The final results showed a linear relationship between information processing (simulated wiring or placement) time and the information complexity of the input (see Figure 7.6). Using the Hick-Hyman law, this relationship can be expressed as

$$IP = -2.328 + 0.7325H$$

where   IP = information processing time (s)
   H = amount of information (bits)

Thus, as the number of alternatives in performing the task increased, so also increased the informational loading on the central processing unit of the human operator, and the corresponding time for task performance. Note that in this relatively real-world case of a complex task, the intercept may not always be a positive value corresponding to simple reaction time.

Fitts' Law and Information Processing of Movement

EXAMPLE 7.3

Information theory was applied to the modeling of human movement by Fitts (1954) who developed the *index of difficulty* to predict movement time. The index of difficulty was defined as a function of the distance of movement and target size in a series of positioning movements to and from identical targets:

$$ID = \log_2(2D/W)$$

where   ID = index of difficulty (bits)
   D = distance between target centers
   W = target width

Movement time was found to follow the Hick-Hyman law, now termed *Fitts' law*:

$$MT = a + b(ID)$$

where MT = movement time (s)
   a = intercept
   b = slope

In a particularly successful application of Fitts' law, Langolf, Chaffin, and Foulke (1976) modeled human movement by different limbs across a wide range of distances, including very small targets visible only with the assistance of a microscope. Their results (see Figure 4.14) yielded slopes of 105 ms/bit for the arm, 45 ms/bit for the wrist, and 26 ms/bit for the finger. The inverse of the slope is interpreted, according to information theory, as the motor system bandwidth. In this case, the bandwidths were 38 bits/s for the finger, 23 bits/s for the wrist, and 10 bits/s for the arm. This decrease in information processing rates was explained as the result of added processing for the additional joints, muscles, and motor units. Interestingly, these results are identical to Gilbreth's classification of movements (see Section 4.2).

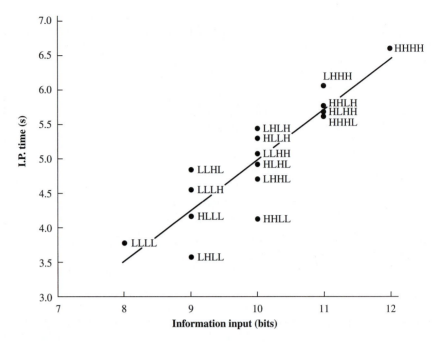

Figure 7.6   Hick-Hyman law illustrated in a wiring task.
(Reprinted from *Applied Ergonomics*, Vol. 19, Bishu and Drury, Information Processing
in Assembly Tasks—A Case Study, pp. 90–98, 1988, with permission from Elsevier
Science.)

RESPONSE EXECUTION

Response execution depends primarily on human movement. More details on
the musculoskeletal system, motor control, and manual work can be found in
Chapter 4. Note that the *Fitts' Tapping Task* (see Figure 7.7) is a simple extension
of the Hick-Hyman law with respect to movement and also an example of a speed-
accuracy trade-off with respect to the size of the target and movement time. Specific applications of responses with respect to controls and to the operation of
machines and other equipment are discussed in Chapter 5.

ATTENTION RESOURCES

Attention resources, or more simply *attention*, refer to the amount of cognitive
capacity devoted to a particular task or processing stage. This amount could vary
considerably from very routine, well-practiced assembly tasks with low attentional demands to air-traffic controllers with very high attentional demands. Furthermore, this cognitive capacity can be applied in a very directed manner, such
as a spotlight on a particular part of the human information processing system,
termed *focused attention*. Or it can be applied in a much more diffuse manner to
various parts or even all the human information processing system, termed

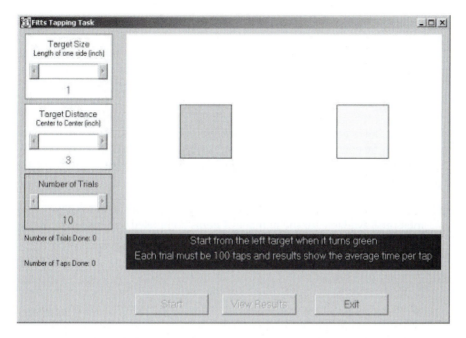

Figure 7.7 Fitts' Tapping Task from DesignTools.

divided attention. An example of focused attention on working memory would occur while an operator was trying to remember a looked-up processing code while entering it into a computer-controlled machine tool. Focused attention can be improved by reducing the number of competing sources of information or demands on the human information processing system or separating these sources in as distinct a manner as possible.

On the other hand, an inspector, sorting apples on a conveyor line, divides his attention between visual perception of blemishes and apple sizes, decision making on the nature of the blemish and the size of the apple, with reference to memory and the images stored there from training, and hand movements to remove the damaged apples and sort the good ones into appropriate bins by size. This latter case of performing various tasks simultaneously is also termed *multi-tasking* or *timesharing*. Because our cognitive resources for attention are relatively limited, timesharing between several tasks will probably result in a decrease in performance for one or more of these tasks as compared to a single task baseline. Again, it can be difficult to improve task performance in such situations, but similar strategies as discussed for focused attention are also utilized. The number and difficulty of tasks should be minimized. The tasks should be made as dissimilar as possible in terms of the demands placed on a processing stage of Figure 7.1. Whereas a purely manual assembly task with auditory instructions can be managed, a musician tuning an instrument would have a much more difficult time listening to verbal comments. One approach that is fairly successful in explaining timesharing performance with multiple tasks is the multiple-resource model of Wickens (1984).

An extension of multiple-resource modes of attention relates to the measurement of *mental workload* or the demands placed on the human information processor. One definition uses the ratio of resources required to the resources available, where time is one of the more important of a number of required resources. In the examples mentioned above, simple assembly may be time-consuming, but is not especially demanding of cognitive resources. On the other hand, air-traffic control, at peak times, can be very demanding. It can be very difficult to actually quantify these demands placed on the operator. Some of the approaches used to do so include these:

- *Primary task* measures may be time required to perform the task divided by total time available, or the number of items completed per unit time. The problem with this approach is that some tasks are better timeshared than others.

- *Sondary task* measures utilize the concept of a reserve capacity that, if not being directed to the performance of the primary task, will be used by a secondary task (choice reaction time), which can be controlled and more easily measured. The problem with this approach is that the secondary task typically seems artificial and intrusive, and it is difficult to identify how the operator prioritizes the performance of both tasks.

- *Physiological measures* (e.g., heart rate variability, eye movement, pupil diameter, electroencephalograms) are thought to respond to the stress imposed by mental workload; although they typically don't interfere with the primary task performance, the equipment needed to measure them may do so.

- *Subjective measures* are thought to aggregate all aspects of mental workload in a simple overall rating (or a weighted average of several scales). Unfortunately subjective reports don't always accurately reflect true performance; also, motivation may strongly affect the ratings.

For a more detailed discussion of mental workload and the advantages and disadvantages of various means to measure it, see Wickens (1984), Eggemeier (1988), and Sanders and McCormick (1993).

A final example of attentional resources relates to the ability of an operator (e.g., a visual inspector) to maintain attention and remain alert over prolonged periods of time. Termed *sustained attention* or *vigilance*, the concern is how to minimize the vigilance decrement that occurs after as little as 30 min and increases up to 50 percent with increasing time (Giambra and Quilter, 1987; see Figure 7.8). Unfortunately, there are few documented countermeasures that work well for industrial tasks. The basic approach is to try to maintain a high level of arousal, which then maintains performance, following the *Yerkes-Dodson (1908) inverted-U* curve (see Figure 7.9). This can be done by providing more frequent rest breaks, providing task variation, providing operators with more feedback on

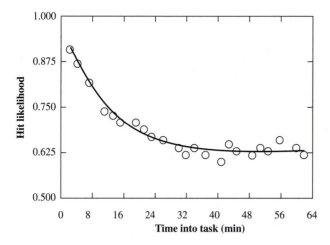

Figure 7.8 Vigilance decrement over time.
(From: Giambra and Quilter, 1987. Reprinted with permission from *Human Factors*, Vol. 29, No. 6, 1987. Copyright 1987 by the Human Factors and Ergonomics Society. All rights reserved.)

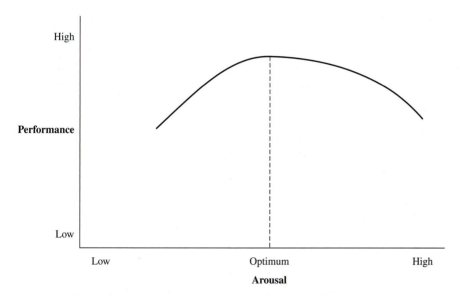

Figure 7.9 The Yerkes-Dodson law showing the inverted-U relationship between performance and the level of arousal.

their detection performance, and using appropriate levels of stimulation, either internal (e.g., caffeine) or external (e.g., music or white noise), or even through the introduction of false signals. However, the latter change of detection criteria will also increase the rate of false alarms (see the discussion of signal detection

Perception Considerations	Yes	No
1. Are key signals enhanced?	❑	❑
2. Are overlays, special patterns, or grazing light used to enhance defects?	❑	❑
Are both top-down and bottom-up processing used simultaneously?	❑	❑
a. Are high-level concepts used to process low-level features?	❑	❑
b. Is data-driven information used to identify sensory features?	❑	❑
4. Is better training used to increase sensitivity of signal detection?	❑	❑
5. Are incentives provided to change the response bias and increase hits?	❑	❑

Memory Considerations	Yes	No
1. Is short-term memory load limited to 7±2 items?	❑	❑
2. Is chunking utilized to decrease memory load?	❑	❑
3. Is rehearsal utilized to enhance recall?	❑	❑
4. Are numbers separated from letters in lists or chunks?	❑	❑
5. Are similar-sounding items separated?	❑	❑
6. Are mnemonics and associations used to enhance long-term memory?	❑	❑

Decision and Response Selection	Yes	No
1. Are a sufficient number of hypotheses examined?	❑	❑
2. Are a sufficient number of cues utilized?	❑	❑
3. Are later cues given equal weight to early cues?	❑	❑
4. Are undesirable cues filtered out?	❑	❑
5. Are decision aids utilized to assist in the process?	❑	❑
6. Are a sufficient number of responses evaluated?	❑	❑
7. Are potential losses and gains weighted appropriately?	❑	❑
8. Are speed-accuracy trade-offs considered?	❑	❑
9. Are the stimuli and responses compatible?	❑	❑

Attentional Resource Considerations	Yes	No
1. Is there task variety?	❑	❑
2. Is performance feedback provided to the operator?	❑	❑
3. Does the operator have internal stimulation (e.g., caffeine)?	❑	❑
4. Does the operator have external stimulation (e.g., music, incentives)?	❑	❑
5. Are rest breaks provided?	❑	❑

Figure 7.10　Cognitive Work Evaluation Checklist.

theory) with consequent costs. Increasing the prominence of the signal will assist in the detection performance (e.g., making the signal brighter, larger, or with greater contrast through special illumination). Overlays that act as special patterns to enhance differences between the defective part and the rest of the object may also be useful. Finally, selecting inspectors with faster eye fixation time and better peripheral vision will also help in inspection performance (Drury, 1982).

To assist the industrial engineer in evaluating and redesigning cognitive tasks, the above details on the human information processing system have been summarized in the Cognitive Work Evaluation Checklist (see Figure 7.10).

7.3 CODING OF INFORMATION: GENERAL DESIGN PRINCIPLES

As mentioned in the introduction to Chapter 4, many, if not most, industrial functions or operations will be performed by machines, because of the greater force, accuracy, and repeatability considerations. However, to ensure that these machines are performing satisfactorily at the desired specifications, there will always be the need for a human monitor. This operator will then receive a variety of information input (e.g., pressure, speed, temperature, etc.) that has to be presented in a manner or form that will be both readily interpretable and unlikely to result in an error. Therefore, there are a number of design principles that will assist the industrial engineer in providing the appropriate information to the operator.

TYPE OF INFORMATION TO BE PRESENTED

Information to be presented can be either static or dynamic, depending on whether it changes over time. The former includes any printed text (even scrolling text on a computer screen), plots, charts, labels, or diagrams that are unchanging. The latter is any information that is continually updated such as pressure, speed, temperature, or status lights. Either of the two categories can also be classified as

- Quantitative—presenting specific numerical values (e.g., 50°F, 60 rpm)
- Qualitative—indicating general values or trends (e.g., up, down, hot, cold)
- Status—reflecting one of a limited number of conditions (e.g., on/off, stop/caution/go)
- Warnings—indicating emergencies or unsafe conditions (e.g., fire alarm)
- Alphanumeric—using letters and numbers (e.g., signs, labels)
- Representational—using pictures, symbols, and color to code information (e.g., "wastebasket" for deleted files)
- Time-phased—using pulsed signals, varying in duration and intersignal interval (e.g., Morse code or blinking lights)

Note that one informational display may incorporate several of these types of information simultaneously. For example, a stop sign is a static warning using alphanumeric letters, an octagonal shape, and the red color as representations of information.

DISPLAY MODALITY

Since there are five different senses (vision, hearing, touch, taste, smell), there could be five different *display modalities* for information to be perceived by the human operator. However, given that vision and hearing are by far the most developed senses and most used for receiving information, the choice is generally limited to those two. The choice of which of the two to use depends on a variety of factors, with each sense having certain advantages as well as certain disadvantages. The

Table 7.1   When to Use Visual or Auditory Forms of Presentation

Use visual forms if:	Use auditory forms if:
The message is long and complex	The message is short and simple
The message deals with spatial references	The message deals with events in time
The message needs to be referred to later	The message is transient
No immediate action is needed	Immediate action is needed
Hearing is difficult (noise) or overburdened	Vision is difficult or overburdened
The operator can be stationary	The operator is moving about

Adapted from Deatherage, 1972.

detailed comparisons given in Table 7.1 may aid the industrial engineer in selecting the appropriate modality for the given circumstances.

Touch or tactile stimulation is useful primarily in the design of controls, which is discussed further in Section 5.3. Taste has been used in a very limited range of circumstances, primarily added to give medicine a "bad" taste and prevent children from accidentally swallowing it. Similarly, odors have been used in the ventilation system of mines to warn miners of emergencies or in natural gas to warn the homeowner of a leaking stove.

SELECTION OF APPROPRIATE DIMENSION

Information can be coded in a variety of dimensions. Select a dimension appropriate for the given conditions. For example, if lights are to be used, one can then select brightness, color, and frequency of pulsing as dimensions to code information. Similarly, if sound is to be used, one can select dimensions such as loudness, pitch, and modulation.

LIMITING ABSOLUTE JUDGMENTS

The task of differentiating between two stimuli along a particular dimension depends on either a *relative judgment*, if a direct comparison can be made of the two stimuli, or an *absolute judgment*, if no direct comparison can be made. In the latter case the operator must utilize working memory to hold one stimulus and make the comparison. As discussed previously, the capacity of working memory is limited to approximately 7 ± 2 items by Miller's rule. Therefore, an individual will be able to identify, at most, five to nine items based on absolute judgment. Research has shown that this holds true for a variety of dimensions: five levels for pure tones, five levels for loudness, seven levels for size of the object, five levels for brightness, and up to a dozen colors. On the other hand, individuals have been able to identify up to 300,000 different colors on relative basis, when comparing them two at a time. If multiple dimensions are used (e.g., brightness and color), then the range can be increased to some degree, but less than would be expected from the combination (direct product) of the two coding dimensions (Sanders and McCormick, 1993).

INCREASING DISCRIMINABILITY OF CODES

When selecting a coding scheme, one must consider that with regards to relative judgment, it is apparent that there has to be some minimum difference between

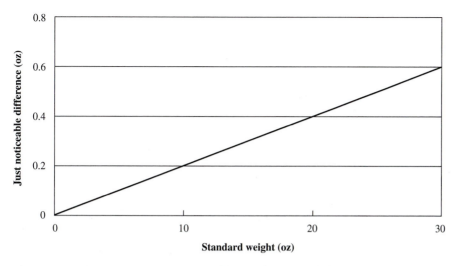

Figure 7.11   Weber's law showing the relationship between just noticeable difference and the level of stimuli which, for this example, is standard weight.

the two codes or stimuli before they can be distinguished as different. This difference is termed the *just noticeable difference* (JND) and is found to vary depending on the level of the stimuli. For example (see Figure 7.11), if an individual is given a 10-oz (0.283-kg) weight, the JND is roughly 0.2 oz (0.056 kg). If the weight is increased to 20 oz (0.566 kg), the JND increases to 0.4 oz (0.113 kg), and so on. This relationship is termed *Weber's law* and can be expressed as

$$k = \text{JND}/S$$

where k = Weber's fraction or slope
          S = standard stimulus

The application of this law is quite evident in an industrial setting. Consider a lighting source with a three-way bulb (100-200-300 W). The brightness change from 100 to 200 W is quite noticeable, whereas the change from 200 to 300 W is much less noticeable. Thus, for a change in a high-intensity signal to be noticeable, the change must be quite large. Although, Weber's law was formulated for relative thresholds, Fechner (1860) extended it to develop psychological scales for measuring a wide range of sensory experiences to form the basis for the science of psychophysics.

COMPATIBILITY OF CODING SCHEMES

Compatibility refers to the relationship of stimuli and responses that are consistent with human expectations, resulting in decreased errors and faster response time. This can occur at several levels: conceptual, movement, spatial, and modality. *Conceptual compatibility* refers to how meaningful the codes are to the individuals using them. Red is an almost universal code for danger or stop. Similarly pictorial realism is very useful, for example, a symbol for a female on a door

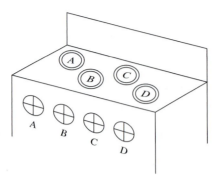

Figure 7.12 Spatial compatibility of controls and burners on a stove top.

indicating women's toilets. *Movement compatibility* refers to the relationship between the movement of controls and displays and is discussed further in Section 5.3. *Spatial compatibility* refers to the physical arrangement of controls and displays. The classic example (see Figure 7.12) is the optimum arrangement of knobs to control the burners on a stove from Chapanis and Lindenbaum (1959). *Modality compatibility* refers to using the same stimulus modality for both the signal and response. For example, verbal tasks (responses to verbal commands) are performed best with auditory signals and spoken responses. Spatial tasks (moving a cursor to a target) are performed best with a visual display and a manual response.

REDUNDANCY FOR CRITICAL SITUATIONS

When several dimensions are combined in a redundant manner, the stimulus or code will be more likely to be interpreted correctly and fewer errors will be made. The stop sign is a good example with three redundant codes: the word *stop*, the universal red color, and the unique (among traffic signs) octagonal shape. Note that these dimensions are all in the visual modality. Using two different modalities will improve the response as compared to two different dimensions within a modality. Thus, for an emergency evacuation of a plant in case of a fire, using both an auditory signal (a siren) and a visual signal (flashing red lights) will be more effective than either modality alone. The trade-off, as discussed earlier in this chapter, is a reduction in the number of potential codes available and a reduction in the amount of information that can be presented.

MAINTAINING CONSISTENCY

When coding systems have been developed by different people in different situations, it is important to maintain consistency. Otherwise, especially under times of stress, it is likely that operators will respond instinctively to previous habits and errors will occur. Therefore, as new warnings are added to a factory having an existing warning system, it is important that the coding system duplicate the existing scheme when the same information is being transmitted, even though the old system might not have had the optimum design in the first place. For example, the color yellow, typically meaning proceed with caution, should have the same meaning for all displays.

7.4   DISPLAY OF VISUAL INFORMATION: SPECIFIC DESIGN PRINCIPLES

FIXED SCALE, MOVING POINTER DESIGN

There are two major alternative analog display designs: a fixed scale with a moving pointer and a moving scale with a fixed pointer (see Figure 7.13). The first is the preferred design because all major compatibility principles are maintained: increasing values on the scale go from left to right, and a clockwise (or left to right) movement of the pointer indicates increasing values. With a moving scale and fixed pointer, one of these two compatibility principles will always be violated. Note that the display itself can be circular, semicircular, a vertical bar, a horizontal bar, or an open window. The only situation in which the moving scale and fixed pointer design has an advantage is for very large scales, which cannot be fully shown on the fixed scale display. In that case an open window display can accommodate a very large scale in back of the display with only the relevant portion showing. Note that the fixed scale and moving pointer design can display very nicely quantitative information as well as general trends in the readings. Also, the same displays can be generated with computer graphics or electronics without a need for traditional mechanical scales.

DIGITAL DISPLAYS FOR PRECISION

When precise numeric values are required and the values remain relatively static (at least long enough to be able to be read), then a digital display or counter should be utilized (see Figure 7.13). However, once the display changes rapidly, then counters become difficult to use. Also, counters are not good for identifying trends. Thus, digital counters were never very successful as a "high-tech" feature for automobile speedometers. A more detailed comparison in Table 7.2 shows the advantages and disadvantages of using moving pointers, moving scales and counters.

DISPLAY BASIC FEATURES

Figure 7.14 illustrates some of the basic features utilized in a dial design. The scale range is clearly depicted with an orderly numerical progression with numbered major markers at 0, 10, 20, and so on, and minor markers at every unit. An intermediate marker at 5, 15, 25, and so on, helps identify readings better. A progression by 5s is less good but still satisfactory. The pointer has a pointed tip just meeting, but not overlapping, the smallest scale markers. Also, the pointer should be close to the surface of the scale to avoid parallax and erroneous readings.

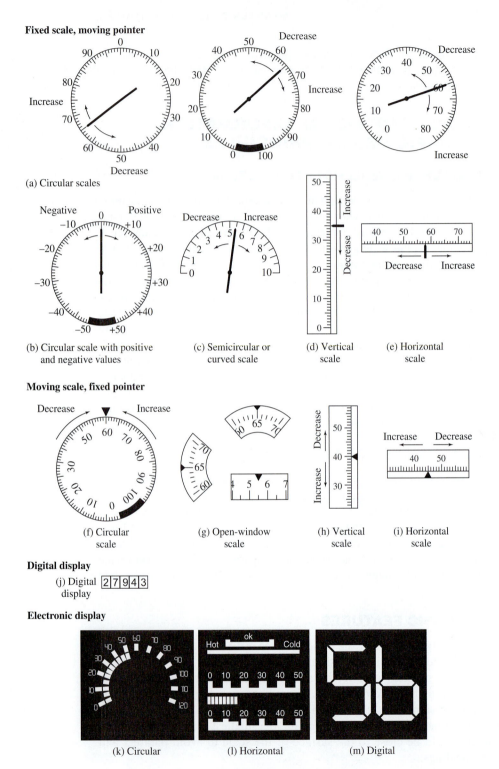

Figure 7.13 Types of displays for presenting quantitative information.

Table 7.2 Comparison of Pointers, Scales, and Counters

Indicator	Quantitative reading	Qualitative reading	Setting	Tracking
		Service rendered		
Moving pointer...	Fair	Good (changes are easily detected)	Good (easily discernible relation between setting knob and pointer)	Good (pointer position is easily controlled and monitored)
Moving scale . . .	Fair	Poor (may be difficult to identify direction and magnitude)	Fair (may be difficult to identify relation between setting and motion)	Fair (may have ambiguous relationship to manual-control motion)
Counter	Good (minimum time to read and results in minimum error)	Poor (position change may not indicate qualitative change)	Good (accurate method to monitor numerical setting)	Poor (not readily monitored)

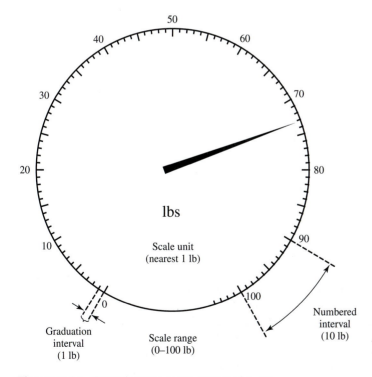

Figure 7.14 Basic features in the design of a dial.

PATTERNS FOR A PANEL OF DIALS

Usually a panel of dials is used to indicate the status of a series of pressure lines or valves as in the control room of a power plant. In such cases, the operator is primarily a monitor, performing check readings to ascertain whether readings (and the status of the system) are normal. Although the actual dials may be quantitative, the operator is primarily determining whether any one dial is indicating a condition out of a normal range. Thus, the key design aspect is to align all normal states and all dial pointers in the same direction, such that any change or deviant reading will stand out from the others. This pattern is accentuated by having extended lines between dials (see Figure 7.15).

MINIMIZING INFORMATIONAL LOADING ON THE WORKER

Human error in reading display information increases as the amount of information per unit area increases. Always consider Miller's rule in limiting the amount of information presented. Proper coding of the information improves readability of the displays and decreases the number of errors. In general, color, symbols or geometric figures, and alphanumerics are the best coding methods. These require little space and allow easy identification of information.

INDICATOR LIGHTS TO GET THE ATTENTION OF THE WORKER

Indicator or warning lights are especially good in attracting attention to potentially dangerous situations. Several basic requirements should be incorporated into their use. They should indicate to the operator what is wrong and what action should be taken. In poor background conditions with poor contrasts, a red (typically a serious situation), a green, and a yellow (less-serious condition) light have advantages over a white light. In terms of size and intensity, a good rule is

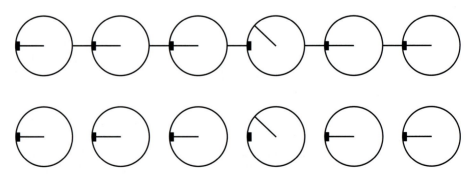

Figure 7.15   Panel of dials designed for check reading.

Table 7.3 Recommended Coding of Indicator Lights

Diameter	State	Color			
		Red	**Yellow**	**Green**	**White**
12.5 mm	Steady	Failure; Stop action; Malfunction	Delay; Inspect	Circuit energized; Go ahead: Ready; Producing	Functional; In position; Normal (on)
25 mm or larger	Steady	System or subsystem in stop action	Caution	System or subsystem in go-ahead state	
25 mm or larger	Flashing	Emergency condition			

to make it twice the size (at least 1 degree of visual arc) and brightness of other panel indicators, and place it not more than 30 degrees off the operator's line of sight. A flashing light, with a flash rate between 1 to 10 per second, will especially attract attention. Immediately after the operator takes action, the flashing should stop, but the light should remain on until the improper condition has been completely remedied. Further details on the coding of indicator lights are given in Table 7.3.

PROPER SIZE ALPHANUMERIC CHARACTERS FOR LABELS

For the most effective alphanumeric coding use recommended illumination levels (see Section 6.1) and consider the following information on letter height, stroke width, width-height ratio, and font. Based on a viewing distance of 20 in (51 cm), the letter or numeral height should be at least 0.13 in (0.325 cm) and the stroke width at least 0.02 in (0.055 cm), to give a stroke width-height ratio of 1:6. This creates a preferred visual angle of 22 arc minutes as recommended by ANSI/HFS 100 (1988) or a point size of 10 as typically found in newspapers (one point equals 1/72 in or 0.035 mm). For distances other than 20 in (51 cm), the value can be scaled so as to yield a comparable visual angle (e.g., for a distance of 40 in, sizes would be doubled).

The above recommendations refer to dark letters on a white background, the preferred format for reading for well-illuminated areas (e.g., typical work areas with windows or good lighting). White letters on a dark background are more appropriate for dark areas (e.g., nighttime conditions) and have a narrower stroke width (1:8 to 1:10 ratio) due to a spreading effect of the white letters. The font refers to the style of type such as Roman, with serifs or the special embellishments on the ends of the strokes, or Gothic, without serifs. In general, a mix of uppercase and lowercase letters is easier for extended reading. However, for special emphasis and attracting attention, as in labels, boldface or uppercase letters with a width-height ratio of about 3:5 are useful (Sanders and McCormick, 1993).

7.5 DISPLAY OF AUDITORY INFORMATION: SPECIFIC DESIGN PRINCIPLES

AUDITORY SIGNALS FOR WARNINGS

As discussed previously, there are special characteristics of the auditory system that warrant using an auditory signal for warnings. Simple reaction time is considerably quicker to auditory than visual signals (e.g., consider the starter pistol to start races). An auditory signal places much higher attentional demands on the worker than a visual signal. Since hearing is omnidirectional and sound waves penetrate barriers (to some degree, depending on thickness and material properties), auditory signals are especially useful if workers are at an unknown location and moving about.

TWO-STAGE AUDITORY SIGNALS

Since the auditory system is limited to short and simple messages, a two-stage signal should be considered when complex information is to be presented. The first stage should be an attention-demanding signal to attract attention, while the second stage would be used to present more precise information.

HUMAN ABILITIES AND LIMITATIONS

Since human auditory sensitivity is best at approximately 1,000 Hz, use auditory signals with frequencies in the range of 500 and 3,000 Hz. Increasing signal intensity will serve two purposes. First it will increase the attention-demanding quality of the signal and decrease response time. Second, it will tend to better differentiate the signal from background noise. On the other hand, one should avoid excessive levels (e.g., well above 100 dB) as these will tend to cause a startle response and perhaps disrupt performance. Where feasible, avoid steady-state signals so as to avoid adaption to the signal. Thus, *modulation* of the signal (i.e., turning the signal on and off in a regular cycle) in the frequency range of 1 to 3 Hz will tend to increase the attention-demanding quality of the signal.

ENVIRONMENTAL FACTORS

Since sound waves can be dispersed or attenuated by the working environment, it is important to take environmental factors into account. Use signal frequencies below 1,000 Hz when the signals need to travel long distances (i.e., more than 1,000 ft), because higher frequencies tend to be absorbed or dispersed more easily. Use frequencies below 500 Hz when signals need to bend around obstacles or pass through partitions. The lower the signal frequency, the more similar sound waves become to vibrations in solid objects, again with lower absorption characteristics.

DISSOCIATING THE SIGNAL FROM NOISE

Auditory signals should be as separate as possible from other sounds, whether useful auditory signals or unneeded noise. This means the desired signal should be as different as possible from other signals in terms of frequency, intensity, and modulation. If possible, warnings should be placed on a separate communication channel to increase the sense of *dissociability* and increase the attention-demanding qualities of the warning.

The above principles for designing displays, both auditory and visual, are summarized as an evaluative checklist in Figure 7.16. If purchased equipment has dials or other displays that don't correspond to these design guidelines, then there is the possibility for operator error and potential loss. If at all possible, those problems should be corrected or the displays replaced.

7.6 HUMAN-COMPUTER INTERACTION: HARDWARE CONSIDERATIONS

KEYBOARDS

The standard computer keyboard used today is based on the typewriter key layout patented by C. L. Sholes in 1878. Termed a *QWERTY keyboard*, because of the sequence of the first six leftmost keys in the third row, it has the distinction of allocating some of the most common English letters to the weakest fingers. One potential explanation is that the most commonly used keys were separated from each other so that they would not jam upon rapid sequential activation. Other more scientifically based alternative layouts, which allocate the letters more proportionately, have been developed, with the 1936 *Dvorak keyboard* being the most notable. However, scientific studies have shown the Dvorak layout to achieve at most a 5 percent improvement over the QWERTY layout, which is probably not a large enough improvement to justify switching and retraining millions of operators.

In certain special circumstances such as stenotyping and mail sorting, a *chord keyboard* may be more appropriate. Whereas in a standard keyboard, individual characters are keyed in sequentially, a chord keyboard requires the simultaneous activation of two or more keys. The basic trade-off is that with such activation, fewer keys are required and considerably more (estimates of 50 percent to 100 percent) information can be entered. It also has the distinct advantage of small size and one-handed operation, allowing the other hand to perform other tasks. However, for general use, the standard keyboard is more than sufficient without the need for additional specialized training.

Many keyboards will have a separate numeric keypad. Early research on telephones established that users preferred that numerals increase from left to right and then from top to bottom, which resulted in the standard layout for telephones. An alternate layout was developed for calculators, in which the numbers increase from the bottom to the top. Most research has shown that the telephone layout is slightly but significantly faster and more accurate than the calculator layout. Unfortunately, the difference was not large enough for the ANSI/HFS 100 standard (Human Factors Society, 1988) to favor one over the other, resulting in, perhaps, the worst situation in performance, with an operator having to alternate

General Principles	Yes	No
1. Is the number of absolute judgments limited to 7±2 items?	❏	❏
2. Is the difference between coding levels well above the JND?	❏	❏
3. Is the coding scheme compatible with human expectations?	❏	❏
4. Is the coding scheme consistent with existing schemes?	❏	❏
5. Is redundancy utilized for critical situations?	❏	❏

Visual Displays	Yes	No
1. Is the message long and complex?	❏	❏
2. Does the message deal with spatial information?	❏	❏
3. Does one need to refer to the message later?	❏	❏
4. Is hearing overburdened or is noise present?	❏	❏
5. Is the operator in a stationary location?	❏	❏
6. For general purpose and trends, is a fixed-scale, moving-pointer display being used?	❏	❏
a. Do scale values increase from left to right?	❏	❏
b. Does a clockwise movement indicate increasing values?	❏	❏
c. Is there an orderly numerical progression with major markers at 0, 10, 20, etc.?	❏	❏
d. Are there intermediate markers at 5, 15, 25, etc. and minor markers at each unit?	❏	❏
e. Does the pointer have a pointed tip just meeting the smallest scale markers?	❏	❏
f. Is the pointer close to the surface of the scale to avoid parallax?	❏	❏
7. For a very large scale, is an open-window display being used?	❏	❏
8. For precise readings, is a digital counter being used?	❏	❏
9. For check reading of a panel of dials, are pointers aligned and a pattern utilized?	❏	❏
10. For attentional purposes, are indicator lights being used?	❏	❏
a. Do the lights flash (1 to 10/sec) to attract attention?	❏	❏
b. Are the lights large (1 degree of visual arc) and bright?	❏	❏
c. Does the light remain on until the improper condition has been remedied?	❏	❏
11. Are alphanumeric characters of proper size?	❏	❏
a. Are they at least a 10 point font at a distance of 20 inches (22 min of visual arc)?	❏	❏
b. In a well-illuminated area, are the letters dark on a light background? Do they have a stroke width-to-height ratio of approximately 1:6?	❏	❏
c. In a dark area, are the letters white on a dark background? Do they have a stroke width-to-height ratio of approximately 1:8 for nighttime use?	❏	❏
d. Are both uppercase and lowercase letters utilized?	❏	❏
e. For special emphasis, are capitals or boldface utilized?	❏	❏

Auditory Displays	Yes	No
1. Is the message short and simple?	❏	❏
2. Does the message deal with events in time?	❏	❏

Figure 7.16 Display design checklist.

3. Is the message a warning or is immediate action required? ❑ ❑
4. Is vision difficult or overburdened? ❑ ❑
5. Is the operator moving about? ❑ ❑
6. Is a two-stage signal being utilized? ❑ ❑
7. Is the frequency of the signal in the range of 500 to 3,000 Hz for best auditory sensitivity? ❑ ❑
8. Is the sound level of the signal well above background noise? ❑ ❑
9. Is the signal being modulated (1 to 3 Hz) to attract attention? ❑ ❑
10. If the signal is traveling over 1,000 ft or around obstacles, is the frequency below 500 Hz? ❑ ❑
11. If a warning, is a separate communication channel being used? ❑ ❑

Figure 7.16 (*continued*)

between both layouts in an office environment (e.g., a telephone next to a computer keyboard).

For any keyboard the keys should be relatively large, spaced center to center 0.71 to 0.75 in (18 to 19 mm) horizontally and 0.71 to 0.82 in (18 to 21 mm) vertically. Smaller keys, which are becoming more common as PCs become smaller and smaller, become a disadvantage to individuals with large fingers with a definite decrease in speed and an increase in errors. The keys will have a preferred displacement between 0.08 and 0.16 in (2.0 and 4.0 mm) with the key force not to exceed 0.33 lb (1.5 N). Actuation of a key should be accompanied by either tactile or auditory feedback. Traditionally, a slight upward slope (0 to 15°) has been advocated. However, more recent research shows that a slight downward slope of $-10°$ may actually provide a more neutral and favorable wrist posture. Also split keyboards have been shown to reduce the ulnar deviation commonly found while using traditional one-piece keyboards. As mentioned in Chapter 5, armrests provide shoulder/arm support and reduce the shoulder muscle activity. These are recommended in place of the more commonly seen wrist rest, which may actually increase the pressure in the carpal tunnel and increase operator discomfort.

POINTING DEVICES

The primary device for data entry is the keyboard. However, with the growing ubiquitousness of graphical user interfaces and depending on the task performed, the operator may actually spend less than half the time using the keyboard. Especially for window and menu-based systems, some type of a cursor-positioning or a pointing device better than the cursor keys on a keyboard is needed. A wide variety of devices have been developed and tested. The *touch screen* either uses a touch-sensitive overlay on the screen or senses the interruption of an infrared beam across the screen as the finger approaches the screen. This approach is quite natural with the user simply touching the target directly on the screen. However, the finger can obscure the target and, in spite of the fairly large targets required, accuracy can be poor. The *light pen* is a special stylus linked to the computer by an electrical cable that senses the electron scanning beam at the particular location on the screen. The user has a similar natural pointing response as with a touch screen, but usually with greater accuracy.

A *digitizing tablet* is a flat pad placed on the desktop, again linked to the computer. Movement of a stylus is sensed at the appropriate position on the tablet, which can either be absolute (i.e., the tablet is a representation of the screen) or relative (i.e., only direct movement across the tablet is shown). Further complications are tablet size versus accuracy trade-offs and optimum control-response ratios. Also, the user needs to look back at the screen to receive feedback. Both displacement and force joysticks (currently termed *track sticks* or *track points*) can be used to control the cursor and have a considerable background of research on types of control systems, types of displays, control-response ratios, and tracking performance. The *mouse* is a handheld device with a roller ball in the base to control position and one or more buttons for other inputs. It is a relative positioning device and requires a clear space next to the keyboard for operation. The *trackball*, an upside-down mouse without the mouse pad, is a good alternative for work surfaces with limited space. More recently, *touchpads*, a form of digitizing tablets integrated into the keyboard, have become popular especially for notebook PCs.

A number of studies have compared these pointing devices, showing clear speed-accuracy trade-offs, i.e., the fastest devices (touch screens and light pens) being quite inaccurate. Keyboard cursor keys were very slow and probably not acceptable. Touchpads were a bit faster than joysticks, but were less preferred by users. Mice and trackballs were generally similarly good in both speed and accuracy and probably indicate why mice are so ubiquitous. However, there is a tendency for users to overgrip a mouse by a factor of 2 to 3 (i.e., use two to three times more than the minimum force required), with potential risk of incurring an injury. Use of a drag-lock, similar to what is found on trackballs, would reduce this risk. For a more detailed review of cursor-positioning devices, please refer to Greenstein and Arnaut (1988) or Sanders and McCormick (1993).

MONITOR SCREENS

The center of the monitor screen should be placed at the normal line of sight, which is roughly about 15 degrees below the horizontal. For a 15-in screen, at a typical 16-in reading distance, the edges are only slightly beyond the recommended ±15 degree cone of primary visual field. The implication is that within this optimal cone, no head movements are needed and eye fatigue is minimized. Thus, the top of the screen should not be above the horizontal plane through the eyes.

A 16-in reading distance, however, may not be optimal. A comfortable reading distance is a function not only of the size of the displayed characters, but also of the person's ability to maintain focus and alignment of the eyes. The mean resting focus (measured in the dark or in the absence of a stimulus with a laser optometer) is roughly 24 in. The implication is that the eye may be under greater stress when viewing characters at distances larger or smaller than 24 in, because then there will be compromise between the "pull" of the stimulus and the tendency of the eye to regress toward the individual's resting position. However, there is a large variation in individual resting focus distances. Therefore, office workers who view computer screens for extended periods of time may wish to have their eyes measured for their resting focus. Then if they are not able to set the monitor screen at the appropriate distance (e.g., excessive short or long distances), they can have their eyes fitted with special "computer-viewing" lenses (Harpster et al., 1989).

The monitor, preferably tiltable, should be placed relatively vertical such that the angle formed by the line of sight and the line normal to surface of the display is relatively small, but definitely less than 40 degrees. Tilting the screen upward should be avoided due to an increased likelihood of developing specular reflections from overhead lighting, leading to glare and decreased visibility. The screen should have minimum flicker, uniform luminance, and glare control if needed (polarizing or micromesh filters). Further details on hardware and furniture requirements for computer workstations can be found in the ANSI/HFS 100 standard (Human Factors Society, 1988).

NOTEBOOKS AND HANDHELD PCS

Portable PCs, or laptop or notebook computers, are becoming very popular, accounting for 34 percent of the U.S. PC market in 2000. Their main advantage over a desktop is reduced size (and weight) and portability. However, with the smaller size, there are distinct disadvantages: smaller keys and keyboard, keyboard attached to screen, and lack of a peripheral cursor-positioning device. The lack of adjustability in placing the screen has been found to give rise to excessive neck flexion (much beyond the recommended 15 degrees), increased shoulder flexion, and elbow angles greater than 90 degrees, which have accelerated feelings of discomfort as compared to using a desktop PC. Adding an external keyboard and raising the notebook computer or adding an external monitor helps alleviate the situation.

Even smaller handheld computers, termed *personal digital assistants*, have been developed but are too new to have had detailed scientific evaluations performed. Being pocket-sized, they offer much greater portability and flexibility, but at an even greater disadvantage for data entry. Decrements in accuracy and speed, when entering text via the touchscreen, have been found. Alternate input methods such as handwriting or voice input may be better.

7.7 HUMAN-COMPUTER INTERACTION: SOFTWARE CONSIDERATIONS

The typical industrial or methods engineer will not be developing programs but will, most likely, be using a variety of existing software. Therefore that person should be aware of current software features or standards that allow best human interaction with the computer and minimize the number of errors that could occur through poor design.

Most current interactive computing software utilizes the *graphical user interface* (GUI), identified by four main elements: windows, icons, menus, and pointers (sometimes collectively termed WIMP). *Windows* are the areas of the screen that behave as if they were independent screens in their own right. They typically contain text or graphics and can be moved around or resized. More than one window can be on a screen at once, allowing users to switch back and forth between various tasks or information sources. This leads to a potential problem of windows overlapping each other and obscuring vital information. Consequently, there needs to be a layout policy with windows being tiled, cascaded, or picture-in-a-picture (PIP). Usually windows have features that increase their usefulness such as scrollbars, allowing the user to move the contents of the window up and down or from left to right. This makes the window behave as if it were a real window onto a much larger world, where new information is brought into view by manipulating the scrollbars. There is usually a title bar attached to the top of the window, identifying it to the user, and there may be special boxes in the corners of the window to aid in resizing and closing.

Icons are small or reduced representations of windows or other entities within the interface. By allowing icons, many windows can be available on the screen at the same time, ready to be expanded to a useful size by clicking on the icon with a pointer (typically a mouse). The icon saves space on the screen and serves as a reminder containing the dialog. Other useful entities represented by icons include a wastebasket for deleting unwanted files, programs, applications, or files accessible to the user. Icons can take many forms: they can be realistic representations of the objects they stand for or they can be highly stylized, but with appropriate reference to the entity (known as compatibility) so that users can easily interpret them.

The *pointer* is an important component of the WIMP interface, since the selection of an appropriate icon requires a quick and efficient means of directly manipulating it. Currently the mouse is the most common pointing device, although joysticks and trackballs can serve as useful alternatives. A touchscreen, with the finger serving as a pointer, can serve as a very quick alternative and even redundant backup/safety measure in emergency situations. Different shapes of the cursor are often used to distinguish different modes of the pointer, such as an arrow for simple pointing, cross-hairs for drawing lines, and a paintbrush for filling in outlines. Pointing cursors are essentially icons or images and thus should have a hot spot that indicates the active pointing location. For an arrow, the tip is the obvious hot spot. However, cutesy images (e.g., dogs and cats) should be avoided because they have no obvious hot spot.

Menus present an ordered list of operations, services, or information that is available to the user. This implies that the names used for the commands in the menu should be meaningful and informative. The pointing device is used to indicate the desired option, with possible or reasonable options highlighted and impossible or unreasonable actions dimmed. Selection often requires an additional action by the user, usually clicking a button on a mouse or touching the screen with the finger or a pointer. When the number of possible menu items increases beyond a reasonable limit (typically 7 to 10), the items need to be grouped in separate windows with only the title or a label appearing on a menu bar. When the title is clicked, the underlying items pop up in a separate window known as a pull-down menu. To facilitate finding the desired item, it is important to group menu items by functionality or similarity. Within a given window or menu, the items should be ordered by importance and frequency of use. Opposite functions, such as SAVE and DELETE, should be clearly kept apart to prevent accidental misselection.

Other menulike features include buttons, isolated picture-in-picture windows within a display that can be selected by the user to invoke specific actions, toolbars, a collection of buttons or icons, and dialog boxes that pop up to bring important information to the user's attention such as possible errors, problems, or emergencies.

Other principles in screen design include simple usability considerations: orderly, clean, clutter-free appearance, expected information located where it should be consistently from screen to screen for similar functions or information. Eye-tracking studies indicate that the user's eyes typically first move to the upper left center of the display and then move quickly in a clockwise direction. Therefore, an obvious starting point should be located in the left upper corner of the screen, permitted the standard left-to-right and top-to-bottom reading pattern found in Western cultures. The composition of the display should be visually pleasing with balance, symmetry, regularity, predictability, proportion, and sequentiality. Density and grouping are also important features.

The appropriate use of uppercase and mixed-case fonts is important, with special symbols thrown in as necessary. Any text should be brief and concise with familiar words with minimal jargon. Simple action terms expressed in a positive mode are much more effective than some negative statements or standard military jargon. Color is appropriate to draw attention but should be used sparingly and limited to no more than eight colors. Note that a relatively high proportion of the population suffers from deficiencies in color vision.

The user should always feel under control and have the ability to easily exit screens or modules and undo previous actions. Feedback should appear for any actions, and the progress should be indicated for any long transactions. Above all, the main consideration in any display should be simplicity. The simpler the design, the quicker the response. Further information on software interface design can be found in Mayhew (1992), Galitz (1993), and Dix et al. (1998). As a convenience to the purchaser or user of software, the above desired features have been summarized in the Graphical User Interface Features Checklist (see Figure 7.17).

SUMMARY

This chapter presented a conceptual model of the human as an information processor along with the capacities and limitations of such a system. Specific details were given for properly designing cognitive work so as not to overload the human with regard to information presented through auditory and visual displays, to information being stored in various memories, and to information being processed as part of the final decision-making and response-selection step. Also, since the computer is the common tool associated with information processing, issues and design features with regard to the computer workstation were addressed. With manual work activities, the physical aspects of the workplace and tools, and the working environment having been addressed in Chapters 4, 5, and 6, the cognitive element is the final aspect of the human operator at work, and the analyst is now ready to implement the new method.

QUESTIONS

1. How can the informational content of a task be quantified?
2. What is redundancy? Give a good everyday example of redundancy.
3. Explain the five stages of the human information processing model.
4. How do information processing stages act to prevent an information overload of the human operator?
5. What are the four possible outcomes explained by signal detection theory?
6. Give an example of a task to which signal detection theory can be applied. What effect would a shift in the criterion have on task performance?
7. What is the meaning of sensitivity in signal detection theory? What techniques can be used to increase sensitivity in an inspection task?
8. What techniques can be used to improve memory?
9. What are some of the biases that may negatively affect a person's decision making?
10. What is compatibility? Give two everyday examples of compatibility.
11. Compare and contrast the different types of attention.
12. What is the inverted-U curve in attention?
13. Under what conditions are auditory displays best used?
14. What is the difference between absolute and relative judgment? What is the limitation in absolute judgment?
15. What is the just noticeable difference and how does it relate to the level of the stimulus?
16. Why is redundancy utilized for critical stimuli?
17. Why is a fixed-scale, moving-pointer display preferred?
18. What is the purpose of using patterns for a set of dials in a control room?
19. What key features are used to increase attention in a visual display?
20. What key features are used to increase attention in an auditory display?
21. What are the trade-offs between the different types of pointing devices?
22. What are the main components of a good graphical user interface?

PROBLEMS

1. *a.* What is the amount of information in a set of eight signal lights if each light has an equal probability of occurrence?

Windows Features	Yes	No
1. Does the software use movable areas of the screen termed *windows*?	❑	❑
2. Is there a layout policy for the windows (i.e., are they tiled, cascaded, or picture-in-picture)?	❑	❑
3. Are there scrollbars to allow the contents of windows to be moved up or down?	❑	❑
4. Are there meaningful titles identifying the windows?	❑	❑
5. Are there special boxes in the corners of the windows to resize or close them?	❑	❑

Icon Features	Yes	No
1. Are reduced versions of frequently used windows, termed *icons*, utilized?	❑	❑
2. Are the icons easily interpretable or realistic representations of the given feature?	❑	❑

Pointer Features	Yes	No
1. Is a pointing device (mouse, joystick, touchscreen) utilized to move icons?	❑	❑
2. Is the pointer or cursor easily identifiable with an obvious active area or hot spot?	❑	❑

Menu Features	Yes	No
1. Are meaningful menus (list of operations) with descriptive titles provided?	❑	❑
2. Are menu items functionally grouped?	❑	❑
3. Are menu items limited to a reasonable number (7 to 10)?	❑	❑
4. Are buttons available for specific common actions?	❑	❑
5. Are toolbars with a collection of buttons or icons used?	❑	❑
6. Are dialog boxes used to notify the user of potential problems?	❑	❑

Other Usability Considerations	Yes	No
1. Is the screen design simple, orderly, and clutter-free?	❑	❑
2. Are similar functions located consistently from screen to screen?	❑	❑
3. Is the starting point for the screen action the upper left-hand corner?	❑	❑
4. Does the screen action proceed left to right and top to bottom?	❑	❑
5. Is any text brief and concise and does it use both uppercase and lowercase fonts?	❑	❑
6. Is color used sparingly for attention (i.e., limited to eight colors)?	❑	❑
7. Does the user have control over exiting screens and undoing actions?	❑	❑
Is feedback provided for any action?	❑	❑

Figure 7.17 Graphical User Interface Features Checklist.

    b. The probabilities of the lights are changed as shown below. Calculate the amount of information and the redundancy in this configuration.

Stimulus	1	2	3	4	5	6	7	8
Probability	0.08	0.25	0.12	0.08	0.08	0.05	0.12	0.22

2. A large state university uses three-digit mail stops to code mail on campus. The initial step in sorting this mail is to sort according to the first digit (there are 10 possible), which signifies a general campus zone. This step is accomplished by pushing a key with corresponding number, which dumps the letter in an appropriate bin. A typical mail sorter can sort 60 envelopes per minute, and it takes a minimum of 0.3 s to just push the key with cognitive processing involved.

a. Assuming that the mail is distributed evenly over the campus zones, what is the mail sorter's bandwidth?

b. After a while the sorter notices that campus mail is distributed as follows. If the mail sorter used this information, how many pieces of mail could the sorter possibly handle in 1 min?

Zone	% Distribution
0	25
1	15
2	25
3–9	5

3. Bob and Bill are two weather forecasters for AccurateWeather. Bob is a veteran forecaster, while Bill is fresh out of school. The following are the records (in number of predictions) on both forecasters' ability to predict whether it will rain in the next 24 h.

a. Which forecaster would you hire? Why?

b. Who is a more conservative forecaster? Why?

c. How would a conservative versus a risky forecaster be beneficial for different geographic regions?

Bob said:	True result Rain	True result No rain
Rain	268	56
No rain	320	5,318

Bill said:	True result Rain	True result No rain
Rain	100	138
No rain	21	349

4. The Dorben Electronics Co., manufacturer of resistors, screens potential quality control inspectors before they are hired. Dorben has developed the following preemployment test. Each potential employee is presented with the same set of 1,000 resistors of which 500 are defective. The results for two applicants are as follows: (1) of the 500 good resistors, applicant 1 labeled 100 as defective, and of the 500 bad resistors, applicant 1 labeled 200 as defective; (2) of the 500 good resistors, applicant 2 labeled 50 as defective, and of the 500 bad resistors, applicant 2 labeled 300 as defective.

a. Treat picking a defective resistor as defective as a hit. Fill in the following table.

	Applicant 1	Applicant 2
Hit rate	_____	_____
False-alarm rate	_____	_____
Miss rate	_____	_____
Correct rejection rate	_____	_____
d'	_____	_____

b. Given that Dorben places a very large emphasis on quality control (i.e., they don't wish to sell a defective product at any cost), which applicant would better fulfill its goal? Why?

c. Given that Dorben wishes to hire the most efficient inspector (i.e., most correct), which applicant would the company hire?

5. The following performance data were collected under similar conditions on two inspectors removing defective products from the assembly line. Comment on the relative performance of the two inspectors. Which one is better at finding defects? Which would you hire if the cost of releasing a defective product were be high? Which does an overall better job? (Hint, consider d'.)

		Case 1	Case 2
JRS	Hit rate	0.81	0.41
	False alarm rate	0.21	0.15
ABD	Hit rate	0.84	0.55
	False alarm rate	0.44	0.31

6. The following response-time data (in msec) were obtained on Farmer Brown and his son Big John while operating a tractor using the right foot to control the clutch, brake, and accelerator. The foot is normally kept on the rest position. The location and sizes of the pedals are shown below as well as some sample response times (in msec) for activating a given control from the rest position.

a. What is the index of difficulty value for each pedal?

b. Plot the response times. What law can be used to explain the relationship between response times and the difficulty in activating a given control?

c. What is the simple reaction time for Farmer Brown?

d. What is the bandwidth for Big John?

e. Which farmer is the better tractor operator? Why?

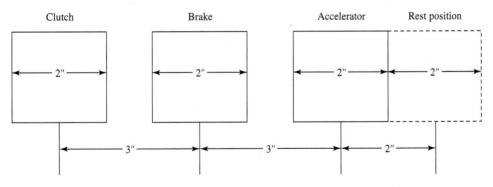

	Accelerator	Brake	Clutch
Farmer Brown	300	432	510
Big John	270	428	480

7. Disregarding the digit 0, which keyboard is fastest for entering digits using one finger? Assume that the home position is digit 5. (Hint: Calculate the index of difficulty.)

												#B			
				Scale	.5 in								7	8	9
#A					↑								4	5	6
1	2	3	4	5	6	7	8	9					1	2	3

8. Pianists must often hit keys in very quick succession. The figure below displays a typical piano keyboard.
   a. Compare the index of difficulty for striking C, C#, F, and F#. Assume one is starting from the A key and distances are measured center to center.
   b. If starting from the A key, it takes 200 ms to strike the C key and 500 ms to strike the F key, what is the bandwidth of a typical pianist?

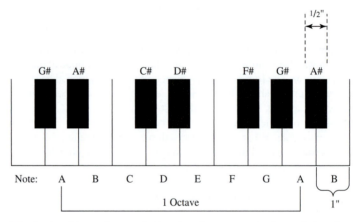

9. The knob with numbers and directional arrows on it shown below is used on refrigerators to control temperature. In which direction, clockwise or counterclockwise, would you turn the knob to make the refrigerator cooler? Why? How would you improve the control to avoid confusion?

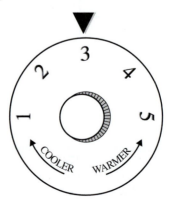

10. The dial shown below represents a pressure gage. The operational range is 50 psi. The operator must read the scale to the nearest 1 psi. Critically evaluate the dial, indicating the poor design practices. Then redesign the dial, following recommended design practices.

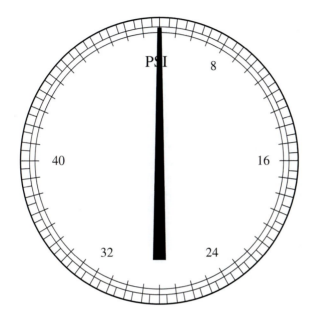

11. The scale shown below represents a scale used to measure weight. The maximum weight possible is 2 lb, and the scale must be read to the nearest 0.1 lb. Critically evaluate the dial, indicating the poor design practices. Then redesign the dial, following recommended design practices.

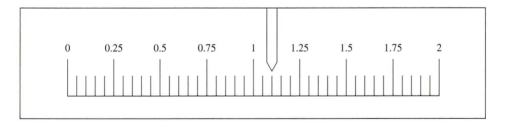

12. Design an EXIT sign which is to be displayed in a public auditorium. Explain the ergonomics principles that need to be considered for this sign.

REFERENCES

ANSI (American National Standards Institute). *ANSI Standard for Human Factors Engineering of Visual Display Terminal Workstations*. ANSI/HFS 100-1988. Santa Monica, CA: Human Factors Society, 1988.

Bishu, R. R., and C. G. Drury. "Information Processing in Assembly Tasks—A Case Study." *Applied Ergonomics*, 19 (1988), pp. 90–98.

Chapanis, A., and L. Lindenbaum. "A Reaction Time Study of Four Control-Display Linkages." *Human Factors*, 1 (1959), pp. 1–7.

Deatherage, B. H. "Auditory and Other Sensory Forms of Information Presentation." In H. P. Van Cott and R. Kinkade (Eds.), *Human Engineering Guide to Equipment Design*. Washington, DC: Government Printing Office, 1972.

Dix, A., J. Finlay, G. Abowd, and R. Beale. Human-Computer Interaction. 2d ed. London: Prentice-Hall, 1998.

Drury, C. "Improving Inspection Performance." In *Handbook of Industrial Engineering*. Ed. G. Salvendy. New York: John Wiley & Sons, 1982.

Drury, C. G., and J. L. Addison. "An Industrial Study on the Effects of Feedback and Fault Density on Inspection Performance." *Ergonomics*, 16 (1973), pp. 159–169.

Eggemeier, F. T. "Properties of Workload Assessment Techniques." In *Human Mental Workload*. Eds. P. Hancock and N. Meshkati. Amsterdam: North-Holland, 1988.

Fechner, G. *Elements of Psychophysics*. New York: Holt, Rinehart and Winston, 1860.

Fitts, P. "The Information Capacity of the Human Motor System in Controlling the Amplitude of Movement." *Journal of Experimental Psychology*, 47 (1954), pp. 381–391.

Galitz, W. O. *User-Interface Screen Design*. New York: John Wiley & Sons, 1993.

Giambra, L., and R. Quilter. "A Two-Term Exponential Description of the Time Course of Sustained Attention." *Human Factors*, 29 (1987), pp. 635–644.

Green, D., and J. Swets. *Signal Detection Theory and Psychophysics*. Los Altos, CA: Peninsula Publishing, 1988.

Greenstein, J. S., and L. Y. Arnaut. "Input Devices." In *Handbook of Human-Computer Interaction*. Ed. M. Helander. Amsterdam: Elsevier/North-Holland, 1988.

Harpster, J .L., A. Freivalds, G. Shulman, and H. Leibowitz. "Visual Performance on CRT Screens and Hard-Copy Displays." *Human Factors*, 31 (1989), pp. 247–257.

Helander, J. G., T. K. Landauer, and P. V. Prabhu, (Eds.). *Handbook of Human-Computer Interaction*, 2d ed. Amsterdam: Elsevier, 1997.

Hick, W. E. "On the Rate of Gain of Information." *Quarterly Journal of Experimental Psychology*, 4 (1952), pp. 11–26.

Human Factors Society. *American National Standard for Human Factors Engineering of Visual Display Terminal Workstations, ANSI/HFS 100-1988*. Santa Monica, CA: Human Factors Society, 1988.

Hyman, R. "Stimulus Information as a Determinant of Reaction Time." *Journal of Experimental Psychology*, 45 (1953), pp. 423–432.

Langolf, G., D. Chaffin, and J. Foulke. "An Investigation of Fitts' Law Using a Wide Range of Movement Amplitudes." *Journal of Motor Behavior*, 8, no. 2 (June 1976), pp. 113–128.

Mayhew, D. J. *Principles and Guidelines in Software User Interface Design*. Englewood Cliffs, NJ: Prentice-Hall, 1992.

Miller, G. "The Magical Number Seven, Plus or Minus Two: Some Limits on Our Capacity for Processing Information." *Psychological Review*, 63 (1956), pp. 81–97.

Sanders, M. S., and E. J. McCormick. *Human Factors in Engineering and Design.* 7th ed. New York: McGraw-Hill, 1993.

Stroop, J. R. "Studies of Interference in Serial Verbal Reactions." *Journal of Experimental Psychology*, 18 (1935), pp. 643–662.

Wickens, C. *Engineering Psychology and Human Performance.* Columbus, OH: Merrill, 1984.

Wickens, C. D. "Processing Resources in Attention." In *Varieties of Attention.* Eds. R. Parasuraman and R. Davies. New York: Academic Press, 1984.

Wickens, C. D., S. E. Gordon, and Y. Liu. *An Introduction to Human Factors Engineering.* New York: Longman, 1997.

Yerkes, R. M., and J. D. Dodson. "The Relation of Strength of Stimulus to Rapidity of Habit Formation." *Journal of Comparative Neurological Psychology*, 18 (1908), pp. 459–482.

SELECTED SOFTWARE

DesignTools (available from the McGraw-Hill text website at www.mhhe.com/niebels-freivalds), New York: McGraw-Hill, 2002.

WEBSITES

Examples of bad ergonomic design—http://www.baddesigns.com/

Workplace and Systems Safety

- Accidents result from a sequence of events with multiple causes.

- Examine accidents by using job safety analysis.

- Detail the accident sequence or system failure by using the fault tree analysis.

- Increase system reliability by adding backups and increasing component reliability.

- Consider trade-offs of various corrective actions by using cost-benefit analysis.

- Be familiar with OSHA safety requirements.

- Control hazards by

  ◆ Eliminating them completely, if possible
  ◆ Limiting the energy levels involved
  ◆ Using isolation, barriers, and interlocks
  ◆ Designing fail-safe equipment and systems
  ◆ Minimizing failures through increased reliability, safety factors, and monitoring

W orkplace safety is an extension of the concept of providing a good, safe, comfortable working environment for the operator, as discussed in Chapter 7. The primary goal here is not to increase production through more efficient working conditions or improved worker morale, but specifically to decrease the number of accidents, which potentially lead to injuries and loss of property. Traditionally, of foremost concern to the employer have been compliance with existing state and federal safety regulations and avoidance of a safety inspection by federal compliance officers (such as OSHA) with commensurate citations, fines, and penalties. However, more recently, the bigger driving force for implementing safety has been the escalating medical costs. Therefore, it only makes sense to implement a thorough safety program to reduce overall costs. Key issues of OSHA safety legislation and workers' compensation are introduced in this chapter, along with a general theories for accident prevention and hazard control. However, details for correcting specific hazards are not addressed here, as there are numerous traditional textbooks on safety that cover these details (Asfahl, 2004; Banerjee, 2003; Goetsch, 2005; Hammer and Price, 2001; National Safety Council, 2000). These books will also address the setup and maintenance of safety management organizations and programs.

8.1 BASIC PHILOSOPHIES OF ACCIDENT CAUSATION AND PREVENTION

Accident prevention is the tactical, sometimes relatively short-term, approach to controlling workers, materials, tools and equipment, and the workplace for the purpose of reducing or preventing the occurrence of accidents. This is in contrast to *safety management,* which is a relatively long-term strategic approach for the overall planning, education, and training of such activities. A good accident prevention process (see Figure 8.1) is an orderly approach very similar to the methods engineering program introduced in Chapter 2.

The first step in the accident prevention process is the identification of the problem in a clear and logical form. Once the problem is identified, the safety engineer the needs to collect data and analyze them so as to understand the causation of the accident and identify possible remedies to prevent it or, if not completely prevent it, at least to reduce the effects or severity of the accident. In many cases, there may be several solutions, and the safety engineer will need to select one of these solutions. Then the remedy will have to be implemented and monitored to ensure that it is truly effective. If it is not effective, the engineer may need to repeat this process and attempt another, perhaps better remedy. This monitoring effectively closes the feedback cycle and ensures a continuous improvement process for accident prevention.

DOMINO THEORY

In identifying the problem, it is important to understand some of the theories of accident causation and the sequence of steps in an accident itself. One such

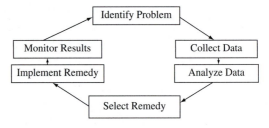

Figure 8.1 Accident prevention process.
(*Adapted from:* Heinrich, Petersen, and Roos, 1980)

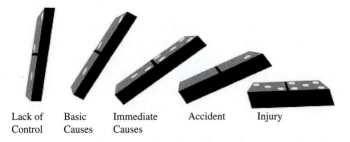

Lack of   Basic   Immediate   Accident   Injury
Control   Causes   Causes

Figure 8.2 The domino theory of an accident sequence.
(*Adapted from:* Heinrich, Petersen, and Roos, 1980)

theory is the *domino theory,* developed by Heinrich, Petersen, and Roos (1980) based on a series of theorems developed in the 1920s which formed the individual dominos (see Figure 8.2):

1. Industrial injuries (or loss of damages) result from accidents, which involve contact with an energy source and the consequent release of that energy.
2. Accidents are the result of immediate causes such as
   a. Unsafe acts by people
   b. Unsafe conditions in the workplace
3. The immediate causes result from more basic causes:
   a. The unsafe acts from personal factors such as lack of knowledge, skill, or simply the lack of motivation or care
   b. The unsafe conditions due to job factors, such as inadequate work standards, wear and tear, poor working conditions, due to either the environment or lack of maintenance
4. The basic causes result from an overall lack of control or proper management.

This domino (the first in the sequence) is essentially the lack of a properly implemented or maintained safety program, which should include elements to properly identify and measure job activities, establish standards proper standards for those jobs, measure worker performance on those jobs, and correct worker performance as needed.

Heinrich, Petersen, and Roos (1980) further postulated that the injury is simply the natural consequence of the previous events having taken place, similar to dominos falling in a chain reaction. As a proactive preventive measure, one could simply remove one of the previous dominos, thereby preventing the rest from falling and stopping the sequence prior to injury. They also emphasized that it was important to try to remove a domino as far upstream as possible, that is, to implement the corrective procedure as early as possible, at the root causes. The implication is that if effort is put into only preventing the injury, similar accidents will still occur in the future with potential for property damage and other types of injuries.

As an adaptation of the domino theory, Heinrich, Petersen, and Roos (1980) also emphasized the concept of *multiple causation;* that is, behind each accident or injury there may be numerous contributing factors, causes, and conditions. These combine in a rather random fashion, such that it might be difficult to identify which, if any, of the factors was the major cause. Therefore, rather than try to find just one major cause, it would be best to try to identify and control as many causes as possible, so as to get the biggest overall effect on controlling or preventing the accident sequence. As an example, among unsafe acts caused by the human, which Heinrich, Petersen, and Roos (1980) claim amount to 88% of all accidents, there could be (1) horseplay, (2) operating equipment improperly, (3) intoxication or drugs, (4) purposefully negating safety devices, or (5) not stopping a machine before cleaning or removing a stuck piece. Among unsafe conditions, which amount to 10% of all accidents (the remaining 2% are unpreventable "acts of God"), there could be (1) inadequate guards, (2) defective tools or equipment, (3) poorly designed machines or workplaces, (4) inadequate lighting, or (5) inadequate ventilation.

Figure 8.3 demonstrates the effects of various corrective actions taken along the domino sequence as well as multiple causation for a scenario in which sparks created by a grinder could ignite solvent fumes and cause an explosion and fire, with resulting burns to the operator. The injury is defined by burns to the operator. The accident leading to the injury is an explosion and fire. The sequence could be stopped by having the operator wear a fire-protective suit. The accident still happens, but a severe injury is avoided. Obviously, this is not the best control method as fire still could occur with other consequences to property. Moving one domino backward, the fire was caused by sparks from the grinder igniting

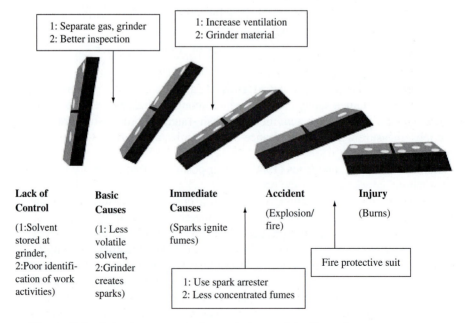

Figure 8.3 A domino sequence for a grinder spark igniting a fire.

volatile fumes in the grinding area. The sequence could be stopped at this stage by using a spark arrester or decreasing the concentration of the fumes through better ventilation. This still is a risky control measure as the spark arrester may not stop all sparks, and the ventilator may fail or slow during power brownouts. Moving backward another domino, more basic causes could include a couple of different factors (note the multiple causation) such as having such a volatile solvent and the fact that the grinder wheel acting on the casting creates sparks. The sequence could be stopped here by having a less volatile solvent or by installing a softer grinding wheel made of a different material that would not create sparks. Again these might not be most effective control measures with extremely hot weather increasing the vaporization of even a rather stable solvent and the grinding wheel perhaps creating sparks with harder castings. Furthermore they can have other, less positive consequences, such as a softer grinding wheel being less effective in smoothing the rough edges on the castings. The final domino of lack of control has probably a multitude of factors: poor identification of work activities that allowed the use of solvent in the grinding area, storage of a solvent in a work area, poor safety inspections, lack of awareness of the grinding operator, etc. At this stage, simply separating the dangerous elements, that is, removing the solvent from the grinding area, is the simplest, cheapest, and most effective solution.

Although, strictly speaking, the Heinrich, Petersen, and Roos (1980) *accident-ratio triangle*, which establishes the foundation for a major injury (see Figure 8.4a), is not an accident causation model, it emphasizes the necessity of moving backward in the accident progression sequence. For each major injury, most likely there were at least 29 minor injuries and 300 no-injury accidents, with untold hundreds or thousands of unsafe acts leading up to the base of the triangle. Therefore, rather than just reactively focus on the major injury or even the minor injuries, it makes sense for the safety engineer to look proactively, further back at the no-injury accidents and unsafe acts leading up to those accidents, as a field of opportunities to reduce potential injury and property damage costs and have a much more significant and effective total loss control program. This accident-ratio triangle was later modified by Bird and Germain (1985) to include property damage and revised numbers (see Figure 8.4b). However, the basic philosophy remained the same.

BEHAVIOR-BASED SAFETY MODELS

More recent accident causation models have focused on behavioral aspects of the human operator. The basis for this approach lies in early crisis research of Hill (1949) followed by the quantification of these crises or more modest situational factors into *life change units* (LCUs) by Holmes and Rahe (1967) (see Table 8.1). The basic premise of the theory is that situational factors tax a person's capacity to cope with stress in the workplace (or life in general), leaving the person more likely to suffer an accident as the amount of stress increases. It was found that 37 percent of individuals who accumulated between 150 and 199 LCUs in 2 years had illnesses. As the LCUs increased from 200 to 299, 51 percent had illnesses,

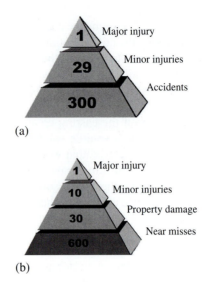

Figure 8.4   (a) Heinrich accident ratio triangle, (b) Bird and Germain (1985) accident ratio triangle.
(*Adapted from:* Heinrich, Petersen, and Roos, 1980)

and for those exceeding 300 LCUs, 79 percent had illnesses. This theory may help explain apparently accident-prone individuals and the need for having stressed workers avoid difficult or dangerous tasks.

Another behavioral accident causation model is the *motivation-reward-satisfaction model* presented by Heinrich, Petersen, and Roos (1980). It expands on Skinnerian concepts (Skinner, 1947) of *positive reinforcement* to achieve certain goals. In terms of safety, worker performance is dependent on the worker's motivation as well as the worker's ability to perform. In the main positive feedback cycle (see Figure 8.5), the better the worker performs; the better the worker is rewarded, the more the worker is satisfied, the greater the worker's motivation to perform better. This positive feedback could be applied both to safety performance and to worker productivity (which is the basis for wage incentive systems discussed in Chapter 17).

The most current and popular variation of behavior-based safety training is the *ABC model*. At the center of the model is behavior (the B part) of the worker, or what the worker does as part of the accident sequence. The C part is the consequence of the worker's behavior, or the events taking place after the behavior, leading to a potential accident and injury. The first A parts are antecedents (sometimes termed *activators*) or events that take place before the behavior occurs. Typically, this will start out as a negative process, in which the safety engineer tries to correct unpleasant consequences and determine what behaviors and antecedents lead to these consequences. For example, an operator takes a short-cut across a moving conveyor—a behavior. The antecedent may be break time as the operator tries to beat the lunchtime rush to get into the cafeteria line first. The

Table 8.1 Table of Life Change Units

Rank	Life Event	Mean Value
1	Death of spouse	100
2	Divorce	73
3	Marital separation	65
4	Jail term	63
5	Death of close family member	63
6	Personal injury or illness	53
7	Marriage	50
8	Fired at work	47
9	Marital reconciliation	45
10	Retirement	45
11	Changes in family member's health	44
12	Pregnancy	40
13	Sex difficulties	39
14	Gain of new family member	39
15	Business readjustment	39
16	Change in financial state	38
17	Death of close friend	37
18	Change to different line of work	36
19	Change in number of arguments with spouse	35
20	Mortgage over critical amount	31
21	Foreclosure of mortgage or loan	30
22	Change in work responsibilities	29
23	Son or daughter leaving home	29
24	Trouble with in-laws	29
25	Outstanding personal achievement	28
26	Wife beginning or stopping work	26
27	Begin or end school	26
28	Change in living conditions	25
29	Revision of personal habits	24
30	Trouble with boss	23
31	Change in work-hours, conditions	20
32	Change in residence	20
33	Change in schools	20
34	Change in recreation	19
35	Change in church activities	19
36	Change in social activities	18
37	Mortgage or loan under critical amount	17
38	Change in sleeping habits	16
39	Change in number of family get-togethers	15
40	Change in eating habits	15
41	Vacation	13
42	Christmas	12
43	Minor violations of the law	11

Source: Heinrich, Petersen, and Roos, 1980.

consequences are typically positive for the operator with more time to eat lunch, but in this particular instance are negative with an injury as the operator slipped on the conveyor. One approach in changing the behavior would be to post warnings on the dangers of jumping across the conveyor and to issue fines for violations.

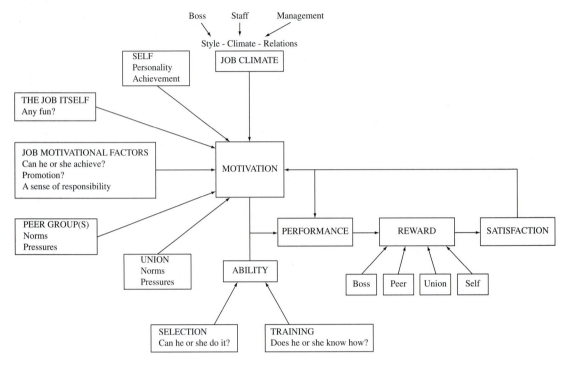

Figure 8.5   The motivation reward satisfaction model.
Source: Heinrich, Petersen, and Roos, 1980

However, this a negative approach that would require major enforcement action. That is, changing antecedents can get behavior started, but in many cases are not sufficient to maintain that behavior, especially if the approach focuses on the negative. A better approach would be to use the motivation-reward-satisfaction model and focus on positive consequences. This could be achieved by staggering lunch breaks for employees so that all would enjoy a relaxing, unrushed lunch break. It is also important to realize that the most effective activators are correlated with the most effective consequences—those that are positive, immediate, and certain.

Generally, behavior-based approaches are quite popular and effective as an accident prevention method, especially considering that the large majority (up to 88 percent) of accidents are due to unsafe acts and behaviors on the part of workers. Unfortunately this approach focuses solely on people and not on physical hazards. So there should also be mechanisms and procedures in place for ensuring safe workplace conditions. Finally, one should be careful that such programs do not become convoluted from the original purpose of promoting safety. From personal experience, a manufacturing company had implemented a positive reinforcement program of providing safety incentives for production workers: all workers in a department achieving a particular safety goal, for example, a month without a recordable injury, were provided a free lunch in the cafeteria. If this record was extended to six months, they received a steak dinner at a popular restaurant; and if they reached one year, then they received a $200 gift certificate. Of course, if there was a recordable injury, they had to start over from scratch. As result, once the stakes got high, injured workers were strongly encouraged by fellow workers not to report the injury to the plant nurse, circumventing the original intent!

8.2 ACCIDENT PREVENTION PROCESS

IDENTIFYING THE PROBLEM

In identifying the problem, the same quantitative exploratory tools (such as Pareto analysis, fish diagram, Gantt chart, job-worksite analysis guide) discussed in Chapter 2 for methods engineering can also be used in the first step in the accident prevention process. Another tool that is effective in identifying whether one department is significantly more hazardous than another is the *chi-square analysis*. This analysis is based on the chi-square goodness of fit test between a sample and a population distribution in the form of categorical data in a contingency table. Practically, this is expressed as a difference between m observed and expected cell counts of injuries (or accidents or dollars):

$$\chi^2 = \sum_{i}^{m} (E_i - O_i)^2 / E_i$$

where E_i = expected value = $H_i \times O_T / H_T$
O_i = observed value
O_T = total of observed values
H_i = hours worked
H_T = total of hours worked
m = number of areas compared

If the resulting χ^2 is greater than $\chi^2_{\alpha,m-1}$, the critical χ^2 at an error level of α and with $m - 1$ degrees of freedom, then there is a significant difference between the expected and observed values in injuries. Example 8.1 demonstrates an application for safety while more details on the statistical procedure can be found in Devore (2003).

Chi-Square Analysis of Injury Data   EXAMPLE 8.1

Dorben Co. has three main production departments: processing, assembly, and packing/shipping. It is concerned with the apparent high number of injuries in processing and would like to know if this is a significant deviation from the rest of the plant. Chi-square analysis comparing the number of injuries in 2006 (shown in Table 8.2) with an expected number based on the number of exposure hours is the appropriate way to study the problem.

The expected number of injuries in processing is found from

$$E_i = H_i \times O_T / H_T = 900,000 \times 36/2,900,000 = 11.2$$

JOB HAZARD ANALYSIS NUMBER _____

JOB
DESCRIPTION _MACHINING OF COUPLING_ PREPARED BY _____
ISSUING
DEPARTMENT _MACHINING_   REVIEWED BY _____
LOCATION _____—_____   DATE _4-12-_

KEY JOB STEPS	POTENTIAL HEALTH AND INJURY HAZARD	SAFE PRACTICES, APPAREL, AND EQUIPMENT
1) PICK UP UNFINISHED COUPLING	FORWARD LEAN WITH HIGH LS/S, COMP FORCE	TILT CARTON FORWARD
2) PLACE COUPLING IN FIXTURE	SHOULDER FLEXION & TORQUE DUE TO LOAD	DECREASE REACH DISTANCE LOWER FIXTURE HEIGHT
3) TIGHTEN FIXTURE WITH WRENCH	SHOULDER FLEXION WITH EXERTION	LOWER FIXTURE HEIGHT
4) BLOW OUT DEBRIS WITH COMPRESSED AIR	INHALE DUST	WEAR DUST MASK
LOOSEN FIXTURE WITH WRENCH (3)	SHOULDER FLEXION WITH EXERTION	LOWER FIXTURE HEIGHT
REMOVE COUPLING FROM FIXTURE (2)	SHOULDER FLEXION & TORQUE DUE TO LOAD	DECREASE REACH DISTANCE LOWER FIXTURE HEIGHT
5) SMOOTH ROUGH EDGES WITH GRINDER	VIBRATION	WEAR GEL GLOVES
6) PLACE FINISHED COUPLING IN CARTON	FORWARD LEAN WITH HIGH LS/S, COMP FORCE	TILT CARTON FORWARD

JHA No. _____
Page _____ of _____

Figure 8.7 Job safety analysis.

SELECT A REMEDY—RISK ANALYSIS AND DECISION MAKING

Once JSAs have been completed and a variety of solutions have been suggested, the safety engineer will need to choose one for implementation. This can be done by using a variety of decision-making tools in the fourth step of the accident prevention process, select a remedy. Most of these tools are just as appropriate for selecting a new method for improved productivity and are presented in Chapter 9. However, one of these tools, *risk analysis,* is more suitable for safety because it calculates the potential risk for an accident or injury and the reduction of risk due to modifications. According to Heinrich, Petersen, and Roos (1980), the analysis is based on the premise that the risk for injury or loss cannot be completely eliminated; that only a reduction in risk or potential loss can be achieved. Furthermore, any modifications should consider maximum cost effectiveness.

According to the method (Heinrich, Petersen, and Roos, 1980), the potential loss increases with (1) increased likelihood or probability that the hazardous event will occur, (2) increased exposure to the hazardous conditions, and (3) increased possible consequences of the hazardous event. Numerical values are assigned to each of the above three factors (see Table 8.2), and then an overall risk score is computed as a product of these factors (see Table 8.3). Note that these numerical values are rather arbitrary, and consequently the final risk score is also rather arbitrary. This, however, doesn't negate the method; it still serves as method to provide good relative comparison between different safety features or controls.

As an example of risk analysis, consider an event that is conceivable but rather unlikely with a value of 0.5, with a weekly exposure and value of 3, and

Table 8.3 Risk Analysis Factor Values

Likelihood	Values		Exposure	Values
Expected	10		Continuous	10
Possible	6		Daily	6
Unusual	3		Weekly	3
Remote	1		Monthly	2
~ Conceivable	0.5		Few/Year	1
~ Impossible	0.1		Yearly	0.5

Possible Consequences	Value
Catastrophe (many fatalities, 10^8 damage)	100
Disaster (few fatalities, 10^7 damage)	40
Very serious (fatality?, 10^6 damage)	15
Serious (serious injuries, 10^5 damage)	7
Important (injuries, 10^4 damage)	3
Noticeable (first aid, 10^3 damage)	1
Adapted from: Heinrich, Petersen, and Roos (1980).	

Table 8.4 Risk Analysis and Cost-Effectiveness

Risk Situation	Value
Very high risk, discontinue operations	400
High risk, immediate correction	200–400
Substantial risk, correction needed	70–200
Possible risk, attention needed	20–70
Risk?, perhaps acceptable	< 20

Source: Heinrich, Petersen, and Roos (1980).

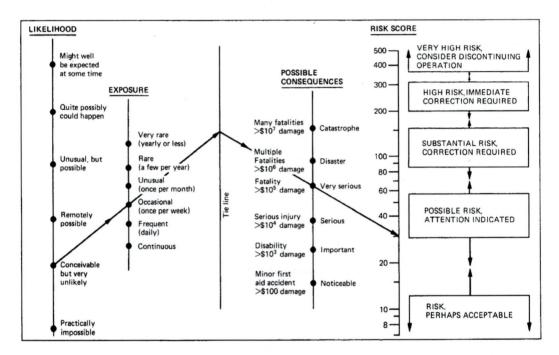

Figure 8.8 Risk analysis calculation.
Source: Heinrich, Petersen, and Roos, 1980.

very serious consequences with a value of 15. The resulting product yields a risk score of 22.5 (= 0.5 × 3 × 15) which corresponds to a rather low risk, with possible attention needed, but not urgent attention. See Table 8.4. This same result can be achieved by using Figure 8.8 and a tie line to connect the two halves of the chart. The cost-effectiveness of two different remedies for the above risky event can be compared by using Figure 8.9. Remedy A reduces the risk by 75 percent but costs $50,000 while remedy B reduces the risk by only 50 percent, but also costs only $500. In terms of cost-effectiveness, remedy A is of doubtful merit and may have difficulty receiving financial support, while remedy B may well be justified because of its lower cost.

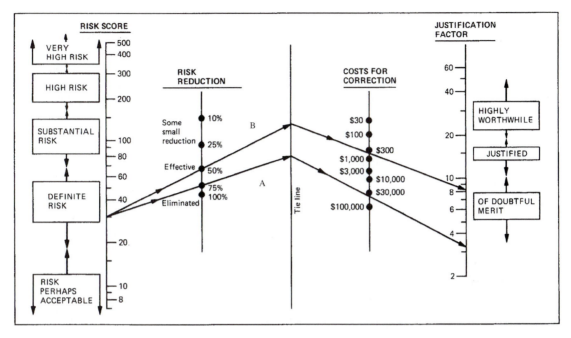

Figure 8.9 Risk analysis and cost-effectiveness.
Source: Heinrich, Petersen, and Roos, 1980.

After an appropriate cost-effective remedy has been selected, the remedy needs to be implemented in the fifth step of the accident prevention method. This should occur at several levels. The safety engineer with appropriate technicians will install the appropriate safety devices or equipment. However, for a completely successful implementation, the individual operators and supervisors must also buy into the new approach. If they don't follow the correct procedures with the new equipment, any potential safety benefits may be lost. As an aside, this also presents an opportunity to discuss the *3 E's* approach: engineering, education, and enforcement. The best remedy almost always is an engineering redesign. This will ensure strict safety and doesn't rely strictly on operator compliance. The next-best remedy is education; however, this does rely on operator compliance and may not always succeed, especially if workers do not follow the correct procedures. Lastly, there is enforcement of strict rules and use of personal protective equipment. This presumes worker noncompliance, requires strict checkups, instills resentment with negative reinforcement, and should be used as a last resort.

MONITORING AND ACCIDENT STATISTICS

The sixth and final step in the accident prevention process is the monitoring of the situation to evaluate the effectiveness of the new method. This provides feedback on the process and closes the loop by restarting the cycle in case the situation is not improving. Typically, numerical data provide a solid benchmark for monitoring any changes. These could be insurance costs, medical costs, or simply numbers of injuries and/or accidents. However, any of these numbers should be normalized to the worker exposure hours so that the numbers can be compared across locations and industries. Furthermore, OSHA recommends expressing injury statistics as *incidence rate* (IR) per 100 full-time employees per year:

$$\text{IR} = 200{,}000 \times I/H$$

where I = number of injuries in given time period
         H = employee hours worked in same time period

For OSHA record-keeping purposes, the injuries should be OSHA-recordable, or more than simple first-aid injuries. However, research has shown that there are considerable similarities between minor and major injuries (Laughery and Vaubel, 1993). Similarly, the *severity rate* (SR) monitors the number of lost-time (LT) days:

$$\text{SR} = 200{,}000 \times \text{LT}/H$$

In addition to simply recording and monitoring the incidence rates as they change from month to month, the safety engineer should apply statistical control charting principles and look for long-term trends. The control chart (see Figure 8.10) is based on a normal distribution of the data and establishing a lower control limit (LCL) and a upper control limit (UCL) as defined by

$$\text{LCL} = \bar{x} - ns$$

$$\text{UCL} = \bar{x} + ns$$

where $\bar{x}$ = sample mean
         s = sample standard deviation
         n = level of control limits

For example, for the case in which we would expect $100(1-\alpha)$ percent of the data to fall between the upper and lower control limits, n would simply be the standard normal variable $z_{\alpha/2}$. For $\alpha = 0.05$, n becomes 1.96. However, for many

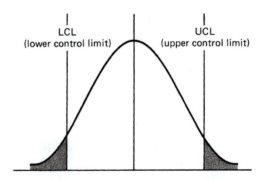

Figure 8.10   Statistical control limits.
Source: Heinrich, Petersen, and Roos, 1980.

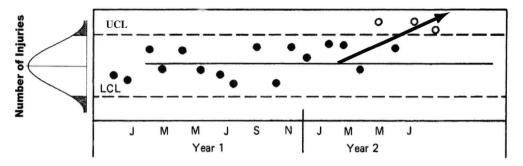

Figure 8.11 Red flagging with control chart.
Adapted from: Heinrich, Petersen, and Roos, 1980.

situations a higher lever of control is needed with $n = 3$ or even $n = 6$ (the Motorola six-sigma control level). For tracking accidents or injuries, the control chart is rotated sideways, and monthly data are plotted on the chart (see Figure 8.11). Obviously the lower control limit is of less concern (other than a nice pat on the back) than the upper control limit. Should several consecutive months fall above the upper control limit, this should be a red flag or signal to the safety engineer that there is a problem and a serious effort should be put into finding the cause. In addition to the *red flagging,* an alert safety engineer should have noticed the upward trend beginning several months previously and started a control action earlier. This trend analysis could be easily performed using a moving linear regression over varying multiple-month periods.

8.3 PROBABILITY METHODS

The accident causation models discussed previously, especially the domino theory, implied a very deterministic response. That is far from the case. Grinding without safety glasses or walking under an unsupported roof in a coal mine at a given moment does not ensure an automatic accident and injury. However, there is a chance that an accident will occur, and the likelihood of that happening can be defined with a numeric probability.

Probability is based on Boolean logic and algebra. Any event is defined by a binary approach; that is, at any given moment, there are only two states—the event either exists and is true (T), or it does not exist and is false (F). There are three operators that define interactions between events:

1. AND, the intersection between two events, with symbol ∩ or • (the dot sometimes is omitted)
2. OR, the union between events, with symbol ∪ or +
3. NOT, the negation of an event, with the symbol ⁻.

The interaction of two events *X* and *Y,* using these operators, follows a specific pattern termed the *truth tables* (see Table 8.5). Interactions between more than

Table 8.5　Boolean Truth Tables

X	Y	X·Y	X +Y		− (Not)
				X	$\overline{X}$
T	T	T	T		
T	F	F	T	T	F
F	T	F	T	F	T
F	F	F	F		

Table 8.6　Boolean Algebra Simplifications

- Basic laws:

$$X \cdot X = X \qquad\qquad X\overline{X} = 0$$
$$X + X = X \qquad\qquad X + \overline{X} = 0$$
$$XT = X \qquad\qquad XF = 0$$

- Distributive laws:

$$XY + XZ = X(Y + Z) \qquad (X+Y)(X+Z) = X + YZ$$
$$XY + X\overline{Y} = X \qquad\qquad X + Y\overline{X} = X + Y$$
$$X + XY = X \qquad\qquad X(X + Y) = X$$
$$X(\overline{X} + Y) = XY \qquad\qquad (X+Y)(X+\overline{Y}) = X$$

two events result in more complicated expressions, which necessitate an ordered processing to evaluate the resultant overall probability of the final accident or injury. The specific order or precedence that must be followed is as follows: ( ),⁻, •, +. Also, certain groupings of events tend to appear repeatedly, so that if one recognizes these patterns, simplification rules can be applied to quicken the evaluation procedure. The most common rules are given in Table 8.6.

The *probability of an event P(X)* is defined as the number of times event X occurs out of the total number occurrences:

$$P(X) = \#X/\#\text{Total}$$

and $P(X)$ must necessarily lie between 0 and 1. The probability of ORed events $X \cup Y$, is defined as

$$P(X + Y) = P(X) + P(Y) - P(XY)$$

if the events are not mutually exclusive, and as

$$P(X + Y) = P(X) + P(Y)$$

if the events are mutually exclusive. Two events are defined as *mutually exclusive* if the two events do not intersect, that is, $X \cap Y = 0$. Thus necessarily X and $\overline{X}$ are mutually exclusive. For the union of more than two events, an alternate expression, based on reverse logic, that is much easier to evaluate is more typically used:

$$P(X + Y + Z) = 1 - [1 - P(X)][1 - P(Y)][1 - P(Z)]$$

The probability of ANDed events is defined as

$$P(XY) = P(X)P(Y) \tag{2}$$

if the two events are independent, and as

$$P(XY) = P(X)P(Y/X) = P(Y)P(X/Y) \tag{3}$$

if the two events are not independent. Two events are defined as *independent* if the occurrence of one event doesn't affect the occurrence of another event. Mathematically this is determined by equating Equations (2) and (3) and removing $P(X)$ from both sides, yielding

$$P(Y) = P(Y/X) \qquad (4)$$

if the two events are independent. Rearrangement of Equation (3) also yields a commonly used expression that is termed *Bayes' rule:*

$$P(Y/X) = P(Y)P(X/Y)/P(X) \qquad (5)$$

Note that two events cannot be both mutually exclusive and independent, because being mutually exclusive necessarily defines as the one event defining the other one, that is, being dependent. Example 8.2 demonstrates these various calculations as well as independent and nonindependent events. More details on basic probability can be found in Brown (1976).

Independent and Not Independent Events	EXAMPLE 8.2

Consider the number of occurrences of A being true (or 1) out of the total number of occurrences in Table 8.7a. This determines the probability of A:

$$P(X) = \#X/\#\text{Total} = 7/10 = 0.7$$

Note that the probability of $\overline{A}$ is the number of occurrences of being false (or 0) out of the total number of occurrences:

$$P(\overline{X}) = \#\overline{X}/\#\text{Total} = 3/10 = 0.3$$

Also $P(\overline{X})$ can be found from

$$P(\overline{X}) = 1 - P(X) = 1 - 0.7 = 0.3$$

Similarly the probability of Y is

$$P(Y) = \#Y/\#\text{Total} = 4/10 = 0.4$$

The probability of $A \cap B$ is the number of occurrences of both A and B being true out of the total number of occurrences:

$$P(XY) = \#XY/\#\text{Total} = 3/10 = 0.3$$

Table 8.7 Independent or Not Independent Events

(a) *X* and *Y* are not independent				(b) *X* and *Y* are independent			
	X				*X*		
Y	0	1	Total	*Y*	0	1	Total
0	2	4	6	0	2	3	5
1	1	3	4	1	4	6	10
Total	3	7	10	Total	6	9	15

The conditional probability of X given that Y has occurred (or is true) is defined as the number of occurrences of X from the $Y = 1$ row:

$$P(X/Y) = \#X/\#\text{Total } Y = 3/4 = 0.75$$

Similarly,

$$P(Y/X) = \#Y/\#\text{Total } X = 3/7 = 0.43$$

Note also Bayes' rule:

$$P(Y/X) = P(Y)P(X/Y)/P(X) = 0.4 \times 0.75/0.7 = 0.43 = P(Y/X)$$

Finally, for two events to be independent, Equation (4) must be true. But for Table 8.7a, we found that $P(Y) = 0.4$ while $P(Y/X) = 0.43$. Therefore events X and Y are not independent. However in Table 8.7b, we find

$$P(X) = \#X/\#\text{Total} = 9/15 = 0.6$$
$$P(X/Y) = \#X/\#\text{Total } Y = 6/10 = 0.6$$

Since both expressions are equal, events X and Y are independent. The same equivalence is found for $P(Y)$ and $P(Y/X)$:

$$P(Y) = \#Y/\#\text{Total} = 10/15 = 0.67$$
$$P(Y/X) = \#Y/\#\text{Total } X = 6/9 = 0.67$$

8.4 RELIABILITY

The term *reliability* defines the probability of the success of a system, which necessarily must depend on the reliability or the success of its components. A system could be either a physical product with physical components or an operational procedure with a sequence of steps or suboperations that need to be completed correctly for the procedure to succeed. These components or steps can be arranged in combinations using two different basic relationships: series and parallel arrangements. In a *series* arrangement (see Figure 8.12a), every component must succeed for the total system T to succeed. This can be expressed as the intersection of all components

$$T = A \cap B \cap C = ABC$$

which if independent (in most cases), yields a probability of

$$P(T) = P(A)P(B)P(C)$$

or if not independent, yields

$$P(T) = P(A)P(B/A)P(C/AB)$$

In a *parallel* arrangement, the total system succeeds if any one component succeeds. This can be expressed as an union of the component

$$T = A \cup B \cup C = A + B + C$$

(a) Series system

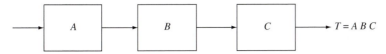

(b) Parallel system

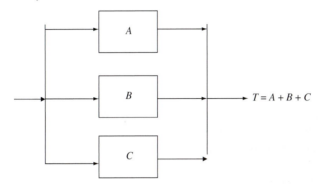

Figure 8.12 Series and parallel components.

which if mutually exclusive (typically), yields a probability of

$$P(T) = 1 - [1 - P(A)][1 - P(B)][1 - P(C)]$$

Two examples (Examples 8.3 and 8.4) demonstrate the calculation of system probabilities.

Reliability of a Two-Stage Amplifier   EXAMPLE 8.3

Consider two prototypes of a two-stage amplifier with backup components. Prototype 1 (see Figure 8.13) has a backup for the full amplifier, while prototype 2 has a backup for each stage. Which has the better system reliability, given that all the components are independent but identical with the same reliability of 0.9?

The best approach is to write all possible paths for system success. For prototype 1 there are two possible paths, either *AB* or *CD*. Written as an expression, system success is

$$T = AB + CD$$

Expressed as a probability, this expression becomes

$$P(T) = P(AB) + P(CD) - P(AB)P(CD)$$

where

$$P(AB) = P(A)P(B) = 0.9 \times 0.9 = 0.81 = P(CD)$$

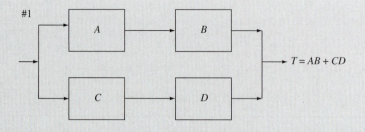

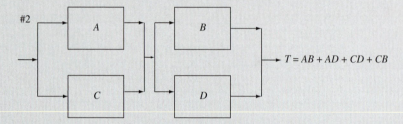

Figure 8.13　Two prototypes of a two-stage amplifier.

The overall system reliability becomes

$$P(T) = 0.81 + 0.81 - 0.81 \times 0.81 = 0.964$$

For prototype 2 there are four possible paths: *AB* or *AD* or *CB* or *CD*. Written as an expression, system success is

$$T = AB + AD + CB + CD$$

which simplifies to

$$T = (A + C)(B + D)$$

[Note that complicated probability expressions will need to be simplified; otherwise there may be an incorrect calculation. There are two basic *distributive laws* that form the basis for all simplification rules. These are:

$$(X + Y)(X + Z) = X + YZ$$
$$XY + XZ = X(Y + Z)$$

The others in Table 8.6 can be derived by substituting $\overline{X}$ for X, Y, or Z above.]
Now substituting values for the variables yields

$$P(A + C) = P(A) + P(C) - P(A)P(C) = 0.9 + 0.9 - 0.9 \times 0.9 = 0.99$$

The overall system reliability becomes

$$P(T) = 0.99 \times 0.99 = 0.9801$$

Therefore prototype 2 is the better amplifier with higher system reliability.

Reliability of a Four-Engine Airplane

EXAMPLE 8.4

Consider an airplane with four independent but identical engines (see Figure 8.14). The airplane can obviously fly with all four engines working, with any three engines working, and also with two engines working, as long as there is one engine working on each side of the plane; that is, two engines working on one side would cause the plane to be so unbalanced as to crash. What is the overall reliability of the airplane given that the reliability of each engine is 0.9?

Writing out all the possible engine scenarios results in these expressions:

$$4 \text{ engines} \Rightarrow ABCD$$

$$3 \text{ engines} \Rightarrow ABC + ABD + BCD + ACD$$

$$2 \text{ engines} \Rightarrow AC + AD + BC + BD$$

$$T = ABCD + ABC + ABD + BCD + ACD + AC + AD + BC + BD$$

The expression must be simplified. Note that the three-engine combinations are roughly redundant to the two-engine combinations:

$$AC + ABC = AC(1 + B) = AC$$

Similarly, the four-engine combination is redundant to any of the two-engine combinations, resulting in a final expression for system reliability of

$$T = AC + AD + BC + BD$$

This expression further simplifies to

$$T = (A + B)(C + D)$$

The probability for each expression in parentheses is

$$P(A + B) = P(A) + P(B) - P(A)P(B) = 0.9 + 0.9 - 0.9 \times 0.9 = 0.99$$

and the total system probability becomes

$$P(T) = 0.99 \times 0.99 = 0.9801$$

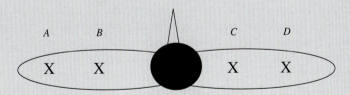

$$T = ABCD + ABC + ABD + BCD + ACD + AC + AD + BC + BD$$

Figure 8.14 Reliability of a four-engine airplane.

Examples 8.2 and 8.3 indicate one of the basic principles for increasing system reliability—increasing redundancy by adding components in parallel to the original component. Thus two or more components or operations are providing the same function. Note that if two or more operations or components are required

(a) Basic event mapping

Y\X	0 ($\bar{X}$)	1 (X)
0 ($\bar{Y}$)	Neither X nor Y occurs	X occurs, but Y does not
1 (Y)	Y occurs, but X does not	Both X and Y occur

(b) $T = X + Y$

Y\X	0	1
0		X
1	Y	XY

(c) $T = XY$

Y\X	0	1
0		
1		XY

(d) $T = \bar{X}$

Y\X	0	1
0		
1		

Figure 8.15 Basics of Karnaugh mapping.

to prevent an accident, then those elements are in series and do not provide redundancy. System reliability can also increased by increasing the reliability of the individual components. If the reliability of each engine in Example 8.3 is increased to 0.99, the total system reliability increases to 0.9998 as opposed to the original 0.9801. Note, however, that there is a trade-off—increasing the reliabilities of individual components or increasing the number of parallel components will necessarily increase the cost of the system. At some point, the increasing costs may not merit the marginal increase in overall system reliability. This decision point may vary considerably, depending on whether the system of interest is a simple consumer product as opposed to commercial airlines.

As the Boolean expressions for system reliability become more complex, the simplification process can become correspondingly more complex and tedious. At that point, the use of *Karnaugh maps* will provide a definitive solution to the problem. A Boolean algebra map is developed so as to represent the spatial representation of all possible events. Each event represents either a row or a column in a gridlike matrix. At its simplest, with two events, the matrix is 2×2, with both conditions (either true or false) of one event Y represented as rows, while both conditions of another event X are represented as columns (see Figure 8.15a). Then the expression $X + Y$ is represented as three cells of the matrix (see Figure 8.15b), the expression XY is represented as one cell of the matrix (see Figure 8.15c), and $\bar{X}$ is a vertical column under 0 (see Figure 8.15d). For more events, the matrix will increase in size; for example, with 4 variables, the matrix will have 16 cells, and for 6 variables the matrix will have 36 cells. Beyond six variables, it will become difficult to handle the matrix, and a computerized approach will become more practical. Also, with more complicated expressions, the expression must be written in the form of a simple sum of products.

Each term or product is then marked appropriately on the matrix. Undoubtedly there will be overlapping areas, or cells that are marked several times.

A simplified expression is then written of nonoverlapping groups of cells, each having a unique characterization of those cells. If done correctly, each group is mutually exclusive, and the addition of the probabilities for each group is quite straightforward. Note that it is better to identify the largest possible areas so that fewer calculations are needed later. These areas will consist of a certain number of cells as determined by powers of 2, that is, 1, 2, 4, 8, etc. This is best demonstrated by Example 8.4, the previous airplane example but using Karnaugh maps. More details on Karnaugh maps can be found in Brown (1976).

Reliability of a Four-Engine Airplane Using Karnaugh Maps | EXAMPLE 8.5

Consider the same airplane of Example 8.4 with four independent but identical engines and the event expression

$$T = ABCD_1 + ABC_2 + ABD_3 + CDB_4 + CDA_5 + AD_6 + BC_7 + AC_8 + BD_9$$

Each combination of events can be delineated on a Karnaugh map (see Figure 8.16a) as indicated by the appropriate numbers. Many of the events or areas are overlapping, and only the nonoverlapping areas should be evaluated for the probability. Four such nonoverlapping areas are shown in Figure 8.16b, although there could be many other possible combinations of nonoverlapping areas. In this case the largest possible combination of areas was selected: one of four cells, two two-cell combinations, and one remaining cell. The resulting expression is

$$T = AD_1 + AC\overline{D}_2 + \overline{A}BC_3 + \overline{A}B\overline{C}D_4$$

with a value of

$$P(T) = (0.9)(0.09) + (0.9)(0.9)(0.1) + (0.1)(0.9)(0.9)$$
$$+ (0.1)(0.9)(0.1)(0.9) = 0.9801$$

Obviously the nine cells could be individually identified and calculated, but that would entail a much greater effort. As noted previously, larger areas will be characterized by powers of 2, that is, 1, 2, 4, 8, etc. To simplify the calculations even further, a reverse logic can also be used to define the nonmarked areas. This value is then subtracted from 1 to obtain the true probability of the event of interest (see Figure 8.16c).

$$\overline{T} = \overline{AB}_1 + \overline{BCD}_2 + A\overline{BCD}_3$$

This results in a probability of

$$P(T) = 1 - P(\overline{T}) = 1 - (0.1)(0.1) - (0.9)(0.1)(0.1)$$
$$- (0.9)(0.1)(0.1)(0.9) = 0.9801$$

and the same value is calculated in the direct method.

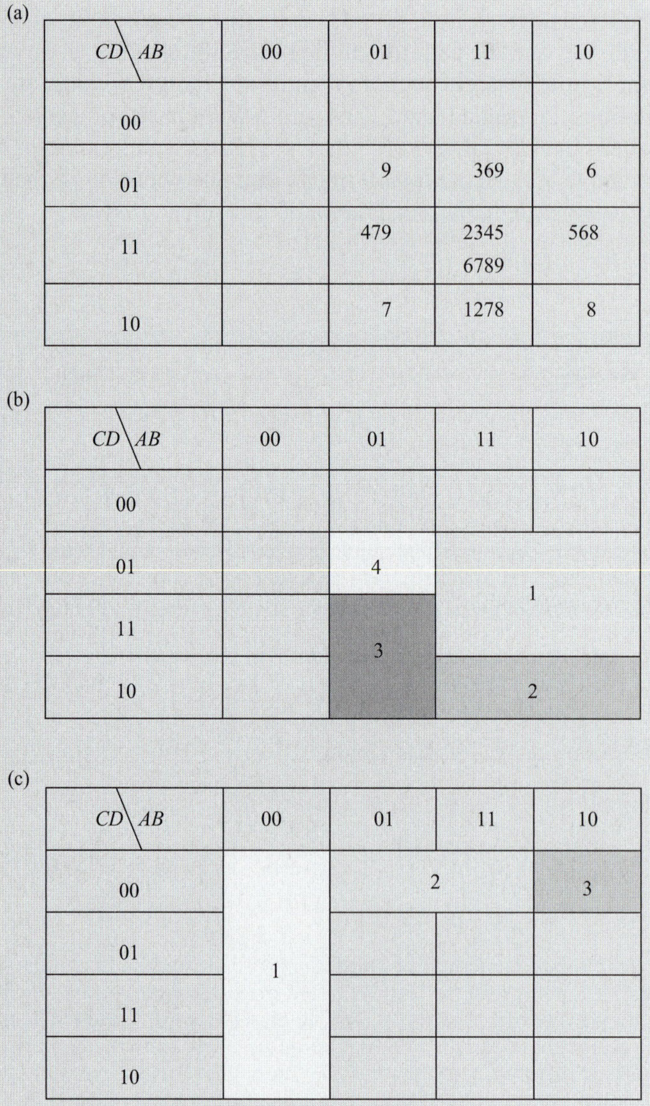

(a)

CD\AB	00	01	11	10
00				
01		9	369	6
11		479	2345 6789	568
10		7	1278	8

(b)

CD\AB	00	01	11	10
00				
01		4		
11		3		1
10			2	

(c)

CD\AB	00	01	11	10
00			2	3
01	1			
11				
10				

Figure 8.16   Reliability of a four-engine airplane using Karnaugh maps.

8.5   FAULT TREE ANALYSIS

Another approach to examining accident sequences or system failures uses
fault tree analysis. This is a probabilistic deductive process using a graphical
model of parallel and sequential combinations of events, or faults, leading to
the overall undesired event, for example, an accident. It was developed by Bell

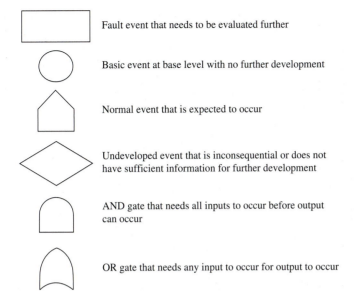

☐	Fault event that needs to be evaluated further
◯	Basic event at base level with no further development
⌂	Normal event that is expected to occur
◇	Undeveloped event that is inconsequential or does not have sufficient information for further development
⌒	AND gate that needs all inputs to occur before output can occur
⌒	OR gate that needs any input to occur for output to occur

Figure 8.17 Fault tree symbols.

Laboratories in the early 1960s to assist the U.S. Air Force in examining missile failures and later used by NASA to ensure overall system safety for the manned space program. These events can be of various types and identified by different symbols (see Figure 8.17). In general, there are two main categories: *fault events,* identified by rectangles, that are to be expanded further, starting from the top head event; and *basic events,* identified by circles, at the bottom of the fault tree that cannot be developed any further. Formally, there can also be house-shaped symbols indicating that "normal" events can be expected to occur and diamond-shaped symbols indicating inconsequential events or events with insufficient data for further analysis. The events are linked with gates that involve the same Boolean logic described previously (see Figure 8.17 for the symbols). An *AND gate* requires that all the inputs occur for the output to occur. An *OR gate* needs at least one of the inputs to occur for the output to occur. Obviously this implies that several of or all the inputs could occur as well, but only one is needed. It also helps to define the input events to the OR gate to cover all possible ways that the output event could occur. There could also be some situations in which the gates may need to be modified by labels indicating certain situations such as a conditional AND: event *A* must occur before event *B* occurs or an exclusive OR: either event *A* or event *B* must occur for the output to occur but not both events. The first case can be alleviated by defining the AND gate such that the second event is conditioned on the first happening, and a third event is conditioned on the first two happening. The second case is a special case of mutually exclusive events and must be handled as such.

The development of a fault tree starts with identifying all the events that are deemed undesirable for normal operation. These events need to be separated into mutually exclusive groups according to similar causes with one head event for each group. For example, in a grinding operation, there could be several mutually exclusive fault events leading to different head events or accidents: getting a chip in the eye, having sparks from the grinder start a fire, the operator losing control of the casting while pushing on it and having his or her fingers scraped by grinder, etc. Next, the relationships between the various causal events and head event are established through a combination of the AND and OR gates. This is continued until basic fault events are reached, which cannot be developed any further. Then probabilities are assigned to each of the basic events, and an overall probability of the head event is calculated using AND and OR expressions. In the final step, appropriate controls are attempted along with estimated reductions in probabilities, leading to a decrease in the probability of the final head event. A simple fault tree analysis is shown in Example 8.6. Obviously, the cost of these controls or modifications has to be considered, and this will be examined through cost-benefit analysis discussed in the next section.

EXAMPLE 8.6	Fault Tree Analysis of a Fire

The grinding shop of Dorben Co. has had several relatively small fires that were quickly put out. However, the company is concerned that a fire could get out of control and burn down the whole plant. One way to analyze the problem is to use the fault tree approach with a major fire as the head event. There are three requirements (well actually four, but neglect oxygen which is ubiquitous): (1) combustible material with a probability of 0.8, (2) an ignition source, and (3) the probability of a small fire getting out of control with a value of 0.1. There could be several ignition sources: (1) an operator smoking in spite of No Smoking signs, (2) sparks from the grinding wheel, and (3) an electrical short in the grinder. The company estimates the probabilities of these events as 0.01, 0.05, and 0.02, respectively. All the events in the first set are required and thus are connected with an AND gate. Note that the second event is conditional on the first, and the third event is conditional on the first two events. For the ignition set of events, any one input is sufficient for ignition to occur, and thus the events are connected with an OR gate. The complete fault tree is shown in Figure 8.18a.

An alternate approach would be to draw a sequence of events, similar to components for a product or operations for a system, as shown in Figure 8.18b. The three main events—combustible, ignition, getting out of control—are drawn in series, since the path must go through all three to have the plant burn down. The ignition sources can be considered in parallel, since the path has only to go through any one of them to have ignition.

The expression for the final head event or system success (in this case, the plant burning down shouldn't really be considered a success, but that is the general term used for a system) is

$$T = ABC$$

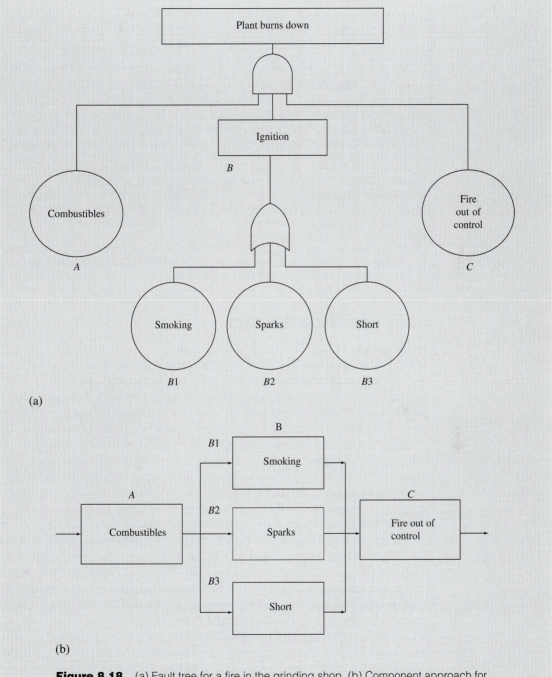

(a)

(b)

Figure 8.18 (a) Fault tree for a fire in the grinding shop. (b) Component approach for a fire in the grinding shop.

where $B = B1 + B2 + B3$. The probabilities are calculated as

$$P(B) = 1 - [1 - P(B1)][1 - P(B2)][1 - P(B3)]$$
$$= 1 - (1 - 0.01)(1 - 0.05)(1 - 0.02) = 0.0783$$
$$P(T) = P(A)P(B)P(C) = (0.8)(0.0783)(0.1) = 0.0063$$

Thus, there is approximately a 0.6 percent chance of the plant burning down.

In spite of the relatively low probability, the company would still like to reduce this probability. Sparks are a natural part of grinding, and shorts are unpredictable occurrences. So neither is a likely avenue for control measures. Two more reasonable approaches would be to enforce the No Smoking ban with the severe punishment of immediate firing. However, even if the probability of smoking went down to zero, the resulting overall probability is still 0.0055 with only a 12 percent reduction. That may not be worth the antagonization of workers by the severe penalties. On the other hand, removing the unnecessary combustibles, perhaps oily rags, from the grinding area may significantly reduce the probability from 0.8 to 0.1. It would not be completely reduced to zero, because there still may be wooden shipping cartons for the castings. In this case, the resulting overall probability goes to 0.00078, which almost a factor-of-10 reduction and is probably more cost-effective.

COST-BENEFIT ANALYSIS

As mentioned in the previous section, a fault tree is very useful in studying a safety problem and understanding the relative contributions of various causes to the head event. However, for this approach to be ultimately effective, the cost of any controls or modifications to the system or workplace also need to be considered, and this provides the basis for *cost-benefit analysis*. The cost part is easy to understand; it is simply the money being spent to retrofit an old machine, to purchase a new machine or a safety device, or to train workers in a safer method, whether it is in a lump sum or prorated over some useful life. The benefit part is a bit more difficult to understand because it is typically a reduction in accident costs or lost production costs or money saved in reduced injuries and medical costs over a period of time. Consider, for example, the medical costs associated with injuries from a 200-ton press collected over a 5-year period (see Table 8.8). The severity levels are based on workers' compensation categories (see Section 8.6), and the costs and probabilities are derived from the company's medical records. The severities range from relative minor skin laceration treatable with first aid to permanent partial disabilities, such as the amputation of a hand, to permanent total disabilities from a major crushing injury or even death. The total expected costs for an injury using the press can be calculated by the summation of the relative expected costs for each level of injury severity based on the product of the respective costs and the proportion for each type of injury. This resultant total expected cost of approximately $10,000 multiplied by the probability of the head event yields a measure of *criticality* associated with using a press. Note that this criticality is usually expressed for a given exposure time (for example, 200,000 worker-hours) or

Table 8.8 Example Expected Costs for an Injury from a Press

Severity	Cost ($)	Probability	Expected Cost ($)
First aid	100	0.515	51.50
Temporary total disability	1,000	0.450	450.00
Permanent partial disability	50,000	0.040	2,000.00
Permanent total disability	500,000	0.015	7,500.00
			10,001.50

certain amount of production. The benefit part of the cost-benefit analysis will then be a reduction in this criticality, obtained from either reducing the probability of the head event or decreasing the severity and resulting costs of any injuries. Example 8.7 demonstrates the use of cost-benefit analysis to find an appropriate redesign of a coffee mill to prevent finger injuries. More details on fault tree and cost-benefit analyses can be found in Bahr (1997), Brown (1976), Cox (1998), and Ericson (2005).

Fault Tree and Cost-Benefit Analyses of Coffee Mill Finger Lacerations

EXAMPLE 8.7

With the rise in popularity of speciality coffees, many consumers have purchased coffee mills in order to grind their own coffee beans for fresher coffee. As a result, there has been an increase of finger lacerations from the rotor blade due to the inadvertent activation of the coffee mill with the fingers still in the mill. Possible contributing factors to such an accident and injury are shown in the fault tree of Figure 8.19 with estimated probabilities for each event. This assumes a simple coffee mill with a switch to activate the rotor on the side of the mill. Working downward from the head event, the rotor must be in motion and the finger must be in the path of the rotor, indicating an AND gate. The reasons for the finger being in the path of the rotor, whether to remove the ground coffee or to clean the container, could be varied and for this example are not developed further. For the rotor to be in motion, the power must be connected and the circuit closed, again indicating an AND gate. The circuit could be closed either normally or abnormally, indicating an OR gate. Note that for either scenario, there is the presumption that the finger is in the container and in the path of the rotor. The normal closure could also include possibilities that the switch failed in a closed position or that the switch was assembled in a closed position, again indicating an OR gate. The abnormal closure could be due to a variety of conditions—broken wire, incorrect wiring, conductive debris, or water in the container—only one of which needs to occur for the circuit to short and close, thus indicating an OR gate. The calculated probabilities are as follows:

$$P(C_1) = 1 - (1 - 0.001)(1 - 0.01)(1 - 0.001)(1 - 0.01) = 0.0199$$

$$P(C_2) = 1 - (1 - 0.001)(1 - 0.01)(1 - 0.001) = 0.1$$

$$B = C_1 + C_2$$

$$P(B) = 1 - (1 - 0.0199)(1 - 0.1) = 0.12$$

$$P(A) = P(B)(1) = 0.12$$

$$P(H) = P(A)(0.2) = 0.12(0.2) = 0.024$$

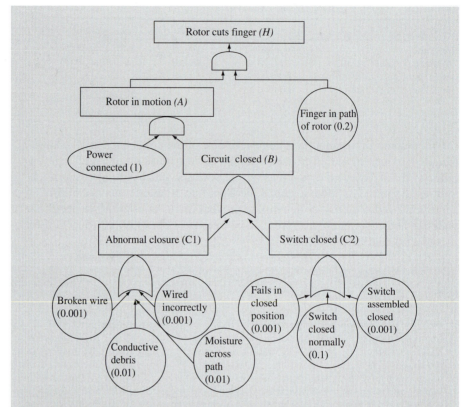

Figure 8.19   Fault tree for a finger injury using a coffee mill.

Assuming an expected $200 cost for finger injuries ranging from simple laceration to complete amputation (developed similarly as in Table 8.8), the resulting criticality C of a coffee mill finger injury is then

$$C = P(H)(\$200) = (0.024)(200) = \$4.80$$

Now comes the interesting part in examining alternative redesigns and safety measures in order to reduce the likelihood of incurring a finger injury. An obvious redesign found on most coffee mills is an interlock switch in the cover of the coffee mill (interlocks are discussed in greater detail in Section 8.8). The basic premise is that the finger cannot be in the bowl and the switch activated simultaneously. This would reduce the probability of "switch closed normally" from 0.1 to 0.0. However, the other failure modes could still occur, and the probability of the head event would not go completely to 0.0 but instead reduces to 0.0048 with a new criticality of $0.96.

The resulting decrease in criticalities from $4.80 to $0.96 is a benefit of $3.84. However, there is an increased associated cost of approximately $1.00 per coffee mill to insert a switch in the cover of the coffee mill versus the simpler switch in the side of the mill. Therefore, the cost-benefit (C/B) ratio is

$$C/B = \$1.00/\$3.84 = 0.26$$

An alternative, cheaper approach could be to apply a warning sticker to the side of the coffee mill stating, "Always Disconnect Power Before Removing Coffee or Cleaning Bowl" at a minimal cost of $0.10 per mill. The probability of a "power connected" event would be reduced, perhaps to 0.3, but not to 0.0, because consumers will be likely to forget or ignore the warning. The resulting probability of the head event is reduced to 0.0072, and the criticality to $1.44. The new benefit is $3.36, yielding a cost-benefit ratio of

$$C/B = \$0.10/\$3.36 = 0.03$$

This approach, on the surface, appears to be much more cost-effective. However, the probability of "power connected" is probably greatly underpredicted, as most consumers will forget to unplug the power before entering the bowl. Therefore, this would not be the preferred solution. Note that if an additional $1.00 were applied to each mill for additional quality control to catch all the wiring and switch errors before shipping, reducing each of those probabilities to 0.0, the resulting cost-benefit ratio at 1.25 is much larger than the installation of the interlock switch in the cover.

8.6 SAFETY LEGISLATION AND WORKERS' COMPENSATION

BASICS AND TERMINOLOGY

In the United States, safety legislation, as well as the rest of the legal system, is based on a combination of *common law, statute law,* and *administrative law.* Common law was derived from unwritten customs and typical usage in England, but adjusted and interpreted by the courts through judicial decisions. Statute law is written law enacted by legislators and enforced by the executive branch. Administrative law is established by the executive branch or government agencies. However, since common law came first, many of our legal terms and principles are derived from that. Thus, the ancient terms of *master, servant,* and *stranger* eventually came to represent an *employer,* an *employee,* and a *guest* or *visitor,* respectively. *Liability* is the obligation to provide compensation for damages or injury, while *strict liability* is a higher level of liability, in which the plaintiff need not prove negligence or fault. The *plaintiff* is the person, typically injured, originating a suit in court. The *defendant* is the entity defending the suit, typically an employer or the manufacturer of a product. *Negligence* is the failure to exercise a reasonable amount of care in preventing injury. Higher forms of negligence include *gross negligence,* with failure to show the slightest care, and *negligence per se,* with no proof needed. Any resulting awards to the plaintiff fall into two categories: *compensatory damages* for medical costs, lost wages, and other direct losses on the part of the plaintiff, and *punitive damages* in the form of additional monetary amounts specifically to punish the defendant.

Under the English common law system, and later under statute law, the employer did have some legal obligations to provide a safe workplace, protect employees against injury, and pay for injuries and damages that could result if the employer failed to fulfill those obligations. These obligations also extended to customers and the general public, for example, visitors to the workplace. However, in practice, these legal obligations didn't amount to much as the burden of proof fell on the worker to prove in court that the employer's negligence had been the sole cause of his or her injury. Several factors made it especially difficult for the employee to prove his or her case. First, the doctrine of *privity* required a direct relationship, as in the form of a contract, between the two contesting parties. Therefore, any workers not having a direct contract with the employer would likely have little success in court. Second, the *assumption of risk* concept implied that a worker who was aware of the hazards of the job, but continued working there, assumed the risks and could not recover damages in case of injury even though it occurred through no fault of her or his own. Third, fellow employee negligence or contributory negligence by the worker himself or herself severely limited the worker's case. Finally, there was always the fear of loss of jobs for the worker or fellow employees, which generally restricted legal actions against employers. In addition, any legal action took many years, delaying the compensation needed for medical expenses, and resulted in inconsistent and relatively insufficient compensation, with much of the money going to the lawyers involved. As a result there arose demands for workers' compensation legislation that would correct these inequities and force employers to take corrective action to safeguard their employees.

WORKERS' COMPENSATION

In the United States, the first workers' compensation laws were enacted in 1908 for federal employees and eventually were enacted in all 50 states and the U.S. territories. These all operate on the general principle of recompensing workers for medical expenses and lost wages without establishing fault. Typically they will have set amounts to be paid for given conditions and even occupations, which could vary among the states. Approximately 80 percent of the U.S. workforce is covered, with some notable exceptions: independent agricultural workers, domestic workers, some charity organizations, railroad and maritime workers, and smaller independent contractors. To ensure that worker benefits do not end if the employer were to go bankrupt, companies are required to obtain insurance. This could be on an exclusive basis through state funds set up for this purpose or on a competitive basis through private insurance companies. In some cases, very large, financially secure companies or entities can be self-insured.

There are three main requirements for a worker to claim compensation:

1. The injury must have resulted from an accident.
2. The injury must have arisen out of employment.
3. The injury must have occurred during the course of employment.

Injuries not considered accidents include those caused by intoxication, those that are self-inflicted, or those that arose out of a heated argument. Also anything that could have occurred normally, such as a heart attack, would not be covered, unless the work were considered so stressful as to be contributing to the heart attack. Injuries arisen out of employment apply to work assigned by a supervisor or work normally expected of that employee. A typical exception that would negate compensation is doing "government work" or using company equipment for personal use. Injuries during the course of employment apply to normal work-hours and not to commuting time to and from work, unless the company provides transportation.

Workers' compensation is typically broken down into four categories of disability: (1) temporary partial, (2) temporary total, (3) permanent partial, and (4) permanent total. *Temporary partial disabilities* are ones in which the worker receives a minor injury and full recovery is expected. The worker can still perform most duties but may suffer some lost time and/or wages. *Temporary total disabilities* are ones in which the worker is incapable of performing any work for a limited time, but full recovery is expected. This category accounts for the majority of workers' compensation cases. *Permanent partial disabilities* are ones in which the worker will not fully recover from injuries but can still perform some work. This category accounts for the majority of workers' compensation costs and is further subdivided into schedule and nonschedule injuries. A *schedule injury* receives a specific payment for a specified time according to a schedule, as shown in Table 8.9. Note that there may be considerable differences in payments among states, as shown between federal and Pennsylvania workers. A *nonschedule injury* is of less specific nature, such as disfigurement, with payments prorated to a schedule injury. *Permanent total disabilities* are sufficiently serious that they will prevent the employee from ever working in regular employment. Again, there may be considerable differences in what constitutes total disability, but in many states this is accorded by the loss of sight in both eyes or the loss of both arms or both legs. In approximately one-half of the states, the compensation is for the duration of the disability or the injured worker's lifetime. In the other half, the duration is limited to 500 weeks. The compensation is some percentage of wages, the majority being two-thirds. In case of the worker's death, benefits are paid to the widow

Table 8.9 Lost-Limb Award Schedule (in weeks) for a Permanent Partial Disability

Amputation or 100% Loss of Use	Federal	Pennsylvania
Arm	312	410
Leg	288	410
Hand	244	325
Foot	205	250
Eye	160	275
Thumb	75	100
First finger (little)	46 (15)	50 (28)
Great toe (other)	38 (16)	40 (16)
Hearing—one ear (both)	52 (200)	60 (260)

for life or until remarriage and to children until the age of 18 or until the maximum period of payments (for example, 500 weeks).

There may be some other important conditions depending on the states. In some cases, the companies may be able to require the injured worker to see a company physician and to perform suitable lighter-duty employment. If the worker refuses, workers' compensation benefits can be terminated. In most cases, though, the workers' compensation cases are settled quickly and amicably, and the worker receives a direct settlement. In some situations, the case can be contested with the employee and employer settling either directly or through the workers' compensation legal system. From the worker's standpoint, it is generally a positive trade-off—she or he accepts a guaranteed lesser amount of compensation in place of the uncertain option to sue the employer for negligence. However, the worker does not give up the rights to sue a third party, such as the manufacturer of faulty equipment or defective tools, the architect or contractor of faulty construction, or even an inspection agency that certified the safety of a building or machines.

From the company's perspective, it is important to try to decrease workers' compensation costs as much as possible. This can be done through a variety of means. First and foremost, implement a safety program so as to reduce workplace hazards and to train operators in the proper procedures. Second, and almost equally important, implement a proper medical management program. This means hiring a good occupational nurse and selecting a knowledgeable local physician to visit the plant and understand the various jobs. This will aid in proper diagnoses and the assignment of workers to light-duty jobs. It is also very important to get the injured employees back to work as quickly as possible, even if only on a light-duty job. Third, review the employment classification of each employee based on the job she or he performs. It makes no sense to have a misclassification result in increased premiums, for example, an office worker misclassified as a grinder operator. Fourth, conduct a thorough payroll audit. Overtime is charged as straight time on workers' compensation. Therefore, double overtime wages of $20 an hour would greatly inflate the company's costs compared to straight $10 an hour charges. Fifth, compare self-insurance and various group insurance programs for lowest cost, and use a deductible. Sixth, check your *mod ratio* frequently. This is the ratio of actual losses to losses expected of similar employers, with 1.00 being average. By implementing a good safety program and reducing accidents, injuries, and consequently workers' compensation claims, the mod ratio will drop significantly. A mod ratio of 0.85 means a 15 percent savings on premiums. Through proper management, workers' compensation costs can be controlled.

8.7 OCCUPATIONAL SAFETY AND HEALTH ADMINISTRATION (OSHA)

OSHA ACT

The Occupational Safety and Health Act of 1970 was passed by Congress "to assure so far as possible every working man and woman in the Nation safe and healthful working conditions and to preserve our human resources." Under the act, the Occupational Safety and Health Administration was created to

1. Encourage employers and employees to reduce workplace hazards and to implement new, or improve existing, safety and health programs.
2. Establish "separate but dependent responsibilities and rights" for employers and employees for the achievement of better safety and health conditions.
3. Maintain a reporting and record-keeping system to monitor job-related injuries and illnesses.
4. Develop mandatory job safety and health standards, and enforce them effectively.
5. Provide for the development, analysis, evaluation, and approval of state occupational safety and health programs.

Since the act can intimately affect the design of the workplace, methods analysts should be knowledgeable regarding the details of this act. The *general-duty clause* of the act states that each employer "must furnish a place of employment which is free from recognized hazards that cause or are likely to cause death or serious physical harm to employees." Furthermore, the act brings out that it is the employers' responsibility to become familiar with standards applicable to their establishments and to ensure that employees have and use personal protective gear and equipment for safety.

OSHA standards fall into four categories: general industry, maritime, construction, and agriculture. All OSHA standards are published in the Federal Register, which is available in most public libraries, in a separate book of regulations (OSHA, 1997), and on the Web (http://www.osha.gov/). OSHA can begin standards-setting procedures on its own initiative or on the basis of petitions from the Secretary of Health and Human Services (HHS), the National Institute for Occupational Safety and Health (NIOSH), state and local governments, nationally recognized standards-producing organizations such as the ASME, and employer or labor representatives. Of these groups, NIOSH, an agency of HHS, is quite active in making recommendations for standards. It conducts research on various safety and health problems and provides considerable technical assistance to OSHA. Especially important are the investigation of toxic substances by NIOSH and its development of criteria for the use of such substances in the workplace.

OSHA also provides free on-site consultation services for employers in all 50 states. This service is available on request, and priority is given to smaller businesses, which are generally less able to afford private-sector consultations. These consultants help employers identify hazardous conditions and determine corrective measures. A listing of such consultants is found on OSHA's Web page (http://www.osha.gov/dcsp/smallbusiness/consult_directory.html).

The act also requires employers of 11 or more employees to maintain records of occupational injuries and illnesses on the *OSHA 300 log*. An *occupational injury* is defined as "any injury such as a cut, fracture, sprain or amputation which results from a work-related accident or from exposure involving a single incident in the work environment." An *occupational illness* is "any abnormal condition or disorder, other than one resulting from an occupational injury, caused by exposure to environmental factors associated with employment." Occupational illnesses include acute and chronic illnesses that may be caused by inhalation, absorption, ingestion, or direct contact with toxic substances or harmful agents. Specifically, they must be recorded if the result is death, loss of one or more workdays, restriction in motion or ability to do the work that had been done, loss of consciousness, transfer to another job, or medical treatment other than first aid.

WORKPLACE INSPECTIONS

To enforce its standards, OSHA is authorized to conduct workplace inspections. Consequently, every establishment covered by the act is subject to inspection by OSHA compliance safety and health officers. The act states that "upon presenting appropriate credentials to the owner, operator, or agent in charge," an OSHA compliance officer is authorized to enter without delay any factory or workplace to inspect all pertinent conditions, equipment, and materials therein and to question the employer, operator, or employees.

OSHA inspections, with few exceptions, are concluded without advance notice. In fact, alerting an employer in advance of an OSHA inspection can bring a fine of up to $1,000 and/or a 6-month jail term. Special circumstances under which OSHA may give notice of inspection to an employer include those where

1. Imminently dangerous situations exist that require correction as soon as possible.
2. Inspections necessitate special preparation or must take place after regular business hours.
3. Prior notice ensures that the employer and employee representatives or other personnel will be present.
4. The OSHA area director determines that advance notice would produce a more thorough or more effective inspection.

Upon inspection, if an imminently dangerous situation is found, the compliance officer asks the employer to abate the hazard voluntarily and to remove endangered employees from exposure. Notice of the imminent danger must also be posted. Before the OSHA inspector leaves the workplace, he or she will advise all affected employees of the hazard.

At the time of the inspection, the employer is asked to select an employer representative to accompany the compliance officer during the inspection. An authorized employee representative is also given the opportunity to attend the opening conference and to accompany the compliance officer during the inspection. In those plants with a union, the union ordinarily designates the employee representative to accompany the compliance officer. Under no circumstances may the employer select the employee representative for the inspection. The act does not require an employee representative for each inspection; however, where there is no authorized employee representative, the compliance officer must consult with a reasonable number of employees concerning safety and health matters in the workplace.

After the inspection tour, a closing conference is held between the compliance officer and the employer or the employer representative. Subsequently, the compliance officer reports the findings to the OSHA office, and the area director determines what citations, if any, will be issued and what penalties, if any, will be proposed.

CITATIONS

Citations inform the employer and employees of the regulations and standards alleged to have been violated, and the proposed time set for their abatement. The employer will receive citations and notices of proposed penalties by certified mail. The employer must post a copy of each citation at or near the place where a violation has occurred, for three days or until the violation is abated, whichever is longer.

The compliance officer has the authority to issue citations at the worksite, following the closing conference. To do so, he or she must first discuss each apparent violation with the area director and must receive approval to issue the citations.

The six types of violations that may be cited, and the penalties that may be imposed, are as follows:

1. *De minimis* (no penalty). This type of violation has no immediate relationship to safety or health, for example, number of toilets.
2. *Nonserious violation.* This type of violation has a direct relationship to job safety and health, but probably would not cause death or serious physical harm. A proposed penalty of up to $7,000 for each violation is discretionary. A penalty for a nonserious violation may be decreased considerably depending on the employer's good faith (demonstrated efforts to comply with the act), history of previous violations, and size of business.
3. *Serious violation.* This is a violation in which there is substantial probability that death or serious harm could result, stemming from a hazard about which the employer knew or should have known. A mandatory penalty of up to $7,000 is assessed for each violation.
4. *Willful violation.* This is a violation that the employer intentionally and knowingly commits. The employer either knows that his or her actions constitute a violation, or is aware that a hazardous condition exists and has made no reasonable effort to eliminate it. Penalties of up to $70,000 may be proposed for each willful violation. If an employer is convicted of a willful violation that has resulted in the death of an employee, there may also be imprisonment for up to 6 months. A second conviction doubles these maximum penalties.
5. *Repeated violation.* A repeated violation occurs when a violation of any standard, regulation, rule, or order is reinspected and another violation of the previously cited section is found. If, on reinspection, a violation of the previously cited standard, regulation, rule, or order is found, but it involves

another piece of equipment and/or a different location in the establishment or worksite, it may be considered a repeated violation. Each repeated violation can bring a fine of up to $70,000. If there is a finding of guilt in a criminal proceeding, then up to 6 months' imprisonment and a $250,000 fine for an individual or a $500,000 fine for a corporation may be imposed.

6. *Imminent danger.* This is a situation in which there is reasonable certainty that a danger exists that can be expected to cause death or serious physical harm either immediately or before the danger can be eliminated through normal enforcement procedures. An imminent danger violation may result in a cessation of the operation or even complete plant shutdown.

Other violations for which citations and proposed penalties may be issued are as follows:

1. Falsifying records, reports, or applications, on conviction, can bring a fine of $10,000 and 6 months in jail.
2. Violating the posting requirements can bring a civil penalty of up to $7,000.
3. Failing to abate or correct a violation can bring a civil penalty of up to $7,000 for each day the violation continues beyond the prescribed abatement date.
4. Assaulting, interfering with, or resisting an inspector in his or her duties can result in a fine of up to $5,000 and imprisonment for up to 3 years.

OSHA ERGONOMICS PROGRAM

In 1990, the high incidences and severity of work-related musculoskeletal disorders found in the meatpacking industry led OSHA to develop ergonomics guidelines to be used in protecting meatpackers from these hazards (OSHA, 1990). The publication and dissemination of these guidelines were meant to be a first step in assisting the meatpacking industry in implementing a comprehensive safety and health program that would include ergonomics. Although the guidelines were initially meant to be advisory in nature, they were eventually developed into new industrywide ergonomics standards. The guidelines were meant to provide information so that the employers could determine if they have ergonomics-type problems, identify the nature and location of those problems, and implement measures to reduce or eliminate them.

The ergonomics program for meatpacking plants is divided into five sections: (1) management commitment and employee involvement, (2) worksite analysis, (3) recommended hazard prevention and controls, (4) medical management, and (5) training and education. Detailed examples tailored for the meatpacking industry are also provided.

Commitment and involvement are essential elements in any sound safety and health program. Commitment by management is especially important in providing both the motivating force and the necessary resources to solve the problems. Similarly, employee involvement is necessary to maintain and continue the program. An effective program should have a team approach, with top management as the team leader, using the following principles:

1. A written program for job safety, health, and ergonomics, with clear goals and objectives to meet these goals, endorsed and advocated by the highest levels of management.
2. A personal concern for employee health and safety, emphasizing the elimination of ergonomics hazards
3. A policy that places the same emphasis on health and safety as on production.
4. Assignment and communication of the responsibility of the ergonomics program to the appropriate managers, supervisors, and employees.
5. A program ensuring accountability from these managers, supervisors, and employees for carrying out these responsibilities.
6. Implementation of a regular review and evaluation of the ergonomics program. This might include trend analyses of injury data, employee surveys, "before and after" evaluations of workplace changes, logs of job improvements, etc.

Employees can be involved via the following:

1. A complaint or suggestion procedure for voicing their concerns to management without fear of reprisal
2. A procedure for prompt and accurate recording of the first signs of work-related musculoskeletal disorders, so that prompt controls and treatment can be implemented
3. Ergonomics committees that receive reports of, analyze, and correct ergonomics problems
4. Ergonomics teams with the required skills to identify and analyze jobs for ergonomics stress

An effective ergonomics program includes four major program elements: worksite analysis, hazard control, medical management, and training and education. Worksite analysis identifies existing hazards and conditions, as well as operations and workplaces where such hazards may develop. The analysis includes a detailed tracking and statistical analysis of injury and illness records, to identify patterns of work-related musculoskeletal disorder development. The first step in implementing the analysis program should be a review and analysis of medical records, insurance records, and OSHA 300 logs using chi-square analysis and tracking incidence rates. Next, baseline screening surveys can be conducted to identify jobs that put employees at risk of developing work-related musculoskeletal disorders. The survey is typically performed with a questionnaire to identify potential ergonomics risk factors in the job process, workplace, or work method, as well as the location and severity of the potential musculoskeletal problems for the individual worker, using the body discomfort charts of Chapter 5. Then a physical worksite analysis should be conducted with a walk-through of the plant and videotaping and analysis of critical jobs, using the work design checklists and analyses tools presented in earlier chapters. Finally, as in any methods program, periodic reviews should be conducted. These may uncover previously missed risk factors or design deficiencies. Trends of injuries and illnesses should be calculated and examined at regular intervals, as a quantitative check on the effectiveness of the ergonomics program.

Hazard control involves the same engineering controls, work practice controls, personal protective equipment, and administrative controls as discussed throughout this book. Engineering controls, where feasible, are the OSHA-preferred method of control.

Proper medical management, including the early identification of signs and the effective treatment of symptoms, is necessary to reduce the risk of developing work-related musculoskeletal disorders. A physician or occupational nurse with experience in musculoskeletal disorders should supervise the program. The person should conduct periodic, systematic workplace walk-throughs to remain knowledgeable about the jobs, identify potential light-duty jobs, and maintain close contact with employees. This information will allow the health providers to recommend assignments of recovering workers to restricted-duty jobs with minimal ergonomic stress on the injured muscle and tendon groups.

Health care providers should participate in the training and education of all employees, including supervisors, on different types of work-related musculoskeletal disorders, means of prevention, causes, early symptoms, and treatments. This demonstration will assist in the early detection of work-related musculoskeletal disorders prior to the development of more severe conditions. Employees should be encouraged to report early signs and symptoms of work-related musculoskeletal disorders, for timely treatment without fear of retribution by management. Written protocols for health surveillance, evaluation, and treatment will assist in maintaining properly controlled procedures.

Training and education are critical components of an ergonomics program for employees potentially exposed to ergonomics hazards. Training allows managers, supervisors, and employees to understand the ergonomics problems associated with their jobs, as well as the prevention, control, and medical consequences of those problems.

1. General training on work-related musculoskeletal disorder risk factors, symptoms, and hazards associated with the job should be given annually to those employees who are potentially exposed.
2. Job-specific training on tools, knives, guards, safety, and proper lifting should be given to new employees prior to their being placed on a full-time job.
3. Supervisors should be trained to recognize the early signs of work-related musculoskeletal disorders and hazardous work practices.
4. Managers should be trained to be aware of their health and safety responsibilities.
5. Engineers should be trained in the prevention and correction of ergonomics hazards through workplace redesign.

A rough-draft version of the guidelines for general industry, as a precursor to an ergonomics standard, was released in 1990, and the final version was signed early in 1992. It contained primarily the same information as found in the guidelines for the meatpacking industry. However, there was considerable negative reaction from industry, and with the Republicans gaining control of Congress in 1992, the ergonomics standard was effectively shelved for the time being.

8.8 HAZARD CONTROL

This section presents the basic principles in controlling hazards. A *hazard* is a condition with the potential of causing injury or damage while *danger* is the relative exposure to or potential consequences of that hazard. Thus, an unprotected worker on scaffold is exposed to a hazard and has the danger of serious injury. If the worker wears a safety harness, there is still a hazard, but the danger of the hazard has been reduced considerably.

Hazards can occur in several general categories: (1) due to inherent properties such as high voltage, radiation, or caustic chemicals; (2) due to potential failure, either of the operator (or some other person) or of the machine (or some other equipment); or (3) due to environmental forces or stresses, for example, wind, corrosion, etc. The general approach is to first completely eliminate the hazard and prevent the accident, and then, if not successful, to reduce the hazard level to the point that, should the accident still happen, the potential injury or damage is minimized. Elimination of a hazard can be achieved through good design and proper procedures, for example, use of noncombustible materials and solvents, rounding edges on equipment, automating corrosive dips (that is, removing the operator from the hazardous environment, building an overpass at railroad and highway intersections, etc.).

If the hazard cannot be completely eliminated, then a second-level approach is to limit the hazard level. For example, an electric power drill in a wet environment has the potential for electrocution. Using a cordless drill would reduce the power level for serious injury, although some shock may still occur. Of course, the trade-off is a reduced torque level and drilling effectiveness. Using a pneumatic drill would completely eliminate the electrocution hazard, but may have limited use, especially for homeowners who don't have compressed air available, may increase the cost of the drill, and, may introduce a new hazard with the release of high-pressure air. The safest solution would be to use a mechanical hand drill, with no minimal hazards due to energy. However, the effectiveness of the tool would be significantly reduced and could introduce musculoskeletal fatigue (a completely separate hazard). Another example of limiting the hazard level is the use of governors on school buses to limit the maximum speed of the vehicles.

If the hazard level cannot be limited due to the inherent nature of electromechanical equipment or power tools, the next approach is to use isolation, barriers, and interlocks to minimize the contact between the energy source and the human operator. Isolation and barriers impose either a distance or a physical impediment between the two. Placing a generator or a compressor outside the plant will limit normal daily contact between operators and the energy source. Only maintenance workers at irregular intervals will have some exposure to the energy source. Fixed machine guards or pulley enclosures are good examples of barriers. The third item, an *interlock,* is a more complex approach or device that prevents incompatible events from occurring at the wrong time. At the most basic level it could simply be a *lockout* (as in OSHA 1910.147, Lockout/Tagout) of a dangerous area to prevent unauthorized individuals from entering this area or a *lock in* of a switch in an energized status so that it can't be accidentally turned off. The more typical active interlock is a mechanism that ensures that a given event doesn't occur at the same time as another event. The previously mentioned switch in the cover of a coffee mill (Example 8.7) is a good example of an interlock that prevents the user's finger from being in the bowl at the same time that the switch is activated.

Another approach is the use of *fail-safe designs.* Systems are designed such that, in the case of failure, they go to the lowest energy level. This can be accomplished through simple, passive devices such as fuses and circuit breakers, which upon experiencing high current levels physically open the circuit and drop the current immediately to zero. This can also be accomplished operationally through the design of valves that either fail in an open position so as to maintain fluid flow (that is, the valve disk is forced away from the seat by the flow) or fail in a closed position to stop current flow (that is, the valve disk is forced into the seat by the flow). Another good example of an operational design is the *deadman control* on a lawnmower or a waverunner. In the first case, the operator holds down a lever to keep the blade turning; if the operator trips and loses control of the lever, the blade stops through either release of the clutch or cutting of the engine. In the second case, the engine key is attached to the operator's wrist by a leash; if the operator is thrown from the craft, the key is pulled out, stopping the engine.

Another approach in hazard control is failure minimization. Rather than allow the system to fail completely even in a fail-safe mode, this approach decreases the probability of system failure. This can be accomplished by increasing safety factors, monitoring system parameters more closely, and replacing key components regularly or providing redundancy for these components. *Safety factor* is defined as the ratio of strength to stress and should obviously be well above 1. Given that there can be considerable variance in the strength of material, for example, two-by-fours used in construction and also some variance in environmental stresses, for example, snow in northern locales, it would make sense to increase the safety factor appropriately to account for these variances and minimize the collapse of the building. Monitoring of key temperatures and pressures, with appropriate adjustments or compensations, can help forestall the system reaching critical levels. Automobile tire wear markers are one commonly used parameter monitoring system. The OSHA requirement of a buddy system for work in hazardous areas is another such example. Regular replacement of the above tires even before the wear markers are exposed is an example of regular replacement of components, or using more two-by-fours (set at 12-in distances, rather than 16 in) would be an example of redundancy in the system.

Finally, in case the system does finally fail, the organization must provide for personal protective equipment, escape and survival equipment, and rescue equipment so as to minimize the resulting injuries and costs. Fire protective clothing, helmets, safety shoes, earplugs, etc., are common examples of personal protective equipment to minimize injury. The storage of self-rescuers at regular locations in mines provides miners extra oxygen in case of methane leaks or fires using up the ambient oxygen and extra time until rescuers can reach them. Similarly, large companies will have their own firefighting equipment to save time compared to relying on local fire departments.

8.9 GENERAL HOUSEKEEPING

General safety considerations related to the building include adequate floor-loading capacity. This is especially important in storage areas, where overloading may cause serious accidents. The danger signs of overloading include cracks in walls or ceilings, excessive vibration, and displacement of structural members.

Aisles, stairs, and other walkways should be investigated periodically to ensure that they are free of obstacles, are not uneven, and are not covered with oil or other material that could lead to slips and falls. In many old buildings, stairs should be inspected, since they are the cause of numerous lost-time accidents. Stairs should have a slope of 28 to 35 degrees, with tread widths of 11 to 12 in (28 to 30.5 cm) and riser heights of 6.5 to 7.5 in (16.5 to 19 cm). All stairways should be equipped with handrails, should have at least 10 fc (100 lx) of illumination, and be painted in light colors.

Aisles should be plainly marked and straight, with well-rounded corners or diagonals at turn points. If aisles are to accommodate vehicle travel, they should be at least 3 ft wider than twice the width of the broadest vehicle. When traffic is only one-way, then 2 ft wider than the broadest vehicle is adequate. In general, aisles should have at least 10 fc (100 lx) of illumination. Color should be used throughout, to identify hazardous conditions (see Table 8.10). More details on aisles, stairs, and walkways can be found in OSHA 1910.21-1910.24, *Walking and Working Surfaces.*

Most machine tools can be satisfactorily guarded to minimize the probability of a worker being injured while operating the machine. The problem is that many older machines are not properly guarded. In these instances, immediate action should be taken to see that a guard is provided and that it is workable and routinely used. An alternate approach is to provide a two-button operation, such as that shown for the press operations in Figure 8.20. Note that the two hand buttons are spread well apart, so the operator's hands are in a safe position when the press starts. These buttons should not require high levels of force; otherwise repetitive-motion injuries are likely to occur. In fact newer buttons can be activated through skin capacitance rather than relying on mechanical pressure. A better alternative may be to automate the process, completely freeing the operator from the nip point or using a robotic manipulator in place of the operator. Further details on machine guarding can be found in OSHA 1910.211-1901.222, *Machinery and Machine Guarding.*

A quality control and maintenance system should be incorporated in the tool room and the tool cribs, so that only reliable tools in good working condition are released to workers. Examples of unsafe tools that should not be released to operators include power tools with broken insulation, electrically driven power tools lacking grounding plugs or wires, poorly sharpened tools, hammers with

Table 8.10 Color Recommendations

Color	Used for:	Examples
Red	Fire protection equipment, danger, and as a stop signal	Fire alarm boxes, location of fire extinguishers and fire hose, sprinkler piping, safety cans for flammables, danger signs, emergency stop buttons
Orange	Dangerous parts of machines, other hazards	Inside of movable guards, safety starting buttons, edges of exposed parts of moving equipment
Yellow	Designating caution, physical hazards	Construction and material handling equipment, corner markings, edges of platforms, pits, stair treads, projections. Black stripes or checks may be used in conjunction with yellow
Green	Safety	Location of first-aid equipment, gas masks, safety deluge showers
Blue	Designating caution against starting or using equipment	Warning flags at starting point of machines, electrical controls, valves about tanks and boilers
Purple	Radiation hazards	Container for radioactive materials or sources
Black and white	Traffic and housekeeping markings	Location of aisles, direction signs, clear floor areas around emergency equipment

mushroomed heads, cracked grinding wheels, grinding wheels without guards, and tools with split handles or sprung jaws.

There are also potentially dangerous materials and hazardous chemicals to be considered. These materials can cause a variety of health and/or safety problems and typically fall into one of three categories: corrosive materials, toxic materials, and flammable materials. Corrosive materials include a variety of acids and caustics that can burn and destroy human tissue upon contact. The chemical action of corrosive materials can take place by direct contact with the skin or through the inhalation of fumes or vapors. To avoid the potential danger resulting from the use of corrosive materials, consider the following measures:

1. Be sure that the material handling methods are completely foolproof.
2. Avoid any spilling or spattering, especially during initial delivery processes.
3. Be sure that operators exposed to corrosive materials have used and are using correctly designed personal protective equipment and waste disposal procedures.
4. Ensure that the dispensary or the first-aid area is equipped with the necessary emergency provisions, including deluge showers and eye baths.

Toxic or irritating materials include gases, liquids, or solids that poison the body or disrupt normal processes by ingestion, absorption through the skin, or inhalation. To control toxic materials, use the following methods:

1. Completely isolate the process from workers.
2. Provide adequate exhaust ventilation.

Figure 8.20 Two-button press operation. © Morton Beebe/CORBIS.

3. Provide workers with reliable personal protective equipment.
4. Substitute a nontoxic or nonirritating material, wherever possible.

More details on toxic materials can be found in OSHA 1910.1000-1910.1200, *Toxic and Hazardous Substances*.

Furthermore, per OSHA regulations, the composition of every chemical compound must be ascertained, its hazards determined, and appropriate control measures established to protect employees. This information must be clearly presented to workers with clear labels and material safety data sheets (MSDSs). Further information on this process (termed *HAZCOM*) can be found in OSHA 1910.1200, *Hazard Communications*.

Flammable materials and strong oxidizing agents present fire and explosion hazards. The spontaneous ignition of combustible materials can take place when there is insufficient ventilation to remove the heat from a process of slow oxidation. To prevent such fires, combustible materials need to be stored in a well-ventilated, cool, dry area. Small quantities should be stored in covered metal containers. Some combustible dusts, such as sawdust, are not ordinarily known to be explosive. However, explosions can occur when such dusts are in a fine enough state to ignite. To avoid explosions, prevent ignition by providing adequate ventilation exhaust systems and by controlling the manufacturing processes to minimize the generation of dust and the liberation of gases and vapors. Gases and vapors may be removed from gas streams by absorption in liquids or solids, adsorption on solids, condensation, and catalytic combustion and incineration. In absorption, the gas or vapor becomes distributed in the collecting liquid found in absorption towers, such as bubble-cap plate columns, packed towers, spray towers, and wet-cell washers. The adsorption of gases and vapors uses a variety of solid adsorbents such as charcoal with an affinity for certain substances such as benzene, carbon tetrachloride, chloroform, nitrous oxide, and acetaldehyde. Further information on flammable materials can be found in OSHA 1910.106, *Flammable and Combustible Liquids*.

In case the flammable materials ignite, suppression of the resulting fire is based on the relatively simple principles of the fire triangle (although the actual implementation may not always be so simple). There are three required components, or legs of a triangle, to a fire: oxygen (or oxidizer in chemical reactions), fuel (or reducing agent in chemical reactions), and heat or ignition. Removal of any one component will suppress the fire (or collapse the triangle). Spraying of water on a house fire cools the fire (removes heat) and also dilutes the oxygen. Using foam on fire (or covering with a blanket) removes oxygen from the fire. Spreading out the logs in a campfire removes fuel. More practically, in a plant, there will be both fixed extinguishing systems such as water sprinklers and portable fire extinguishers. These are categorized by types and sizes. The four basic types are class A, for ordinary combustibles and could use water or foam; class B, for flammable liquids typically using foam; class C, for electrical equipment using nonconductive foams; and class D for oxidizable metals. Further information on fire suppression can be found in OSHA 1910.155-1910-165, *Fire Protection*.

SUMMARY

This chapter covered the basics on safety, including the accident prevention process, starting with various theories on accident causation; the use of probability in understanding system reliability, risk management, and fault tree analysis; the use of cost-benefit analysis and other tools for decision making; various statistical tools for monitoring the success of the safety program; basic hazard control; and federal safety regulations relating to industry. Only the basics of hazard control were presented here. Specific details on specific workplace hazards can be found in numerous traditional safety textbooks such as by Asfahl (2004), Banerjee (2003), Goetsch (2005), Hammer and Price (2001), National Safety Council (2000), and Spellman (2005). However, there should be sufficient information for the industrial engineer to start a safety program with the goal of providing a safe working environment for the employees.

QUESTIONS

1. What is the difference between accident prevention and safety management?
2. What are the steps in the accident prevention process?
3. Describe the "dominoes" of the domino theory? What is the key aspect to this approach?
4. How does multiple causation affect accident prevention?
5. Compare and contrast life-change-unit, motivation-reward-satisfaction, and ABC models. What is the common link for all these models?
6. Explain the significance of using chi-square analysis in accident prevention.
7. What is the purpose of risk analysis in accident prevention?
8. What is red flagging?
9. Discuss the difference between independent and mutually exclusive events.
10. In what ways can the reliability of a system be improved?
11. Compare and contrast AND and OR gates.
12. What is criticality and what role does it play in cost-benefit analysis?
13. Compare and contrast common law and statute law.
14. What is the difference between liability and strict liability?
15. What is the difference between negligence, gross negligence, and negligence per se?
16. What is the difference between compensatory and punitive damages?
17. Prior to workers' compensation, what three common law conditions were used by employers to disqualify an injured worker from receiving benefits?
18. What are the three main requirements to obtain workers' compensation?
19. Compare and contrast the four categories of disabilities recognized under workers' compensation.
20. What is the difference between schedule and nonschedule injuries?
21. What is a third-party suit?
22. What are some ways that a company could try to keep its workers' comp costs down?
23. Why is the OSHA general duty clause so important?
24. What types of citations can be issued by OSHA?
25. What are the key elements of OSHA's proposed ergonomics program?
26. What is the difference between a hazard and a danger?
27. What is the general approach used in hazard control?
28. Explain why a deadman switch is a good example of a fail-safe design. Give an example where it might be used.
29. What is a safety factor?
30. What is the fire triangle? Explain how its principles are used in fire extinguishers.

PROBLEMS

1. For the injury data given in the table below:
   a. What are the incidence and severity rates for each department?
   b. Which department has significantly more injuries than the others?
   c. Which department has a significantly higher severity rate than the others?
   d. As a safety specialist, which department would you tackle first? Why?

Department	Injuries	Lost Days	Hours Worked
Casting	13	3	450,000
Spruing	2	0	100,000
Shearing	5	1	200,000
Grinding	6	3	600,000
Packing	1	3	500,000

2. An engineer was adjusting the gearbox of a large steam engine while it was in operation and dropped a wrench into the path of the gears. As a result, the engine was thrown out of alignment and badly wrecked. Luckily the engineer only suffered minor lacerations from ejected pieces of metal. It was hypothesized by the engineer that the metal-handle wrench simply slipped out of his oily hands.

   a. Use the domino and multiple-causation theories to examine this accident scenario.

   b. Use job safety analysis to suggest control measures (and indicate relative effectiveness for each) that could have prevented the injuries and damage. What is your ultimate recommendation?

3. Given $P(A) = 0.6$, $P(B) = 0.7$, $P(C) = 0.8$, $P(D) = 0.9$, $P(E) = 0.1$ and independent events, determine $P(T)$ for

   a. $T = AB + AC + DE$

   b. $T = A + ABC + DE$

   c. $T = ABD + BC + E$

   d. $T = A + B + CDE$

   e. $T = ABC + BCD + CDE$

4. Perform a cost-benefit analysis using fault tree analysis on a stairway. Assume that over the last year, three accidents were caused by slippery surfaces, five were caused by inadequate railings, and three were caused by someone negligently leaving tools or other obstacles on the steps. The average cost for each accident was $200 (total for first aid, lost time, etc.). Assume that you have been allocated $1,000 to improve the safety of the stairway (although you don't need to spend all the money). Three alternatives could be employed:

   1. New surfaces, which will reduce accidents caused by slippery surfaces by 70 percent and cost $800

   2. New railings, which will reduce accidents caused by inadequate railings by only 50 percent (since not all pedestrians use railings!) and cost $1,000

   3. Signs and educational programs, which are estimated to reduce both railing and obstacle-related accidents, each by 20 percent (people forget easily) but cost only $100

   (To calculate basic event probabilities, assume the stairway is used 5 times per hour, 8 h/day, 5 days/week, 50 weeks/year.)

   a. Draw a fault tree of the situation.

   b. Evaluate all alternatives (or combinations) to determine the best allocation of the $1,000.

5. Widgets are painted and cured with heat from a drier in the paint shop. Three components are required to start a fire in the paint shop and cause major damage: fuel,

ignition, and oxygen. Oxygen is always present in the atmosphere and thus has a probability of 1.0. Ignition can occur either from a spark due to static electricity (with a probability of 0.01) or from overheating of the drier mechanism (0.05). Fuel is provided by the volatile vapors that can arise from three sources: paint vapors during the drying process (0.9), paint thinner used to thin the paint (0.9), and solvent used for the cleanup of equipment (0.3). Property damage from a fire could amount to as much as $100,000. You have three choices of solutions in minimizing the likelihood of a major fire:

1. Spend $50 to move cleanup operations to a different room, which would reduce the probability of having solvent vapors in the paint area to 0.0.

2. Spend $3,000 on a new ventilation system which reduces the probability of having vapor from each of the three fuels to 0.2.

3. Spend $10,000 for a new type of paint and spraying system that doesn't release volatile vapors, reducing the probability of paint and paint thinner vapors to 0.0.

   a. Draw a fault tree and recommend the most cost-effective solution.

   b. Consider the domino theory as applied to this scenario. Name each of the dominoes in the proper order, and provide two other possible solutions that would apply to this scenario.

6. a. Draw the fault tree given by the Boolean expression $T = AB + CDE + F$.

   b. The severity of T is 100 lost workdays per accident. The probabilities of the basic events are $A = 0.02$, $B = 0.03$, $C = 0.01$, $D = 0.05$, $E = 0.04$, and $F = 0.05$. What is the expected loss associated with the head event T as given?

   c. Compare two alternatives from a cost-benefit standpoint. Which would you recommend?

   (i) 100 to reduce C and D to 0.005

   (ii) $200 to reduce F to 0.01

7. Calculate the system reliability. Assume events are independent.

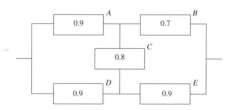

8. One of the early propeller planes (a Ford Tri-Motor) had three engines, one directly in the middle and one on each wing. In this configuration, the plane can fly with any two engines or just the middle one alone. Assuming the reliability of each engine is 0.9, what is the reliability of the overall plane?

9. NASA uses four identical onboard computers (that is, three are redundant) to control the space shuttle. If the reliability of any computer is 0.9, what is the overall system reliability?

10. Pick one of the OSHA 1910 standards. Then examine the OSHA Review Commission decisions regarding this standard. Are there any similarities between the cases? Who typically won? Were the original citations and/or fines issued by OSHA modified by the court?

REFERENCES

Asfahl, C. R. *Industrial Safety and Health Management*. New York: Prentice-Hall, 2004.

Bahr, N. J. *System Safety Engineering and Risk Assessment: A Practical Approach*. London: Taylor & Francis, 1997.

Banerjee, S. *Industrial Hazards and Plant Safety*. London: Taylor & Francis, 2003.

Bird, F., and G. Gemain. *Practical Loss Control Leadership*. Loganville, GA: International Loss Control Institute, 1985.

Brown, D. B. *Systems Analysis and Design for Safety*. New York: Prentice-Hall, 1976.

Cox, S. *Safety, Reliability and Risk Management: An Integrated Approach*. Butterworth Heinemann, 1998.

Devore, J. L. *Probability and Statistics for Engineering and the Sciences*. 6th ed. New York: Duxbury Press, 2003.

Ericson, C. A. *Hazard Analysis Techniques for System Safety*. New York: Wiley-Interscience, 2005.

Goetsch, D. L. *Occupational Safety and Health for Technologists, Engineers, and Managers*. 5th ed. Upper Saddle River, NJ: Pearson Prentice Hall, 2005.

Hammer, W., and D. Price. *Occupational Safety Management and Engineering*. 5th ed. New York: Prentice-Hall, 2001.

Heinrich, H.W., D. Petersen, and N. Roos. *Industrial Accident Prevention*. 5th ed. New York: McGraw-Hill, 1980.

Holmes, T. H., and R. H. Rahe. "The Social Readjustment Rating Scale." *Journal of Psychosomatic Research*. 11 (1967), pp. 213–218.

Laughery, K., and K. Vaubel. "Major and Minor Injuries at Work: Are the Circumstances Similar or Different?" *International Journal of Industrial Ergonomics*. 12 (1993), pp. 273–279.

National Safety Council. *Accident Prevention Manual for Industrial Operations*. 12th ed. Chicago: National Safety Council, 2000.

OSHA. *Code of Federal Regulations—Labor*. 29 CFR 1910. Washington, DC: Office of the Federal Register, 1997.

OSHA. *Ergonomics Program Management Guidelines for Meatpacking Plants*. OSHA 3123. Washington, DC: The Bureau of National Affairs, Inc., 1990.

Skinner, B. F. "Superstition' in the Pigeon". *Journal of Experimental Psychology*. 38 (1947), pp. 168–172.

Spellman, F. R. *Safety Engineering: Principles and Practices*. Lanham, MD: Government Institutes, 2005.

WEBSITES

NIOSH homepage—http://www.cdc.gov/niosh/
OSHA consultants—http://www.osha.gov/dcsp/smallbusiness/consult_directory.html
OSHA homepage—http://www.oshrc.gov
OSH Review Commission homepage—http://www.oshrc.gov

CHAPTER 9

Proposed Method Implementation

KEY POINTS

- Decide among alternative methods, using value engineering, cost-benefit analysis, crossover charts, and economic analyses.

- *Sell* the new method; people are resistant to change.

- Establish sound base rates by using reliable job evaluations.

- Accommodate workers of all abilities.

resenting and installing the proposed method is the fifth step in the systematic development of a work center to produce a product or perform a service. However, the analyst must first choose which proposed method to present. Several alternative methods may be feasible, some more effective than others, some more costly than others. A variety of decision-making tools are presented in this chapter to assist the analyst in selecting the best alternative. Obviously, many factors may comprise the definition of *best*, and these tools will help the analyst in weighing these factors appropriately.

Selling the proposed method is the next, and probably the most important, element in the presentation procedure. This step is as important as any of the preceding steps, since a method not sold is usually not installed. No matter how thorough the data gathering and analysis and the ingenuity of the proposed method, the value of the project is zero unless it is installed.

Humans naturally resent the attempts of others to influence their thinking. When someone approaches with a new idea, the instinctive reaction is to put up a defense against it and resist any changes. We feel that we must protect our own individuality, preserve the sanctity of our own egos. All of us are just egotistical enough to convince ourselves that our ideas are better than those of anyone else. It is natural for us to react in this manner, even if the new idea is to our own advantage. If the idea has merit, there is a tendency to resent it because we did not think of it first.

The presentation of the proposed method should include the decision making that went into choosing the final design and should emphasize the savings in material and labor that would be achieved with it. Next the quality and reliability improvements obtained by installing the improved method should be emphasized. Finally the payback time for capital investment should be addressed. Without reasonable cost recovery the project will not go forward.

Once the proposed method has been properly presented and sold, it can be installed. Installation, like presentation, requires sales ability. During installation, the analyst must continue selling the proposed method to engineers and technicians on their own level, to subordinate executives and supervisors, and to labor and organized labor representatives.

9.1   DECISION-MAKING TOOLS

DECISION TABLES

Decision tables are a structured approach for taking out the subjectivity in making decisions, that is, determining which of several alternative methods changes should be implemented. The tables essentially consist of condition–action statements, similar to if–then statements in computer programs. *If* the right condition or combination of conditions exists, *then* specified actions are taken. Thus, the tables can unambiguously describe complex, multirule, multivariable decision systems.

Such decision tables, also known as *hazard action tables*, are frequently utilized in safety programs to specify certain actions for given hazard conditions (Gausch, 1972). The hazard may be identified by two different variables: frequency, how often the accident is likely to occur; and severity, how severe the loss will be. Frequency may be categorized as extremely remote, remote, reasonably probable, and highly probable, while severity may have levels of negligible, marginal, critical, and catastrophic. This results in a hazard action table with five plans, as shown in Table 9.1.

Consider the right column marked with an asterisk. The analyst would conclude that a condition that is highly probable and would result in a catastrophe, with possible death and severe injury to personnel, should immediately be eliminated by shutting down the operation. Obviously, this is a simplified example that can be envisioned mentally. However, if one has 20 states for each of two different variables, then there are 400 categories of conditions, which cannot be remembered easily. Overall, decision tables emphasize making better-quality decisions through better decision analysis techniques and less time pressure; that is, the action plans can be worked out ahead of time, rather than having to deal with instantaneous pressures, possibly resulting in errors.

VALUE ENGINEERING

A simple way to expand the evaluation of alternatives is to apply numbers and form a payoff matrix. This is often termed *value engineering* (Gausch, 1974). Each solution may have different values with respect to the desired benefits. A weight is determined for each benefit (0 to 10 is a reasonable range), and then a value (0 to 4, with 4 being best) is assigned to reflect how well each solution produces the desired

Table 9.1 Hazard Action Table

Frequency	Severity			
	Negligible	**Marginal**	**Critical**	**Catastrophic**
Extremely remote				
Remote				
Reasonably probable				(*)
Probable				
Actions				
Forget it				
Long-range study				
Correct (1 year)				
Correct (90 days)				
Correct (30 days)				
Shutdown				

Source: Heinrich, Petersen, and Roos, 1980.

benefit. The assigned value is multiplied by the appropriate weight, and the products are summed for the final score. The highest sum is the most appropriate solution.

Note that benefits will have different relative weights for different companies, different departments within a company, or even different points in time for the same department. Also note that the Evaluating Alternatives step of Muther's Systematic Layout Planning (see Section 3.8) is a form of value engineering.

COST-BENEFIT ANALYSIS

A more quantitative approach to deciding between different alternatives is a *cost-benefit analysis*. This approach requires five steps:

1. Determine what is changed due to better design, that is, increased productivity, better quality, decreased injuries, and so on.
2. Quantify these changes (benefits) into monetary units.
3. Determine the cost required to implement the changes.
4. Divide the cost by the benefit for each alternative, to create a ratio.
5. The smallest ratio determines the desired alternative.

Step 2 is probably the most difficult to assess and quantify. It is not always possible to assign dollar values; sometimes it may be percent changes, injuries, or other values. Example 9.1 may help in understanding all three decision-making tools. Other examples of cost-benefit analyses as related to less definable benefits, such as health and safety issues, are presented in Brown (1976).

EXAMPLE 9.1	Cutoff Operation

The Dorben Co. manufactures simple, small knife blades inserted into a plastic handle. One of the operations in the formation of the blade is the cutting off of knife blades from a thin strip of stainless steel via a foot-pedal-operated press. Using tweezers, the worker procures a rubber nib from a parts bin and inserts it over the blade to protect it. After press activation, the cutoff blade is placed on a holder plate for later assembly into the handle (a good example of the effective therblig *preposition*!). Because of the small blade size, a stereoscope is used to assist in the operation. The operators have complained about wrist, neck, back, and ankle pain. Possible method changes include (1) replacement of the mechanical pedal with a foot-operated electric switch, to reduce ankle fatigue; (2) better adjustment of the position of the stereoscope, to reduce neck fatigue; (3) implementation of a video projection system, for heads-up viewing; (4) use of a gravity feed bin for the nibs, to improve productivity; and (5) replacement of the tweezers with a vacuum-operated stylus, to both improve productivity and eliminate a potential cumulative trauma disorder (CTD) causing pinch grip.

Assume the productivity improvements shown in Table 9.2, based on an MTM-2 analysis (see Section 13.1) and injury reductions based on the CTD Risk Index shown in Figure 5.25.

Company policy authorizes methods engineers to proceed, with no further authorization needed, if condition 1 and either condition 2 or 3 are met: (1) implementation costs are less than $200 (i.e., petty cash), (2) productivity increases are more than 5 percent, (3) injury risks decrease more than 33 percent. In terms of decision tables, the situation could be structured as shown in Table 9.3.

In terms of value engineering, weights of 6, 4, and 8 can be assigned to the three factors of interest: increased productivity, decreased injury rates, and low-cost solutions (see Table 9.4). Each solution is rated from 0 to 4 for each of the factors. The resulting product sums are 28, 36, 18, 58, and 42, and the gravity feed bin change at 58 is clearly the best solution.

For a cost-benefit analysis, anticipated benefits could be quantified by both increases in productivity and decreases in injury rates. Assume that the company profits $645 for each 1 percent increase in productivity over the course of the year. Similarly, a decrease in workers' compensation and medical costs due to a decrease in CTD injuries can be considered a benefit. The company has averaged one CTD case leading to surgery every 5 years of operation. Assuming one CTD surgery case costs the company $30,000,

Table 9.2 Expected Changes in Productivity, Injury Risk Potential, and Cost for Various Method Changes in Cutoff Operation

Work design and methods changes	Δ Productivity (%)	Δ CTD Risk (%)	Cost ($)
1. Foot-operated electric switch	0	−1[*]	175
2. Adjust stereoscope	0	−2	10
3. Video projection system	+1[**]	−2	2,000
4. Gravity feed bin	+7	−10	40
5. Vacuum stylus	+1[**]	−40	200

[*]The current CTD Risk Index does not address lower extremities. However, there is reason to believe that the lower force for the electric switch will have some beneficial effect.
[**]Can't be quantified from MTM-2, but some benefit is expected.

Table 9.3 Decision Table for Cutoff Operation

Methods Changes	Conditions				Action
	1	**2**	**3**	**Policy**	
1. Electric switch					—
2. Adjust stereoscope					—
3. Video projection system					—
4. Gravity feed bin					Proceed
5. Vacuum stylus					Proceed

Table 9.4 Value Engineering Analysis of the Cutoff Operation

Evaluating Alternatives

Plant: Dorben Co.	Alternatives	A	B	C	D	E	
Project: Cutoff Operation		Electric switch	Adjust stereoscope	Video projection	Gravity feed bin	Vacuum stylus	
Date: 6-12-97							
Analyst: AF							

Factor/Consideration	Wt	Ratings and weighted ratings										Comments
		A		**B**		**C**		**D**		**E**		
Increase in productivity	6	0	0	0	0	1	6	3	18	1	6	
Decrease in injuries	4	1	4	1	4	1	4	2	8	3	12	
Low-cost solution	8	3	24	4	32	1	8	4	32	3	24	
Totals		28		36		18		58		42		

Remarks:
Gravity feed bin is the most justifiable methods change.

the expected loss per year is $6,000. For each 1% decrease in risk, the company bene-fits $60 per year. The increases in productivity and decreases in injury rates can be quantified as shown in Table 9.5.

From Table 9.5, it is obvious that for any ratio less than 1 (methods changes 2, 4, 5, and 6), there are more benefits than the costs required to implement the methods change. On the other hand, methods change 4 is by far the most cost-effective. Inter-estingly, a combination of methods changes 2, 4, and 5 (alternative 6) may be worth considering for the comparatively low total dollar amounts expended.

Table 9.5 Cost-Benefit Analysis for Cutoff Operation

Method Changes	Benefit ($) Productivity	Injury rates	Total	Cost ($)	Cost benefit
1. Electric switch	0	60	60	175	2.92
2. Adjust stereoscope	0	120	120	10	0.08
3. Video projection system	645	120	765	2,000	2.61
4. Gravity feed bin	4,515	600	5,115	40	0.01
5. Vacuum stylus	645	2,400	3,045	100	0.03
6. Methods changes 2, 4, 5	5,160	3,120	8,280	150	0.02

CROSSOVER CHARTS

Crossover (or *break-even*) *charts* are very useful in comparing the payback times of alternative methods changes. One may use general-purpose equipment with low capital costs but higher setup costs, while the other may use special equipment at a higher capital cost but with lower setup costs. At some production quantity, the two methods are equal, and this is the *crossover point*. This relates to the most prevalent mistake made by planners. Large amounts of money are tied up in fixtures that show large savings while in use, but they are seldom used. For example, a savings of 10 percent in direct labor costs on a job in constant use would probably justify greater expense in tools than an 80 or 90 percent savings on a small job that appears on the production schedule only a few times a year (a good example of the Pareto analysis in Section 2.1).

The economic advantage of lower labor costs is the controlling factor in determining the tooling; consequently, jigs and fixtures may be desirable even where only small quantities are involved. Other considerations, such as improved interchangeability, increased accuracy, or reducing labor troubles, may provide the dominant reason for elaborate tooling, although this is usually not the case.

MULTIPLE-CRITERIA DECISION MAKING

Decision making in the presence of multiple, often conflicting, criteria can be approached by a relatively new process called *multiple-criteria decision making* (MCDM), developed by Saaty (1980). For example, assume that an analyst has four alternatives to consider a_1, a_2, a_3, a_4, which would be applied to four possible states of the product or market S_1, S_2, S_3, S_4. Also assume that this analyst estimates the following outcomes for the various alternatives and states of the market:

Alternatives	States of product or market S_1	S_2	S_3	S_4	Total
a_1	0.30	0.15	0.10	0.06	0.61
a_2	0.10	0.14	0.18	0.20	0.62
a_3	0.05	0.12	0.20	0.25	0.62
a_4	0.01	0.12	0.35	0.25	0.23
Total	0.46	0.53	0.83	0.76	

If the outcomes represent profits or returns, and the state of the market will be S_2, then the analyst will definitely decide on alternative a_1. If the outcomes represent scrap or some other factor that the analyst wishes to minimize, then alternative a_3 is chosen. (Although a_4 also has an outcome of 0.12, the analyst chooses a_3, since there is less variability in outcome under this alternative than with a_4.) Seldom should decisions be made under an assumed certainty. Usually, some risk is involved in predicting the future state of the market. Assume that the analyst is able to estimate the following probability values associated with each of the four states of the market:

$$
\begin{array}{ll}
S_1 & \dots \dots \dots 0.10 \\
S_2 & \dots \dots \dots 0.70 \\
S_3 & \dots \dots \dots 0.15 \\
S_4 & \dots \dots \dots 0.05 \\
\hline
& 1.00
\end{array}
$$

Crossover Analysis of Fixture and Tooling Costs

EXAMPLE 9.2

The production engineer in a machining department has devised two alternative methods involving different tooling for a job being machined in the shop. Data on the present and proposed methods are shown in Table 9.6. Which method would be more economical in view of the activity? The base pay rate is $9.60 per hour. The estimated activity is 10,000 pieces per year. The fixtures are capitalized and depreciated in 5 years. A cost analysis reveals that a total unit cost of $0.077, represented by alternative 2, is the most economical in the long run.

A crossover chart (see Figure 9.1) allows the analyst to decide which method to use for given quantity requirements. The present method is the best for quantities up to about 7,700 per year:

$$(0.137 + 0.0006)x + 0 = (0.097 + 0.001)x + 300$$
$$x = 300/(0.137 - 0.098) = 7,692 \approx 7,700$$

Alternative method 1 is better for quantities between 7,700 and 9,100 per year:

$$(0.097 + 0.001)x + 300 = (0.058 + 0.007)x + 600$$
$$x = 300/(0.098 - 0.065) = 9,090 \approx 9,100$$

and method 2 is best for quantities above 9,100 per year. Note that in this latter approach, fixture costs were absorbed up front, while tool costs were considered as expendable supplies.

Table 9.6 Fixture and Tooling Costs

Method	Standard time (min)	Fixture cost	Tool cost	Average tool life	Unit direct labor cost	Unit fixture cost	Unit tool cost	Unit total cost
Present method	0.856 each	None	$6	10,000 pieces	$0.137	None	$0.0006	$0.1376
Alternative 1	0.606 each	$300	20	20,000 pieces	0.097	$0.006	0.0010	0.104
Alternative 2	0.363 each	600	35	5,000 pieces	0.058	0.012	0.0070	0.077

Figure 9.1
A crossover (break-even) chart for fixture and tooling costs.

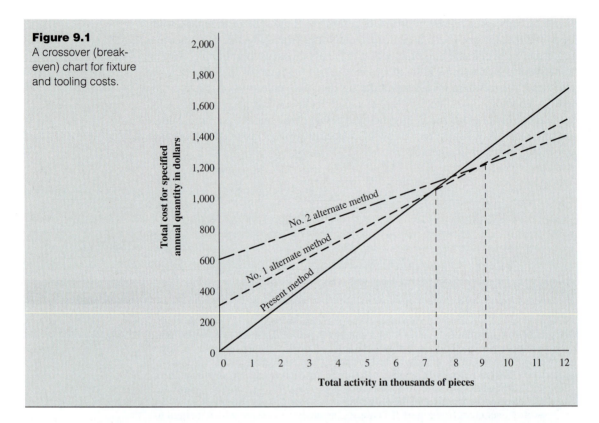

EXAMPLE 9.3
Crossover Analysis of Competing Methods

Insufficient volume may make it impractical to consider many alternative proposals that may offer substantial savings over existing methods. For example, an operation done on a drill press involved a 0.5-in hole reamed to the tolerance of 0.500 to 0.502 in. The activity of the job was estimated to be 100,000 pieces. The time study department established a standard of 8.33 h per thousand to perform the reaming operation, and the reaming fixture cost $2,000. Since a base rate of $7.20 per hour was in effect, the money rate per thousand pieces was $60.

Now, assume that a methods analyst suggests broaching the inside diameter, because the part can be broached at the rate of 5 h per thousand. This would be a savings of 3.33 h per thousand pieces, or a total savings of 333 h. At the $7.20 base rate, this would mean a direct labor savings of $2,397.60. However, it would not be practical to go ahead with this idea, since the tool cost for broaching is $2,800. Thus, the change would not be sound unless the labor savings could be increased to $2,800 to offset the cost of the new broaching tools.

Since the labor savings in a new broaching setup would be 3.33 × $7.20 per thousand, 116,800 pieces would have to be ordered before the change in tooling would be justified.

$$\frac{\$2,800 \times 1,000}{\$7.20 \times 3.33} = 116,783 \text{ pieces}$$

However, if the broaching method had been used originally instead of the reaming procedure, it would have paid for itself in

$$\frac{\$2,800 - \$2,000}{\$7.20 \times 3.33/M} = 33,367 \text{ pieces}$$

With production requirements of 100,000 pieces, the labor savings would be 3.33 × $7.20 × 66.6 thousand (the difference between 100,000 and 33,400) = $1,596.80 over the present reaming method. Had a motion analysis been made in the planning stage, this savings might have been realized. Figure 9.2 illustrates these relationships with the customary crossover chart.

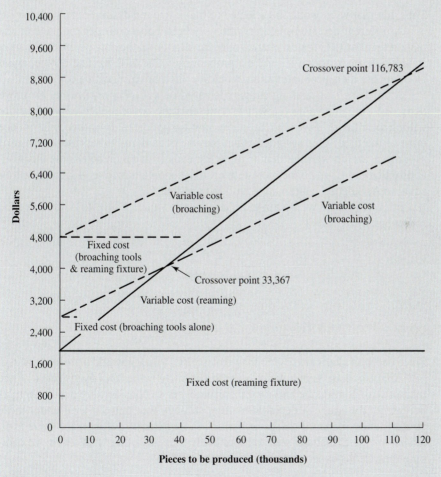

Figure 9.2 Crossover chart illustrating the fixed and variable costs of two competing methods.

A logical decision-making strategy would be to calculate the expected return under each decision alternative and then select the largest value to maximize or the smallest value to minimize. Here,

$$E(a) = \sum_{j=1}^{n} P_j C_{ij}$$
$$E(a_1) = 0.153$$
$$E(a_2) = 0.145$$
$$E(a_3) = 0.132$$
$$E(a_4) = 0.15$$

Thus, alternative a_1 would be selected to maximize the desired result.

A different decision-making strategy would be to consider the state of the market that has the greatest chance of occurring. From the data given, this market state would be S_2, since it carries a probability value of 0.70. The choice, again based on the most probable future, would be alternative a_1, with a 0.15 return.

A third decision-making strategy is based on a *level of aspiration*. Here, we assign an outcome value C_{ij}, which represents the consequence of what we are willing to settle for if we are reasonably sure we will get at least this consequence most of the time. This assigned value may be considered to represent a level of aspiration, which we shall denote as A. For each a_j, we then determine the probability that the C_{ij} in connection with each decision alternative is greater than or equal to A. Select the alternative with the greatest $P(C_{ij} \geq A)$.

For example, if we assign the consequence value of 0.10 to A, we have the following:

$$(C_{ij} \geq 0.10)$$
$$a_1 = 0.95$$
$$a_2 = 1.00$$
$$a_3 = 0.90$$
$$a_4 = 0.90$$

Since decision alternative a_2 has the greatest $P(C_{ij} \geq A)$, it would be recommended.

Analysts may be unable to assign probability values to various states of the market with confidence and may therefore want to consider any one of them as being equally likely. A decision-making strategy that may be used under these circumstances is based on the *principle of insufficient reason*, since there is no reason to expect that any state is more likely than any other state. Here, we compute the various expected values based on

$$E(a) = \frac{\sum_{j=1}^{n} C_{ij}}{n}$$

In our example, this would result in

$$E(a_1) = 0.153$$
$$E(a_2) = 0.155$$
$$E(a_3) = 0.155$$
$$E(a_4) = 0.183$$

Based on this choice, alternative four would be proposed.

A second strategy analysts may consider when making decisions under uncertainty is based on the *criterion of pessimism*. When one is pessimistic, one anticipates the worst. Therefore, in a maximization problem, the minimum consequence would be selected for each decision alternative. The analyst compares these minimum values and selects the alternative that has the maximum of the minimum values. In our example:

Alternative	Min C_{ij}
a_1	0.06
a_2	0.10
a_3	0.05
a_4	0.01

Here, alternative a_2 would be recommended, since its minimum value of 0.10 is a maximum when compared to the minimum values of the other alternatives.

The *plunger criterion* is a third decision-making strategy that analysts may want to consider. This criterion is based on an optimistic approach. If one is optimistic, one expects the best, regardless of the alternative chosen. Therefore, in a maximizing problem, the analyst would select the maximum C_{ij} for each alternative and would then select the alternative with the largest of these maximum values. Thus

Alternative	Min C_{ij}
a_1	0.30
a_2	0.20
a_3	0.25
a_4	0.35

Here, decision alternative a_4 would be recommended, because of its maximum value of 0.35.

Most decision makers are neither completely optimistic nor completely pessimistic. Instead, a coefficient of optimism X is established, where

$$0 \leq X \leq 1$$

Then a Q_i is determined for each alternative, where

$$Q_i = (X)(\text{Max } C_{ij}) + (1 - X)(\text{Min } C_{ij})$$

The alternative recommended is the one associated with the maximum Q_i for maximization, and with the minimum Q_i for minimization.

A final decision-making approach based on uncertainty is the *minimax regret criterion*. This criterion involves the calculation of a *regret matrix*. For each alternative, based on a state of the market, analysts calculate a regret value. This regret value is the difference between the payoff actually received and the payoff that could have been received if the decision maker had been able to foresee the state of the market.

To construct the regret matrix, the analyst selects the maximum C_{ij} for each state S_j and then subtracts the C_{ij} value of each alternative associated with that state. In our example, the regret matrix would be

Alternative	States			
	S_1	S_2	S_3	S_4
a_1	0	0	0.25	0.19
a_2	0.20	0.01	0.17	0.05
a_3	0.25	0.03	0.15	0
a_4	0.29	0.03	0	0

The analyst then selects the alternative associated with the minimum of the maximum regrets (minimax).

Alternative	Min r_{ij}
a_1.	0.25
a_2.	0.20
a_3.	0.25
a_4.	0.29

Based on the minimax regret criterion, a_2 would be selected, in view of its minimum regret of 0.20.

Such decision making is very common in manual materials handling (see Chapter 4), where there is always a trade-off between worker safety and worker productivity. The greater the focus on worker safety, for example, through the reduction in loads and the corresponding biomechanical stresses on the lower back, the worse the productivity of the load handled. To maintain job productivity at the desired level, a reduction of load weight requires an increased task frequency, with corresponding heavier physiological demands. A metabolic evaluation will lead to the conclusion that the infrequent lifting of heavy loads is preferable to the frequent lifting of lighter loads. However, from a biomechanical point of view, the load weight should be minimized, regardless of frequency, resulting in a conflict. This problem was examined by Jung and Freivalds (1991) using MCDM for a critical range of task frequencies of 1 lift/min to 12 lifts/min (see Figure 9.3). For infrequent tasks (< 7/min), the biomechanical stress predominates; for higher-frequency tasks (> 7/min), the

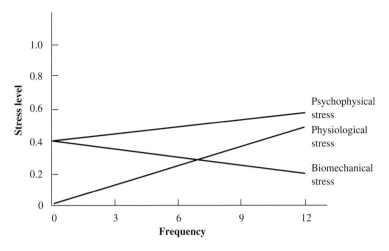

Figure 9.3 Unacceptability of stress levels used in reconciling the conflicting guidelines. (*From*: Jung and Freivalds, 1991)

physiological stress predominates. At about 7 lifts/min, however, both stresses participate equally in determining the overall stress level to the worker. Thus, depending on the alternatives and the effect of each alternative with regard to specific attributes of interest, different solutions can be obtained. Analysts should become familiar with these decision-making strategies and should use those that are most appropriate to their organizations.

ECONOMIC DECISION TOOLS

The three most frequently used appraisal techniques for determining the desirability of investing in a proposed method are (1) the return on sales method, (2) the return on investment or payback method, and (3) the discounted cash flow method.

The *return on sales* method involves computing the ratio of (1) the average yearly profit brought about through using the method to (2) the average yearly sales or increase in dollar value added to the product, based on the pessimistic estimated life of the product. However, while this ratio provides information on the effectiveness of the method and the resulting sales efforts, it does not consider the original investment required to get started on the method.

The *return on investment* method gives the ratio of (1) the average yearly profit brought about through using the method, based upon the pessimistic estimated life of the product, to (2) the original investment. Of two proposed methods that would result in the same sales and profit potential, management would prefer to use the one requiring the least investment of capital. The reciprocal of the return on investment is often referred to as the *payback* method. This gives the time that it would take to realize a full return on the original investment.

The *discounted cash flow* method computes the ratio of (1) the present worth of cash flow, based on a desired percentage return, to (2) the original investment.

This method calculates the rate of flow of money in and through the company and the *time value of money*. The time value of money is important. Because of interest earned, a dollar today is worth more than a dollar at any later date. For example, at 15-percent compound interest, $1 today is worth $2.011 five years from now. Expressing it another way, $1 received 5 years from now would be worth about 50 cents today. Interest may be thought of as the return obtainable by the productive investment of capital.

The following applies to the present-value concept:

Single payment		
– Future-worth factor	(given P, find F)	$F = P(1 + i)^n$
– Present-worth factor	(given F, find P)	$P = F(1 + i)^{-n}$
Uniform series		
– Sinking fund factor	(given F, find R)	$R = \dfrac{Fi}{(i + 1)^n - 1}$
– Capital recovery	(given P, find R)	$R = \dfrac{Pi(1 + i)^n}{(1 + i)^n - 1}$
– Future-worth factor	(given R, find F)	$F = R[(1 + i)^n - 1]/i$
– Present-worth factor	(given R, find P)	$P = \dfrac{R[(1 + i)^n - 1]}{i(1 + 1)^n}$

where i = interest rate for a given period
   n = number of interest periods
   P = present sum of money (present worth of principal)
   F = future sum of money at end of n periods from present date; equivalent to P with interest i
   R = end-of-period payment or receipt in a uniform series continuing for coming n periods; the entire series equivalent to P at interest i

An assumed return rate I is the basis of the cash flow computation. All cash flows following the initial investment for the new method are estimated and adjusted to their present worth, based on the assumed return rate. The total estimated cash flows for the pessimistic estimated life of the product are then summed as a profit (or loss) in terms of cash on hand today. Finally, this total is compared to the initial investment.

Estimates of the product demand 10 years hence may deviate considerably from reality. Thus, the element of chance is introduced, and the probabilities of success tend to diminish with the increased length of the payoff period. The results of any study are only as valid as the reliability of the input data. Constant follow-up can determine the validity of the assumptions. The analyst should not hesitate to alter decisions if the original data prove to be invalid. Sound financial analysis is intended to facilitate the decision-making process, not replace good business judgment.

Economic Justification of a Proposed Method

EXAMPLE 9.4

An example should clarify the use of the three methods for appraising the potential of a proposed method.

   Investment for proposed method: $10,000
   Desired return on investment: 10 percent
   Salvage value of jigs, fixtures, and tools: $500
   Estimated life of the product for which the proposed method will be used: 10 years

Present worth of cash flow:

$$
\begin{array}{ll}
(3,000)(0.9091) = \$2,730 & (3,800)(0.5645) = 2,140 \\
(3,800)(0.8264) = 3,140 & (3,000)(0.5132) = 1,540 \\
(4,600)(0.7513) = 3,460 & (2,200)(0.4665) = 1,025 \\
(5,400)(0.6830) = 3,690 & (1,400)(0.4241) = 595 \\
(4,600)(0.6209) = 2,860 & (500)(0.3855) = 193 \\
& \overline{\$21,337}
\end{array}
$$

Salvage value of tools:

$$(500)(0.3855) = \$193$$

Total present worth of anticipated gross profit and tool salvage value: $21,566. Ratio of present worth to original investment:

$$\frac{12,566}{10,000} = 2.16$$

The new method satisfactorily passes all three appraisal methods (see Table 9.7). A 61-percent return on sales and a 32.3-percent return on capital investment represent very attractive returns. The return of the $10,000 capital investment will take place in 3.09 years, and the cash flow analysis reveals that the original investment will be recovered in 4 years while earning 10 percent. During the 10-year anticipated life of the product, $11,566 more than the original investment will be earned.

Table 9.7 Comparison of Economic Justification Methods

End of year	Increase in sales values due to proposed method	Cost of production with proposed method	Gross profit due to proposed method
1	$5,000	$2,000	$3,000
2	6,000	2,200	3,800
3	7,000	2,400	4,600
4	8,000	2,600	5,400
5	7,000	2,400	4,600
6	6,000	2,200	3,800
7	5,000	2,000	3,000
8	4,000	1,800	2,200
9	3,000	1,600	1,400
10	2,000	1,500	500
Totals	$53,000	$20,700	$32,300
Average	$ 5,300	$ 2,070	$ 3,230

$$\text{Return on sales} = \frac{3,230}{5,300} = 61\% \qquad \text{Return on investment} = \frac{3,230}{10,000} = 32.3\%$$

$$\text{Payback} = 1/0.323 = 3.09 \text{ years}$$

9.2 INSTALLATION

After a proposed method has been approved, the next step is installation. The analyst should stay with the job during installation, to ensure that all details are carried out in accordance with the proposed plan. The analyst should verify that the work center being established is equipped with the facilities proposed, the planned working conditions are provided, the tooling is done in accordance with recommendations, and the work is progressing satisfactorily. A mechanic may make a slight change without considering the consequences, result is in less than anticipated benefits. The analyst should "sell" the new method to the operator, supervisor, and setup person along the way so that these employees will be more accepting of the new method.

Once the new work center has been installed, the analyst must check all aspects to see whether they conform to the specifications established. In particular, the analyst must verify that the *reach* and *move* distances are the correct length, the tools are correctly sharpened, the mechanisms function soundly, stickiness and sluggishness have been worked out, safety features are operative, material is available in the quantities planned, working conditions associated with the work center are as anticipated, and all parties have been informed of the new method.

Once every aspect of the new method is ready for operation, the supervisor assigns the operator who will be working with the method. The analyst should then stay with the operator as long as necessary to ensure that the operator is familiar with the new assignment. This period may be a matter of a few minutes, several hours, or even several days, depending on the complexity of the assignment and the flexibility and adaptability of the worker.

Once the operator begins to get a feel for the method and can work along systematically, the analyst can proceed with other work. However, the installation phase should not be considered complete until the analyst has checked back several times during the first few days after installation to ensure that the proposed method is working as planned. Also check with line supervisors to ensure that they spot-check and monitor the new method.

RESISTANCE TO CHANGE

It is not unusual for workers to resist changes in the method. Although many may think of themselves as open-minded, most people are quite comfortable with their present job or workplace, even if it might not be the most comfortable or pleasant. Their fear of change and the impact it might have on their jobs, pay, and security overrides other concerns (see Maslow's hierarchy of human needs, Section 18.3). Worker reactions to change can be quite obstinate and perplexing, as experienced by Gilbreth in the following classic example. While doing a motion study in a bed manufacturing plant, he noticed a large middle-aged woman ironing bed sheets in an obviously very inefficient and fatiguing manner. To iron each sheet, she picked up a large and heavy iron, sat down, and pressed down hard on the sheet, about 100 times for each sheet. She was fatigued and had back pain.

With a few work design changes using counterweights to support the iron (the modern-day equivalent of a tool balancer), he considerably reduced the physical workload. However, the reaction of the woman was completely opposite of that expected. Instead of being the only worker physically capable of doing the job (and receiving praise from her supervisor), she was only one of many. Her status was lost and she was dead set against the change (DeReamer, 1980).

Thus, it is important to "sell" the new method to the operators, supervisors, mechanics, and others. Employees should be notified well in advance about any method changes that will affect them. The resistance to change is directly proportional to the magnitude of the change and the time available to implement the change. Therefore, large changes should be made in small steps. Don't change the whole workstation, chair or stool, and tooling all at once. Start with the chair, maybe change the tools next, and then finally change the workstation.

Explain the reasons for the change. People resist what they don't understand. Instead of just replacing the worker's pistol-grip tool for a horizontal work surface with an in-line tool, explain that this tool will weigh less and require less upper arm motion, that is, it will be more comfortable to use.

As a general rule for dealing with emotion, it is better to emphasize positives, for example, "This new tool is going to be much easier to use," and de-emphasize negatives, such as, "That tool is heavy and unsafe." Get the worker to participate directly in the process of the methods change or work design. Workers have a good record of following their own suggestions, and where workers have been involved in the decision making, there has typically been less resistance to the changes. One successful approach is to form worker committees or ergonomic teams (see Section 18.5).

Threats of force, that is, management reprisals for not making the change, may be counterproductive, setting up counteremotions to resist the change. In addition, people often resist the social aspects of change, rather than the technical aspects. Therefore, if it can be shown that other employees are using the same device, an operator will be much more likely to go along with the change.

The last step in methods engineering, after appropriate standards have been set, is to maintain the method, that is, to determine if the anticipated productivity gains are being realized. Here, the industrial engineers must be very careful to determine whether any effect is truly due to the new method or due to the *Hawthorne effect*, solely.

HAWTHORNE EFFECT

This is an often quoted study emphasizing the need for worker involvement in methods changes or production planning leading to increased motivation and productivity. Actually, this was a very poorly designed series of studies from which it is difficult to draw any true scientific effect or conclusion, other than the need to be very careful when assuming that productivity improvements are due strictly to the methods changes. The initial, less well-known study was a joint project between the National Research Council and Western Electric Co. to examine the effects of illumination on productivity. The study was conducted at the very large Hawthorne plant (some 40,000 workers) near Chicago, from 1924 to 1927. In general, management found that employees were reacting to changes in illumination in the way they assumed they were expected to react; that is, when illumination was increased, they were expected to produce more, and they did. When the illumination was decreased, they were expected to produce less and did so. A follow-up study tested this point even further. Lightbulbs were changed and workers were allowed to assume that the illumination had been increased. In fact, the lightbulbs were replaced with ones of exactly the same wattage. However, the workers commented favorably on the increased illumination and responded with increased productivity. Thus, physiological effects were being confounded by psychological effects (Homans, 1972).

Based on these results, Western Electric decided to conduct a further series of studies on mental attitudes and workers' effectiveness. These are the more famous Hawthorne studies, conducted in conjunction with the Harvard School of Business Administration from 1927 to 1932 (Mayo, 1960). Six female operators were placed in a separate room and subjected to various experimental conditions: (1) special group incentive for the six workers, as opposed to the more than 100 workers in a given department; (2) inclusion of two 5- or 10-min rest breaks; (3) shorter workdays; (4) shorter workweeks; and (5) lunches and/or beverages provided by the company. Any changes were discussed with the group of six ahead of time, and any that were seriously objected to were discarded. Also, later, the effects of these changes on the workers were discussed in a rather formal interview process.

With the workers welcoming these opportunities to vent their feelings, these structured interviews degenerated into open-ended gripe sessions. Interestingly, production generally increased during this span of 5 years (except for minor changes in product or start/stop of vacation periods), regardless of test conditions. In addition, absences and sick days decreased considerably for the six operators, as compared to their coworkers. Western Electric, already known for its concern for worker welfare, attributed this somewhat surprising result to an overall increase in concern for the worker, with a resulting increase in social satisfaction (Pennock, 1929–1930).

Unfortunately, these conclusions are oversimplified, as other effects were also confounded in the study. There were large variations in supervisory practices, different and inconsistent measures of productivity were utilized throughout the 5 years of the study, and significant changes in methods were introduced (Carey, 1972). For example, a drop delivery system was implemented for the experimental group for the purpose of counting the relays produced. However, based on principles of motion economy (in Section 4.3), this also tended to increase production rates.

Regardless of the controversy, the Hawthorne studies have three main implications: (1) the basic rule of experimentation—"the act of measuring something changes it"—was confirmed; (2) proper human relations can act as a strong motivator (see Section 10.3); and (3) it is very difficult to tease apart confounded factors in an uncontrolled study. Therefore, the analyst should be very careful about jumping to conclusions about the effectiveness of methods changes on productivity. Part of the productivity changes could be due to the improved method, while part could be due to improved morale or motivation of the affected operator. Also, any productivity measurements, even though innocuous on the surface, could have unintended effects, if the workers become aware of them and perform as expected.

9.3 JOB EVALUATION

This is the sixth step in the systematic procedure of applying methods engineering. Every time a method is changed, the job description should be altered to reflect the conditions, duties, and responsibilities of the new method. When a new method is introduced, a job evaluation should be done, so that a qualified operator may be assigned to the work center and an appropriate base rate provided.

A job evaluation should start with an accurate title, a detailed job description identifying the job's specific duties and responsibilities, and the minimum requirements of the worker performing the job. The worker should be enlisted in accurately defining the job's responsibilities. A combination of personal interviews and questionnaires, along with direct observation, results in a concise definition of the job and the duties that it entails. The mental and physical functions required to perform the work should also be included, and such definitive words as *direct*, *examine*, *plan*, *measure*, and *operate* should be used. The more accurate the description, the better, with Figure 9.4 illustrating a detail example for a clerical job. These job descriptions are useful supervisory tools that can aid in the selection, training, and promotion of employees, and in the assessment of work distribution.

Essentially the job evaluation is a procedure by which an organization ranks its jobs in order of their worth or importance and should provide the following:

1. A basis for explaining to employees why one job is worth more (or less) than another job
2. A reason to employees whose rates of pay are adjusted because of a change in method
3. A basis for assigning personnel with specific abilities to certain jobs
4. Criteria for a job when new personnel are employed or promotions are made
5. Assistance in the training of supervisory personnel
6. A basis for determining where opportunities for methods improvement exist

JOB EVALUATION SYSTEMS

The majority of job evaluation systems in use today are a variation or combination of four principal systems: the classification method, the point system, the factor comparison method, and the ranking method. The *classification method*,

JOB TITLE ___Shipping and Receiving Clerk___ DEPT. ___Shipping___

MALE <u>X</u> FEMALE ___ DATE _____ TOTAL POINTS <u>280</u> CLASS <u>5</u>

JOB DESCRIPTION

Directs and assists in loading and unloading, counting, and receiving or rejecting pur-chased parts and supplies, and later delivers to proper departments.

Examines receivals out of line with purchase orders. Maintains file on all purchase orders and/or shipping orders and keeps open orders up-to-date. Maintains daily and weekly shipping reports and monthly inventory reports.

Assists in packing of all foreign and domestic shipments. Makes up request for inspec-tion form on certain materials received and rejection form for all items rejected.

Job requires thorough knowledge of packing, shipping and receiving routine, plant layout, shop supplies, and finished parts. Needs to have a knowledge of simple office routine. Ability to work with other departments, as a service department, and to deal effectively with venders. Job requires considerable accuracy and dependability. The effects of poor decisions include damaged receivables and shipments, inaccurate inven-tories, and extra material handling. Considerable lifting of weights up to 100 lbs. is involved. Works in conjunction with two class 4 packers and shippers.

Job Evaluation	Degree	Points
Education	1	15
Experience and training	2	50
Initiative and ingenuity	3	50
Analytical ability	3	50
Personality requirements	2	30
Supervisory responsibility	1	25
Responsibility for loss	1	10
Physical application	6	25
Mental or visual application	1	5
Working conditions	5	<u>20</u>
		280

Figure 9.4 Job analysis for a shipping and receiving clerk.

sometimes called the *grade description plan*, consists of a series of definitions designed to differentiate jobs into wage groups. Once the grade levels have been defined, analysts study each job and assign it to the appropriate level, on the basis of the complexity of its duties and responsibilities and its relation to the description of the several levels. The U.S. Civil Service Commission uses this plan extensively.

To use this method of job evaluation, the analyst uses the following procedure:

1. Prepare a grade description scale for each type of job, such as machine operations, manual operations, skilled (craft) operations, or inspection.
2. Write the grade descriptions for each grade in each scale, using such factors as
   a. Type of work and complexity of duties
   b. Education necessary to perform job
   c. Experience necessary to perform job
   d. Responsibilities
   e. Effort demanded
3. Prepare job descriptions for each job. Classify each job by *slotting* (placing in a specific category) the job description into the proper grade description.

In the *point system*, analysts directly compare all the attributes of a job with the attributes in other jobs, using the following procedure:

1. Establish and define the basic factors common to most jobs, indicating the elements of value in all jobs.
2. Specifically define the degrees of each factor.
3. Establish the points to be accredited to each degree of each factor.
4. Prepare a description of each job.
5. Evaluate each job by determining the degree of each factor contained in it.
6. Sum the points for each factor to get the total points for the job.
7. Convert the job points into a wage rate.

The *factor comparison* method of job evaluation usually has the following elements:

1. Determine the factors establishing the relative worth of all jobs.
2. Establish an evaluation scale that is usually similar to a point scale, except that the units are in terms of money. For example, a $2,000 per month benchmark job might attribute $800 to the responsibility factor, $400 to education, $600 to skill, and $200 to experience.
3. Prepare job descriptions.
4. Evaluate key jobs, factor by factor, by ranking each job from the lowest to the highest.
5. Pay wages on each key job, based on various factors. The money allocation automatically fixes the relationships among jobs for each factor, and therefore establishes the ranks of jobs for each factor.
6. Evaluate other jobs, factor by factor, on the basis of the monetary values assigned to the various factors in the key jobs.
7. Determine a wage by adding the money values of the various factors.

Both the point system and the factor comparison method are more objective and thorough in their evaluations of the various jobs involved; both plans study the basic factors common to most jobs that influence their relative worth. Of the two plans, the point system is the more commonly used and is generally considered the more accurate method for occupational rating.

The *ranking method* arranges jobs in order of importance, or according to relative worth. Here, the entire job is considered, including the complexity and degree of difficulty of the duties, the requirements for specific areas of knowledge, the required skills, the amount of experience required, and the level of authority and responsibility assigned to the job. This method became popular in the United States during World War II, because of its simplicity and ease of installation. Generally speaking, the ranking method is less objective than the other techniques necessitating greater knowledge of all jobs, and for this reason has not been used extensively in recent years. The following steps apply to the ranking method:

1. Prepare job descriptions.
2. Rank jobs (usually, departmentally first) in the order of their relative importance.
3. Determine the class or grade for groups of jobs, using a bracketing process.
4. Establish the wage or wage range for each class or grade.

FACTOR SELECTION

Under the factor comparison method, most companies use five factors. In some point programs, 10 or more factors may be used. However, it is preferable to use a small number of factors. The objective is to use only as many factors as are necessary to provide a clear-cut difference among the jobs. The elements of any job may be classified according to

1. What job demands the employee meets in the form of physical and mental factors
2. What the job takes from the employee in the form of physical and mental fatigue
3. The responsibilities that the job demands
4. The conditions under which the job is done

Other factors may include education, experience, initiative, ingenuity, physical demands, mental and/or visual demands, working conditions, hazards, equipment responsibility, process, materials, products, and the work and safety of others.

These factors are present in varying degrees in all jobs, and any job under consideration falls under one of the several degrees of each factor. The various factors are equally important. To recognize these differences in importance, the analyst assigns weights or points to each degree of each factor, as shown in Table 9.8. Figure 9.5 illustrates a completed job rating, with substantiating data.

For example, education may be defined as a requirement with first-degree education requiring only the ability to read and write, with second-degree education requiring the use of simple arithmetic, which would be characteristic of two years of high school. Third-degree education could be equivalent to four years of

Table 9.8 Points Assigned to Factors and Key to Grades

Factors	1st degree	2nd degree	3rd degree	4th degree	5th degree
Skill					
1. Education 14		28	42	56	70
2. Experience. 22		44	66	88	110
3. Initiative and ingenuity 14		28	42	56	70
Effort					
4. Physical demand 10		20	30	40	50
5. Mental and/or visual demand. 5		10	15	20	25
Responsibility					
6. Equipment or process. 5		10	15	20	25
7. Material or product 5		10	15	20	25
8. Safety of others 5		10	15	20	25
9. Work of others 5		10	15	20	25
Job conditions					
10. Working conditions. 10		20	30	40	50
11. Unavoidable hazards 5		10	15	20	25

Source: National Electrical Manufacturers Association.

high school while fourth-degree education may be equivalent to four years of high school plus four years of formal trades training. Fifth-degree education could be equivalent to four years of technical university training.

Experience appraises the time that an individual with the specified education usually requires to learn to perform the work satisfactorily from the standpoint of both quality and quantity. Here, first degree could involve up to 3 months; second degree, 3 months to 1 year; third degree, 1 to 3 years; fourth degree, 3 to 5 years; and fifth degree, over 5 years. In a similar manner, each degree of each factor is identified with a clear definition and with specific examples, when applicable.

PERFORMANCE EVALUATION

Considerable judgment is needed to evaluate each job with respect to the degree of each factor required for the plan. Consequently, it is usually desirable to have a committee consisting of a union representative, the department supervisor, the department steward, and a management representative (usually from human resources) perform the evaluation.

Committee members should assign their degree evaluations independently of the other members and should evaluate all jobs for the same factor before proceeding to the next factor. The correlation among different evaluators should be reasonably high, such as 0.85 or higher. The members should then discuss any differences, until there is agreement on the level of the factor.

JOB CLASSIFICATION

After all jobs have been evaluated, the points assigned to each job should be tabulated. Next, the number of labor grades within the plant should be determined. Typically, the number of grades runs from 8 (typical of smaller plants and lesser skilled industries) to 15 (typical of larger plants and higher skilled industries)

JOB RATING-SUBSTANTIATING DATA
DORBEN MFG. CO.
UNIVERSITY PARK, PA.

JOB TITLE: Machinist (General)   CODE: 176   DATE: Nov. 12

FACTORS	DEG.	POINTS	BASIS OF RATING
Education	3	42	Requires the use of fairly complicated drawings, advanced shop mathematics, variety of precision instruments, shop trade knowledge. Equivalent to four years of high school or two years of high school plus two to three years of trades training.
Experience	4	88	Three to five years installing, repairing, and maintaining machine tools and other production equipment.
Initiative and ingenuity	3	42	Rebuild, repair, and maintain a wide variety of medium-size standard automatic and hand-operated machine tools. Diagnose trouble, disassemble machine and fit new parts, such as antifriction and plain bearings, spindles, gears, cams, etc. Manufacture replacement parts as necessary. Involves skilled and accurate machining using a variety of machine tools. Judgment required to diagnose and remedy trouble quickly so as to maintain production.
Physical demand	2	20	Intermittent physical effort required tearing down, assembling, installing, and maintaining machines.
Mental or visual demand	4	20	Concentrated mental and visual attention required. Laying out, setup, machining, checking, inspecting, fitting parts on machines.
Responsibility for equipment or process	3	15	Damage seldom over $900. Broken parts of machines. Carelessness in handling gears and intricate parts may cause damage.
Responsibility for material or product	2	10	Probable loss due to scrapping of materials or work, seldom over $300.
Responsibility for safety of others	3	15	Safety precautions are required to prevent injury to others; fastening work properly to face plates, handling fixtures, etc.
Responsibility for work of others	2	10	Responsible for directing one or more helpers a great part of time. Depends on type of work.
Working conditions	3	30	Somewhat disagreeable conditions due to exposure to oil, grease, and dust.
Unavoidable hazards	3	15	Exposure to accidents, such as crushed hand or foot, loss of fingers, eye injury from flying particles, possible electric shock, or burns.

REMARKS: Total 307 -- assign to job class 4.

Figure 9.5   Job rating and substantiating form.

(see Figure 9.6). For example, if the point range of all the jobs within a plant is from 110 to 365, the grades shown in Table 9.9 could be established. Similar ranges are not necessary for the various labor grades. Increasing the point ranges might be desirable for more highly compensated jobs.

The jobs falling within the various labor grades are then reviewed relative to one another, to ensure fairness and consistency. For example, it would not be appropriate for a class A machinist to be in the same grade level as a class B machinist.

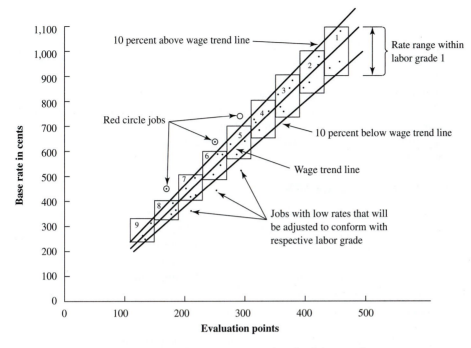

Figure 9.6 Evaluation points and base rate range for nine labor grades.

Table 9.9 Labor Grades

Grade	Score range (points)	Grade	Score range (points)
12	100–139	6	250–271
11	140–161	5	272–293
10	162–183	4	294–315
9	184–205	3	316–337
8	206–227	2	338–359
7	228–249	1	360 and above

Next, hourly rates are assigned to each of the labor grades. These rates are based on area rates for similar work, company policy, and the cost-of-living index. Frequently, analysts establish a rate range for each labor grade. The total performance of each operator determines his or her pay rate within the established range, and *total performance* refers to quality, quantity, safety, attendance, suggestions, and so on.

JOB EVALUATION PROGRAM INSTALLATION

After plotting area rates against the point values of the various jobs, the analyst develops a rate versus point value trend line, which may or may not be a straight line. Regression techniques are helpful in developing this trend line. Several points will be either above or below it. Points significantly above the trend line represent employees whose present rate is higher than that established by the job evaluation plan; points significantly below the trend line represent employees whose present rate is less than that prescribed by the plan.

Employees whose rates are less than called for by the plan should receive immediate increases to the new rate. Employees whose rates are higher than called for by the plan (such rates are referred to as *red circle rates*) are not given a rate decrease. However, they are also not given an increase at their next review, unless the cost-of-living adjustment results in a rate higher than their current pay. Finally, any new employee would be paid the new rate advocated by the job evaluation plan.

POTENTIAL CONCERNS

Although a point job evaluation system is probably the most favorable approach to bringing both equity and objectivity to the determination of individual compensation, there can be some problems. If the job description is not worded carefully, some employees may refuse to perform certain work, simply because these tasks are not included in the job description. A point plan can also create unnecessary and undesirable power relationships within the company since they provide an obvious pecking order that can interfere with cooperation and group decision making.

Another problem is that individuals will recognize that they can increase their job evaluation points by increasing their responsibility, by adding unnecessary work, or by adding another employee. These additions may be unnecessary and may really add only additional overhead costs, in addition to the cost increase of the higher-paid job resulting from the point additions.

One of the principal concerns heard in the courtrooms and legislative hearings deals with the principle of "equal pay for equal work." A point job evaluation system is based on this concept. However, decisions have to be made on how much weight to give to initiative and experience, found in many salaried jobs (many held by women) as opposed to physical demand and hazardous conditions found in many production jobs (many held by men). Also, the analyst must realize that there is no inherent worth to any job: it is worth what is provided in the marketplace. If there is a lack of nurses exacerbated with an aging population, salaries will rise. Conversely, if there is an over abundance of programmers, salaries will drop.

It is also important that a regular follow-up of the plan be done, so that it is adequately maintained. Jobs change, so it is necessary to review all jobs periodically and make adjustments when necessary. Finally, employees must understand the essential fairness of the job evaluation plan and attempt to work with the system.

9.4 AMERICANS WITH DISABILITIES ACT

While implementing a new method and performing job evaluations, the analyst must consider the implications of the Americans with Disabilities Act (ADA).

The ADA was passed in 1990 to "outlaw discrimination in employment against a qualified individual with a disability." This is an important consideration for all employers with 15 or more employees, as it may entail considerable workplace redesign or other accommodations. The ADA covers such employment practices as recruitment, hiring, promotion, training, pay, layoffs, firing, leave, benefits, and job assignments, the last of which would be the concern of methods analysts. The ADA protects any individual "with a physical or mental impairment that substantially limits a major life activity." *Substantial* implies something more than minor, while *major life activity* includes hearing, seeing, speaking, breathing, walking, manually feeling or manipulating, learning, or working. Temporary injuries of limited duration are not covered.

The individual with a disability must be qualified to perform the "essential functions" of the job with or without "reasonable accommodations." Essential functions are basic job duties that an employee must be able to perform. They can be determined from the job analysis techniques presented earlier in this chapter. *Reasonable accommodation* is any change or adjustment to a job or work environment that allows the individual to perform the essential functions of the job and enjoy the benefits and privileges that all employees enjoy. Those accommodations could include physical modification of the tools, equipment, or workstation; job restructuring; modification of work schedules; and modification of training materials or policies, etc. The purpose of any modifications would be to make them usable and accessible. One consideration is that any such change will be ergonomically beneficial to all workers. Many of the work design principles (Chapters 4 to 7) should be useful here also.

A reasonable accommodation is one that does not place an undue hardship on the employer, that is, one that is not unduly costly, extensive, substantial, or disruptive, or that fundamentally alters the nature or operation of the business. Variables that affect cost are the company size, financial resources, and its operational nature or structure. Unfortunately, there is no specific or quantitative definition of the cost factor. Most likely, that will evolve through the legal system as various cases of discrimination are brought forward in the courts. For further information on the definitions, legal aspects, and accessibility modifications guidelines, the methods engineer should consult the ADA (1991).

9.5 FOLLOW-UP

The eighth and last step in a methods engineering program is Follow-Up (please refer back to Figure 2.1). The seventh step, Establish Time Standards, is not strictly part of methods changes and will not be discussed in this section on methods. However, it is a necessary part of any successful work center and will be covered in great detail in Chapters 10 to 16.

Following up the method, in the short term, includes ensuring that the installation was correct, so that the operators can be trained in the proper work practices and can achieve the desired productivity levels. It also includes economic analyses to verify that the projected savings are truly achieved. If follow-up is not done, management may question the need for such changes and may be less willing to support similar methods changes in the future. Finally, it is important to keep everyone sold on the method, so that the operators don't slide back to the old patterns of movement, supervisors do not slack off in the enforcement of new procedures, and management does not waiver in its commitment to the overall program.

Following up the new method is a critical aspect of maintaining a smoothly running and efficient work center. Otherwise, several years later, another methods engineer will be examining the present method and asking the same questions— "Why?" and "What is the purpose of this operation?"—which were asked as part of this operation analysis. Thus, it is very important to close the feedback loop and maintain a continuous improvement cycle, as shown in Figure 2.1.

9.6 SUCCESSFUL METHODS IMPLEMENTATIONS

As an example of an effective methods program to manufacturing, one Ohio company realized a 17,496-ton annual savings. By forming a mill section into a ring and welding it, the company could replace an original rough-forged ring weighing 2,198 lb. The new mill section blank weighed only 740 lb. The saving of 1,458 lb of high-grade steel, amounting to twice the weight of the finished piece, was brought about by the simple procedure of reducing the excess material that previously had to be cut away. Another more detailed change is shown in example 9.5.

Methods improvements can also be implemented in less traditional operations. An analytic laboratory in a New Jersey plant applied the principles of operation analysis in laying out new workbenches in the form of a cross, so that each chemist had an L-shaped worktable. This arrangement allowed each chemist to reach any part of the workstation by taking only one stride. The new workbench consolidated equipment, thus saving space and eliminating the duplication of facilities. One glassware cabinet serviced two chemists. A large four-place fume head allowed multiple activity in an area that was formerly a bottleneck. All utility outlets were also relocated for maximum efficiency.

EXAMPLE 9.5	Methods Change for Auto-Starter

A representative case history, which follows operation analysis, is the production of an auto-starter, a device for starting AC motors by reducing the voltage through a transformer. A subassembly of the auto-starter is the arc box. This part sits in the bottom of the auto-starter and acts as a barrier between the contacts so there are no short-circuits. The present design consists of the components shown in Table 9.10.

In assembling these components, the operator places a washer, lock washer, and nut on one end of each rod. Next, the worker inserts the rods through the three holes in the first barrier. Then the operator places one spacer on each of the rods (three in all) and adds another barrier. Operators repeat this until six barriers are on the rods, separated by the tubing.

Table 9.10 Auto-Starter Components

Asbestos barriers with three drilled holes	6
Spacers of insulating tubing 2 in long	15
Steel rods threaded at both ends	3
Pieces of hardware	18
Pieces total	42

It was suggested that the six barriers be made with two slots, one at each end, and that two strips of asbestos be made for supporting the barriers, with six slots in each. These would be slipped together and placed in the bottom of the auto-starter as needed. The manner of assembly would be the same as that used in putting together the separator in an eggbox. A total of 15 suggested improvements occurred after the analysis was completed, with the results shown in Table 9.11.

Table 9.11 Improvements for Auto-Starter

Old method	New method	Savings
42 parts	8	34
10 workstations	1	9
18 transportations	7	11
7,900 ft of travel	200	7,700
9 storages	4	5
0.45-h time	0.11-h time	0.34-h
$1.55 costs	$0.60	$0.95

Methods improvements can be used to streamline service organizations. A state government division developed an operation analysis program that resulted in an estimated annual savings of more than 50,000 h. This was brought about by combining, eliminating, and redesigning all paperwork activities; improving the plant layout; and developing paths of authority.

Methods improvement is also effective in office environments. One industrial engineering department of a Pennsylvania company was given the problem of simplifying the paperwork necessary for shipping molded parts manufactured in one of its plants to an outlying plant for assembly. The department developed a new method that reduced the average daily shipment of 45 orders from 552 sheets of paper forms to 50 sheets. The annual savings in paper alone was significant.

More recently, methods improvement was tied in very successfully to an ergonomics program. A central Pennsylvania manufacturer of automobile car- peting was cited by OSHA for excessive work-related musculoskeletal injuries. In lieu of the developing ergonomics standard, the company was cited under the General Duty Clause of the 1970 OSHAct for not providing a safe working environment for its employees. As part of the 4-year abatement program, the company hired an occupational nurse to provide proper medical management and an outside ergonomics consultant to assist on the workplace redesigns necessary to reduce the injury rates. A detailed worksite analysis identified or confirmed from the medical records the critical jobs and workstations. Successful methods changes implemented included water jet cutters in place of manual trimming of the carpets with knives, a decrease in the number of times the carpets were handled manually with excessive pinch grips, and changes in gluing and sewing operations. In addition, the working conditions through the employees' perspective were analyzed with a health and safety survey, and the workers were provided with various levels of ergonomics training. As a result, the number of OSHA-recordable musculoskeletal injuries dropped from a high of 55 in the first year, to 35 in the second year, to 17 in the third year, and to an eventual low of 8 in the fifth year. However, more important, the ergonomics program was considered so successful by OSHA that the abatement program oversight was terminated after the second year!

SUMMARY

Increased output and improved quality are the primary outcomes of methods and work design changes, but methods changes also distribute the benefits of improved production to all workers and help develop better working conditions and a safer working environment, so the worker can do more work at the plant, do a good job, and still have enough energy to enjoy life. The examples of effective implementation of methods changes clearly demonstrate the need to follow an orderly approach, as presented in Figure 2.1. The methods engineer should note that it is not sufficient to use sophisticated mathematical algorithms or the latest software tools to develop the ideal method. It is necessary to sell the plan both to management and to the workers themselves. Additional interpersonal techniques and strategies for dealing with people to better sell the method are presented in Chapter 18.

QUESTIONS

1. Compare and contrast a decision table with value engineering.
2. How does one define the benefits especially related to health and safety in cost-benefit analysis?
3. What are the principal concerns of management with regard to a new method that is relatively costly to install?
4. What is meant by the discounted cash flow method?
5. What is meant by the payback method? How is it related to the return on investment method?
6. What is the relationship between return on capital investment and the risk associated with the anticipated sales of the product for which a new method will be used?
7. What two specific subjects should be emphasized in writing the job description?
8. Is time a common denominator of labor cost? Why or why not?
9. What is a job evaluation?
10. Which four methods of job evaluation are being practiced in this country today?
11. Explain in detail how a point plan works.

12. Which factors influence the relative worth of a job?

13. What is the weakness of using historical records as a means of establishing standards of performance?

14. Explain why a range of rates, rather than just one rate, should be established for every labor grade.

15. What are the principal negative considerations that should be understood prior to the installation of a point job evaluation system?

16. What are the principal benefits of a properly installed job evaluation plan?

17. Which three considerations constitute a successful job evaluation plan?

18. How does ADA enter into a methods change?

PROBLEMS

1. How much capital could be invested in a new method if it is estimated that $5,000 would be saved the first year, $10,000 the second year, and $3,000 the third year? Management expects a 30 percent return on invested capital.

2. You have estimated the life of your design to be 3 years. You expect that a capital investment of $20,000 will be required to get it into production. You also estimate, based on sales forecasts, that the design will result in an after-tax profit of $12,000 the first year and $16,000 the second year, and a $5,000 loss the third year. Management has asked for an 18 percent return on capital investment. Should the company go ahead with the investment to produce the new design? Explain.

3. In the Dorben Company, a materials handling operation in the warehouse is being done by hand labor. Annual disbursements for this labor and for related expenses (social security, accident insurance, and other fringe benefits) are $8,200. The methods analyst is considering a proposal to build certain equipment to reduce this labor cost. The first cost of this equipment will be $15,000. It is estimated that the equipment will reduce annual disbursements for labor and labor extras to $3,300. Annual payments for power, maintenance, and property taxes plus insurance are estimated to be $400, $1,100, and $300, respectively. The need for this particular operation is anticipated to continue for 10 years. Because the equipment is specially designed for the particular purpose, it will have no salvage value. It is assumed that the annual disbursements for labor, power, and maintenance will be uniform throughout the 10 years. The minimum rate of return before taxes is 10 percent. Based on an annual cost comparison, should the company proceed with the new material handling equipment?

4. A job evaluation plan based on the point system uses the following factors:

   a. Education: maximum weight 100 points; four grades
   b. Effort: maximum weight 100 points; four grades
   c. Responsibility: maximum weight 100 points; four grades

   A floor sweeper is rated as 150 points, and this position carries an hourly rate of $6.50. A class 3 milling machine operator is rated as 320 points, which results in a money rate of $10.00 per hour. What grade of experience would be given to a drill press operator with an $8.50 per hour rate and point ratings of grade 2 education, grade 1 effort, and grade 2 responsibility?

5. A job evaluation plan in the Dorben Company provides for five labor grades, of which grade 5 has the highest base rates and grade 1 the lowest. The linear plan

involves a range of 50 to 250 points for skill, 15 to 75 points for effort, 20 to 100 points for responsibility, and 15 to 75 points for job conditions. Each of the four factors has five degrees. Each labor grade has three money rates: low, mean, and high.

    a. If the high money rate of labor grade 1 is $8 per hour and the high money rate of labor grade 5 is $20 per hour, what would be the mean money rate of labor grade 3?

    b. What degree of skill is required for a labor grade of 4, if second-degree effort, second-degree responsibility, and first-degree job conditions apply?

6. In the Dorben Company, the analyst installed a point job evaluation plan covering all indirect employees in the operating divisions of the plant. Ten factors were used in this plan, and each factor was broken down into five degrees. In the job analysis, the shipping and receiving clerk position was shown as having second-degree initiative and ingenuity, valued at 30 points. The total point value of this job was 250 points. The minimum number of points attainable in the plan was 100, and the maximum was 500.

    a. If 10 job classes prevailed, what degree of initiative and ingenuity would be required to elevate the job of shipping and receiving clerk from job class 4 to job class 5?

    b. If job class 1 carried a rate of $8 per hour and job class 10 carried a rate of $20 per hour, what rate would job class 7 carry? (*Note*: Rates are based on the midpoint of job class point ranges.)

7. An ergonomist suggests replacing two mail clerks, each of whom sorts 3,000 pieces/h and gets paid $10 per hour, with a machine vision system that can sort 6,000 pieces/h but costs $1/per hour to maintain and $50,000 to purchase. How many hours would it take to pay off the system?

REFERENCES

ADA. *Americans with Disabilities Act Handbook*. EEOC-BK-19. Washington, DC: Equal Employment Opportunity Commission and U.S. Dept. of Justice, 1991.

Brown, D. B. *Systems Analysis & Design for Safety*. Englewood Cliffs, NJ: Prentice-Hall, 1976.

Carey, A. "The Hawthorne Studies: A Radical Criticism." In *Concepts and Controversy in Organizational Behavior*. Ed. W. R. Nord. Pacific Palisades, CA: Goodyear Publishing Co., 1972.

DeReamer, R. *Modern Safety and Health Technology*. New York: John Wiley & Sons, 1980.

Dunn, J. D., and F. M. Rachel. *Wage and Salary Administration: Total Compensation Systems*. New York: McGraw-Hill, 1971.

Ellig, Bruce R. *Compensation Issues of the Eighties*. Amherst, MA: Human Resource Development Press, Inc., 1988.

Fleischer, G. A. "Economic Risk Analysis." In *Handbook of Industrial Engineering*. 2d ed. Ed. Gavriel Salvendy. New York: John Wiley & Sons, 1992.

Gausch, J. P. "Safety and Decision-Making Tables." *ASSE Journal*, 17 (November 1972), pp. 33–37.

Gausch, J. P. "Value Engineering and Decision Making." *ASSE Journal*, 19 (May 1974), pp. 14–16.

Grant, E., W. Ireson, and R. S. Leavenworth. *Principles of Engineering Economy*. 6th ed. New York: Ronald Press, 1976.

Heinrich, H. W., D. Petersen, and N. Roos. *Industrial Accident Prevention*. 5th ed. New York: McGraw-Hill, 1980.

Homans, G. "The Western Electric Researches." In *Concepts and Controversy in Organizational Behavior*. Ed. W. R. Nord. Pacific Palisades, CA: Goodyear Publishing Co., 1972.

Jung, E. S., and A. Freivalds. "Multiple Criteria Decision-Making for the Resolution of Conflicting Ergonomic Knowledge in Manual Materials Handling." *Ergonomics*, 34, no. 11 (November 1991), pp. 1351–1356.

Lawler, Edward E. *What's Wrong with Point-Factor Job Evaluation?* Amherst, MA: Human Resource Development Press, Inc., 1988.

Livy, B. *Job Evaluation: A Critical Review*. New York: Halstead, 1973.

Lutz, Raymond P. "Discounted Cash Flow Techniques." In *Handbook of Industrial Engineering*. 2d ed. Ed. Gavriel Salvendy. New York: John Wiley & Sons, 1992.

Mayo, E. *The Human Problems of an Industrial Civilization*. New York: The Viking Press, 1960.

McCormick, Ernest J. "Job Evaluation." In *Handbook of Industrial Engineering*. Ed. Gavriel Salvendy. New York: John Wiley & Sons, 1982.

Milkovich, George T., Jerry M. Newman, and James T. Brakefield. "Job Evaluation in Organizations." In *Handbook of Industrial Engineering*. 2d ed. Ed. Gavriel Salvendy. New York: John Wiley & Sons, 1992.

OSHA. *Ergonomics Program Management Guidelines for Meatpacking Plants*. Washington, DC: Bureau of National Affairs, 1990.

Otis, Jay, and Richard H. Leukart. *Job Evaluation: A Sound Basis for Wage Administration*. Englewood Cliffs, NJ: Prentice-Hall, 1954.

Pennock, G. A. "Industrial Research at Hawthorne." *The Personnel Journal*, 8 (June 1929–April 1930), pp. 296–313.

Risner, Howard. *Job Evaluation: Problems and Prospects*. Amherst, MA: Human Resource Development Press, Inc., 1988.

Saaty, T. L. *The Analytic Hierarchy Process*. New York: McGraw-Hill, Inc., 1980.

Salvendy, Gavriel, and Douglas W. Seymour. *Prediction and Development of Industrial Work Performance*. New York: John Wiley & Sons, 1973.

Thuesen, H. G., W. J. Fabrycky, and G. J. Thuesen. *Engineering Economy*. 5th ed. Englewood Cliffs, NJ: Prentice-Hall, 1977.

Wegener, Elaine. *Current Developments in Job Classification and Salary Systems*. Amherst, MA: Human Resource Development Press, Inc., 1988.

SELECTED SOFTWARE

DesignTools (available from the McGraw-Hill text website at www.mhhe.com/niebel-freivalds), New York: McGraw-Hill, 2002.

Time Study

- Use time study to establish time standards.

- Use both audio and visual break points to divide the operation into elements.

- Use continuous timing to obtain a complete record of times.

- Use snapback timing to avoid clerical errors.

- Perform a time check to confirm the validity of the time study.

The seventh step in the systematic process of developing the efficient work center is the establishment of time standards. These can be determined by using estimates, historical records, and work measurement procedures. In past years, analysts relied more heavily on estimates as a means of establishing standards. However, experience has shown that no individual can establish consistent and fair standards simply by looking at a job and judging the amount of time required to complete it.

With the historical records method, production standards are based on the records of similar, previously performed jobs. In common practice, the worker punches in on a time clock or data collection hardware every time he or she begins a new job and then punches out after completing the job. This technique tells how long it actually took to do a job, but not how long it should have taken. Some jobs carry personal, unavoidable, and avoidable delay time to a much greater extent than they should, while other jobs do not carry their appropriate share of delay time. Historical records have consistently deviated by as much as 50% on the same operation of the same job.

Any of the work measurement techniques—stopwatch (electronic or mechanical) time study, pre-determined time systems, standard data, time formulas, or work sampling studies—represents a better way to establish fair production standards. All these techniques are based on establishing an actual allowed time standard for performing a given task, with due allowance for fatigue and for personal and unavoidable delays.

Accurately established time standards make it possible to increase the efficiency of the equipment and the operating personnel while poorly established standards, although better than no standards at all, lead to high costs, labor dissension, and possibly even the failure of the enterprise. This may mean the difference between the success or failure of a business.

10.1   A FAIR DAY'S WORK

The fundamental principle in industry is that an employee is entitled to a fair day's pay in return for which the company is entitled to a fair day's work. A fair day's work can be defined as the amount of work that can be produced by a qualified employee when working at a standard pace and effectively utilizing her or his time where work is not restricted by process limitations. This definition does not clarify what is meant by qualified employees, standard pace, and effective utilization of flexibility. For example, the term *qualified employee* can be further defined as a representative average of those employees who are fully trained and able satisfactorily to perform any and all phases of the work involved, in accordance with the requirements of the job under consideration.

Standard pace can be defined as the effective rate of performance of a conscientious, self-paced, qualified employee when working neither fast nor slow and giving due consideration to the physical, mental, or visual requirements of the specific job. One interpretation specifies that as a worker walking without load, on smooth, level ground at a rate of 3 mi/h.

There is also some uncertainty over the definition of *effective utilization*. Typically this could be the maintenance of a standard pace while performing essential elements of the job during all portions of the day except that which is required for reasonable rest and personal needs, under circumstances in which the job is not subject to process, equipment, or other operating limitations.

In general, a fair day's work is one that is fair to both the company and the employee. This means that the employee should give a full day's work for the time that he or she gets paid, with reasonable allowances for personal delays, unavoidable delays, and fatigue. The worker is expected to operate in the prescribed method at a pace that is neither fast nor slow, but one that may be considered representative of all-day performance by the experienced, cooperative employee. Time study is one method of determining a fair day's work.

10.2   TIME STUDY REQUIREMENTS

Certain fundamental requirements must be realized before the time study is made. For example, whether the standard is required on a new job or on an old job in which the method or part of the method has been altered, the operator should be thoroughly acquainted with the new technique before the operation is studied. Also, the method must be standardized for the job before the study begins. Unless all details of the method and working conditions have been standardized, the time standards will have little value and will become a continual source of mistrust, grievances, and internal friction.

Analysts should tell the union steward, the department supervisor, and the operator that the job is to be studied. Each of these parties can then take the appropriate steps necessary to allow a smooth, coordinated study. The operator should verify that she or he is performing the correct method and should become acquainted with all details of that operation. The supervisor should check the method to make sure that feeds, speeds, cutting tools, lubricants, and so forth conform to standard practice, as established by the methods department. Also the supervisor should investigate the amount of material available so that no shortages take place during the study. The union steward should then make sure that only trained, competent operators are selected, should explain why the study is being made, and should answer any pertinent questions raised by the operator.

ANALYST'S RESPONSIBILITY

All work involves varying degrees of skill, as well as physical and mental effort. There are also differences in the aptitude, physical application, and dexterity of the workers. It is easy for the analyst to observe an employee at work and measure the actual time taken to perform a task. It is considerably more difficult to evaluate all variables and determine the time required for a qualified operator to perform the task.

The time study analyst should ensure that the correct method is being used, accurately record the times taken, honestly evaluate the performance of the operator, and refrain from any operator criticism. The analyst's work must be completely dependable and accurate. Inaccuracy and poor judgment will not only affect the operator and the company financially, but may also result in loss of confidence by the operator and the union. The time study analyst should always be honest, tactful, pleasing, patient, and enthusiastic, and should always use good judgment.

SUPERVISOR'S RESPONSIBILITY

The supervisor should notify the operator in advance that his or her work assignment is to be studied. The supervisor should see that the proper method established by the methods department is being utilized, and that the operator selected is competent and has adequate experience on the job. Although the time study analyst should have a practical background in the area of work being studied, analysts cannot be expected to know all specifications for all methods and processes. Therefore, the supervisor should verify that the cutting tools are properly ground, the correct lubricant is being used, and a proper selection of feeds, speeds, and cut depths is being made. The supervisor should also make certain that operators use the prescribed method, conscientiously assisting and training all employees in perfecting this method. Once the time study is completed, the supervisor should sign the original copy, indicating compliance with the study.

UNION'S RESPONSIBILITY

Most unions recognize that standards are necessary for the profitable operation of a business and that management continues to develop such standards using accepted work measurement techniques. Furthermore, every union steward knows that poor time standards cause problems for both labor and management.

Through training programs, the union should educate all its members in the principles, theories, and economic necessity of time study practice. Operators cannot be expected to be enthusiastic about time study if they know nothing about it. This is especially true in view of its background (see Chapter 1).

The union representative should make certain that the time study includes a complete record of the job conditions, that is, work method and workstation layout. The representative should ascertain that the current job description is accurate and complete and also encourage the operator to cooperate with the time study analyst.

OPERATOR'S RESPONSIBILITY

Every employee should be sufficiently interested in the welfare of the company to support the practices and procedures inaugurated by management. Operators should give new methods a fair trial and should cooperate in helping to work out any bugs. The operator is closer to the job than anyone else, and she or he can make a real contribution to the company by helping to establish ideal methods.

The operator should assist the time study analyst in breaking the job down into elements, thus ensuring that all details of the job are specifically covered. The operator should also work at a steady, normal pace while the study is being made and should introduce as few foreign elements and extra movements as possible. The worker should use the exact method prescribed, as any action that artificially lengthens the cycle time could result in a too liberal standard.

10.3 TIME STUDY EQUIPMENT

The minimum equipment required to conduct a time study program includes a stopwatch, time study board, time study forms, and pocket calculator. Videotape equipment can also be very useful.

STOPWATCH

Two types of stopwatches are in use today: the traditional decimal minute watch (0.01 min) and the much more practical electronic stopwatch. The decimal minute watch, shown in Figure 10.1, has 100 divisions on its face, and each division is equal to 0.01 min; that is, a complete sweep of the long hand requires 1 min. The small dial on the watch face has 30 divisions, each of which is equal to 1 min. Therefore, for every full revolution of the sweep hand, the small hand moves one division, or 1 min. To start this watch, move the side slide toward the crown. Depressing the crown moves both the sweep hand and the small hand back to zero. Releasing the crown puts the watch back into operation, unless the side slide is moved away from the crown. Moving the side slide away from the crown stops the watch.

Electronic stopwatches cost approximately $50. These watches provide resolution to 0.001 s and an accuracy of ± 0.002 percent. They weigh about 4 oz and

Figure 10.1 Decimal minute watch.

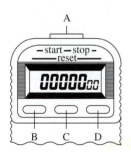

Figure 10.2 Electronic stopwatch. (A) Start/stop, (B) memory retrieval, (C) mode (continuous/snapback), and (D) other functions.

are about $4 \times 2 \times 1$ in (see Figure 10.2). They permit timing any number of individual elements, while also counting the total elapsed time. Thus, they provide both *continuous* and *snapback timing* (button C), with none of the disadvantages of mechanical watches. To operate the watch, press the top button (button A). Each time the top button is pressed, a numerical readout is presented. Pressing the memory button (button B) causes previous readouts to be retrieved. A slightly fancier version incorporates the watch into an electronic time study board (see Figure 10.3).

With mechanical watches costing $150 and up and electronic watches falling in price, the mechanical watches are quickly disappearing from use. On the other hand, more general purpose personal digital assistants (PDAs) are becoming more popular.

VIDEOTAPE CAMCORDERS

Videotape camcorders are ideal for recording operators' methods and elapsed time. By taking pictures of the operation and then studying them one frame at a time, analysts can record exact details of the method used and can then assign normal time values. They can also establish standards by projecting the film at the same speed at which the pictures were taken and then performance-rating the

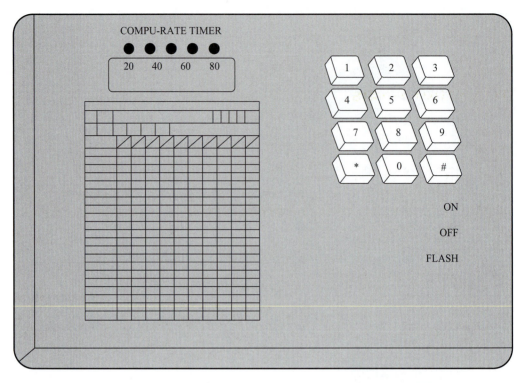

Figure 10.3 Computer-assisted electronic stopwatch.

operator. Because all the facts are there, observing the videotape is a fair and accurate way to rate performance. Then, too, potential methods improvements that would seldom be uncovered with a stopwatch procedure can be revealed through the camera eye. Another advantage of videotapes is that with MVTA software (discussed in Time Study Software below), time studies can be done almost automatically. More recently with the arrival of digital camcorders and PC editing software, time studies can be done practically online. Videotapes are also excellent for training the novice time study analyst, since sections can easily be rewound and repeated until sufficiently mastered.

TIME STUDY BOARD

When the stopwatch is being used, analysts find it convenient to have a suitable board to hold the time study form and the stopwatch. The board should be light, so as not to tire the arm, and yet strong and sufficiently hard to provide a suitable backing for the time study form. Suitable materials include ¼-in plywood or smooth plastic. The board should have both arm and body contacts, for comfortable fit and ease of writing while it is being held. For a right-handed observer, the watch should be mounted in the upper right-hand corner of the board. A spring clip to the left would hold the time study form. Standing in the proper position, the time study analyst can look over the top of the watch to the workstation and follow the operator's movements, while keeping both the watch and the time study form in the immediate field of vision.

TIME STUDY FORMS

All the details of the study are recorded on a time study form. The form provides space to record all pertinent information concerning the method being studied, tools utilized, and so on. The operation being studied is identified by such information as the operator's name and number, operation description and number, machine name and number, special tools used and their respective numbers, department where the operation is performed, and prevailing working conditions. Providing too much information concerning the job being studied is better than too little.

Figure 10.4 illustrates a time study form that has been developed by the authors. It is sufficiently flexible to be used for practically any type of operation. On this form, analysts would record the various elements of the operation horizontally across the top of the sheet, and the cycles studied would be entered vertically, row by row. The four columns under each element are R for *ratings*; W for *watch time* or *watch readout*; OT for *observed time*, that is, the differential time between successive watch times; and NT for *normal time*.

TIME STUDY SOFTWARE

There are several software packages available for the time study analyst. TimStudy, by the Royal J. Dossett Corp., uses a custom-built Datawriter to collect the data electronically and then upload them directly to a desktop PC for analysis. CITS/APR, by C-Four, uses more versatile handheld PCs to collect data, and allows for much more detailed data analysis because a customized spreadsheet interface links directly into Excel. Unfortunately, both utilize specialized hardware, not readily available. More recently programs using more general-purpose personal digital assistants (PDAs) have gained popularity: Palm CITS by C-Four, QuickTimes™ by Applied Computer Services, Inc. and WorkStudy+™ 3.0 by Quetech, Ltd. A simple, user-friendly program for the Palm, QuickTS, has also been supplied for users of this textbook (see Figure 10.5). Any of these software products will allow the analyst to eliminate much of the drudgery of clerical transcription and to improve the accuracy of computations.

For those analysts performing time studies from videotapes, an interesting option is the Multimedia Video Task Analysis (MVTA, Nexgen Ergonomics). MVTA interfaces directly to a VCR through a graphical interface and enables users to interactively identify break points in the video record while analyzing at any desired speed (real time, slow/fast motion, or frame by frame in forward/reverse direction). MVTA then automatically produces time study reports and computes frequency of occurrence of each event as well as postural analysis for work design.

Time Study Observation Form

Study No.: 2-85	Date: 3-1-	Page 1 of 1
Operation: DIE CASTING	Operator: B. JONES	Observer: A F

Element No. and Description		1 REMOVE PART FROM DIE, LUBRICATE DIE, INSPECT				2 PLACE PART IN FIXTURE, TRIM ASIDE PART																					
Note	Cycle	R	W	OT	NT	R	W	OT	NT	R	W	OT	NT	R	W	OT	NT	R	W	O	NT	R	W	OT	NT
	1	90		30	270	90		23	207																
	2	100		27	270	100		21	210																
	3	90		31	279	90		23	207																
	4	85		35	298	100		20	200																
	5	100		28	280	100		20	200																
	6	110		25	275	110		18	198																
	7	90		31	279	90		24	216																
	8	100		28	280	85		24	204																
	9	90		32	288	90		23	207																
	10	110		26	286	105		19	200																
	11																								
	12																								
	13																								
	14																								
	15																								
	16																								
	17																								
	18																								

Summary					
Total OT	2.93	2.15			
Rating	—	—			
Total NT	2.805	2.049			
No. Observations	10	10			
Average NT	.281	.205			
% Allowance	17	17			
Elemental Std. Time	.329	.240			
No. Occurrences	1	1			
Standard Time	.329	.240			

Total Standard Time (sum standard time for all elements): **.569**

Foreign Elements					Time Check			Allowance Summary	
Sym	W1	W2	OT	Description					
A					Finishing Time	3:48.00		Personal Needs	5
B					Starting Time	3:42.00		Basic Fatigue	4
C					Elapsed Time	6.00		Variable Fatigue	8
D					TEBS	.60		Special	—
E					TEAF	.32		Total Allowance %	17
F					Total Check Time	.92		Remarks:	
G					Effective Time	5.08			
Rating Check					Ineffective Time	0			
Synthetic Time			%		Total Recorded Time	6.00			
Observed Time					Unaccounted Time	0			
					Recording Error %	0			

Figure 10.4 Snapback study of a die casting operation (rating elements every cycle).

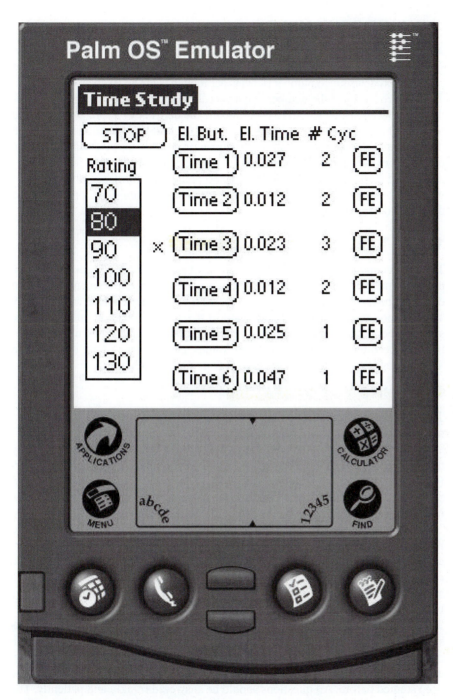

Figure 10.5 The QuickTS time study program on a Palm PDA.

TRAINING EQUIPMENT

A simple inexpensive piece of equipment that can assist in the training of time study analysts is the metronome used by music students. This device can be set to provide a predetermined number of beats per minute, such as 104 beats per minute. This happens to equal the number of cards dealt per minute when dealing at a standard pace (see Chapter 11). By synchronizing the delivery of a card at a four-hand bridge table such that a card is delivered with each beat of the metronome, we can demonstrate standard pace. Many metronomes, especially the electronic ones, can be set to provide a louder beep (or light blink) to occur every 3, 4, or 5 beats to create a more realistic element. To illustrate 80 percent performance, the instructor need only set the metronome to 83 beats per minute and then synchronize the card dealing and beep accordingly.

10.4   TIME STUDY ELEMENTS

The actual conduct of a time study is both an art and a science. To ensure success, analysts must be able to inspire confidence in, exercise judgment with, and develop a personable approach to everyone with whom they come in contact. They should thoroughly understand and perform the various functions related to the study: selecting the operator, analyzing the job and breaking it down into its elements, recording the elapsed elemental values, performance-rating the operator, assigning appropriate allowances, and working up the study itself.

CHOOSING THE OPERATOR

The first step in beginning a time study is to select the operator with the assistance of the departmental or line supervisor. In general, an operator who is average or somewhat above average in performance gives a more satisfactory study than a low-skilled or highly superior operator. The average operator usually performs the work consistently and systematically. That operator's pace will tend to be approximately in the standard range (see Chapter 11), thereby making it easier for the time study analyst to apply a correct performance factor.

Of course, the operator should be completely trained in the method, should like the work, and should demonstrate an interest in doing a good job. The operator should be familiar with time study procedures and practices, and should have confidence in both time study methods and the analyst. The operator should also be cooperative enough to follow through willingly with suggestions made by both the supervisor and the time study analyst.

The analyst should approach the operator in a friendly manner and demonstrate an understanding of the operation to be studied. The operator should have the opportunity to ask questions about the timing technique, method of rating, and application of allowances. In some instances, the operator may never have been studied before. All questions should be answered frankly and patiently. The operator should be encouraged to offer suggestions, and when the operator does so, the analyst should receive them willingly, thus showing respect for the skill and knowledge of the operator.

RECORDING SIGNIFICANT INFORMATION

Analysts should record the machines, hand tools, jigs or fixtures, working conditions, materials, operations, operator name and clock number, department, study date, and observer's name. Space for such details is provided under Remarks on the time study form. A sketch of the layout may also be helpful. The more pertinent information is recorded, the more useful the time study becomes over the years. It becomes a resource for establishing standard data and developing formulas (see Chapter 12). It will also be useful for methods improvement, operator evaluation, tool evaluation, and machine performance evaluation.

When machine tools are used, the analyst should specify the name, size, style, capacity, and serial or inventory number as well as the working conditions. Dies, jigs, gages, and fixtures should be identified by their numbers and with short descriptions. If the working conditions during the study are different from the normal conditions for that job, they will affect the performance of the operator. For example, in a drop forge shop, if a study were made on an extremely hot day, the working conditions would be poorer than usual, and operator performance would reflect the effect of the intense heat. Consequently, an allowance (see Chapter 11) would be added to the operator's normal time. If the working conditions improve, the allowance can be diminished. Conversely, if the working conditions become poorer, the allowance should be raised.

POSITIONING THE OBSERVER

The observer should stand, not sit, a few feet to the rear of the operator, so as not to distract or interfere with the worker. Standing observers are better able to move around and follow the movements of the operator's hands as the operator goes through the work cycle. During the course of the study, the observer should avoid any conversation with the operator, as this could distract the worker or upset the routines.

DIVIDING THE OPERATION INTO ELEMENTS

For ease of measurement, the operation should be divided into groups of motions known as *elements*. To divide the operation into its individual elements, the analyst should watch the operator for several cycles. However, if the cycle time is over 30 min, the analyst can write the description of the elements while making the study. If possible, the analyst should determine the operational elements before the start of the study. Elements should be broken down into divisions that are as fine as possible and yet not so small that reading accuracy is sacrificed. Elemental divisions of around 0.04 min are about as fine as can be read consistently by an experienced time study analyst. However, if the preceding and succeeding elements are relatively long, an element as short as 0.02 min can be readily timed.

To identify endpoints completely and develop consistency in reading the watch from one cycle to the next, consider both sound and sight in the elemental breakdown. For example, the *break points* of elements can be associated with such sounds as a finished piece hitting the container, a facing tool biting into a casting, a drill breaking through the part being drilled, and a pair of micrometers being laid on a bench.

Each element should be recorded in its proper sequence, including a basic division of work terminated by a distinctive sound or motion. For example, the element "up part to manual chuck and tighten" would include the following basic divisions: reach for part, grasp part, move part, position part, reach for chuck wrench, grasp chuck wrench, move chuck wrench, position chuck wrench, turn chuck wrench, and release chuck wrench. The termination point of this element would be the chuck wrench being dropped on the head of the lathe, as evidenced by the accompanying sound. The element "start machine" could include reach for lever, grasp lever, move lever, and release lever. The rotation of the machine, with the accompanying sound, would identify the termination point so that readings could be made at exactly the same point in each cycle.

Frequently, different time study analysts in a company adopt a standard elemental breakdown for given classes of facilities, to ensure uniformity in establishing break points. For example, all single-spindle bench-type drill press work may be broken down into standard elements, and all lathe work may be composed of a series of predetermined elements. Having standard elements as a basis for operation breakdown is especially important in the establishment of standard data (see Chapter 12).

Some additional suggestions may help in breaking elements down:

1. In general, keep manual and machine elements separate, since machine times are less affected by ratings.
2. Likewise, separate constant elements (those elements for which the time does not deviate within a specified range of work) from variable elements (those elements for which the time does vary within a specified range of work).
3. When an element is repeated, do not include a second description. Instead, in the space provided for the element description, give the identifying number that was used when the element first occurred.

10.5 START OF STUDY

At the start of the study, record the time of day (on a whole minute) from a "master" clock while simultaneously starting the stopwatch. (It is assumed that all data are recorded on the time study form.) This is the *starting time* ① as shown in Figure 10.6. One of two techniques can be used for recording the elemental times during the study. The *continuous timing* method, as the name implies, allows the stopwatch to run for the entire duration of the study. In this method, the analyst reads the watch at the break point of each element, and the time is allowed to continue. In the *snapback* technique, after the watch is read at the break point of each element, the watch time is returned to zero; as the next element takes place, the time increments from zero.

Time Study Observation Form	Study No.: 1-3		Date: 3-22-	Page 1 of 1
	Operation: MACHINING		Operator: J. SMITH	Observer: AF

Element No. and Description		1 FEED BAR TO STOP ③ ④				2 INDEX, FEED CUTTING TOOL TO BAR				3 TURN 1½" 550 RPM ⑤				4 WITHDRAW TOOL AND BAR SET DOWN												
Note	Cycle	R	W	OT	NT	R	W	OT	NT	R	W	OT	NT	R	W	OT	NT	R	W	O	NT	R	W	OT	NT
	1	85		19	162	105		12	126	100		60	600	90		17	153								
	2	90		22	198	105		13	137	100		60	600	100		16	160								
	3	100		17	170	105		11	116	100		60	600	105		17	179								
	4																								
	5				⑩																				
	6																								
	7																								
	8																								
	9																								
	10																								
	11																								
	12																								
	13																								
	14																								
	15																								
	16																								
	17																								
	18																								

Summary									
Total OT	.58			.36		1.80		.50	
Rating ③→	—			—		—		—	
Total NT	.530			.379		1.800		.492	
No. Observations	3			3		3		3	
Average NT ⑪	.177			.126		.600		.164	
% Allowance	10			10		10		10	
Elemental Std. Time	.195			.139		.660		.180	
No. Occurrences	1			1		1		1	
Standard Time	.195			.139		.660		.180	

							Total Standard Time (sum standard time for all elements):	**1.174**

Foreign Elements					Time Check			Allowance Summary	
Sym	W1	W2	OT	Description	Finishing Time ⑥→	9:22.00		Personal Needs	5
A	0	35	35	CHECK DIMENSIONS	Starting Time ①→	9:16.00		Basic Fatigue	4
B	↑⑥↗				Elapsed Time ⑨→	6.00		Variable Fatigue	1
C					TEBS ②→	1.86		Special	—
D					TEAF ⑦→	.60		Total Allowance %	10
E					Total Check Time	2.46 ←⑧		Remarks: MACHINE CYCLE	
F					Effective Time	3.24 ←⑫		(ELEMENT #3) TIME = .60 min	
G					Ineffective Time	.35 ←⑬			
Rating Check					Total Recorded Time ⑭→	6.05			
Synthetic Time				%	Unaccounted Time ⑮→	.05			
Observed Time					Recording Error % ⑯→	.8%			

Figure 10.6 Summary of steps in performing and computing a time study.

When recording the watch readings, note only the necessary digits and omit the decimal point, thus giving as much time as possible to observing the performance of the operator. If using a decimal minute watch and if the break point of the first element occurs at 0.08 min, record only the digit 8 in the W (watch time) column. Other example recordings are shown in Table 10.1.

Table 10.1   Recording Watch Readings for Continuous Timing

Consecutive reading of watch in decimal minutes	Recorded reading
0.08	8
0.25	25
1.32	132
1.35	35
1.41	41
2.01	201
2.10	10
2.15	15
2.71	71
3.05	305
3.17	17
3.25	25

SNAPBACK METHOD

The snapback method has both advantages and disadvantages compared to the continuous technique. Some time study analysts use both methods, believing that studies of predominantly long elements are more adapted to snapback readings, while short-cycle studies are better suited to the continuous method.

Since elapsed element values are read directly in the snapback method, no clerical time is needed to make successive subtractions, as for the continuous method. Thus, the readout can be inserted directly in the OT (*observed time*) column. Also, elements performed out of order by the operator can be readily recorded without special notation. In addition, proponents of the snapback method state that delays are not recorded. Also, since elemental values can be compared from one cycle to the next, a decision could be made as to the number of cycles to study. However, it is actually erroneous to use observations of the past few cycles to determine how many additional cycles to study. This practice can lead to studying entirely too small a sample.

Among the disadvantages of the snapback method is that it encourages the removal of individual elements from the operation. These cannot be studied independently, because elemental times depend on the preceding and succeeding elements. Consequently, omitting such factors as delays, foreign elements, and transposed elements could allow erroneous values in the readings accepted. One of the traditional objections to the snapback method was the amount of time lost while snapping the hand back to zero. However, this problem has been eliminated with the use of electronic watches. Also, short elements (0.04 min and less) are more difficult to time with this method. Finally, the overall time must be verified by summing the elemental watch readings, a process that is more prone to error.

Figure 10.4 illustrates a time study of a die casting operation using the snapback method.

CONTINUOUS METHOD

The continuous method of recording elemental values is superior to the snapback method for several reasons. The most significant is that the resulting study presents a complete record of the entire observation period; as a result, it appeals to the operator and the union. The operator is able to see that no time has been left out of the study, and all delays and foreign elements have been recorded. Since all the facts are clearly presented, this technique of recording times is easier to explain and sell.

The continuous method is also better adapted to measuring and recording very short elements. With practice, a good time study analyst can accurately catch three successive short elements (less than 0.04 min), if they are followed by an element of about 0.15 min or longer. This is possible by remembering the watch readings of the break points of the three short elements and then recording their respective values while the fourth, longer element is taking place.

On the other hand, more clerical work is involved in calculating the study if the continuous method is used. Since the watch is read at the break point of each element while the hands of the watch continue their movements, it is necessary to make successive subtractions of the consecutive readings to determine the elapsed elemental times. For example, the following readings might represent the break points of a 10-element study: 4, 14, 19, 121, 25, 52, 61, 76, 211, 16. The elemental values of this cycle would be 4, 10, 5, 102, 4, 27, 9, 15, 35, and 5. Figure 10.7 illustrates a completed time study of the same die casting operation using the continuous method.

ADDRESSING DIFFICULTIES

During the time study, analysts may observe variations from the element sequence originally established. Occasionally, analysts may miss specific break points. These difficulties complicate the study; the less often they occur, the easier it is to calculate the study.

When missing a reading, the analyst should immediately indicate an "M" in the W column. In no case should the analyst approximate and endeavor to record the missed value. This practice can destroy the validity of the standard established for the specific element. If the element were then to be used as a source of standard data, appreciable discrepancies in future standards might result. Occasionally, the operator omits an element; this is handled by drawing a horizontal line through the space in the W column. This should happen infrequently, since it is usually a sign of an inexperienced operator or a lack of standardization in the method. Of course, the operator can inadvertently omit an element, for example, forgetting to "vent cope" in making a bench mold. If elements are omitted repeatedly, the analyst should stop the study and investigate the necessity of performing the omitted elements. This should be done in cooperation with the supervisor and the operator, so that the best method can be established. The observer is expected to be on the alert constantly for better ways to perform the elements; as ideas come to mind, the observer should jot them down in the note section of the time study form, for future evaluation.

Time Study Observation Form

		Study No.: 2-85		Date: 3-1		Page 1 of 1
		Operation: DIE CASTING		Operator: B. JONES		Observer: A F

Element No. and Description		1 REMOVE PART FROM DIE, LUBRICATE DIE, INSPECT				2 PLACE PART IN FIXTURE, TRIM ASIDE PART																						
Note	Cycle	R	W	OT	NT	R	W	OT	NT	R	W	OT	NT	R	W	OT	NT	R	W	O	NT	R	W	OT	NT			
	1	90	90	30	270	90	113	23	207																			
	2	100	40	27	270	100	61	21	210																			
	3	90	92	31	279	90	25	23	207																			
	4	85	50	35	298	100	70	20	200																			
	5	100	98	28	280	100	318	20	200																			
	6	110	43	25	275	110	61	18	198																			
	7	90	92	31	279	90	416	24	216																			
	8	100	44	28	280	85	68	24	204																			
	9	90	500	32	288	90	23	23	207																			
	10	110	49	26	286	105	68	19	200																			
	11																											
	12																											
	13																											
	14																											
	15																											
	16																											
	17																											
	18																											

Summary

Total OT	2.93	2.15				
Rating	—	—				
Total NT	2.805	2.049				
No. Observations	10	10				
Average NT	.281	.205				
% Allowance	17	17				
Elemental Std. Time	.329	.205				
No. Occurrences	1	1				
Standard Time	.329	.205				

Total Standard Time (sum standard time for all elements): **.569**

Foreign Elements					Time Check			Allowance Summary	
Sym	W1	W2	OT	Description	Finishing Time	3:48.00		Personal Needs	5
A					Starting Time	3:42.00		Basic Fatigue	4
B					Elapsed Time		6.00	Variable Fatigue	8
C					TEBS	.60		Special	—
D					TEAF	.32		Total Allowance %	17
E					Total Check Time	.92		Remarks:	
F					Effective Time	5.08			
G					Ineffective Time	0			
Rating Check					Total Recorded Time		6.00		
Synthetic Time				%	Unaccounted Time		0		
Observed Time					Recording Error %		0		

Figure 10.7 Continuous timing study of a die casting operation (rating every cycle).

The observer may also see elements performed out of sequence. This happens fairly frequently when a new or inexperienced employee is studied on a long-cycle job made up of many elements. Avoiding such disturbances is one of the prime reasons for studying a competent, fully trained employee. However, when elements are performed out of order, the analyst should immediately go to the element being performed and draw a horizontal line through the middle of its W space. Directly below this line, the analyst should write the time the operator began the element, and above it, the completion time. This procedure should be repeated for each element performed out of order, as well as for the first element performed back in the normal sequence.

During a time study, the operator may encounter unavoidable delays, such as an interruption by a clerk or supervisor, or tool breakage. The operator may also intentionally cause a change in the order of work by going for a drink of water or stopping to rest. Such interruptions are referred to as *foreign elements*.

Foreign elements can occur either at the break point or during the course of an element. The majority of foreign elements, particularly those controlled by the operator, occur at the termination of an element. If a foreign element occurs during an element, it is signified by letters (A, B, C, etc.) in the NT column of this element. If the foreign element occurs at the break point, it is recorded in the NT column of the work element that follows the interruption (⑤ in Figure 10.6). The letter A is used to signify the first foreign element, the letter B to signify the second, and so on.

As soon as the foreign element has been properly designated, the analyst should write a short description of it in the lower left-hand corner of the space. The time that the foreign element begins is entered in the W1 block of the foreign element section, and the time it ends is entered in the W2 block. These values can then be subtracted when the time study is calculated, to determine the exact duration of the foreign element. This value is then entered in the OT column of the foreign element section. Figure 10.6 illustrates the correct handling of a foreign element.

Occasionally, a foreign element is of such short duration that it is impossible to record the foreign element in the fashion outlined. Typical examples of this would be dropping a wrench on the floor and quickly picking it up, wiping one's brow with a handkerchief, or turning to speak briefly to the supervisor. In such cases, where the foreign element may be 0.06 min or less, the most satisfactory method of handling the interruption is to allow it to accumulate in the element and immediately circle the reading, indicating that a "wild" value has been encountered. A short comment should be entered in the note section across from the element in which the interruption occurred, justifying the circled number. Cycle 7 in Figure 10.8 illustrates the correct handling of a wild value.

CYCLES IN STUDY

Determining how many cycles to study to arrive at an equitable standard is a subject that has caused considerable discussion among time study analysts as well as union representatives. Since the activity of the job, as well as its cycle time, directly influences the number of cycles that can be studied from an economic standpoint, the analyst cannot be completely governed by sound statistical practice that demands a certain sample size based on the dispersion of individual

Time Study Observation Form

Study No.: 14 Date: 3/15/ Page 1 of 2
Operation: DIE CASTING Operator: RAIN BOW Observer: P. ROCHE

Element No. and Description	Cycle	1 PICK UP & PLACE CASTING IN FIXTURE, DEPRESS 2 PARTS		2 OPEN FIX-TURE, GET PART, TURN 90°, PLACE IN 2ND FIXTURE		3 ENGAGE FEED, OPEN FIXTURE, REMOVE PART		5-1 CLEAN WORKSTATION		5-2 PUNCH-IN		5-3 SET UP STOPS IN FIXTURE		5-4 PUNCH OUT TALLY PRODUCTION	
Note	Cycle	R	W	R	W	R	W	R	W	R	W	R	W	R	W
	1							132	132	182	50	415	233	550	135
	2	62	12	78	16	88	10								
	3	604	16	21	17	30	9								
	4	43	13	59	16	70	11								
	5	828	15 (B)	44	21	58	9								
	6	71	13	91	20	905	14								
DROPPED CASTING	7	30	(29)	46	16	57	11								
	8	70	13	88	18	1002	14								
	9	15	13	32	17	40	8								
	10	52	12	68	16	78	10								
	11	92	14	112	20	24	12								
	12	38	14	56	18	66	10								
	13	81	15	1200	19	11	11								
	14	25	14	41	16	50	9								
	15	63	13	80	17	91	11								
	16	1305	14	24	19	34	10								
	17	50	16	69	19	83	14								
	18														

Summary

	Elem 1	Elem 2	Elem 3	Elem 5-1	Elem 5-2	Elem 5-3	Elem 5-4
Total OT	2.07	2.85	1.74	1.32	.50	2.33	1.35
Rating	110	110	110	110	110	110	110
Total NT	2.277	3.135	1.914	1.452	.550	2.563	1.485
No. Observations	15	16	16	1	1	1	1
Average NT	.152	.196	.120	1.452	.550	2.563	1.485
% Allowance	12	12	12	12	12	12	12
Elemental Std. Time	.170	.219	.134	1.626	.616	2.867	1.663
No. Occurrences	1	1	1				
Standard Time	.170	.219	.134				

Total Standard Time (sum standard time for all elements): .523

Foreign Elements

Sym	W1	W2	OT	Description
A	670	813	143	TALK TO SUPERVISOR
B				
C				
D				
E				
F				
G				

Rating Check
Synthetic Time   %
Observed Time

Time Check

Finishing Time	2:39.00
Starting Time	2:25.00
Elapsed Time	14.00
TEBS	0
TEAF	.17
Total Check Time	.17
Effective Time	12.10
Ineffective Time	1.43
Total Recorded Time	14.00
Unaccounted Time	0
Recording Error %	0

Allowance Summary

Personal Needs	5
Basic Fatigue	4
Variable Fatigue	3
Special	—
Total Allowance %	12

Remarks: STANDARD TIME PER PIECE WITHOUT SET UP TIME

Figure 10.8 Time study with overall rating.

element readings. General Electric Company has established Table 10.2 as an approximate guide to the number of cycles to observe.

A more accurate number can be established by using statistical methods. Since time study is a sampling procedure, the observations can be assumed to be distributed normally about an unknown population mean with an unknown variance.

Table 10.2 Recommended Number of Observation Cycles

Cycle time (min)	Recommended number of cycles
0.10	200
0.25	100
0.50	60
0.75	40
1.00	30
2.00	20
2.00–5.00	15
5.00–10.00	10
10.00–20.00	8
20.00–40.00	5
40.00–above	3

Source: Information taken from the Time Study Manual of the Erie Works of the General Electric Company, developed under the guidance of Albert E. Shaw, manager of wage administration.

Using the sample mean $\bar{x}$ and sample standard deviation s, the normal distribution for a large sample leads to the confidence interval

$$\bar{x} \pm \frac{zs}{\sqrt{n}}$$

where

$$s = \sqrt{\frac{\sum\limits_{i=1}^{i=n} (x_i - \bar{x})^2}{n - 1}}$$

However, time studies often involve only small samples ($n < 30$); therefore, a t distribution must be used. The confidence interval equation is then

$$\bar{x} \pm t\,\frac{s}{\sqrt{n}}$$

The $\pm$ term can be considered an error term expressed as a fraction of $\bar{x}$:

$$k\bar{x} = ts/\sqrt{n}$$

where k = an acceptable fraction of $\bar{x}$.
     Solving for n yields

$$n = \left(\frac{ts}{k\bar{x}}\right)^2$$

It is also possible to solve for n before making the time study by interpreting historical data of similar elements, or by actually estimating $\bar{x}$ and s from several snapback readings with the highest variation.

| EXAMPLE 10.1 | Calculation of Required Number of Observations |

A pilot study of 25 readings for a given element showed that $\bar{x} = 0.30$ and $s = 0.09$. A 5% desired acceptable fraction k and an alpha = 0.05 for 24 degrees of freedom (25 minus 1 degree of freedom for estimating one of the parameters) yield $t = 2.064$. (See Table A3–3, Appendix 3, for values of t.) Solving the last equation yields

$$n = \left(\frac{0.09 \times 2.064}{0.05 \times 0.30}\right)^2 = 153.3 \approx 154 \text{ observations}$$

To ensure the required confidence, always round up.

10.6 EXECUTION OF STUDY

This section provides an overview of the main steps needed to execute the time study. More details on rating operator performance and adding allowances are presented in Chapter 11.

| EXAMPLE 10.2 | Statistical Calculation of an Outlier |

Consider the seventh cycle of element 1 (circled) in Figure 10.8, when the operator drops the casting and has an excessively long element. The 1.5IQR criterion is based on descriptive statistics using a box plot and considers any value outside the 1.5t interquartile range (range between the first and third quartiles) to be an outlier (Montgomery and Runger, 1994). For element 1, the mean value is 0.138, the standard deviation is 0.01265, the first quartile is 0.13, and the third quartile is 0.15. Therefore, the value of 1.5IQR is

$$1.5\text{IQR} = 1.5 \times (0.15 - 0.13) = 0.3$$

Since the circled value of 0.25 is well beyond the mean plus the 1.5IQR value of 0.168, this can be considered an outlier and discarded from the calculation of the standard time.

Using the three-sigma rule (or four-sigma, Montgomery, 1991) for a 95 percent confidence interval with t at 14 degrees of freedom of 2.145 yields a value of

$$ts/\sqrt{n} = 2.145 \times 0.01265/\sqrt{15} = 0.0701$$

The three-sigma rule yields the critical value of 0.21 while the four-sigma rule yields the critical value of 0.28. The second value is very close to the 1.5IQR criterion and yields the same conclusion, an outlier.

RATING OPERATOR PERFORMANCE

Since the actual time required to perform each element of the study depends to a high degree on the skill and effort of the operator, it is necessary to adjust upward the time of the good operator and downward the time of the poor operator to a standard level. Therefore, before leaving the workstation, analysts should give a fair and impartial performance rating to the study. On short-cycle, repetitive work, it is customary to apply one rating to the entire study, or an average rating for each element (see Figure 10.8). However, where the elements are long and entail diversified manual movements, it is more practical to evaluate the performance of each element as it occurs. This was done for the die casting study shown in Figures 10.4 and 10.7, in which the elements were over 0.20 min in duration. The time study form includes provisions for both the overall rating and the individual element rating.

In the performance rating system, the observer evaluates the operator's effectiveness in terms of a *qualified operator* performing the same element. The rating value is expressed as a decimal or percentage and is assigned to the observed element, in the R column (③ in Figure 10.6). A qualified operator is defined as a thoroughly experienced operator working under customary conditions at the workstation, at a pace neither too fast nor too slow, but representative of a pace that can be maintained throughout the day.

The basic principle of performance rating is to adjust the mean observed time (OT) for each element performed during the study to the *normal time* (NT) that would be required by the qualified operator to perform the same work:

$$NT = OT \times R/100$$

where R is the performance rating of the operator expressed as a percentage, with 100 percent being standard performance by a qualified operator. To do a fair job of rating, the time study analyst must be able to disregard personalities and other varying factors and consider only the amount of work being done per unit of time, as compared to the amount of work that the qualified operator would produce. Chapter 11 more fully explains the performance-rating techniques in common use.

ADDING ALLOWANCES

No operator can maintain a standard pace every minute of the working day. Three classes of interruptions can take place, for which extra time must be provided. The first is personal interruptions, such as trips to the restroom and drinking fountain; the second is fatigue, which can affect even the strongest individual on the lightest work. The third is unavoidable delays, such as tool breakage, supervisor interruptions, slight tool trouble, and material variations, all of which require that some allowance be made. Since the time study is made over a relatively short period, and since foreign elements should have been removed in determining the normal time, an allowance must be added to the normal time to arrive at a fair standard that can reasonably be achieved by an operator. The time required for a fully qualified, trained operator, working at a standard pace and exerting average effort, to perform the operation is termed the *standard time* (ST) for that operation. The allowance is typically given as a fraction of normal time and is used as a multiplier equal to 1 + allowance:

$$ST = NT + NT \times \text{allowance} = NT \times (1 + \text{allowance})$$

An alternative approach is to formulate the allowances as a fraction of the total workday, since the actual production time might not be known. In that case, the expression for standard time is

$$ST = NT/(1 - \text{allowance})$$

Chapter 11 details the means for arriving at realistic allowance values.

10.7 CALCULATING THE STUDY

After properly recording all the necessary information on the time study form, observing an adequate number of cycles, and performance-rating the operator, the analyst should record the finishing time (⑥ in Figure 10.6) at the same master clock used for the start of the study. For continuous timing, it is very important to verify the final stopwatch reading with the overall elapsed clock reading. These two values should be reasonably close ($\pm$ 2 percent difference). (A sizable discrepancy may mean that an error has occurred, and the time study may need to be repeated.) Finally, the analyst should thank the operator and proceed to the next step, the study computations.

For the continuous method, each watch reading must be subtracted from the preceding reading to get the elapsed time; this value is then recorded in the OT column. Analysts must be especially accurate in this phase, because carelessness at this point can completely destroy the validity of the study. If the elemental performance rating was used, the analyst must multiply the elapsed elemental times by the rating factor and record the result in the NT spaces. Note that since NT is a calculated value, it is typically recorded with three digits.

Elements that have been missed by the observer are signified by an M in the W column and are disregarded. Thus, if the operator happened to omit element 7 of cycle 4 in a 30-cycle study, the analyst would have only 29 values of element 7 with which to calculate the mean observed time. The analyst should disregard not only the missed element but also the succeeding one, since the subtracted value in the study would include the time for performing both elements.

To determine the elapsed elemental time on out-of-order elements, it is merely necessary to subtract the appropriate watch times.

For foreign elements, the analyst must deduct the time required for the foreign element from the cycle time of the applicable element. The analyst can obtain the time taken by the foreign element by subtracting the W1 reading in the foreign element section of the time study form from the W2 reading.

After all elapsed times have been calculated and recorded, the analyst should study them carefully for any abnormality. Values that are extreme can be considered as statistical outliers of the cyclical repetition of an element. To determine if an elemental time is an outlier, one can use either the 1.5IQR criterion (interquartile range, Montgomery and Runger, 1994) or the three-sigma rule (or four-sigma, Montgomery, 1991). Both rules are demonstrated in Example 10.2. These values are then circled and excluded from further consideration in working up the study. For example, in cycle 7 of element 1 in Figure 10.8, the operator dropped a casting. Whereas previously that was considered a "wild" value, it would be just as easily excluded based on the statistical outlier criteria.

Machine elements have little variation from cycle to cycle, while considerably wider variation could be expected in manual elements. When unexplainable time variations occur, the analyst should be quite careful before circling such values. Remember that this is not a performance-rating procedure. By arbitrarily discarding high or low values, the analyst may end up with an incorrect standard.

If elemental rating is used, then after the elemental elapsed time values have been computed, the analyst should determine the normal elemental time by multiplying each elemental value by its respective performance factor. This normal time is then recorded in the NT columns for each element ((10) in Figure 10.6). Next, the analyst determines the mean elemental normal value by dividing the total of the times recorded in the NT columns by the number of observations.

After determining all elemental elapsed times, the analyst should check to ensure that no arithmetic or recording errors have been made. One method of checking for such accuracy is to complete the *time check* of the time study form (see Figure 10.6). To do this, though, the analyst needed to have synchronized the starting and stopping of the watch at a master clock, recording the *starting time* (1) and the *finishing time* (6) on the form. The analyst then sums three quantities: (1) total observed times, known as *effective time* ((12) on the form); (2) total foreign elements time, known as *ineffective time* ((13) on the form); and (3) total of *time elapsed before study* (TEBS (2) on the form) and time elapsed after study (TEAF (7) on the form). The time elapsed before study is the readout when the analyst snaps the watch at the start of the first element. The time elapsed after study is the last readout when the analyst snaps the watch at the very end of the study. These last two quantities are sometimes totaled, forming the *check time* ((8) on the form). The three quantities together equal the *total recorded time* ((14) on the form). The difference between the finishing and starting times on the master clock equals the *actual elapsed time* ((9) on the form). Any difference between the total recorded time and the elapsed time is called *unaccounted time* ((15) on the form). Normally, in a good study, this value would be zero. The unaccounted time divided by the elapsed time is a percentage called the *recording error*. This recording error should be less than 2 percent. If it exceeds 2 percent, the time study should be repeated.

After the normal elemental times have been calculated, the analyst should add the percentage allowance to each element to determine the allowed or standard time. In the time study of Figure 10.8, the normal time for element 1 is multiplied by 1.12 to yield the following elemental standard time for element 1:

$$ST = 0.152 \times (1 + 0.12) = 0.170$$

The nature of the job determines the amount of allowance to be applied, as discussed in Chapter 11. Suffice it to say at this point that 15 percent is the average allowance used for manual elements, and 10 percent is the allowance usually applied to machine elements.

In most cases, each element occurs once within each cycle and the number of occurrences is simply 1. In some cases, an element may be repeated within a cycle (e.g., three passes on a lathe within the same cycle). In that case, the number of occurrences becomes 2 or 3, and the time accrued by that element within the one cycle is doubled or tripled.

The standard times for each element are then summed to obtain the standard time for the entire job, which is recorded in the space labeled *Total Standard Time* on the time study form.

10.8   THE STANDARD TIME

The sum of the elemental times gives the standard in minutes per piece, using a decimal minute watch, or hours per piece, using a decimal hour watch. The majority of industrial operations have relatively short cycles (less than 5 min); consequently, it is sometimes more convenient to express standards in hours per hundred pieces. For example, the standard on a press operation might be 0.085 h per hundred pieces. This is a more satisfactory method of expressing the standard than 0.00085 h per piece or 0.051 min per piece.

The percent efficiency of the operator can be expressed as

$$E = 100 \times H_e/H_c = 100 \times O_c/O_e$$

where E = percent efficiency
H_e = standard hours earned
H_c = clock hours on job
O_e = expected output
O_c = current output

Thus, an operator producing 10,000 pieces during the working day would earn 8.5 h of production and would perform at an efficiency of 8.5/8 = 106 percent.

Once the standard time has been computed, it is given to the operator in the form of an operation card. The card can be either computer-generated or run off on a copier. The operation card serves as the basis for routing, scheduling, instruction, payroll, operator performance, cost, budgeting, and other necessary controls for the effective operation of a business. Figure 10.9 illustrates a typical production operation card.

TEMPORARY STANDARDS

Employees require time to become proficient in any new or different operation. Frequently, time study analysts establish a standard on a relatively new operation on which there is insufficient volume for the operator to reach top efficiency. If the analyst bases operator grading on the usual conception of output (i.e., rating the operator below 100), the resulting standard may seem unduly tight, and the operator will probably be unable to make any incentive earnings (see Chapter 17). On the other hand, if the analyst considers that the job is new and the volume is low, and establishes a liberal standard, then if the size of the order is increased, or if a new order for the same job is received, trouble may occur.

PRODUCTION OPERATION CARD

DESCRIPTION Shower head face

DWG. NO. JB-1102   PART NO. J-1102-1

MADE FROM 2½" diam. 70-30 extruded brass rod

DATE 9-15

Routing 9-11-12--14-12-18

OP. NO.	OPERATION	DEPT.	MACHINE AND SPECIAL TOOLS	SET-UP MINUTES		EACH PC. MINUTES
1	Saw slug	9	J.& L. Air Saw	15 min		.077
2	Forge	11	150 Ton Maxi F-1102	70 min		.234
3	Blank	12	Bliss 72 F-1103	30 min		.061
4	Pickle	14	HCL. Tank	5 min		.007
5	Pierce 6 holes	12	Bliss 74 F-1104	30 min		.075
6	Rough ream and chamfer	12	Delta 17" D.P. F-1105	15 min		.334
7	Drill 13/64" holes	12	Avey D.P. F-1106	15 min		.152
8	Machine stem and face	12	#3 W. & S.	45 min		.648
9	Broach 6 holes	12	Bliss 74½	30 min		.167
10	Inspect	18	F-1109,F-1110,F-1112		Daywork	

Figure 10.9 Typical production operation card. (The "F" numbers refer to the fixtures used with the involved operation.)

Perhaps the most satisfactory method of handling such situations is the issuance of temporary standards. The analyst establishes the standard by giving consideration to the difficulty of the work assignment and the number of pieces to be produced. Then, by using a learning curve (see Chapter 18) for the work, as well as the existing standard data, the analyst can develop an equitable temporary standard for the work. The resulting standard will be considerably more liberal than if the job involved a large volume. When released to the production floor, the standard is clearly marked "temporary," and will include the maximum quantity for which it applies. When temporary standards are released, they should only be in effect for the duration of the contract, or for 60 days, whichever is shorter. Upon expiration, they should be replaced by permanent standards.

EXAMPLE 10.3	Calculation of Hours Earned and Percent Efficiency

The standard time for an operation is 11.46 min per piece. In an 8-h shift, the operator would be expected to produce

$$\frac{8 \text{ h} \times 60 \text{ min/h}}{11.46 \text{ min/piece}} = 41.88 \text{ pieces}$$

However, if the operator produced 53 pieces in a given working day, the standard hours (see Chapter 17) earned would be

$$H_e = \frac{53 \text{ pieces} \times 11.46 \text{ min/piece}}{60 \text{ min/h}} = 10.123 \text{ h}$$

The standard S_h expressed in hours per hundred pieces C is

$$S_h = \frac{11.46 \text{ min/piece} \times 100 \text{ pieces}/C}{60 \text{ min/h}} = 19.1 \text{ h}/C$$

The standard hours earned would be

$$H_e = \frac{19.1 \text{ h}/C \times 53 \text{ pieces}}{100 \text{ pieces}/C} = 10.123 \text{ h}$$

The operator's efficiency would be

$$E = 100 \times 10.123/8 = 126.5\%$$

or more simply

$$E = 100 \times 53/41.88 = 126.5\%$$

SETUP STANDARDS

The elements of work commonly included in setup standards involve all events that take place between completion of the previous job and the start of the present job. The setup standard also includes "teardown" or "put-away" elements, such as punching in on the job, getting tools from the tool crib, getting drawings from the dispatcher, setting up the machine, punching out on the job, removing tools from the machine, returning tools to the crib, and tallying production. Figures 10.8 and 10.10 illustrate a study that involved four setup elements (S-1, S-2, S-3, and S-4).

In establishing setup times, the analyst should use the identical procedure followed in establishing production standards, except that there will be no opportunity to get a series of elemental values for determining the mean times. Also, the analyst cannot observe the operator performing the setup elements in advance; consequently, the analyst is obliged to divide the setup into elements while the study is taking place. However, since setup elements, for the most part, are long-duration, there is a reasonable amount of time to break the job down, record the time, and evaluate the performance as the operator proceeds from one work element to the next.

There are two ways of handling setup times distributed by quantity, or allocated by job. In the first method, they would be distributed over a specific manufacturing quantity, such as 1,000 or 10,000 pieces. This method is satisfactory only when the magnitude of the production order is standard. For example, industries that ship from stock and reorder on a basis of minimum-maximum inventories are able to control their production orders to conform to economical lot sizes. In such cases, the setup time can be equitably prorated over the lot size. For example, suppose that the economical lot size of a given item is 1,000 pieces and that reordering is always done on the basis of 1,000 units. If the standard setup time in a given operation is 1.50 h, then the allowed operation time could be increased by 0.15 h per 100 pieces to take care of the make-ready and put-away elements.

This method is not at all practical if the size of the order is not controlled. In a plant that requisitions on a job-order basis, that is, releases production orders specifying quantities in accordance with customer requirements, it is impossible to standardize on the size of the work orders. For example, this week an order for 100 units may be issued, and next month an order for 5,000 units of the same part may be needed. In the example, the operator would only be allowed 0.15 h to set up the machine for the 100-unit order, which would be inadequate. On the 5,000-unit order, however, the operator would be given 7.50 h, which would be considerably too much time.

It is more practical to establish setup standards as separate standard times (see Figures 10.8 and 10.10). Then, regardless of the quantity of parts to be produced, a fair standard would prevail. In some concerns, the setup is performed by a person other than the operator who does the job. The advantages of having separate setup personnel are quite obvious. Lower-skilled workers can be utilized as operators when they do not have to set up their own facilities. Setups are more readily standardized and methods changes more easily introduced when the responsibility for setup rests with one individual. Also, when sufficient facilities are available, production can be continuous if the next work assignment is set up while the operator is working on the present job.

PARTIAL SETUPS

Frequently, it is not necessary to set up a facility completely to perform a given operation, because some of the tools of the previous operation are required in the

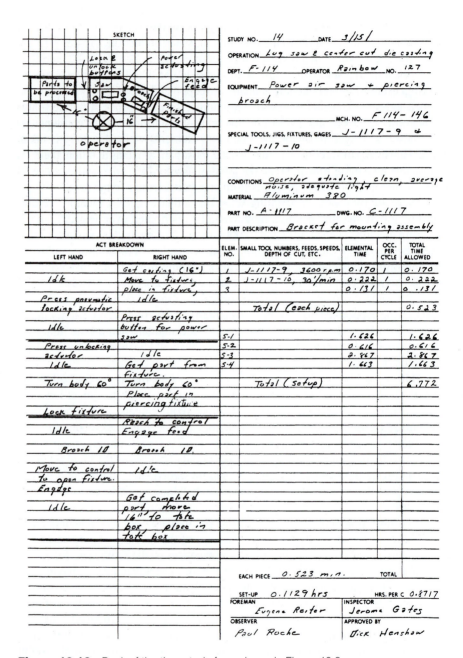

Figure 10.10 Back of the time study form shown in Figure 10.8.
The study indicates a piece time of 0.523 min (0.8717 h/C) and a setup time of
6.772 min (0.1129 h).

job being set up. For example, in hand-screw machine or turret lathe setups, careful scheduling of similar work for the same machine allows partial setups from one job to be used for the next. Instead of having to change six tools in the hex turret, it may only be necessary to change two or three. This savings in setup time is one of the principal benefits of a well-formulated group technology program.

Since the sequence of work scheduled for a given machine seldom remains the same, it is difficult to establish partial setup times to cover all possible variations. For example, the standard for a complete setup for a given No. 4 Warner & Swasey turret lathe might be 0.80 h. However, if this setup is performed after job X, it might take only 0.45 h; after job Y, it may require 0.57 h; while following job Z, 0.70 h may be necessary. The variations possible in partial setup time are so broad that the only practical way to establish their values accurately is to use standard data (see Chapter 12) for each job.

In plants where setup times are less than 1 h and production runs are reasonably long, it is common practice to allow operators the full setup time for each job performed. This is advantageous for several reasons: First, if the plant incorporates wage incentives, operators are considerably more satisfied because of higher earnings, and they plan their work to the best possible advantage. This results in more production per unit of time and in lower total costs. Also, considerable time and paperwork are saved by avoiding having to determine a standard for the partial setup operation and its application in all pertinent cases. In fact, this saving tends to approach the extra amount paid to operators resulting from the difference between the time required to make the complete setup and the time required to make the partial setup.

SUMMARY

To summarize, the steps for performing and computing a typical time study are as follows (see Figure 10.6 for numbers corresponding to these steps):

1. Synchronize at the master clock and record the starting time.
2. Walk to the operation and start the study. The readout at the snap is the time elapsed before the study.
3. Rate operator performance while the element is taking place, and record either the single or the average rating.
4. Snap the watch at the start of the next element. For continuous timing, enter the readout in the W column; for snapback timing, enter the readout in the OT column, as shown.
5. For a foreign element, indicate in the appropriate NT column and record the times in the *Foreign Element* section.
6. Once all elements have been timed, snap the watch at the master clock. Record the finishing time.
7. Record the readout as the time elapsed after the study.
8. Add 2. and 7. to obtain the check time.

9. Subtract 1. from 6. to obtain the elapsed time.
10. Calculate the normal time by multiplying the observed time by the rating.
11. Sum all observed times and the normal times for each element. Find the average normal time.
12. Add all OT totals to obtain the effective time.
13. Add all foreign elements to obtain the ineffective time.
14. Add 8., 12., and 13. to obtain the total recorded time.
15. Subtract 14 from 9. to obtain the unaccounted time. Use the absolute value. (The difference could be either negative or positive, and positive numbers are desired.)
16. Divide 15. by 9. to obtain the percent recording error. Hopefully, this value is less than 2 percent!

QUESTIONS

1. How is a fair day's pay determined?
2. What benchmark is used for standard pace?
3. Why should the supervisor sign the time study?
4. What are the effects of poor time standards?
5. What equipment is needed by the time study analyst?
6. What features of the PDA make it attractive to time study analysts?
7. How can the metronome be used as a training tool for performance rating?
8. What considerations should be given to the choice of the operator to be studied?
9. Why is it essential to record complete information on the tools and facility on the time study form?
10. Why are working conditions important in identifying the method being observed?
11. Why would a time study analyst who is hard of hearing have difficulty in performing a time study?
12. Differentiate between constant and variable elements. Why should they be kept separate when dividing the job into elements?
13. What advantages does the continuous method of watch recording offer over the snapback method?
14. Explain why electronic stopwatches have increased the use of the snapback procedure.
15. Why is the time of day recorded on the time study form?
16. How is a large variation in the sequence of elemental time recordings handled?
17. Explain what a foreign element is and how foreign elements are handled under the continuous method.
18. What factors enter into the determination of the number of cycles to observe?
19. Why is it necessary to rate the operator?
20. When should individual elements of each cycle be rated?
21. Define a qualified operator.

22. Why are allowances applied to the normal time?
23. What is the significance of a circled elapsed time?
24. What steps are taken in the computation of a time study conducted in accordance with the continuous overall performance-rating procedure?
25. How does the walking pace of 3 mi/h agree with your concept of a standard performance?
26. Define standard time.
27. Why is it usually more convenient to express standards as time per hundred pieces, rather than time per piece?
28. Why would temporary standards be established?
29. Which elements of work are included in the setup standard?

PROBLEMS

1. Based on Table 10.2, how many observations should be taken on an operation for which the annual activity is 750 pieces and the cycle time is estimated at 15 min?
2. Take a simple operation that you perform regularly, such as brushing your teeth, shaving, or combing your hair, and estimate the time it takes you to perform that operation. Now measure the time it takes while working at a standard pace. Is your estimate within plus or minus 20 percent of the estimated time?
3. To demonstrate various levels of performance to a group of union stewards, the time study supervisor of XYZ Company is using the metronome while dealing bridge hands. How many times per minute should the metronome beat to demonstrate the following levels of performance? 60 percent, 75 percent, 100 percent, 125 percent
4. You used the General Electric guide sheet to determine the number of observations to study. The guide sheet indicated that 10 cycles were required. After making the study, you used the standard error of the mean to estimate the number of observations needed for a given confidence level. The resulting calculation indicated that 20 cycles should be studied. What would be your procedure? Why?
5. The time study analyst at Dorben Company developed the following snapback stopwatch readings where elemental performance rating was used. The allowance for this element was assigned a value of 16 percent. What would be the standard time for this element?

Snapback reading	Performance factor
28	100
24	115
29	100
32	90
30	95
27	100
38	80
28	100
27	100
26	105

6. What would be the required number of readings if the analyst wanted to be 87 percent confident that the mean observed time was within ±5 percent of the true mean and the following values were established for an element after 19 cycles were observed?

0.09, 0.08, 0.10, 0.12, 0.09, 0.08, 0.09, 0.12, 0.11, 0.12, 0.09, 0.10, 0.12, 0.10, 0.08, 0.09, 0.10, 0.12, 0.09

7. The following data resulted from a time study made on a horizontal milling machine:

Mean manual effort time per cycle: 4.62 min
Mean cutting time (power feed): 3.74 min
Mean performance rating: 115 percent
Machine allowance (power feed): 10 percent
Fatigue allowance: 15 percent
What is the standard time for the operation?

8. A work measurement analyst in Dorben Company took 10 observations of a high-production job. He performance-rated each cycle and then computed the mean normal time for each element. The element with the greatest dispersion had a mean of 0.30 min and a standard deviation of 0.03 min. If it is desirable to have sampled data within ±5 percent of the true data, how many observations should this time study analyst take of this operation?

9. In Dorben Company, the work measurement analyst took a detailed time study of the making of shell molds. The third element of this study had the greatest variation in time. After studying nine cycles, the analyst computed the mean and standard deviation of this element, with the following results:

$$\bar{x} = 0.42 \quad s = 0.08$$

If the analyst wanted to be 90 percent confident that the mean time of the sample was within ±10 percent of the mean of the population, how many total observations should have been taken? Within what percentage of the average of the total population is $\bar{x}$ at the 95 percent confidence level, under the measured observations?

10. Based on the data provided in Figure 10.8, what would be the efficiency of an operator who set up the machine and produced an order of 5,000 pieces in a 40-h workweek?

11. Establish the money rate per hundred pieces from the following data:

Cycle time (averaged measured time): 1.23 min
Base rate: $8.60 per h
Pieces per cycle: 4
Machine time (power feed): 0.52 min per cycle
Allowance: 17 percent on effort time; 12 percent on power feed time
Performance rating = 88 percent

12. The following data resulted from a study taken on a horizontal milling machine:

Pieces produced per cycle: 8
Average measured cycle time: 8.36 min
Average measured effort time per cycle: 4.62 min
Average rapid traverse time: 0.08 min
Average cutting time power feed: 3.66 min
Performance rating: 115 percent
Allowance (machine time): 10 percent
Allowance (effort time): 15 percent

The operator works on the job a full 8-h day and produces 380 pieces. How many standard hours does the operator earn? What is the operator's efficiency for the 8-h day?

13. Express the standard of 5.761 min in hours per hundred pieces. What would the operator's efficiency be if 92 pieces were completed during a working day? What would the efficiency be if the operator set up the machine (standard for setup = 0.45 h) and produced 80 pieces during the 8-h workday?

14. Find the observed, normal, and standard times for the following jobs shown on the CD. Assume standard performance and a 10 percent allowance.

   a. Stamping extrusions

   b. Stamping end couplings

   c. Flashlight assembly

   d. Union assembly

   e. Hospital bed rail assembly

   f. Stitching (garments)

   g. Labeling (garments)

   h. Cut and tack (garments)

REFERENCES

Barnes, Ralph M. *Motion and Time Study: Design and Measurement of Work*. 7th ed. New York: John Wiley & Sons, 1980.

Gomberg, William. *A Trade Union Analysis of Time Study*. 2d ed. Englewood Cliffs, NJ: Prentice-Hall, 1955.

Griepentrog, Carl W., and Gilbert Jewell. *Work Measurement: A Guide for Local Union Bargaining Committees and Stewards*. Milwaukee, WI: International Union of Allied Industrial Workers of America, AFL-CIO, 1970.

Lowry, S. M., H. B. Maynard, and G. J. Stegemerten. *Time and Motion Study and Formulas for Wage Incentives*. 3d ed. New York: McGraw-Hill, 1940.

Montgomery, D. C. *Design and Analysis of Experiments*. 3d ed. New York: John Wiley & Sons, 1991.

Montgomery, D. C., and G. C. Runger. *Applied Statistics and Probability for Engineers*. New York: John Wiley & Sons, 1994.

Mundel, M. E. *Motion and Time Study: Improving Productivity*. 5th ed. Englewood Cliffs, NJ: Prentice-Hall, 1978.

Nadler, Gerald. Work Design: *A Systems Concept*. Rev. ed. Homewood, IL: Richard D. Irwin, 1970.

Rotroff, Virgil H. *Work Measurement*. New York: Reinhold Publishing, 1959.

Smith, George, L. *Work Measurement—A Systems Approach. Columbus*, OH: Grid, 1978.

United Auto Workers. *Time Study—Engineering and Education Departments. Is Time Study Scientific?* Publication No. 325. Detroit, MI: Solidarity House, 1972.

SELECTED SOFTWARE

CITS/APR. C-Four, P.O. Box 808, Pendleton, SC 29670 (http://www.c-four.com/).

DesignTools (available from the McGraw-Hill text website at www.mhhe.com/niebel-freivalds), New York: McGraw-Hill, 2002.

MVTA. Nexgen Ergonomics, 3400 de Maisonneuve Blvd. West, Suite 1430, Montreal, Quebec, Canada H3Z 3B8 (http://www.nexgenergo.com/).

Palm CITS. C-Four, P.O. Box 808, Pendleton, SC 29670 (http://www.c-four.com/).

QuickTimes. Applied Computer Services, Inc., 7900 E. Union Ave., Suite 1100, Denver CO 80237 (http://www.acsco.com/).

QuickTS (included with textbook), New York: McGraw-Hill, 2002.

TimStudy. Royal J. Dosset Corp., 2795 Pheasant Rd., Excelsior, MN 55331.

Work study +3.0, Quetech Ltd., 1-866-222-1022 (http://www.quetech.com)

SELECTED VIDEOTAPES

Time Study Fundamentals, ½-in VHS, C-Four, P.O. Box 808, Pendleton, SC 29670.

Performance Rating and Allowances

KEY POINTS

- Use ratings to adjust observed times to those expected with standard performance.

- Speed rating is the fastest and simplest method.

- Rate the operator before recording the time.

- For a study with long elements, rate each separately.

- For a study with short elements, rate the overall study.

- Use allowances to compensate for fatigue and delays at work.

- Provide a minimum of 9% to 10% constant allowance for personal needs and basic fatigue.

- Add allowances to normal time as a percentage of normal time to obtain standard time.

During a time study, analysts carefully observe the performance of the operator. The performance being executed seldom conforms to the exact definition of *standard*. Thus, some adjustment must be made to the mean observed time to derive the time required for a qualified operator to do the job when working at a standard pace. To arrive at the time required by a qualified worker, time study analysts must increase the time if an above-standard operator has been selected for study and decrease the time if a below-standard operator has been selected for study. Only in this manner can they establish a true standard for qualified operators.

Performance rating is probably the most important step in the entire work measurement procedure. It is also the step most subject to criticism, since it is based entirely on the experience, training, and judgment of the work measurement analyst. Regardless of whether the rating factor is based on the speed or tempo of the output or on the performance of the operator compared to that of a qualified worker, experience and judgment are still the criteria for determining the rating factor. For this reason, analysts must have sufficient training and high personal integrity.

After calculation of the normal time, one additional step must be performed to arrive at a fair standard. This last step is the addition of an allowance to account for the many interruptions, delays, and slowdowns caused by fatigue in every work assignment. For example, when planning a road trip of 1,300 mi, we know that the trip cannot be made in 20 h driving at a speed of 65 mi/h. An allowance must be added for periodic stops for personal needs, driving fatigue, unavoidable stops due to traffic congestion and stoplights, possible detours and the resulting rough roads, car trouble, and so forth. Thus, we may actually estimate that the trip will take 25 h, allowing an additional 5 h for all delays. Similarly, analysts must provide an allowance if the resulting standard is to be fair and readily maintainable by an average worker performing at a steady, normal pace.

11.1 STANDARD PERFORMANCE

Standard performance is defined as the level of performance attained by a thoroughly experienced operator working under customary conditions at a pace neither too fast nor too slow, but representative of one that can be maintained throughout the day. To better define such performance, we use carefully defined benchmark examples that are familiar to us all. Typical benchmarks include dealing 52 cards in 0.50 min or walking 100 ft (3 mi/h or 4.83 km/h) in 0.38 min. However, a specific description should be given of the distance of the four hands dealt with respect to the dealer and of the technique of grasping, moving, and disposing of the cards or whether the ground is level, whether a load is being carried, and how heavy the load is while walking. The more clear-cut and specific the definition of the conditions, the better it is.

The benchmark examples should be supplemented by a clear description of the characteristics of an employee carrying out a standard performance. A representative description of such an employee might be as follows: a worker who is adapted to the work and has attained sufficient experience to perform the job in an efficient manner, with little or no supervision. The worker possesses coordinated mental and physical qualities, enabling him or her to proceed from one element to another without hesitation or delay, in accordance with the principles of motion economy. The worker maintains a good level of efficiency through knowledge and proper use of all tools and equipment related to the job. She or he cooperates and performs at a pace best suited for continuous performance.

Considerable, individual differences can exist between workers. Differences in inherent knowledge, physical capacity, health, trade knowledge, physical dexterity, and training can cause one operator to consistently outperform another. For example, in a random selection of 1,000 employees, the frequency distribution of performance would approximate the normal curve, with over 997 cases on average falling within three-sigma limits. Graphically, this distribution of the 1,000 people would appear as shown in Figure 11.1 (Presgrave, 1957). Based on the ratio of the two extremes (1.39/0.61) the best individual would be more than twice as fast as the slowest individual.

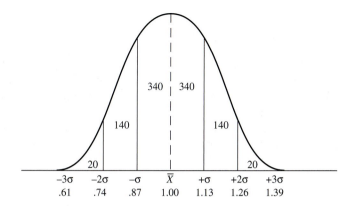

Figure 11.1   Expected distribution of the output of 1,000 people selected at random.

11.2   SOUND RATING CHARACTERISTICS

The first and most important characteristic of any rating system is accuracy. Since the majority of rating methods rely on the judgment of the time study analyst, perfect consistency in rating is impossible. However, rating procedures applied to different operators employing the same method that yield standards deviating less than 5 percent from the mean are considered adequate.

Time study analysts should be similarly accurate. It is not difficult to correct the rating habits of an analyst who rates consistently high or consistently low. However, it is very difficult to correct the rating ability of an analyst who is inconsistent, rating too high one day and too low the next day, and that person should probably not continue to conduct time studies. Inconsistency, more than anything else, destroys the operator's confidence in the time study procedure.

Performance rating should be done only during the observation of elemental times. As the operator progresses from one element to the next, using the prescribed method, an analyst should carefully evaluate the operator's speed, dexterity, false moves, rhythm, coordination, effectiveness, and the other factors influencing output and judge the operator's performance to standard performance. Once the performance has been recorded, it should not be changed. However, this does not imply that the observer always has perfect judgment. If the rating is questioned, the job or operation should be restudied to prove or disprove the recorded evaluation.

The frequency of rating depends on the cycle time. In short-cycle repetitive operations, little deviation in operator performance is expected during the course of the average-length study (15 to 30 min), and it is perfectly satisfactory to evaluate the performance of the entire study. Also, an observer who endeavors to rate each element in such a study will be so busy recording values that he or she will be unable to effectively observe, analyze, and evaluate the operator's performance.

When the study is relatively long (over 30 min) or is made up of several long elements, operator performance may vary during the course of the study. In such studies, analysts should rate each element especially if they are longer than 0.10 min. But in general, the more frequently a study is rated, the more accurate will be the evaluation of operator performance. As examples, the time studies shown in Figures 10.4, 10.6, and 10.7 rate each element, while the time study shown in Figure 10.8 rates the overall study.

11.3   RATING METHODS

SPEED RATING

Speed rating is a performance evaluation method that considers only the rate of work per unit time. In this method, the observer measures the effectiveness of the operator against the concept of a qualified operator doing the same work, and then assigns a percentage to indicate the ratio of the observed performance to standard performance.

This method particularly emphasizes the observer having complete knowledge of the job before doing the study. To illustrate, the pace of machine workers in a plant producing aircraft engine parts would appear considerably slower than the pace of stitchers in the garment industry. The greater precision of aircraft work requires such care that the movements of the various operators might appear unduly slow to one not completely familiar with the work.

In speed rating, analysts first appraise the performance to determine whether it is above or below normal. They then try to place the performance in the precise position on the rating scale that correctly evaluates the numerical difference between the standard and the performance demonstrated. Thus, 100 percent is usually considered normal. A rating of 110 percent indicates that the operator was performing at a speed 10 percent greater than normal, and a rating of 90 percent means that the operator was performing at a speed 90 percent of normal.

All analysts, through years of experience, will eventually develop a mental model of standard performance. However, it may be useful for the novice analyst to consider the performance of several common tasks in order to develop an initial mental model. Two such tasks were suggested by Presgrave (1957): (1) walking at 3 mi/h (4.83 km/h), that is, 100 ft (30.5 m) in 0.38 min, and (2) dealing a deck of 52 cards into four equal piles closely spaced in 0.5 min (note that the opposite thumb feeds the cards to the dealing hand). A guide with specific anchor points for various levels is presented in Table 11.1.

It may also be useful for the novice analyst to begin by rating only by 10s, that is, 80, 90, 100, and then eventually moving onto 5s, and so on. Also, it is very important for any analyst to record the rating in the R column on the time study form before snapping and looking at the watch readout. Otherwise, the analyst may be accused of *rating by the watch.*

Some companies use a speed rating technique normalized to 60 percent standard. This is based on the standard hour approach, that is, producing 60 min of

Table 11.1 Speed Rating Guide

Rating	Verbal anchor points	Walking speed (mi/h)	Cards dealt per 1/2 min
0	No activity	0	0
67	Very slow, clumsy	2	35
100	Steady, deliberate	3	52
133	Brisk, businesslike	4	69
167	Very fast, high degree of dexterity	5	87
200	Upper limit for short period	6	104

work every hour. On this basis, a rating of 80 would mean that the operator was working at a speed of 80/60, or 133 percent, which is 33 percent above normal. A rating of 50 would indicate a speed of 50/60, or 83.3 percent of normal.

Time study analysts use speed rating for elemental, cycle, or overall rating. For example, all the time studies shown in Chapter 10 utilized speed rating.

THE WESTINGHOUSE SYSTEM

One of the oldest used rating systems, then termed *leveling,* was developed by Westinghouse Electric Corporation (Lowry, Maynard, and Stegemerten, 1940). This *Westinghouse rating system* considers four factors in evaluating the performance of the operator: skill, effort, conditions, and consistency.

The system defines *skill* as "proficiency at following a given method" and further relates it to expertise, as demonstrated by the proper coordination of mind and hands. The skill of an operator results from experience and inherent aptitudes, such as natural coordination and rhythm. Skill increases over time, because increased familiarity with the work brings speed, smoothness of motions, and freedom from hesitations and false moves. A decrease in skill is usually caused by some impairment of ability, brought about by physical or psychological factors, such as failing eyesight, failing reflexes, and the loss of muscular strength or coordination. Therefore, a person's skill can vary from job to job, and even from operation to operation on a given job.

There are six degrees of skill: poor, fair, average, good, excellent, and super. Table 11.2 illustrates the characteristics of the various degrees of skill, with their equivalent percentage values. The skill rating is then translated to its equivalent percentage value, which ranges from +15 percent for superskill to −22 percent for poor skill. This percentage is then combined algebraically with the ratings for effort, conditions, and consistency, to arrive at the final rating, or performance-rating factor.

This rating method defines *effort* as a "demonstration of the will to work effectively." Effort is representative of the speed with which skill is applied, and can be controlled to a high degree by the operator. When evaluating the operator's effort, the observer must rate only the "effective" effort, because occasionally an operator will apply misdirected effort to increase the cycle time of the study.

Table 11.2 Westinghouse System Skill Ratings

+0.15	A1	Superskill
+0.13	A2	Superskill
+0.11	B1	Excellent
+0.08	B2	Excellent
+0.06	C1	Good
+0.03	C2	Good
0.00	D	Average
−0.05	E1	Fair
−0.10	E2	Fair
−0.16	F1	Poor
−0.22	F2	Poor

Source: Lowry, Maynard, and Stegemerten (1940), p. 233.

Table 11.3 Westinghouse System Effort Ratings

+0.13	A1	Excessive
+0.12	A2	Excessive
+0.10	B1	Excellent
+0.08	B2	Excellent
+0.05	C1	Good
+0.02	C2	Good
0.00	D	Average
−0.04	E1	Fair
−0.08	E2	Fair
−0.12	F1	Poor
−0.17	F2	Poor

Source: Lowry, Maynard, and Stegemerten (1940), p. 233.

The six effort classes for rating purposes are poor, fair, average, good, excellent, and excessive. Excessive effort is assigned a value of +13 percent, and poor effort, a value of −17 percent. Table 11.3 gives the numerical values for the different degrees of effort and outlines the characteristics of the various categories.

The *conditions* referred to in this performance rating procedure affect the operator and not the operation, and they include temperature, ventilation, light, and noise. Thus, if the temperature at a given workstation is 60°F, yet it is customarily maintained at 68 to 74°F, the conditions will be rated as lower than normal. Factors that affect the operation, such as poor tools or materials, would not be considered when applying the performance factor for working conditions.

The six general classes of conditions, with values ranging from +6 percent to −7 percent, are ideal, excellent, good, average, fair, and poor. Table 11.4 gives the respective values for these conditions.

The last of the four factors that influence the performance rating is operator *consistency*. Unless the analyst uses the snapback method, or makes and records successive subtractions as the study progresses, the consistency of the operator must be evaluated as the study is worked up. Elemental time values that constantly repeat would have perfect consistency. This situation occurs very infrequently, as there always tends to be some variability due to material hardness, tool cutting

Table 11.4 Westinghouse System
Condition Ratings

+0.06	A	Ideal
+0.04	B	Excellent
+0.02	C	Good
0.00	D	Average
−0.03	E	Fair
−0.07	F	Poor

Source: Lowry, Maynard, and Stegemerten (1940), p. 233.

Table 11.5 Westinghouse System
Consistency Ratings

+0.04	A	Perfect
+0.03	B	Excellent
+0.01	C	Good
0.00	D	Average
−0.02	E	Fair
−0.04	F	Poor

Source: Lowry, Maynard, and Stegemerten (1940), p. 233.

edge, lubricant, erroneous watch readings, and foreign elements. Elements that are mechanically controlled will have near-perfect consistency and are rated 100.

The six classes of consistency are perfect, excellent, good, average, fair, and poor. Perfect consistency is rated +4 percent, and poor consistency is rated −4 percent, with the other classes falling in between these values. Table 11.5 summarizes these values.

Once the skill, effort, conditions, and consistency of the operation have been assigned, and their equivalent numerical values established, analysts can determine the overall performance factor by algebraically combining the four values and adding their sum to unity. For example, if a given job is rated C2 on skill, C1 on effort, D on conditions, and E on consistency, the performance factor would be as follows:

Skill	C2	+0.03
Effort	C1	+0.05
Conditions	D	+0.00
Consistency	E	−0.02
Algebraic sum		+0.06
Performance factor		1.06

Many companies have modified the Westinghouse system to include only the skill and effort factors of the overall rating. They contend that consistency is very closely allied to skill, and that conditions are rated average in most instances. If conditions deviate substantially from normal, the study could be postponed, or the effect of the unusual conditions could be taken into consideration in the application of the allowance (see Section 11.10).

The Westinghouse rating system demands considerable training to differentiate the levels of each attribute. It is appropriate for either cycle rating or overall study rating but not for elemental rating because except for very long elements, analysts would not have time to evaluate the dexterity, effectiveness, and physical application of each element. Also in the author's opinion, a rating system that is simple, concise, easily explained, and keyed to well-established benchmarks is more successful than a complex rating system, such as the Westinghouse, requiring involved adjustment factors and computational techniques that may confuse the average shop employee.

SYNTHETIC RATING

In an effort to develop a rating method that would not rely on the judgment of a time study observer and would give consistent results, Morrow (1946) established a procedure known as *synthetic rating*. The synthetic rating procedure determines a performance factor for representative effort elements of the work cycle by comparing actual elemental observed times to times developed through fundamental motion data (see Chapter 13). Thus, the performance factor may be expressed algebraically as

$$P = \frac{F_t}{O}$$

where P = performance or rating factor
F_t = fundamental motion time
O = observed mean elemental time for elements used in F_t

This factor would then be applied to the remainder of the manually controlled elements comprising the study. Again, machine-controlled elements are not rated. A typical illustration of synthetic rating appears in Table 11.6.

For element 1,

$$P = 0.096/0.08 = 120\%$$

and for element 4,

$$P = 0.278/0.22 = 126\%$$

The mean of these is 123 percent, which is the rating factor used for all effort elements.

More than one element should be used to establish a synthetic rating factor, because research has proved that operator performance varies significantly from element to element, especially in complex work. Unfortunately, a major objection to the synthetic rating procedure is the time required to develop the predetermined time system analysis of the given job.

OBJECTIVE RATING

The *objective rating* method, developed by Mundel and Danner (1994), eliminates the difficulty of establishing a standard pace for every type of work. This

Table 11.6 Examples of Synthetic Ratings

Element no.	Observed average time (min)	Element type	Fundamental motion time (min)	Performance factor
1	0.08	Manual	0.096	123
2	0.15	Manual	—	123
3	0.05	Manual	—	123
4	0.22	Manual	0.278	123
5	1.41	Power-fed	—	100
6	0.07	Manual	—	123
7	0.11	Manual	—	123
8	0.38	Power-fed	—	100
9	0.14	Manual	—	123
10	0.06	Manual	—	123
11	0.20	Manual	—	123
12	0.06	Manual	—	123

procedure establishes a single work assignment to which the pace of all other jobs is compared. After the judgment of pace, a secondary factor assigned to the job indicates its relative difficulty. Factors influencing the difficulty adjustment are (1) amount of body used, (2) foot pedals, (3) bimanualness, (4) eye–hand coordination, (5) handling or sensory requirements, and (6) weight handled or resistance encountered.

Numerical values, resulting from experiments, have been assigned for a range of each factor. The sum of the numerical values for each of the six factors comprises the secondary adjustment. The rating R can thus be expressed as follows:

$$R = P \times D$$

where　P = pace rating factor
　　　　D = job difficulty adjustment factor

This performance rating procedure gives consistent results. Comparing the pace of the operation under study to an operation completely familiar to the observer is easier than simultaneously judging all the attributes of an operation and comparing them to a concept of normal for that specific job. The secondary factor does not affect inconsistency, since this factor merely adjusts the rated time by some percentage. Tables of percentage values for the effects of various difficulties in the operation performed are given in Mundel and Danner (1994).

11.4 RATING APPLICATION AND ANALYSIS

The value of a rating is written in the R column of the time study form. Typically, the decimal point is omitted, and a whole number (i.e., percent) value is written to save time. After the stopwatch phase is complete, the analyst multiplies the observed time (OT) by the rating R, scaled by 100, to yield the normal time (NT):

$$NT = OT \times R/100$$

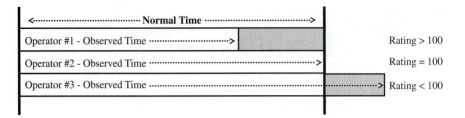

Figure 11.2   Relationships of observed time, ratings, and normal time.

In effect, this rates the operator's performance compared to that of a qualified operator working at the standard performance pace without overexertion, while adhering to the correct method (see Figure 11.2).

The performance rating plan that is easiest to apply, easiest to explain, and gives the most valid results is straight speed rating, augmented by synthetic benchmarks. As has been explained, 100 is considered normal in this procedure, and performance greater than normal is indicated by values directly proportional to 100. The speed rating scale usually covers a range of from 50 to 150. Operators performing outside this 3-to-1 productivity range may be studied, but this is not recommended. The closer the performance is to standard, the better the chance of achieving a fair normal time.

Four criteria determine whether time study analysts using speed rating can consistently establish values within 5 percent of the rating average calculated by a group of trained analysts:

1. Experience in the class of work performed
2. Use of synthetic benchmarks on at least two of the elements performed
3. Selection of an operator who gives performances somewhere between 85 and 115 percent of standard pace
4. Use of the mean value of three or more independent studies and/or different operators

Experience with the class of work performed is critical. The analyst should be sufficiently familiar with the work to understand every detail of the method being used. For example, on an assembly job using a fixture, the observer should be familiar with the difficulty in positioning the components in the fixture, should know the class of fit between all mating parts, and should have a clear understanding of the relationship between time and class of fit. The observer should also know the proper sequence of events and the weights of all the parts handled. This does not necessarily mean that the analyst must have been an actual operator in the work being studied, although this would be desirable. On the other hand, an analyst with 10 years' experience in the metal trades would have considerable difficulty establishing standards in the garment industry.

Whenever more than one operator is available to be studied, the analyst should select the one who is thoroughly experienced on the job, has a reputation of being receptive to time study practice, and consistently performs at a pace near or slightly better than standard. The closer the operator performs to a standard pace, the easier it will be to rate him or her. For example, if 0.50 min is normal for dealing a deck of cards into four bridge hands, performance within ±15 percent of this standard would be fairly easy to identify. However, a performance 50 percent faster or slower than normal would cause considerable difficulty in establishing an accurate rating factor for the performance demonstrated.

The accuracy of the analyst's rating can be checked using synthetic standards. In the time study of Figure 10.8, the average of observed time for element 1 is 2.07/15 = 0.138 min. When 0.138 min is divided by the MTM conversion factor of 0.0006, an observed time of 230 TMUs (see Chapter 13) is obtained. The fundamental motion time for element 1 is 255 TMUs, which yields a synthetic rating of 255/230 = 111 percent. In this particular study, the analyst was on target with a speed rating of 110 percent, and the synthetic rating was used as a validation of that judgment.

For rating the overall study, the analyst should take three or more independent samples before arriving at the standard. These can be made on the same operator at different times of the day, or on different operators. The point is that as the number of samples increases, compensating errors diminish the overall error. A recent example clarifies this point. One author, along with two other trained industrial engineers, reviewed performance rating training films involving 15 different operations. The results are shown in Table 11.7. The standard ratings were not disclosed until all 15 operations had been rated. The average deviation of

Table 11.7 Performance Ratings of Three Different Engineers Observing 15 Different Operations

Operation	Standard rating	Engineer A		Engineer B		Engineer C		Average of Engineers A, B, C	
		Rating	Deviation	Rating	Deviation	Rating	Deviation	Rating	Deviation
1	110	110	0	115	+5	100	−10	108	−2
2	150	140	−10	130	−20	125	−25	132	−18
3	90	110	+20	100	+10	105	+15	105	+15
4	100	100	0	100	0	100	0	100	0
5	130	120	−10	130	0	115	−15	122	−8
6	120	140	+20	120	0	105	−15	122	+2
7	65	70	+5	70	+5	95	+30	78	+13
8	105	100	−5	110	+5	100	−5	103	−2
9	140	160	+20	145	+5	145	+5	150	+10
10	115	125	+10	125	+10	110	−5	120	+5
11	115	110	−5	120	+10	115	0	115	0
12	125	125	0	125	0	115	−10	122	−3
13	100	100	0	85	−15	110	+10	98	−2
14	65	55	−10	70	+5	90	+25	72	+7
15	150	160	+10	140	−10	140	−10	147	−3
Average of 15 operations	112	115.0		112.3		111.3		112.9	
Average deviation	0	+3.0		+0.7		−0.7		+0.9	

all three engineers for the 15 different operations was only 0.9 on a speed rating scale with a range of 50 points to 150 points. Yet, engineer C was 30.0 high on operation 7 and engineer B was 20.0 low on operation 2. When the known ratings were within the 70 to 130 range, the average rating of the three analysts exceeded ± 5 points of the known rating in only one case (operation 3).

11.5   RATING TRAINING

To be successful, analysts must develop track records for setting accurate standards that are accepted by both labor and management. To maintain the respect of all parties, the rates must be consistent. This is especially the case for a speed rating.

In general, when studying operators performing somewhere in the range of 0.70 to 1.30 of standard, good analysts should regularly establish standards within ± 5 percent of the true rate. Thus, if several operators are performing the same job, and different analysts, each studying a different operator, establish time standards on the job, then the resulting standard from each study should be within ± 5 percent of the mean of the group of studies.

To ensure speed rating consistency, both with their own rates and with the rates established by the others, analysts should continually participate in training programs. Such training should be more intense for the neophyte time study analyst. One of the most widely used training methods is the observation of videotapes illustrating diverse operations performed at different productivity levels. Each film has a known level of performance. (Selected training videotapes are listed at the end of this chapter.) After the film is shown, the correct speed rating is compared with the values established independently by the trainees. If the analysts' values deviate substantially from the correct value, then specific information should be put forward to justify the rating. For example, the analyst may have underrated the operator due to an apparently effortless sequence of motions, whereas the operator's smooth, rhythmic blending of movements may really have been an indication of high dexterity and manipulative ability.

As successive operations are reviewed, analysts should plot their ratings against the known values (see Figure 11.3). A straight line indicates perfection, whereas high irregularities on both sides of the line indicate inconsistency as well as an inability to evaluate performance. In Figure 11.3, the analyst rated the first film 75, but the correct rating was 55. The second was rated 80, while the proper rating was 70. In all but the first case, the analyst was within the company's established area of correct rating. Note that, due to the nature of confidence intervals, the ± 5 percent accuracy criterion is valid only around 100 percent, or standard performance. When performance is below 70 percent of standard or above 130 percent of standard, an experienced time study analyst would expect an error much larger than 5 percent.

It is also helpful to plot successive ratings on the x axis and indicate the positive or negative magnitude of deviation from the correct value on the y axis (see Figure 11.4). The closer the time study analyst's rating comes to the x axis, the more nearly correct he or she is.

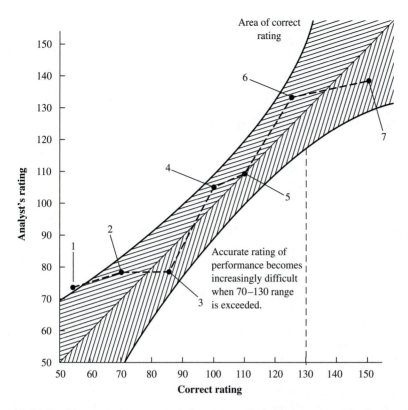

Figure 11.3 Chart showing a record of seven studies with the analyst tending to rate a little high on studies 1, 2, 4, and 6 and a little low on studies 3 and 7. Only study 1 was rated outside the range of desired accuracy.

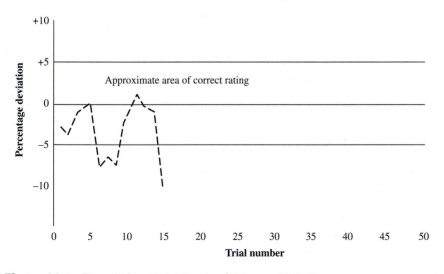

Figure 11.4 Record of an analyst's rating factors on 15 studies.

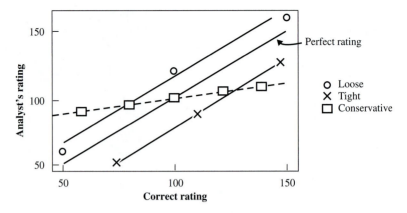

Figure 11.5 Examples of loose, tight, and conservative rater.

A recent statistical study involving the performance rating of 6,720 individual operations by a group of 19 analysts over a period of approximately 2 years produced several interesting observations. Analysts overrated low performance levels and underrated high performance levels. This is typical of novice raters who tend to be conservative raters and are afraid to deviate too far from standard performance. In statistical applications, this tendency is termed *regression to the mean* and results in a relatively flat line compared to the expected line with a slope of 1 (see Figure 11.5). The novice rater who rates higher than the true value for performances below standard performance produces a *loose rate.* The result is a standard time that is too easy for operators to achieve, which means the company would lose money on that operation. For performances above the standard, a novice rater who rates lower than the true value produces a *tight rate,* which results in a time that is difficult for operators to achieve. Some analysts, even after years of experience, consistently tend to rate either tight or loose.

The statistical study also concluded that the operation being examined has an effect on errors in rating performance. Complex operations tend to be more difficult to performance-rate than simpler operations, even for experienced analysts. At low performance levels, overrating is greater for difficult operations than for simple operations, while at high performance levels, underrating is greater for the easy-to-perform operations.

11.6 ALLOWANCES

The watch readings of any time study are taken over a relatively short time. Therefore, normal time does not include unavoidable delays, which may not even have been observed, and other legitimate lost time. Consequently, analysts must make some adjustment to compensate for such losses. The application of these adjustments, or *allowances,* may be considerably broader in some companies than in others. For example, Table 11.8 reveals the items that 42 firms included in their allowances.

Table 11.8 Typical Industrial Allowances

Allowance factor	No. of firms	Percentage of firms
1. Fatigue..........................	39	93
A. General	19	45
B. Rest periods...................	13	31
Did not specify A or B	7	17
2. Time required to learn..............	3	7
3. Unavoidable delay.................	35	83
A. Operator......................	1	2
B. Machine......................	7	17
C. Both operator and machine	21	50
Did not specify A, B, or C.........	6	14
4. Personal needs....................	32	76
5. Setup or preparation operations	24	57
6. Irregular or unusual operations	16	38

Source: Hummel (1935).

Allowances are applied to three parts of the study: (1) the total cycle time, (2) machine time only, and (3) manual effort time only. Allowances applicable to the total cycle time are expressed as a percentage of the cycle time, and compensate for such delays as personal needs, cleaning the workstation, and oiling the machine. Machine time allowances include time for tool maintenance and power variance, while representative delays covered by effort allowances are fatigue and certain unavoidable delays.

Two methods are frequently used for developing standard allowance data. One is direct observation, which requires observers to study two, or perhaps three, operations over a long time. Observers record the duration of and reason for each idle interval. After establishing a reasonably representative sample, observers summarize their findings to determine the percent allowance for each applicable characteristic. Data obtained in this fashion, like those for any time study, must be adjusted to standard performance. Since observers must spend a long time directly observing one or more operations, this method is exceptionally tedious, not only to analysts, but also to the operators. Another disadvantage is the tendency to take too small a sample, which may yield biased results.

The second technique involves work sampling studies (see Chapter 14). This method requires taking a large number of random observations, thus requiring only part-time, intermittent services of the observer. In this method, no stopwatch is used, as observers merely walk through the area under study at random times and note briefly what each operator is doing. The number of delays recorded, divided by the total number of observations during which the operator is engaged in productive work, approximates the allowance required by the operator to accommodate the delays encountered.

Figure 11.6 attempts to provide a scheme for ordering the various types of allowances according to function. The main division is fatigue versus special allowances. *Fatigue allowances,* as the name implies, provide time for the worker to recover from fatigue incurred as a result of the job or work environment, and these

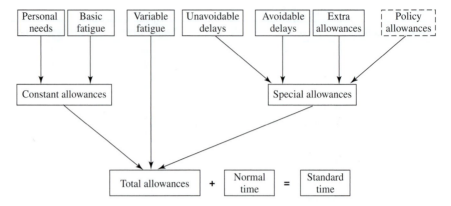

Figure 11.6   Types of Allowances.

are subdivided into *constant* and *variable fatigue allowances. Special allowances* include many different factors related to the process, equipment, and materials and are termed *unavoidable delays, avoidable delays, extra,* and *policy allowances.*

11.7 CONSTANT ALLOWANCES

PERSONAL NEEDS

Personal needs include those cessations in work necessary for maintaining the general well-being of the employee; examples are trips to the drinking fountain and the restroom. The general working conditions and class of work influence the time necessary for personal delays. For example, working conditions involving heavy work performed at high temperatures, such as that done in the pressroom of a rubber-molding department or in a hot-forge shop, would require greater allowance for personal needs than light work performed in comfortable temperature areas.

There is no scientific basis for a numerical percent to give; however, detailed production checks have demonstrated that a 5 percent allowance for personal time, or approximately 24 min in 8 h, is appropriate for typical shop working conditions. Lazarus (1968) reported that of 235 plants in 23 industries, the personal allowance ranged from 4.6 to 6.5 percent. Thus, the 5 percent figure appears to be adequate for the majority of workers.

BASIC FATIGUE

The *basic fatigue allowance* is a constant to account for the energy expended to carry out the work and to alleviate monotony. A value of 4 percent of normal time is considered adequate for an operator who is doing light work, while seated, under good working conditions, with no special demands on the sensory or motor systems (ILO, 1957).

Between the 5 percent personal needs allowance and the 4 percent basic fatigue allowance, most operators are given an initial 9 percent constant allowance, to which other allowances may be added, if necessary.

11.8 VARIABLE FATIGUE ALLOWANCES

BASIC PRINCIPLES

Closely associated with the allowance for personal needs is the allowance for fatigue, although this allowance is usually applied only to the effort portions of the study. Fatigue is not homogeneous in any respect. It ranges from strictly physical to purely psychological or combinations of the two. The result is a decrease in the will to work. The major factors that affect fatigue include working conditions, especially noise, heat, and humidity; the nature of the work, such as posture, muscular exertion, and tediousness; and the general health of the worker. Although heavy manual work, and thus muscular fatigue, is diminishing in industry, due to mechanization, other fatigue components, such as mental stress and tediousness, may be increasing. Because not all fatigue can be eliminated, proper allowance must be made for the working conditions and repetitiveness of the work.

One method of determining the fatigue allowance is to measure the decline in production throughout the working period. The production rate for every quarter of an hour during the course of the working day may be measured. Any decline in production that cannot be attributed to methods changes or personal or unavoidable delays may be attributed to fatigue and expressed as a percentage. Brey (1928) expressed the coefficient of fatigue as follows:

$$F = (T - t) \times 100/T$$

where F = coefficient of fatigue
T = time required to perform operation at end of continuous work
t = time required to perform operation at beginning of continuous work

Many attempts have been made to measure this fatigue through various physical, chemical, and physiological means, none of which have so far been completely successful. Therefore, the International Labour Office (ILO, 1957) has tabulated the effect of various working conditions, to arrive at appropriate allowance factors (see Table 11.9) with a more detailed point system introduced in a later edition (ILO, 1979). These factors include standing versus sitting, abnormal positions, use of force, illumination, atmospheric conditions, required job attention, noise level, mental strain, monotony, and tediousness. To use this table, the analyst would determine the allowance factors for each element of the study and then sum them for a total variable fatigue allowance, which is then added to the constant fatigue allowance.

Table 11.9 ILO Recommended Allowances

A. Constant allowances:
    1. Personal allowance 5
    2. Basic fatigue allowance 4
B. Variable allowances:
    1. Standing allowance 2
    2. Abnormal position allowance:
       a. Slightly awkward 0
       b. Awkward (bending) 2
       c. Very awkward (lying, stretching) 7
    3. Use of force, or muscular energy (lifting, pulling, or pushing):
       Weight lifted, lb:
         5 0
        10 1
        15 2
        20 3
        25 4
        30 5
        35 7
        40 9
        45 11
        50 13
        60 17
        70 22
    4. Bad light:
       a Slightly below recommended 0
       b. Well below 2
       c. Quite inadequate 5
    5. Atmospheric conditions (heat and humidity)—variable 0–100
    6. Close attention:
       a. Fairly fine work 0
       b. Fine or exacting 2
       c. Very fine or very exacting 5
    7. Noise level:
       a. Continuous 0
       b. Intermittent—loud 2
       c. Intermittent—very loud 5
       d. High-pitched—loud 5
    8. Mental strain:
       a. Fairly complex process 1
       b. Complex or wide span of attention 4
       c. Very complex 8
    9. Monotony:
       a. Low 0
       b. Medium 1
       c. High 4
  10. Tediousness:
       a. Rather tedious 0
       b. Tedious 2
       c. Very tedious 5

These ILO recommendations were developed through consensus agreements between management and workers across many industries and have not been directly substantiated. On the other hand, since the 1960s, much work has been done in developing specific standards for the health and safety of the U.S. worker. Here, we examine how well these standards compare with the ILO fatigue allowances.

ABNORMAL POSTURE

Allowances for posture are based on metabolic considerations and can be supported by metabolic models that have been developed for various activities (Garg, Chaffin, and Herrin, 1978). Three basic equations for sitting, standing, and bending can be used to predict and compare the energy expended for various postures. Using an average adult (both male and female) body weight of 152 lb (69 kg) and adding an additional energy expenditure of 2.2 kcal/min for manual hand work (Garg, Chaffin, and Herrin, 1978), we obtain energy expenditures of 3.8, 3.86, and 4.16 kcal/min, for sitting, standing, and bending, respectively. Since sitting is a basic comfortable posture that can be maintained for extended periods, the other postures are compared to sitting. The ratio of standing to sitting energy expenditures is 1.02, or an allowance of 2 percent, while the ratio of bending to sitting energy expenditures is 1.10, or an allowance of 10 percent. The first is identical to the ILO recommendations. The second is slightly larger than the ILO value of 7 percent, but may represent an extreme case of posture, which cannot be maintained for an extended time.

MUSCULAR FORCE

Fatigue, allowances can be formulated on two important physiological principles: muscle fatigue and muscle recovery after fatigue. The most immediate result of muscle fatigue is the significant reduction in muscle strength. Rohmert (1960) quantified these principles as follows:

1. Reduction in maximum strength occurs if the static holding force exceeds 15 percent of maximum strength.
2. The longer the static muscular contraction, the greater the reduction in muscle strength.
3. Individual or specific muscle variations are minimized if forces are normalized to the individual's maximum strength for that muscle.
4. Recovery is a function of the degree of fatigue; that is, a given percent decrease in maximal strength will require a given amount of recovery.

These concepts of fatigue and recovery were further quantified by Rohmert (1973) into a series of curves for rest allowances RA as a function of force and holding time (see Figure 11.7):

$$RA = 1{,}800 \times (t/T)^{1.4} \times (f/F - 0.15)^{0.5}$$

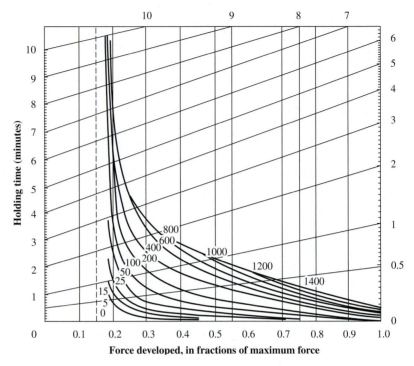

Figure 11.7 Percentage rest allowances for various combinations of holding forces and time.
(*From*: Rohmert, 1973).

where   RA = rest allowance (% of time t)
          t = duration of holding time (min)
          f = holding force (lb)
          F = maximum holding force (lb)
          T = maximum holding time for holding force f (min), defined as

$$T = 1.2/(f/F - 0.15)^{0.618} - 1.21$$

The maximum holding force F can be approximated based on data collected on 1,522 industrial male and female workers (Chaffin, Freivalds, and Evans, 1987). The average of the three basic standardized lifting strengths (arm, leg, torso) is approximately 100 lb (45.5 kg). Using this value of maximum holding force for infrequent lifts (less than one lift every 5 min) of short duration yields the allowances tabulated in Table 11.10. For more frequent lifts (more than one lift every 5 min), metabolic considerations predominate, and the NIOSH lifting guidelines (see Section 4.4) should be utilized in determining the limitation for lifting. Also, loads above 51 lb (23.2 kg) are not allowed by the NIOSH lifting guidelines.

For muscular energy, we consider the formula for determining the amount of rest required for heavy work presented in Section 4.4:

Table 11.10 Comparison of ILO and Calculated Allowances for Use of Muscular Force

Load(lb)	ILO	Calculated
5	0	0
10	1	0
15	2	0
20	3	0.5
25	4	1.3
30	5	2.7
35	7	4.5
40	9	7.0
45	11	10.2
50	13	14.4
60	17	NA
70	22	NA

$$R = (W - 5.33)/(W - 1.33)$$

where R = time required for rest, as percentage of total time
          W = average energy expenditure during work (kcal/min)

This expression can be reformulated into a rest allowance:

$$RA = (\Delta W/4 - 1) \times 100$$

where RA = rest allowance as percentage added to normal time
          ΔW = energy expenditure increment for work = $W - 1.33$ kcal/min

Measuring the heart rate is easier than measuring energy expenditure. The rest allowance can therefore be reformulated with the heart rate as follows:

$$RA = (\Delta HR/40 - 1) \times 100$$

where RA = rest allowance, as percentage added to work time
          ΔHR = difference between working heart rate and resting heart rate

---

Calculation of Rest Allowance for Infrequent Use of Muscular Force

EXAMPLE 11.1

Consider a worker lifting a 40-lb (18.2-kg) load less than once every 5 min. First, calculate the maximum holding time for a load that is 40/100 = 40 percent of an average person's maximum strength capability:

$$T = 1.2/(0.4 - 0.15)^{0.618} - 1.21 = 1.62 \text{ min}$$

Then substitute a short-duration exertion of 0.05 min and a maximum holding time of 1.62 min into the rest allowance equation:

$$RA = 1{,}800 \times (0.05/1.62)^{1.4} \times (0.4 - 0.15)^{0.5}$$
$$= 1{,}800 \times (0.0768) \times (0.5) = 6.96 \approx 7\%$$

The resulting variable rest allowance of 7 percent is added to the typical 9 percent constant allowance for a total allowance of 16 percent.

ATMOSPHERIC CONDITIONS

Modeling the human body and its responses to atmospheric conditions is a very difficult task. Many attempts have been made to combine the physiological manifestations and the changes of several environmental conditions into one simple index (Freivalds, 1987). However, no such index can suffice, and considerable variability in allowances can result. The ILO allowances are based on an outdated concept of cooling power, and they greatly underpredict required rest allowances. Thus, there is considerable deviation in the ILO allowances from true stress levels. This is explained in detail in Freivalds and Goldberg (1988).

A better approach is to consider the more recent guidelines developed by NIOSH (1986), utilizing the wet-bulb globe temperature (WBGT, described in Section 6.3) and working energy expenditures. The resulting fatigue allowances for unacclimatized workers (from Figure 6.13) can be quantified through a least-squares regression, resulting in

$$RA = e^{(-41.5 + 0.0161W + 0.497WBGT)}$$

where W = working energy expenditure (kcal/h)
WBGT = wet-bulb globe temperature (°F)

EXAMPLE 11.2   Calculation of Rest Allowance for Overall Fatigue

Consider the strenuous task shown in fig 4.20 for shoveling coal into a furnace, with an energy expenditure of 10.2 kcal/min. The required rest allowance RA is

$$RA = [(10.2 - 1.33)/4 - 1] \times 100 = 122\%$$

Therefore, to provide adequate time for recovery from fatigue, the worker would need to spend more than 4 h out of an 8-h shift resting. Note that the acceptable ΔW for males is $5.33 - 1.33 = 4$ kcal/min. (For females, substitute the values 4.0 for 5.33 and 1.0 for 1.33.)

NOISE LEVEL

The Occupational Safety and Health Administration (OSHA, 1983) has established permissible noise exposures for workers in industry. The permissible levels depend on the duration of the exposure, as shown in Table 11.11.

Should the total daily exposure consist of exposures to several different noise levels, then the combined exposure is calculated using the equation

$$D = C_1/T_1 + C_2/T_2 + \cdots \leq 1$$

where D = noise dose (decimal value)
C = time spent at specified noise level (h)
T = time permitted (Table 11.11) at specified noise level (h)

and the required relaxation allowance (in %) is simply

$$RA = 100 \times (D - 1)$$

Table 11.11 OSHA-Permissible Noise Levels

Noise level (dBA)	Permissible time (h)
80	32
85	16
90	8
95	4
100	2
105	1
110	0.5
115	0.25
120	0.125
125	0.063
130	0.031

Calculation of a Relaxation Allowance for Atmospheric Conditions

EXAMPLE 11.3

A worker is performing manual assembly at a seated workstation and is expending roughly 200 kcal/h. If the $WBGT_{IN}$ is 88.5°F, then

$$RA = e^{[-41.5 + 0.0161(200) + 0.497(88.5)]} + e^{5.7045} \approx 300\%$$

Based on a relaxation allowance of 300 percent, the worker would need 45 min of rest for each 15 min of work.

Thus, the total exposure to various noise levels cannot exceed a 100-percent dose. For example, a worker may be exposed to 95 dBA noise for 3 h and 90 dBA noise for 5 h. Although each exposure is separately permissible, the combined dose is not:

$$D = 3/4 + 5/8 = 1.375 > 1$$

Therefore, the required relaxation allowance for OSHA compliance is

$$RA = 100 \times (1.375 - 1) = 37.5\%$$

Thus, 90 dBA is the maximum permissible level for an 8-h day, and any sound level above 90 dBA will require a relaxation allowance.

For computation of the noise dose, all sound levels between 80 and 130 dBA must be included in the computations (although continuous levels above 115 dBA are not allowed at all). Since Table 11.11 provides only certain key times, the following computational formula can be used for intermediate noise levels:

$$T = \frac{32}{2^{(L-80)/5}}$$

where L = noise level (dBA)

ILLUMINATION LEVELS

Reconciliation of the ILO (1957) allowances and the IESNA (1995) recommended illumination levels (see Section 6.1) can be approached as follows. For rest allowances, a task that is slightly below recommended guidelines can be considered to be within the same illumination subcategory, perhaps slightly substandard, at the low end of the range, and is assigned a 0 percent allowance. A task that is well below adequate illumination may be defined as being one subcategory beneath its recommended illumination and is assigned a 2 percent allowance. A task with quite inadequate illumination may be defined as being two or more subcategories below its recommended level and receives a 5 percent allowance. These definitions are fairly realistic, as human perceptions of illumination are logarithmic; that is, as illuminance increases, we require a greater intensity difference before a change is noted (IES, 1981).

EXAMPLE 11.4	Calculation of a Relaxation Allowance for Noise

A worker is exposed to the noise levels shown in Table 11.12 during an 8-h workday. The last entry is obtained as follows:

$$T = \frac{32}{2^{(96-80)/5}} = 3.48$$

The noise dose is

$$D = 1/32 + 4/8 + 3/3.48 = 1.393$$

and the rest allowance is

$$RA = 100 \times (1.393 - 1) = 39.3\%$$

Since the 8-h noise exposure dosage exceeds OSHA requirements, the worker must be provided with a 39.3-percent rest allowance. Note that, again, the ILO-recommended values greatly underpredict required allowances.

Table 11.12 Noise Levels Over an 8-h Day

Noise level L (dBA)	Time spent C (h)	Time permitted $T/$(h)
80	1	32
90	4	8
96	3	3.48

The literature contains some evidence that increased task illumination results in better-skilled performance. The most germane performance measure is task completion time under varying illumination conditions. However, it should be kept in mind that performance accuracy is also important. For example, Bennett,

Table 11.13 Modeled Reading Times as a function of illumination

Illumination (fc)	Modeled time (s)	% Change from 75 fc	ILO category	Allowance (%)
75	207.3	—	(Recommended)	0
50	210.0	1.3	Slightly below	0
30	213.9	3.2	Well below	2
20	217.2	4.8	Well below	2
15	219.8	6.0	Inadequate	5
10	223.6	7.9	Inadequate	5

Source: Bennett, Chitlangio, and Pangrekar, 1977.

Chitlangio, and Pangrekar (1977) reported the time to read a 450-word, pencil-written article as a cubic function of task illumination:

$$\text{Time} = 251.8 - 33.96 \log \text{FC} + 6.15 (\log \text{FC})^2 - 0.37 (\log \text{FC})^3$$

where   Time = mean reading time (s)
        FC = task illumination (fc)

The recommended IES (1981) illumination for reading this pencil-written material is category E (from Table 6.2), 50 to 100 fc (500 to 1,000 lx). The weighting factors total zero (from Table 6.3), so the recommended illumination is 75 fc (750 lx), the middle value in category E. Table 11.13 compares modeled reading times as a function of decreasing illumination and allowances. Decreasing the illumination to the next-lower subcategory increased performance times by 3 to 5 percent, which is not too far from the 2 percent ILO allowance for well below recommended lighting. The next-lower illumination subcategory produced times that were 6 to 8 percent greater than that at the recommended levels, which is somewhat greater than the 5 percent allowance for inadequate illumination. Overall, this study (Bennett, Chitlangio, and Pangrekar, 1977) came reasonably close to supporting the ILO allowances.

VISUAL STRAIN

The ILO visual strain rest allowance provides no allowance for fairly fine work, a 2 percent allowance for fine or exacting work, and a 5 percent allowance for very fine or very exacting work. These allowances only refer to the precision of the visual task requirements, without mentioning the other task conditions that have a very large effect on visual requirements: illumination (or luminance), glare, flicker, color, viewing time, and contrast. Therefore, the ILO allowances are only rough approximations. More specific values can be determined by target detectability, as first quantified by Blackwell (1959) in his visibility curves (see Figure 6.3). Four factors have the greatest effect in determining how visible a target in a task will be:

1. *Background luminance of the task.* This is the magnitude of the light reflected from the target's background into the eyes of an observer, measured in foot-lamberts (fL).

2. *Contrast.* This is the difference between the luminance levels of the target and the background. The contrast also needs to be adjusted (divided) by the following factors: real-world conditions (2.5), movement of the target (2.78), and uncertainty in the location (1.5).

3. *Time available for observation.* This ranges from a few milliseconds to several seconds, and it can affect the speed and accuracy of performance.

4. *Size of target, measured as a visual angle in arc minutes.*

Blackwell's visibility curves can be modeled by the following equation (with allowable ranges):

$$\%\text{Det} = 81 \times C^{0.2} \times L^{0.045} \times T^{-0.003} \times A^{0.199}$$

where $\%\text{Det}$ = % targets detected (0–100%)
C = contrast (0.001–1.8)
L = background luminance (1–100 fL)
T = viewing time (0.01–1 s)
A = visual angle (1–64 arc min)

The percentage of targets detected may be used to check visual strain allowances by specifying a percentile range for population description abilities. Since highly used percentile ranges are the 50th and 95th percentiles, these may also be applied to target detection, to define relaxation allowance categories. At least a 95 percent target detection defines a visual task without significant problems and can thus define the ILO fairly fine work category with its associated 0 percent allowance. At least a 50 percent target detection defines the fine or exacting work category with its 2 percent allowance. Finally, less than half of targets detected can define very fine or very exacting work with its associated 5 percent allowance.

It must be stressed that Blackwell's model does not directly define the rest allowance. Instead, it defines absolute target detection ability, which in turn can be used to define a rest allowance. Thus, the rest allowance should be inversely proportional to the expected percentage of detected targets.

In general, small visual angles usually produce the lowest performance levels, whereas viewing time only affects performance for the higher contrast levels.

MENTAL STRAIN

Mental strain is very difficult to measure clearly across many types of tasks. For mental workload, standardized measures of performance have not yet been clearly defined, and variability between individuals on the same task is high. Also underlying any mental strain definition is an understanding of the factors that make a task complex, on which models are lacking. Investigation of the basis and adequacy of these relaxation allowances thus necessarily requires (1) an independent indicator of task complexity and (2) objective evidence of changing work output with fatigue or time on the job. Even given this information, experimental differences in motivation can greatly affect observed results, rendering comparisons between studies useless. The vagueness of the ILO relaxation allowance complicates matters even further: 1 percent for a fairly complex process, 4 percent for a process requiring a complex or wide span of attention, and 8 percent for a very complex process. At best, a controlled study with timed reading or mental arithmetic tasks, such as those by Okogbaa and Shell (1986), can serve as a crude check of these allowances. Both of these tasks could be considered complex and requiring wide attention span and thus deserving of a 4 percent rest allowance. However, reading performance decreased at a rate of 3.5 percent per hour, while arithmetic performance decreased at a rate of approximately 2 percent per hour. Thus, the ILO (1957) guidelines support performance decrements due to mental strain for 1 h, but are inadequate for longer periods and may need to be modified.

| Calculation of Rest Allowance for Visual Strain | EXAMPLE 11.5 |

The inspection of resistors on electronic circuit boards may be considered exacting work, which requires a 2 percent allowance based on ILO guidelines. To confirm this, we use the following calculations. The board is viewed at a distance of 12 in, without magnification, and the stripes on each resistor are 0.02 in wide. The required visual angle is $3{,}438 \times 0.02/12 = 5.73$ arc minutes. The resistor body (task background) has a luminance of 10 fL, and the contrast between the stripe and background is 0.5. The contrast is divided by a factor of $1.5 \times 2.5 = 3.75$ to adjust for real-world detection and uncertain location (Freivalds and Goldberg, 1988). The mean time for eye fixation is 0.2 s. Plugging these values into the detectability equation yields

$$\%\text{Det} = 81 \times (0.5/3.75)^{0.2} \times 10^{0.045} \times 0.2^{-0.003} \times 5.73^{0.199}$$
$$= 81 \times 0.668 \times 1.109 \times 1.005 \times 1.414 = 85.3\%$$

A detectability of 85.3 percent is under 95 percent and would require a 2 percent allowance.

MONOTONY

Assignment of a monotony rest allowance, as defined by ILO (1957), is most appropriate as "the result of repeated use of certain mental faculties, as in mental arithmetic." Tasks with low monotony receive no additional allowance; tasks with medium monotony receive 1 percent, and tasks with high monotony receive a 4 percent allowance. Since the cognitive tasks of Okogbaa and Shell (1986) were performed over 4 h, perhaps they should also receive the monotony allowance. However, even the addition of the maximum 4 percent allowance would only extend the period of adequacy to 2 h. Vigilance tasks present another example of monotonous work. Baker, Ware, and Sipowicz (1962) noted that subjects detected 90 percent of short light interruptions in a lamp after 1 h of continuous testing. By the end of 10 h, the subjects were only detecting about 70 percent of signals, or a drop in performance of 2 percent per hour. Again, the ILO allowance is not sufficient to compensate performance decrements that occur over an entire shift, and better allowances need to be developed.

TEDIOUSNESS

Allowances for task tediousness (or task repetition) are 0 percent for a rather tedious task, 2 percent for a tedious task, and 5 percent for a very tedious task. As defined by ILO (1957), this allowance is applied to elements in which there is "repeated use of certain members of the body, such as fingers, hands, arms or legs." In other words, a tedious task repeatedly utilizes the same physical movements, whereas a monotonous task repeatedly uses the same mental faculties. A methods study used to simplify work and make it more efficient also tends to make it more tedious or repetitive for skilled workers, making it more likely that the workers will be prone to work-related musculoskeletal disorders (see Chapter 5).

Developmental work on risk assessment models for CTD (ANSI, 1995; Seth, Weston, and Freivalds, 1999) has found that the frequency of motions, postures of the hand and wrist, and the forces exerted by the hand are key factors in increasing the risk for CTD. However, these relatively crude models are far from reliable and are not validated over a wide range of jobs and industries. Still, epidemiological data from NIOSH (1989) have indicated that 10,000 damaging wrist motions per shift is a threshold point at which CTD cases increase noticeably, and that at 20,000 motions, the number of cases increases significantly. This would seem to imply that 10,000 motions is a limit for unimpaired performance and for relaxation allowances of up to 100 percent, which is much greater than recommended by ILO (1957). Obviously, most of these models are very much in the developmental stages, and considerable validation must be performed before specific values for allowances can be set.

11.9   SPECIAL ALLOWANCES

UNAVOIDABLE DELAYS

This class of delays applies to effort elements and includes interruptions from the supervisor, dispatcher, time study analyst, and others; material irregularities; difficulty in maintaining tolerances and specifications; and interference delays where multiple machine assignments are made.

As can be expected, every operator experiences numerous interruptions during the course of the workday. The supervisor or group leader may interrupt the operator to give instructions or to clarify certain written information. The inspector may interrupt to point out the reasons for some defective work that passed through the operator's workstation. Interruptions also come from planners, expediters, fellow workers, production personnel, and others.

Unavoidable delays are frequently a result of material irregularities. For example, the material may be in the wrong location; or it may be running too soft or too hard, or too short or too long; or it may have excessive stock on it, as in forgings when the dies begin to wash out, or on castings due to incomplete removal of risers. When material deviates substantially from standard specifications, the customary unavoidable delay allowance may prove inadequate.

The analyst must then restudy the job and allow time for the extra elements introduced by the irregular material.

As discussed in Section 2.3, if more than one machine is assigned to an operator during the workday, one or more machines must wait until the operator completes work on another machine. As more machines are assigned to the operator, machine interference time increases. Example 11.6 shows the application of Wright's formula (from Section 2.3) in calculating the allowance for such machine interference.

Allowance for Machine Interference   EXAMPLE 11.6

In a quilling production, an operator is assigned a package of 60 spindles. The mean machine running time per package, determined by stopwatch study, is 150 min. The standard mean servicing time per package, also developed by time study, is 3 min. Machine interference time as calculated from Wright's formula (see Section 2.3) was found to be 1,160 percent of the mean servicing time.

Thus, we would have

Machine running time	150.00 min
Servicing time	3.00 min
Machine interference time	$11.6 \times 3.0 = 34.80$ min
Standard time for 60 packages	187.80 min
Standard time per package	$\dfrac{187.80}{60} = 3.13$ min

Alternatively, the machine interference time can be calculated as a percentage allowance, to which other allowances can be added:

$$\% \text{ allowance} = \frac{34.8}{153} \times 100 = 22.75\%$$

The amount of interference is also related to the performance of the operator. Thus, the operator demonstrating a low level of effort experiences more machine interference than the operator who, through higher effort, reduces the time spent in attending the stopped machine. The analyst determines the normal interference time as calculated from the methods presented in Chapter 2. If the normal interference time is smaller than the observed machine interference time, then the ratio of the two times will yield a measure of operator performance.

AVOIDABLE DELAYS

It is not customary to provide any allowance for avoidable delays, such as visits with other operators for social reasons, uncalled-for work stoppages, and idleness other than rest to overcome fatigue. While operators may take these delays at the expense of output, no allowance for these cessations of work is provided in the development of the standard.

EXTRA ALLOWANCES

In metal fabrication and related operations, the allowance for personal, unavoidable, and fatigue delays usually approximates 15 percent. However, in certain cases, an extra allowance may be needed for a fair standard. For example, for a substandard lot of raw material, analysts may need to add an extra allowance to account for an unduly high number of rejects. Or, a situation may arise in which, because of the breakage of a jib crane, the operator is obliged to place a 50-lb casting in the chuck of the machine. An extra allowance would be needed for the additional fatigue in manually handling the work.

One extra allowance frequently used, especially in the steel industry, is a percentage added to a portion or all of the cycle time to account for the operator observing the process to maintain efficient progress of the operation. This allowance is frequently referred to as *attention time* allowance and may cover such situations as an inspector observing tin plate coming off the line, a first helper observing the conditions of a molten bath or receiving instructions from the melter, or a crane operator receiving directions from the crane hooker. Without this extra allowance, such operators would find it impossible to make the same earnings as fellow employees.

The time required to clean the workstation and lubricate the operator's machine may be classified as an unavoidable delay. When these elements are the responsibilities of the operator, management must provide an applicable allowance. Analysts often include this time as a total cycle time allowance when these functions are performed by the operator. The type and size of equipment and the material being fabricated have a considerable effect on the time required to do these tasks. One company has established a table of allowances to cover these items (see Table 11.14). Sometimes supervisors give operators 10 or 15 min at the end of the day to perform these elements. When this is done, the established standards would not include any allowance for cleaning and oiling the machine.

A tool maintenance allowance provides time for the operator to maintain tools after the original setup. In setup, first-class tools already properly ground are generally provided. However, in long production runs, the tools may need

Table 11.14 Clean Machine Allowance Chart

	Percent per machine		
Item	Large	Medium	Small
1. Clean machine when lubricant is used	1	¾	½
2. Clean machine when lubricant is not used	¾	½	¼
3. Clean and put away large amounts of tools or equipment	½	½	½
4. Clean and put away small amounts of tools or equipment	¼	¼	¼
5. Shut machine down for cleaning (this percentage is for machines equipped with chip pans, which are stopped at intervals to permit sweeper to clean away large chips)	1	¾	½

to be sharpened periodically and the operator should be given an appropriate allowance.

POLICY ALLOWANCES

A policy allowance is used to provide a satisfactory level of earnings for a specified level of performance under exceptional circumstances. Such allowances could cover new employees, the differently abled, workers on light duty, and others. These are typically decided by management, perhaps with union negotiations.

11.10   APPLYING ALLOWANCES

The fundamental purpose of all allowances is to add enough time to normal production time to enable the average worker to meet the standard when performing at standard performance. There are two ways of applying allowances. The most common is to add a percentage to the normal time, so that the allowance is based on a percentage of the productive time only. It is also customary to express the allowance as a multiplier, so that the normal time (NT) can be readily adjusted to the standard time (ST):

$$ST = NT + NT \times \text{allowance} = NT \times (1 + \text{allowance})$$

where   ST = standard time
        NT = normal time

Thus, if a 10 percent allowance were provided on a given operation, the multiplier would be $1 + 0.1 = 1.1$.

For example, the computation of a total allowance might be

Personal needs	5.0 percent
Basic fatigue	4.0
Unavoidable delay	1.0
Total	10.0 percent

Normal time would then be multiplied by 1.1 to determine the standard time. Using the time study example in Figure 10.6, the average normal time of 0.177 min for element 1 is multiplied by 1.1, corresponding to a 10 percent allowance, to yield a standard time of 0.195 min. Out of a 480-min workday, the operator would work $480/1.1 = 436$ min and would be allowed 44 min rest. This rest would be allocated as lunch and other breaks. Note that based on the principles covered in Chapter 4, frequent, short breaks are preferred over infrequent, long breaks.

Some companies apply the percent allowance to the total working day, since the actual production time might not be known. For the previous example, the multiplier of normal times becomes $100/(100 - 10) = 1.11$ (instead of 1.1), and the standard time for element 1 becomes 0.196. Of the 480-min workday, $480 \times 0.1 = 48$ min (instead of 44 min) would be allocated to rest. Although the difference between the two approaches is not large, it could add up to over a year for several hundred workers. This would then become a policy decision for the company.

SUMMARY

Performance rating is a means of adjusting the observed time on a job, so as to derive the time required for a qualified operator to do the job when working at a standard pace. Since the rating is based entirely on the experience, training, and judgment of the time study analyst, there may be criticism of the method. Consequently, many different rating systems have been developed in an attempt to obtain an "objective" system. However, each rating system still ultimately depends on the subjectivity and honesty of the rater. Thus, it is important to train the time study analyst thoroughly in rating properly and consistently. That this can be done successfully has been borne out by many studies.

After ratings have been used to adjust the observed times, allowances need to be added to account for delays and interruptions. Typical allowances used in industry are 5% for personal needs, 4% for basic fatigue, and additional amounts for variable fatigue. Some guidelines are provided in Table 11.15 for allocating variable fatigue allowances in a more quantitative manner than has typically been done. These guidelines are especially appropriate for abnormal position, use of force, atmospheric, and other work environment conditions. The allowances for visual strain, mental strain, monotony, and tediousness are currently less reliable and need to be developed in greater detail. Finally additional allowances for unavoidable delays and extra allowances (e.g., machine cleaning) may need to be added. Note that the analyst must be accurate and consistent in applying these allowances. Otherwise if the allowances are too high, manufacturing costs are unduly inflated; if the allowances are too low, tight standards result, causing poor labor relations and eventual failure of the system.

QUESTIONS

1. Why has industry been unable to develop a universal conception of standard performance?
2. Which factors enter into large variances in operator performance?
3. What are the characteristics of a sound rating system?
4. During a time study, when should a rating be given? Why is this important?
5. What governs the frequency of performance rating during a given study?
6. Explain the Westinghouse system of rating.
7. Under the Westinghouse rating system, why are "conditions" evaluated?
8. What is synthetic rating? What is its principal weakness?
9. What is the basis of speed rating, and how does this method differ from the Westinghouse system?
10. Which four criteria are fundamental for doing a good job in speed rating?
11. Why is training in performance rating a continuous process?
12. Why should more than one element be used in the establishment of a synthetic rating factor?
13. Would there be any objection to studying an operator who is performing at an excessive pace? Why or why not?

Table 11.15 Revised Table of Allowances

Constant Allowances	
Personal Needs	5
Basic Fatigue	4

Variable Relaxation Allowances	
Posture Allowances	
Standing	2
Awkward (bending, lying, crouching)	10
Illumination Levels	
One level (one IES subcategory) below recommended	1
Two levels below recommended	3
Three levels (full IES category) below recommended	5
Visual Strain (Close Attention)	
Fine or exacting work	2
Very fine or exacting work	5
Mental Strain	
First hour	2
Second hour	4
Each succeeding hour	+2
Monotony	
First hour	2
Second hour	4
Each succeeding hour	+2

Use of Muscular Force or Energy
Infrequent lifting, extended static holds
(<1 lift per 5 min) . . . . $RA = 1{,}800 \times (t/T)^{1.4} \times (f/F - 0.15)^{0.5}$,
where $T = 1.2/(f/F - 0.15)^{0.618} - 1.21$

Frequent lifting
(>1 lift per 5 min) . . . . .Use NIOSH Lifting Guidelines with LI < 1.0
Overall whole body activities . . . . $RA = (\Delta HR/40 - 1) \times 100$ or
$RA = (\Delta W/4 - 1) \times 100$

Atmospheric Conditions . . . .$RA = \exp(-41.5 + 0.0161W + 0.497\ \text{WBGT})$

Noise Level . . . .$RA = 100 \times (D - 1)$, where $D = C_1/T_1 + C_2/T_2 + \cdots$

Repetitiveness (Tediousness)
No established standards yet . . . . .Use CTD risk analysis and keep risk index < 1.0

14. In what ways can an operator give the impression of high effort and yet produce at a mediocre or poor level of performance?

15. What main areas are allowances intended to cover?

16. What are the two methods used in developing standard allowance data? Briefly explain the application of each technique.

17. Give several examples of personal delays. Which percentage allowance seems adequate for personal delays, under typical shop conditions?

18. What are some of the major factors that affect fatigue?
19. Which operator interruptions would be covered by the unavoidable delays allowance?
20. What percentage allowance is usually provided for avoidable delays?
21. When are extra allowances provided?
22. Why is fatigue allowance frequently applied only to the effort areas of the work cycle?
23. Why are allowances based on a percentage of the productive time?
24. What are the advantages of having operators oil and clean their own machines?
25. Give several reasons for not applying an extra allowance to operations, if the major part of the cycle is machine-controlled and the internal time is small compared to the cycle time.

PROBLEMS

1. Rate the overall study for each of the following jobs shown on the website.
   a. Stamping end extrusions
   b. Stamping end couplings
   c. Flashlight assembly
   d. Union assembly
   e. Hospital bed rail assembly
   f. Stitching (garments)
   g. Labeling (garments)
   h. Cut and tack (garments)
2. Develop an allowance factor for an assembly element for which the operator stands in a slightly awkward position, regularly lifts a weight of 15 lb, and has good light and atmospheric conditions. The attention required is fine, the noise level is continuous at 70 dBA, and the mental strain is low, as are the monotony and the tediousness of the work.
3. Calculate the fatigue allowance for an operation for which the operator loads and unloads a 25-lb gray iron casting once every 5 min at a height of 30 in.
4. What would be the allowance for problem 3 if the frequency increased to 5 per minute?
5. In XYZ Co., an all-day study revealed the following noise sources: 0.5 h, 100 dBA; 1 h, less than 80 dBA; 3.5 h, 90 dBA; 3 h, 92 dBA. Calculate the rest allowance.
6. What fatigue allowance should be given to a job if it took 1.542 min to perform the operation at the end of continuous work, but only 1.480 min at the beginning of continuous work?
7. Based on the ILO tabulation, what would the allowance factor be on a work element involving a 42-lb pulling force in inadequate light, in which exacting work was required?
8. Calculate the fatigue allowance for a worker shoveling scrap metal into a bin. The operator's working heart rate is approximately 130 beats/min and resting heart rate is 70 beats/min.
9. Calculate the fatigue allowance for a 200-lb worker monitoring a steel furnace while standing next to it. The WBGT index indicates 92°F.

REFERENCES

Åberg, V., K. Elgstrand, P. Magnus, and A. Lindholm. "Analysis of Components and Prediction of Energy Expenditure in Manual Tasks." *The International Journal of Production Research,* 6, no. 3 (1968), pp. 189–196.

ANSI. *Control of Work-Related Cumulative Trauma Disorders—Part I: Upper Extremities.* ANSI Z-365 Working Draft. Itasca, NY: American National Standards Institute, 1995.

Baker, R. A., J. R. Ware, and R. R. Sipowicz. "Signal Detection by Multiple Monitors." *Psychological Record,* 12, no. 2 (April 1962), pp. 133–137.

Barnes, Ralph M. *Motion and Time Study: Design and Measurement of Work.* 7th ed. New York: John Wiley & Sons, 1980.

Bennett, C. A., A. Chitlangia, and A. Pangrekar. "Illumination Levels and Performance of Practical Visual Tasks." *Proceedings of the 21st Annual Meeting of the Human Factors Society* (1977), pp. 322–325.

Blackwell, H. R. "Development and Use of a Quantitative Method for Specification of Interior Illumination Levels on the Basis of Performance Data." *Illuminating Engineer,* 54 (June 1959), pp. 317–353.

Brey, E. E. "Fatigue Research in Its Relation to Time Study Practice." *Proceedings, Time Study Conference.* Chicago, IL: Society of Industrial Engineers, February 14, 1928.

Chaffin, D. B., A. Freivalds, and S. R. Evans. "On the Validity of an Isometric Biomechanical Model of Worker Strengths." *IIE Transactions,* 19, no. 3 (September 1987), pp. 280–288.

Davis, H. L., T. W. Faulkner, and C. L. Miller. "Work Physiology." *Human Factors,* 11, no. 2 (April 1969), pp. 157–165.

Freivalds, A. "Development of an Intelligent Knowledge Base for Heat Stress Evaluation." *International Journal of Industrial Engineering,* 2, no. 1 (November 1987), pp. 27–35.

Freivalds, A., and J. Goldberg. "Specification of Bases for Variable Relaxation Allowances." *The Journal of Methods-Time Measurement,* 14 (1988), pp. 2–29.

Garg, A., D. B. Chaffin, and G. D. Herrin. "Prediction of Metabolic Rates for Manual Materials Handling Jobs." *American Industrial Hygiene Association Journal,* 39, no. 12 (December 1978), pp. 661–674.

Hancock, Walton M. "The Learning Curve." In *Handbook of Industrial Engineering,* 2d ed. Ed. Gavriel Salvendy. New York: John Wiley & Sons, 1992.

Hummel, J. O. P. *Motion and Time Study in Leading American Industrial Establishments* (MS Thesis). University Park, PA: Pennsylvania State University, 1935.

IESNA. *Lighting Handbook,* 8th ed. Ed. M. S. Rea. New York: Illuminating Engineering Society of North America, 1995, pp. 459–478.

ILO. *Introduction to Work Study.* Geneva, Switzerland: International Labour Office, 1957.

ILO. *Introduction to Work Study.* 3d ed. Geneva, Switzerland: International Labour Office, 1979.

Konz, S., and S. Johnson. *Work Design.* 5th ed. Scottsdale, AZ: Holcomb Hathaway Publishers, 2000.

Lazarus, I. "Inaccurate Allowances Are Crippling Work Measurements." *Factory* (April 1968), pp. 77–79.

Lowry, S. M., H. B. Maynard, and G. J. Stegemerten. *Time and Motion Study and Formulas for Wage Incentives.* 3d ed. New York: McGraw-Hill, 1940.

Moodie, Colin L. "Assembly Line Balancing." In *Handbook of Industrial Engineering,* 2d ed. Ed. Gavriel Salvendy. New York: John Wiley & Sons, 1992.

Morrow, R. L. *Time Study and Motion Economy.* New York: Ronald Press, 1946.

MTM Association. *Work Measurement Allowance and Survey.* Fair Lawn, NJ: MTM Association, 1976.

Mundel, Marvin E., and David L. Danner. *Motion and Time Study: Improving Productivity.* 7th ed. Englewood Cliffs, NJ: Prentice-Hall, 1994.

Murrell, K. F. H. *Human Performance in Industry.* New York: Reinhold Publishing, 1965.

Nadler, Gerald. *Work Design: A Systems Concept.* Rev. ed. Homewood, IL: Richard D. Irwin, 1970.

NIOSH. *Criteria for a Recommended Standard for Occupational Exposure to Hot Environments.* Washington, DC: National Institute for Occupational Safety and Health, Superintendent of Documents, 1986.

NIOSH. *Health Hazard Evaluation—Eagle Convex Glass Co.* HETA 89-137-2005. Cincinnati, OH: National Institute for Occupational Safety and Health, 1989.

Okogbaa, O. G., and R. L. Shell. "The Measurement of Knowledge Worker Fatigue." *IIE Transactions,* 12, no. 4 (December 1986), pp. 335–342.

OSHA, "Occupational Noise Exposure: Hearing Conservation Amendment." *Federal Register,* 48 (1983), Washington, DC: Occupational Safety and Health Administration, pp. 9738–9783.

Presgrave, R. W. *The Dynamics of Time Study.* 4th ed. Toronto, Canada: The Ryerson Press, 1957.

Rohmert, W. "Problems in Determining Rest Allowances, Part I: Use of Modern Methods to Evaluate Stress and Strain in Static Muscular Work." *Applied Ergonomics,* 4, no. 2 (June 1973), pp. 91–95.

Rohmert, W. "Ermittlung von Erholungspausen für statische Arbeit des Mensche." *Internationale Zeitschrift für Angewandte Physiologie einschließlich Arbeitsphysiologie,* 18 (1960), pp. 123–140.

Seth, V., R. Weston, and A. Freivalds. "Development of a Cumulative Trauma Disorder Risk Assessment Model." *International Journal of Industrial Ergonomics,* 23, no. 4 (March 1999), pp. 281–291.

Silverstein, B. A., L. J. Fine, and T. J. Armstrong. "Occupational Factors and Carpal Tunnel Syndrome." *American Journal of Industrial Medicine,* 11, no. 3 (1987), pp. 343–358.

Stecke, Kathryn E. "Machine Interference: Assignment of Machines to Operators." In *Handbook of Industrial Engineering,* 2d ed. Ed. Gavriel Salvendy. New York: John Wiley & Sons, 1992.

SELECTED SOFTWARE

Design Tools (available from the McGraw-Hill text website at www.mhhe.com/niebel-freivalds), New York: McGraw-Hill, 2002.

QuickTS (available from the McGraw-Hill text website at www.mhhe.com/niebel-freivalds), New York: McGraw-Hill, 2002.

SELECTED VIDEOTAPES

Fair Day's Work Concepts. 1/2″ VHS, C-Four, 102 E. Main St., P.O. Box 808, Pendleton, SC 29670-0808 (www.c-four.com)

Workplace Fundamentals. 1/2″ VHS, C-Four, 102 E. Main St., P.O. Box 808, Pendleton, SC 29670-0808 (www.c-four.com)

Workplace Rating Exercises. 1/2″ VHS, C-Four, 102 E. Main St., P.O. Box 808, Pendleton, SC 29670-0808 (www.c-four.com)

Standard Data and Formulas

KEY POINTS

- For common work elements, use standard data, tabular or graphical collections of normal times.

- Keep setup and cyclical elements separate.

- Keep constant and variable elements separate.

- Use formulas to provide quick and consistent normal times for variable elements.

- Keep formulas as clear, concise, and simple as possible.

- After summing normal times, add allowances for new standard times.

Standard time data are elemental times obtained from time studies that have been stored for later use. For example, a regularly repeated setup elemental time should not be remeasured for every operation. The principle of applying standard data was established many years ago by Frederick W. Taylor, who proposed that each elemental time be properly indexed so that it could be used to establish future time standards. When we speak of standard data today, we refer to all the tabulated element standards, plots, nomograms, and tables that allow the measurement of a specific job without the use of a timing device, such as a stopwatch.

Standard data can have several levels of refinement: motion, element, and task. The more refined the standard data element, the broader its range of usage. Thus, motion standard data have the greatest application, but it takes longer to develop such a standard than either element or task standard data. Element standard data are widely applicable and allow the faster development of a standard than motion data. This chapter is devoted to element standard data, and Chapter 13 addresses motion standard data and their use to predetermine standard time systems.

A time study formula is an alternative and, typically, simpler presentation of standard data, especially for variable elements. Such formulas have particular application in nonrepetitive work for which it is impractical to establish standards for each job using an individual time study. Formula construction involves the design of an algebraic expression that establishes a time standard in advance of production by substituting known values peculiar to the job for the variable elements.

Work standards calculated from standard data and formulas are relatively consistent in that the tabulated elements result from many proven stopwatch time studies. Since the values are tabulated, it is only necessary to summate the required elements to establish a standard, and all analysts should arrive at identical performance standards for a given method. Consistency therefore is ensured for standards established by different analysts in a plant, as well as for the various standards computed by a given time study observer.

Standards on new work can usually be computed more quickly through standard data or formulas than by a stopwatch time study. This allows the establishment of standards for indirect labor operations, which is usually impractical if stopwatch studies are required. Typically, one work measurement analyst can establish 25 rates per day using standard data or formulas but only five rates per day using stopwatch methods.

12.1 STANDARD TIME DATA DEVELOPMENT

GENERAL APPROACH

To develop standard time data, analysts must distinguish constant elements from variable elements. A constant element is one whose time remains approximately the same, cycle after cycle. A variable element is one whose time varies within a specified range of work. Thus, the element "start machine" would be a constant, while the element "drill 3/8-in-diameter hole" would vary with the depth of the hole, the feed, and the speed of the drill.

Standard data are indexed and filed as they are developed. Also, setup elements are kept separate from elements incorporated into each piece time, and constant elements are separated from variable elements. Typical standard data for machine operation would be tabulated as follows: (1) setup, (*a*) constants, (*b*) variables; (2) each piece, (*a*) constants, (*b*) variables.

Standard data are compiled from different elements in time studies of a given process over a period of time. Only those studies proved valid through use are included in the data. In tabulating standard data, the analyst must be careful to define the endpoints clearly. Otherwise, there may be a time overlap in the recorded data. For example, in the element "out stock to stop" on a bar feed No. 3 Warner & Swasey turret lathe, the element could include reaching for the feed lever, grasping the lever, feeding the bar stock through the collet to a stock stop located in the hex turret, closing the collet, and reaching for the turret handle. Then again, this element may involve only the feeding of bar stock through the collet to a stock stop. Since standard data elements are compiled from a great number of studies taken by different time study observers, the limits or endpoints of each element should be carefully defined.

Any missing values in a standard data tabulation will need to be measured, typically by a stopwatch time study (see Chapter 10). Sometimes, very short individual elements will be difficult if not impossible to measure separately. However, the analyst can determine their individual values by timing groups of elements collectively and using simultaneous equations to solve for the individual elements, as shown in Example 12.1.

| Calculation of Brief Element Times | EXAMPLE 12.1 |

Element a is "pick up small casting," element b is "place in leaf jig," c is "close cover of jig," d "position jig," e "advance spindle," and so on. These elements are timed in groups, as follows:

$$a + b + c = \text{element 1} = 0.070 \text{ min} = A \qquad (1)$$
$$b + c + d = \text{element 3} = 0.067 \text{ min} = B \qquad (2)$$
$$c + d + e = \text{element 5} = 0.073 \text{ min} = C \qquad (3)$$
$$d + e + a = \text{element 2} = 0.061 \text{ min} = D \qquad (4)$$
$$e + a + b = \text{element 4} = 0.068 \text{ min} = E \qquad (5)$$

First, we add these five equations:

$$3a + 3b + 3c + 3d + 3e = A + B + C + D + E$$

Then let

$$A + B + C + D + E = T$$
$$3a + 3b + 3c + 3d + 3e = T = 0.339 \text{ min}$$

and

$$a + b + c + d + e = \frac{0.339}{3} = 0.113 \text{ min}$$

Therefore,

$$A + d + e = 0.113 \text{ min}$$

Then

$$d + e = 0.113 \text{ min} - 0.07 \text{ min} = 0.043 \text{ min}$$

since

$$c + d + e = 0.073 \text{ min}$$

$$c = 0.073 \text{ min} - 0.043 \text{ min} = 0.030 \text{ min}$$

Likewise,

$$d + e + a = 0.061$$

and

$$a = 0.061 - 0.043 = 0.018 \text{ min}$$

Substituting in Equation (1), we get

$$b = 0.070 - (0.03 + 0.018) = 0.022$$

Substituting in Equation (2), we see that

$$d = 0.067 - (0.022 + 0.03) = 0.015 \text{ min}$$

Substituting in Equation (3), we arrive at

$$e = 0.073 - (0.015 + 0.03) = 0.028 \text{ min}$$

TABULAR DATA

For example, when developing standard data times for machine elements, the analyst may need to tabulate horsepower requirements for various materials in relation to depth of cut, cutting speeds, and feeds. To avoid overloading existing equipment, the analyst should have information on the workload being assigned to each machine for the conditions under which the material is being removed. For example, in the machining of high-alloy steel forgings on a lathe capable of a developed horsepower of 10, it would not be feasible to take a 3/8-in depth of cut while operating at a feed of 0.011 in per revolution and a speed of 200 surface feet per minute. Tabular data, either from the machine tool manufacturer or from empirical studies (see Table 12.1) indicate a horsepower requirement of 10.6 for these conditions. Consequently, the work would need to be planned for a feed of 0.009 in at a speed of 200 surface ft; this would only require a horsepower rating of 8.7. Such tabular data are best stored, retrieved, and accumulated into a final standard time using commercially available spreadsheet programs (e.g., Microsoft Excel).

USING NOMOGRAMS AND PLOTS

Because of space limitations, tabularizing values for variable elements is not always convenient. By plotting a curve or a system of curves in the form of an alignment chart, the analyst can express considerable standard data graphically on one page.

Figure 12.1 illustrates a nomogram for determining turning and facing time. For example, if the problem is to determine the production in pieces per hour to turn 5 linear inches of a 4-in-diameter shaft of medium carbon steel on a machine utilizing 0.015-in feed per revolution and having a cutting time of 55% of the cycle time, the answer could be readily determined graphically. Connecting a recommended cutting speed of 150 ft/min for medium carbon steel, shown on scale 1, to the 4-in diameter of the work, shown on scale 2, results in a speed of 143 rpm, shown on scale 3. The 143-rpm point is connected with the 0.015-in feed per revolution, shown on scale 4. This line extended to scale 5 shows a feed of 2.15 in/min. This feed point connected with the length of cut, shown on scale 6 (5 in), gives the

Table 12.1 Horsepower Requirements for Turning High-Alloy Steel Forgings for Cuts 3/8-inch and 1/2-inch Deep at Varying Speeds and Feeds

Surface feet	3/8-in depth cut (feeds, in/rev)						1/2-in depth cut (feeds, in/rev)					
	0.009	0.011	0.015	0.018	0.020	0.022	0.009	0.011	0.015	0.018	0.020	0.022
150	6.5	8.0	10.9	13.0	14.5	16.0	8.7	10.6	14.5	17.3	19.3	21.3
175	8.0	9.3	12.7	15.2	16.9	18.6	10.1	12.4	16.9	20.2	22.5	24.8
200	8.7	10.6	14.5	17.4	19.3	21.3	11.6	14.1	19.3	23.1	25.7	28.4
225	9.8	11.9	16.3	19.6	21.7	23.9	13.0	15.9	21.7	26.1	28.9	31.8
250	10.9	13.2	18.1	21.8	24.1	26.6	14.5	17.7	24.1	29.0	32.1	35.4
275	12.0	14.6	19.9	23.9	26.5	29.3	15.9	19.4	26.5	31.8	35.3	39.0
300	13.0	16.0	21.8	26.1	29.0	31.9	17.4	21.2	29.0	34.7	38.6	42.5
400	17.4	21.4	29.1	34.8	38.7	42.5	23.2	28.2	38.7	46.3	51.5	56.7

required cutting time, on scale 7. Finally, this cutting time of 2.35 min, connected with the percentage of cutting time, shown on scale 8 (in this case, 55 percent), gives the production in pieces per hour, on scale 9 (in this case, 16).

Figure 12.2 illustrates a plot of forming time in hours per hundred pieces for a certain gage of stock over a range of sizes expressed in square inches. Each of the 12 points in this plot represents a separate time study. The plotted points indicate a straight-line relationship, which can be expressed as a formula:

$$\text{Standard time} = 50.088 + 0.00038 \text{ (size)}$$

Details on developing such formulas are found below in Section 12.2.

Using nomograms or plots has some distinct disadvantages. First, it is easy to introduce an error in reading from the plot, because of the amount of interpolation usually required. Second, there is the chance of outright error through incorrect reading or misalignment of the intersections on the various scales.

12.2 FORMULA CONSTRUCTION FROM EMPIRICAL DATA

IDENTIFY VARIABLES

The first step in formula construction is to identify the dependent and independent variables involved. Since the analyst is concerned with setting time standards, the dependent variable frequently will be time. For example, a formula might be developed for curing bonded rubber parts between 2 and 8 oz in weight. The independent variable is the weight of the rubber, while the dependent variable is the time to cure. The range for the dependent variable would be 2 to 8 oz, while the dependent variable of time would have to be quantified from studies.

ANALYZE ELEMENTS AND COLLECT DATA

After the initial identification is finished, the next step is to collect data for the formula, written from existing studies or taking new studies, to obtain a sufficiently

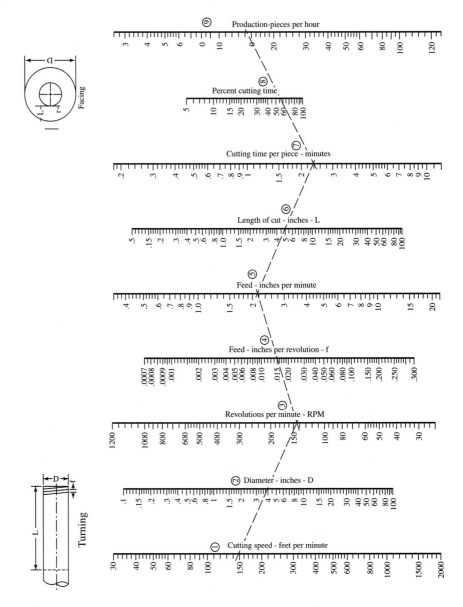

Figure 12.1   Nomogram for determining facing and turning time. (Crobalt, Inc.)

large sample to cover the range of work for the formula. Obviously, variable elements tend to vary in proportion to some characteristics of the work, such as size, shape, or hardness. These elements should be carefully studied to determine which factors influence the time, and to what extent. In general, the constant elements should not deviate substantially.

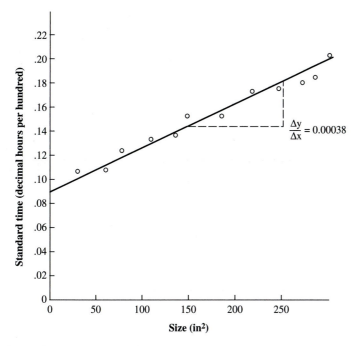

Figure 12.2 Forming time for different stock sizes.

In general, the more studies used, the more data will be available, and the more normal will be the conditions reflected. If desired, statistical procedures such as the power test (Neter et al., 1996) can be used to determine the exact number of studies to be collected.

PLOT DATA AND COMPUTE VARIABLE EXPRESSIONS

Next, the data are posted to a spreadsheet (e.g., Microsoft Excel) for analysis of the constants and variables. The constants are identified and combined and the variables analyzed so as to have the factors influencing time expressed in an algebraic form. By plotting a curve of time versus the independent variable, the analyst may deduce potential algebraic relationships. For example, plotted data may take a number of forms: a straight line, nonlinear increasing trend, a nonlinear decreasing trend, or no obviously regular geometric form. If a straight line, then the relationship is quite straightforward:

$$y = a + bx$$

with the constants a and b determined from least-squares regression analysis. If the plot shows a nonlinear increasing trend, then power relationships of the form x^2, x^3, x^n, or e^x should be attempted. For nonlinear decreasing trends, negative power or negative exponentials should be attempted. For asymptotic trends, log relationships or negative exponentials of the form

$$y = 1 - e^{-x}$$

should be attempted.

Note that adding terms to the model will always produce a better model with a higher percentage of the variance (expressed as r^2) in the data explained. However, the model may not be statistically significantly better; that is, statistically there is no difference in the quality of the predicted value between the two models. Furthermore, the simpler the formula, the better it can be understood and applied. Cumbersome expressions involving many terms to powers should be avoided. The range for each variable should be specifically identified. The limitations of the formula must be noted by describing its applicable range in detail.

There is a formalized procedure for computing the best model, termed the *general linear test*. It computes the decrease in unexplained variance between the simpler model, termed the *reduced model,* and the more complex model, termed the *full model*. The decrease in variance is tested statistically, and the more complex model is used only if the decrease is significant (see Example 12.2). Further details on curve fitting and model development can be found in various statistical textbooks such as Neter et al. (1996) or Rawling (1988).

EXAMPLE 12.2

In the element "strike arc and weld," analysts obtained the following data from 10 detailed studies:

Study number	Size of weld	Minutes per inch of weld
1	1/8	0.12
2	3/16	0.13
3	1/4	0.15
4	3/8	0.24
5	1/2	0.37
6	5/8	0.59
7	11/16	0.80
8	3/4	0.93
9	7/8	1.14
10	1	1.52

Plotting the data resulted in the smooth curve shown in Figure 12.3. A simple linear regression of the dependent variable "minutes" against the independent variable "weld" yields:

$$y = -0.245 + 1.57x \qquad (1)$$

with $r^2 = 0.928$ and sum of squares (SSE) = 0.1354.

Since Figure 12.3 indicates a definite nonlinear trend to the data, adding a quadratic component to the model seems reasonable. Regression now yields

$$y = 0.1 - 0.178x + 1.61x^2 \tag{2}$$

with $r^2 = 0.993$ and SSE $= 0.012$. The increase in r^2 would seem to indicate a definite improvement in the fit of the model. This improvement can be tested statistically with the general linear test

$$F = \frac{[SSE(R) - SSE(F)]/(df_R - df_F)}{SSE(F)/df_F}$$

where   SSE(R) = sum of squares error for reduced (i.e., simpler) model
         SSE(F) = sum of squares error for full (i.e., more complex) model
           df_R = degrees of freedom for reduced model
           df_F = degrees of freedom for full model

Comparing the two models yields

$$F = \frac{(0.1354 - 0.012)/(8 - 7)}{0.012/7} = 71.98$$

Since 71.98 is considerably larger than $F_{(1,7)} = 5.59$, the full model is a significantly better model.

The process can be repeated by adding another term with a higher power (e.g., x^3), which yields the following model:

$$y = 0.218 - 1.14x + 3.59x^2 - 1.16x^3 \tag{3}$$

with $r^2 = 0.994$ and SSE $= 0.00873$. However in this case, the general linear test does not yield a statistically significant improvement:

$$F = \frac{(0.12 - 0.00873)/(7 - 6)}{0.00873/6} = 2.25$$

The F value of 2.25 is smaller than the critical $F_{(1,6)} = 5.99$.

Interestingly, using a simple quadratic model of the form

$$y = 0.0624 + 1.45x^2 \tag{4}$$

with $r^2 = 0.993$ and sum of squares $= 0.0133$ yields the best and simplest model. Comparing this model [Equation (4)] with the second model [Equation (2)] yields

$$F = \frac{(0.133 - 0.012)/(8 - 7)}{0.012/7} = 0.758$$

The F value is not significant, and the extra linear term in x does not yield a better model.

This best-fitting quadratic model can be checked by substituting a 1-in weld, to yield

$$y = 0.0624 + 1.45(1)^2 = 1.51$$

This checks quite closely with the time study value of 1.52 min.

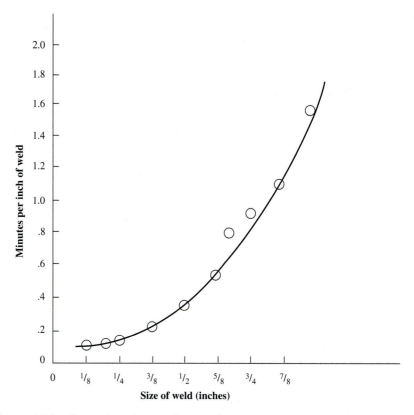

Figure 12.3   Curve plotted on regular coordinate paper takes quadratic form.

At times the analyst may recognize that more than one independent variable is influencing time and that the final expression may consist of a multiple combination of various powers of the independent variables. If that is the case, multivariate regression techniques must be applied. These calculations are quite tedious and specialized statistical software packages, such as MINITAB or SAS, will need to be utilized.

CHECK FOR ACCURACY AND FINALIZE

Upon completion of the formula, analysts should verify it before releasing it for use. The easiest and fastest way to check the formula is to use it to check existing time studies. Any marked differences (roughly 5 percent) between the formula value and the time study value should be investigated. If the formula does not have the expected validity, the analyst should accumulate additional data by taking more stopwatch and/or standard data studies.

The final step in the formula development process is to write the formula report. The analyst should consolidate all data, calculations, derivations, and applications of the formula and present this information in a complete report prior to putting the formula into use. This will make available all the facts regarding the process employed, the operating conditions, and the scope of formula.

12.3 ANALYTICAL FORMULAS

Standard times can be calculated using analytical formulas found in technical handbooks or from information provided by machine tool manufacturers. By finding the appropriate feeds and speeds for different types and thicknesses of materials, analysts can calculate cutting times for different machining operations.

DRILL PRESS WORK

A drill is a fluted end-cutting tool used to originate or enlarge a hole in solid material. In drilling operations on a flat surface, the axis of the drill is at 90 degrees to the surface being drilled. When a hole is drilled completely through a part, the analyst must add the lead of the drill to the length of the hole to determine the entire distance the drill must travel to make the hole. When a blind hole is drilled, the distance from the surface being drilled to the deepest penetration of the drill is the distance that the drill must travel (see Figure 12.4).

Since the commercial standard for the included angle of drill points is 118 degrees, the lead of the drill may be readily found through the expression

$$l = \frac{r}{\tan A}$$

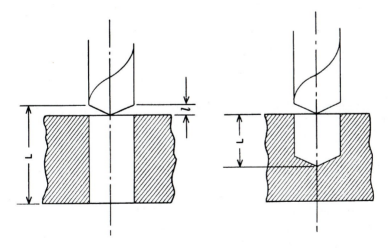

Figure 12.4 Drill travel distance.
Distance L indicates the distance the drill must travel when drilling through (illustration at left) and when drilling blind holes (illustration at right) (lead of drill is shown by distance l).

where l = lead of drill
r = radius of drill
$\tan A$ = tangent of one-half the included angle of drill

To illustrate, calculate the lead of a general-purpose drill with 1-in diameter:

$$l = \frac{0.5}{\tan 59°}$$

$$l = \frac{0.5}{1.6643}$$

$$l = 0.3\text{-in lead}$$

After determining the total length that the drill must move, we divide this distance by the feed of the drill in inches per minute, to find the drill cutting time in minutes.

Drill speed is expressed in feet per minute (ft/min), and feed in thousandths of an inch per revolution (r). To change the feed into inches per minute when the feed per revolution and the speed in feet per minute are known, the following equation can be used:

$$F_m = \frac{3.82 f S_f}{d}$$

where F_m = feed (in/min)
f = feed (in/r)
S_f = surface feet per minute
d = diameter of drill (in)

For example, to determine the feed in inches per minute of a 1-in drill running at a surface speed of 100 ft/min and a feed of 0.013 in/r, we have

$$F_m = \frac{(3.82)(0.013)(100)}{1} = 4.97 \text{ in/min}$$

To determine how long it would take for this 1-in drill running at the same speed and feed to drill through 2 in of a malleable iron casting, we use the equation

$$T = \frac{L}{F_m}$$

where T = cutting time (min)
L = total length drill must move
F_m = feed (in/min)

which should yield

$$T = \frac{2 \text{ (thickness of casting)} + 0.3 \text{ (lead of drill)}}{4.97}$$
$$= 0.464\text{-min cutting time}$$

The cutting time thus calculated does not include an allowance, which must be added to determine the standard time. The allowance should include time for variations in material thickness and for tolerance in setting the stops, both of which affect the cycle cutting time. Personal and unavoidable delay allowances should also be added to arrive at an equitable standard time.

LATHE WORK

Many variations of machine tools are classified as lathes. These include the engine lathe, turret lathe, and automatic lathe (automatic screw machine). All of these lathes are used primarily with stationary tools or with tools that translate over the surface to remove material from the revolving work, which includes forgings, castings, or bar stock. In some cases, the tool revolves while the work is stationary, as on certain stations of automatic screw machine work. For example, a slot in a screw head can be machined in the slotting attachment on the automatic lathe.

Many factors alter speeds and feeds, such as the condition and design of the machine tool, the material being cut, the condition and design of the cutting tool, the coolant used for cutting, the method of holding the work, and the method of mounting the cutting tool.

As in drill press work, feeds are expressed in thousandths of an inch per revolution, and speeds in surface feet per minute. To determine the cutting time for L in of cut, the length of cut in inches is divided by the feed in inches per minute, or

$$T = \frac{L}{F_m}$$

where T = cutting time (min)
   L = total length of cut
   F_m = feed (in/min)

and

$$F_m = \frac{3.82 \, f S_f}{d}$$

where f = feed (in/r)
   S_f = speed (surface ft/min)
   d = diameter of work (in)

MILLING MACHINE WORK

Milling refers to the removal of material with a rotating multiple-tooth cutter. While the cutter rotates, the work is fed past the cutter. This differs from a drill press, for which the work is usually stationary. In addition to machining plane and irregular surfaces, operators use a milling machine for cutting threads, slotting, and cutting gears.

In milling work, as in drill press and lathe work, the speed of the cutter is expressed in surface feet per minute. Feeds or table travel are usually expressed in thousandths of an inch per tooth. To determine the cutter speed in revolutions per minute from the surface feet per minute and the diameter of the cutter, use the following expression:

$$N_r = \frac{3.82 S_f}{d}$$

where N_r = cutter speed (rpm)
   S_f = cutter speed (ft/min)
   d = outside diameter of cutter (in)

To determine the feed of the work in inches per minute into the cutter, use the expression

$$F_m = f n_t N_r$$

where F_m = feed of work into cutter (in/min)
   f = feed of cutter (in per tooth)
   n_t = number of cutter teeth
   N_r = cutter speed (rpm)

The number of cutter teeth suitable for a particular application may be expressed as

$$n_t = \frac{F_m}{F_t N_\rho}$$

where F_t = chip thickness.

To compute the cutting time on milling operations, the analyst must take into consideration the lead of the milling cutter when figuring the total length of cut under power feed. This can be determined by triangulation, as illustrated in Figure 12.5, which shows the slab-milling of a pad.

In this case, to arrive at the total length that must be fed past the cutter, the lead BC is added to the length of the work (8 in). Clearance for removal of the work after the machining cut is handled as a separate element, because greater feed under rapid table traverse is used. By knowing the diameter of the cutter, you can determine AC as being the cutter radius, and you can then calculate the height of the right triangle ABC by subtracting the depth of cut BE from the cutter radius AE, as follows:

$$BC = \sqrt{AC^2 - AB^2}$$

In the preceding example, suppose we assume that the cutter diameter is 4 in and that it has 22 teeth. The feed per tooth is 0.008 in, and the cutting speed is 60 ft/min. We can compute the cutting time by using the equation

$$T = \frac{L}{F_m}$$

where T = cutting time (min)
   L = total length of cut under power feed
   F_m = feed (in/min)

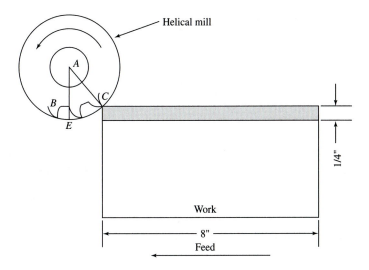

Figure 12.5 Slab-milling a casting 8 in long.

Then L would be equal to (8 inches 1 *BC*) and

$$BC = \sqrt{4 - 3.06} = 0.975$$

Therefore,

$$L = 8.975$$
$$F_m = fn_tN_r$$
$$F_m = (0.008)(22)N_r$$

or

$$N_r = \frac{3.82Sf}{d} = \frac{3.82(60)}{4} = 57.3 \text{ rpm}$$

Then

$$F_m = (0.008)(22)(57.3) = 10.1 \text{ in/min}$$

and

$$T = \frac{8.975}{10.1} = 0.888 \text{ min cutting time}$$

Through knowledge of feeds and speeds, analysts can determine the required cutting or processing time for various work performed in their plants. The illustrations cited in drill press, lathe, and milling work are representative of the techniques used to establish raw cutting times. The necessary applicable allowances must be added to these values to create fair total element allowed values.

12.4   USING STANDARD DATA

For easy reference, constant standard data elements should be tabulated and filed under the machine or the process. Variable data can be either tabulated or expressed as a curve or equation, and then filed under the facility or operation class.

Table 12.2 illustrates standard data for blanking and piercing stock on a punch press. By identifying the job according to the distance that the strip of sheet stock is moved per piece, the analyst can find the standard time for the complete operation.

Table 12.3 illustrates standard setup data for No. 5 Warner & Swasey turret lathes at a specific plant. To determine the setup time with these data, the analyst would visualize the tooling in the square and hex turret and would then refer to the table. For example, if a certain job required a chamfering tool, turning tool, and facing tool in the square turret, and needed two boring tools, one reamer, and a collapsible tap in the hex turret, the setup standard time would be 69.70 min plus 25.89 min, or 95.59 min. To arrive at this solution, the analyst would find the value of the relevant tooling under the square turret column (line 8) and the most time-consuming applicable tooling in the hex turret section, which in this case is tapping. This would give a value of 69.7 min. Since three additional tools are in the hex turret (first bore, second bore, and ream), the analyst would then multiply 8.63 by 3 to get 25.89 min. Finally, adding 25.89 min to 69.70 min yields the total required setup time.

SUMMARY

When properly applied, standard data permit the rapid establishment of accurate time standards before the job is performed. This feature makes their use especially attractive for estimating the cost of new work, for cost quotes, and for subcontracting purposes. The use of standard data also simplifies many administrative problems in plants where there may be restrictions pertaining to such matters as the type of study to be made (continuous or snapback), the number of cycles to be studied,

Table 12.2   Standard Data for Blanking and Piercing Strip Stock Hand Feed with Piece Automatically Removed on Toledo 76 Punch Press

L (distance in inches)	T (time in hours per hundred hits)
1	0.075
2	0.082
3	0.088
4	0.095
5	0.103
6	0.110
7	0.117
8	0.123
9	0.130
10	0.137

Table 12.3 Standard Setup Data for No. 5 Turret Lathes

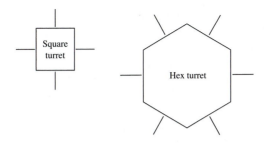

Basic tooling

		Hex turret					
No. **Square turret**	**Partial**	**Chamfer**	**Bore or turn**	**Drill**	**S. tap or ream**	**C. tap**	**C. die**
1. Partial	31.5	39.6	44.5	48.0	47.6	50.5	58.5
2. Chamfer	38.2	39.6	46.8	49.5	50.5	53.0	61.2
3. Face or cutoff	36.0	44.2	48.6	51.3	52.2	55.0	63.0
4. Turn, bore, groove, radius	40.5	49.5	50.5	53.0	54.0	55.8	63.9
5. Face and chamfer	37.8	45.9	51.3	54.0	54.5	56.6	64.8
6. Face and cutoff	39.6	48.6	53.0	55.0	56.0	58.5	66.6
7. Face and turn or turn and cutoff .	45.0	53.1	55.0	56.7	57.6	60.5	68.4
8. Face, turn, and chamfer	47.7	55.7	57.6	59.5	60.5	69.7	78.4
9. Face, turn, and cutoff	48.6	57.6	57.5	60.0	62.2	71.5	80.1
10. Face, turn, and groove	49.5	58.0	59.5	61.5	64.0	73.5	81.6

11. Circled basic tooling from above .... _____
12. Each additional tool in square .... 4.20 × _____ = _____
13. Each additional tool in hex .... 8.63 × _____ = _____
14. Remove and setup three jaws .... 5.9 _____
15. Set up subassembly or fixture .... 18.7 _____
16. Set up between centers .... 11.0 _____
17. Change lead screw .... 6.6 _____

Total setup _____ min

the operator(s) to be studied, and the observer to conduct the study. By using the standard data technique, analysts may not only avoid such details, but may also alleviate some sources of tension between labor and management.

In general, the more refined the element times, the greater the coverage of the data. Consequently, in job shops, it is practicable to have individual element values, as well as grouped or combined values, so that the data for a given facility will have enough flexibility to allow rates to be set for all types of work scheduled for a machine. For short-cycle jobs, fundamental motion data are especially useful for establishing standards.

A time study formula can similarly establish standards in a fraction of the time required for individual studies. One advantage of formulas over standard data is that a less skilled (and less costly) person can plug data into formulas more quickly than adding standard data elements. Also, since columns of figures must be added in the standard data method, there is a greater chance for omission or arithmetic error in setting a standard than in using a formula.

QUESTIONS

1. What is meant by *standard data*?
2. What is the approximate ratio of the time required to set standards by stopwatch methods to the time required using standard data methods?
3. What are the advantages to establishing time standards by using standard data rather than taking individual studies?
4. What are some of the disadvantages of using curves to tabulate standard data?
5. What advantages do formulas have over standard data in establishing time standards?
6. Is the use of time study formulas restricted to machine shop operations where feeds and speeds have been analytically shown to influence times? Explain.
7. What are the characteristics of a sound time study formula?
8. What is the danger of using too few studies in the derivation of a formula?
9. Explain in detail how one would develop a best-fit formula.

PROBLEMS

1. What would be the horsepower requirements of turning a mild steel shaft with 3-in diameter if a cut of 1/4 in with a feed of 0.022 in/r at a spindle speed of 250 rpm were established?
2. How long would it take to turn 6 in of 1-in bar stock on a No. 3 Warner & Swasey turret lathe running at 300 ft/min and feeding at the rate of 0.005 in/r?
3. A plain milling cutter with 3-in diameter and a face width of 2 in is being used to mill a piece of cold-rolled steel 1.5 in wide and 4 in long. The depth of cut is 3/16 in. How long will it take to make the cut if the feed per tooth is 0.010 in and a 16-tooth cutter running at a surface speed of 120 ft/min is used?
4. Compute the times for elements a, b, c, d, and e when $a + b + c$ is timed at 0.057 min; $b + c + d$ is timed at 0.078 min; $c + d + e$ equals 0.097 min; $d + e + a$ is 0.095 min; and $e + a + b$ is 0.069 min.
5. What would be the lead of a 3/4-in-diameter drill with an included angle of 118 degrees?
6. What would be the feed in inches per minute of a 3/4-in drill running at a surface speed of 80 ft/min and a feed of 0.008 in/r?
7. How long would it take the preceding drill to drill through a casting 2.25 in thick?
8. The analyst in Dorben Company made 10 independent time studies in the hand paint spraying section of the finishing department. The product line under study

revealed a direct relation between spraying time and product surface area. The following data were collected:

Study no.	Rating factor	Product surface area	Standard time
1	0.95	170	0.32
2	1.00	12	0.11
3	1.05	150	0.31
4	0.80	41	0.14
5	1.20	130	0.27
6	1.00	50	0.18
7	0.85	120	0.24
8	0.90	70	0.23
9	1.00	105	0.25
10	1.10	95	0.22

Compute the slope and intercept constant, using regression line equations. How much spray time would you allow for a new part with a surface area of 250 in^2?

9. The work measurement analyst in the Dorben Company wants to develop an accurate equation for estimating the cutting of various configurations in sheet metal with a band saw. The data from eight time studies for the actual cutting element provided the following information:

No.	Lineal inches	Standard time
1	10	0.40
2	42	0.80
3	13	0.54
4	35	0.71
5	20	0.55
6	32	0.66
7	22	0.60
8	27	0.61

What would be the relation between the length of cut and the standard time, using the least squares technique?

10. The work measurement analyst in the XYZ Company wishes to develop standard data involving fast, repetitive manual motions, for use in a light assembly department. Because of the shortness of the desired standard data elements, the analyst is obliged to measure them in groups as they are performed on the factory floor. On a certain study, this analyst is endeavoring to develop standard data for five elements, denoted A, B, C, D, and E. Using a fast (0.001) decimal minute watch, the analyst studied a variety of assembly operations and arrived at the following data:

$$A + B + C = 0.131 \text{ min}$$
$$B + C + D = 0.114 \text{ min}$$
$$C + D + E = 0.074 \text{ min}$$
$$D + E + A = 0.085 \text{ min}$$
$$E + A + B = 0.118 \text{ min}$$

Compute the standard data values for each of the elements A, B, C, D, and E.

11. The work measurement analyst in the Dorben Company is developing standard data for prepricing work in the drill press department. Based on the following recommended speeds and feeds, compute the power feed cutting time of 1/2-in high-speed drills with a 118 degree included angle to drill through material that is 1 in thick. Include a 10 percent allowance for personal needs and fatigue.

Material	Recommended speed (ft/min)	Feed (in/r)
Al (copper alloy)	300	0.006
Cast iron	125	0.005
Monel (R)	50	0.004
Steel (1112)	150	0.005

12. The following data were obtained between standard time and the blanking of various leather pieces from cow hides. Derive a formula for this relationship.

Study no.	Area of leather (in^2)	Standard time (min)
1	5.0	0.07
2	7.5	0.10
3	15.5	0.13
4	25.0	0.20
5	34.0	0.24

13. Develop a formula for the relationship between time and area from the following data:

Study no.	Time	Area
1	4	28.6
2	7	79.4
3	11	182.0
4	15	318.0
5	21	589.0

REFERENCES

Brisley, C. L., and R. J. Dossett. "Computer Use and Non-Direct Labor Measurement Will Transform Profession in the Next Decade." *Industrial Engineering*, 12, no. 8 (August 1980), pp. 34–43.

Cywar, Adam W. "Development and Use of Standard Data." In *Handbook of Industrial Engineering*. Ed. Gavriel Salvendy. New York: John Wiley & Sons, 1982.

Fein, Mitchell. "Establishing Time Standards by Parameters." *Proceedings of the Spring Conference of the American Institute of Industrial Engineers*. Norcross, GA: American Institute of Industrial Engineers, 1978.

Lowry, S. M., H. B. Maynard, and B. J. Stegemerten. *Motion and Time Study*. 3d ed. New York: McGraw-Hill, 1940.

Metcut Research Associates. *Machining Data Handbook*. Cincinnati, OH: Metcut Research Associates, 1966.

Neter, J., M. Wasserman, M. H. Kutner, and C. J. Nachstheim. *Applied Linear Statistical Models*. 4th ed. New York: McGraw-Hill, 1996.

Pappas, Frank G., and Robert A. Dimberg. *Practical Work Standards*. New York: McGraw-Hill, 1962.

Rawling, J. O. *Applied Regression Analysis*. Pacific Grove, CA: Wadsworth & Brooks, 1988.

Rotroff, Virgil H. *Work Measurement*. New York: Reinhold Publishing, 1959.

SELECTED SOFTWARE

MINITAB. 3081 Enterprise Dr., State College, PA 16801.

SAS. SAS Institute, Cary, NC 27513.

Predetermined Time Systems

Since the time of Frederick W. Taylor, management has realized the desirability of assigning standard times to the basic elements of work. These times are referred to as *basic motion times, synthetic times*, or *predetermined times*. They are assigned to fundamental motions and groups of motions that cannot be precisely evaluated with ordinary stopwatch time study procedures. They are also the result of studying a large sample of diversified operations with a timing device, such as a motion picture camera or videotape machine, capable of measuring very short elements. The time values are synthetic in that they are often the result of logical combinations of therbligs; they are basic in that further refinement is both difficult and impractical; they are predetermined because they are used to predict standard times for new work resulting from methods changes.

Since 1945, there has been a growing interest in the use of basic motion times as a method of establishing rates quickly and accurately without using the stopwatch or other time recording devices. Today, practicing methods analysts can obtain information from over 50 different systems of established predetermined time. Essentially, these predetermined time systems are sets of motion–time tables with explanatory rules and instructions on the use of the motion–time values. Considerable specialized training is essential to the practical application of these techniques. In fact, most companies require certification before analysts are permitted to establish standards using the Work-Factor, Methods-Time Measurement (MTM), or MOST systems.

To give the reader a good overview of the field of predetermined time systems we will review MTM in some detail, since it is the pioneer, as well as a quicker subsystem of MTM, called MTM-2. In addition, a derivative of MTM, a complete Maynard Operation Sequence Technique (MOST) will be discussed. Derivations of the common predetermined time systems are illustrated in Figure 13.1. MTM-2 and MOST are the most commonly used predetermined time systems, according to 141 industrial engineers surveyed.

13.1   METHODS–TIME MEASUREMENT

MTM-1

Methods-Time Measurement (Maynard, Stegemerten, and Schwab, 1948) gives time values for the fundamental motions of reach, move, turn, grasp, position, disengage, and release. The authors defined MTM as "a procedure which analyzes any manual operation or method into the basic motions required to perform it, and assigns to each motion a pre-determined time standard which is determined by the nature of the motion and the conditions under which it is made."

MTM-1 data are the result of frame-by-frame analyses of motion picture films of diversified areas of work. The data taken from the various films were rated by the Westinghouse technique, tabulated, and analyzed to determine the degree of difficulty caused by variable characteristics. For example, both the distance and the type of reach affect the reach time. Further analysis categorized five distinct cases of reach, each requiring a different time allotment for a given distance:

1. Reach to object in fixed location, or to object in other hand, or to object on which other hand rests.
2. Reach to single object in location that may vary slightly from cycle to cycle.
3. Reach to object jumbled with other objects, requiring search and select as well.
4. Reach to a very small object, or an object for which accurate grasp is required.
5. Reach to indefinite location to get hand in position for body balance, or for next motion, or out of the way.

In addition, they found that move time was influenced by both the distance and the weight of the object being moved, as well as by the specific type of move. The three cases of move are

1. Move object to other hand, or against stop.
2. Move object to approximate or indefinite location.
3. Move object to exact location.

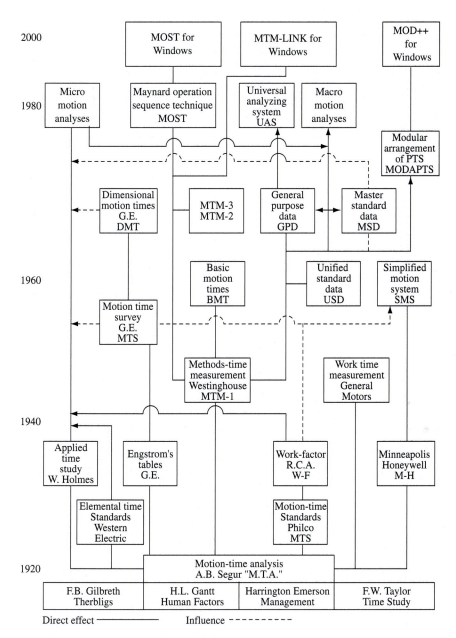

Figure 13.1 Family tree of predetermined time systems.
Courtesy: Standards International, Chicago, Illinois.

Finally, 2 cases of release and 18 cases of position also influence time.

Table 13.1 summarizes the MTM-1 values. The time values of the therblig grasp vary from 2.0 to 12.9 TMU [1 time-measurement unit (TMU) = 0.00001 h], depending on the classification of the grasp.

Table 13.1   Summary of MTM-1 Data

Table I—Reach—R						
Distance Moved (in)	**Time (TMU)**				**Hand in Motion**		**Case and Description**

Distance Moved (in)	A	B	C or D	E	A	B	Case and Description
$1/2$ or less	2.0	2.0	2.0	2.0	1.6	1.6	A Reach to object in fixed
1	2.5	2.5	3.6	2.4	2.3	2.3	location, or to object in other
2	4.0	4.0	5.9	3.8	3.5	2.7	hand or on which other hand
3	5.3	5.3	7.3	5.3	4.5	3.6	rests.
4	6.1	6.4	8.4	6.8	4.9	4.3	B Reach to single object in
5	6.5	7.8	9.4	7.4	5.3	5.0	location which may vary
6	7.0	8.6	10.1	8.0	5.7	5.7	slightly from cycle to cycle.
7	7.4	9.3	10.8	8.7	6.1	6.5	C Reach to object jumbled with
8	7.9	10.1	11.5	9.3	6.5	7.2	other objects in a group so
9	8.3	10.8	12.2	9.9	6.9	7.9	that search and select occur.
10	8.7	11.5	12.9	10.5	7.3	8.6	D Reach to a very small object
12	9.6	12.9	14.2	11.8	8.1	10.1	or where accurate grasp is
14	10.5	14.4	15.6	13.0	8.9	11.5	required.
16	11.4	15.8	17.0	14.2	9.7	12.9	E Reach to indefinite location
18	12.3	17.2	18.4	15.5	10.5	14.4	to get hand in position for
20	13.1	18.6	19.8	16.7	11.3	15.8	body balance or next motion
22	14.0	20.1	21.2	18.0	12.1	17.3	or out of way.
24	14.9	21.5	22.5	19.2	12.9	18.8	
26	15.8	22.9	23.9	20.4	13.7	20.2	
28	16.7	24.4	25.3	21.7	14.5	21.7	
30	17.5	25.8	26.7	22.9	15.3	23.2	

Table II—Move—M							

Distance Moved (in)	Time (TMU)				Wt. Allowance			Case and Description
	A	B	C	Hand in Motion B	Wt. (lb) up to	Factor	Constant (TMU)	
$1/2$ or less	2.0	2.0	2.0	1.7	2.5	0	0	A Move object
1	2.5	2.9	3.4	2.3				to other hand
2	3.6	4.6	5.2	2.9	7.5	1.06	2.2	or against
3	4.9	5.7	6.7	3.6				stop.
4	6.1	6.9	8.0	4.3	12.5	1.11	3.9	
5	7.3	8.0	9.2	5.0				
6	8.1	8.9	10.3	5.7	17.5	1.17	5.6	
7	8.9	9.7	11.1	6.5				
8	9.7	10.6	11.8	7.2				B Move object
9	10.5	11.5	12.7	7.9	22.5	1.22	7.4	to approxi-
10	11.3	12.2	13.5	8.6				mate or
12	12.9	13.4	15.2	10.0	27.5	1.28	9.1	indefinite
14	14.4	14.6	16.9	11.4				location.
16	16.0	15.8	18.7	12.8	32.5	1.33	10.8	
18	17.6	17.0	20.4	14.2				
20	19.2	18.2	22.1	15.6	37.5	1.39	12.5	
22	20.8	19.4	23.8	17.0				
24	22.4	20.6	25.5	18.4	42.5	1.44	14.3	C Move object
26	24.0	21.8	27.3	19.8				to exact
28	25.5	23.1	29.0	21.2	47.5	1.50	16.0	location.
30	27.1	24.3	30.7	22.7				

Table III—Turn and Apply Pressure—T and AP											
	Time TMU for Degrees Turned										
Weight	**30°**	**45°**	**60°**	**75°**	**90°**	**105°**	**120°**	**135°**	**150°**	**165°**	**180°**
Small—0 to 2 lb	2.8	3.5	4.1	4.8	5.4	6.1	6.8	7.4	8.1	8.7	9.4
Medium—2.1 to 10 lb	4.4	5.5	6.5	7.5	8.5	9.6	10.6	11.6	12.7	13.7	14.8
Large—10.1 to 35 lb	8.4	10.5	12.3	14.4	16.2	18.3	20.4	22.2	24.3	26.1	28.2

Apply Pressure Case A—10.6 TMU. Apply Pressure Case B—16.2 TMU

Table IV—Grasp—G		
Case	**Time (TMU)**	**Description**
1A	2.0	**Pick Up Grasp**—Small, medium or large object by itself, easily grasped.
1B	3.5	Very small object or object lying close against a flat surface.
1C1	7.3	Interference with grasp on bottom and one side of nearly cylindrical object. Diameter larger than $1/2''$.
1C2	8.7	Interference with grasp on bottom and one side of nearly cylindrical object. Diameter $1/4''$ to $1/2''$.
1C3	10.8	Interference with grasp on bottom and one side of nearly cylindrical object. Diameter less than $1/4''$.
2	5.6	**Regrasp.**
3	5.6	**Transfer Grasp.**
4A	7.3	Object jumbled with other objects so search and select occur. Larger than $1'' \times 1'' \times 1''$.
4B	9.1	Object jumbled with other objects so search and select occur. $1/4'' \times 1/4'' \times 1/8''$ to $1'' \times 1'' \times 1''$.
4C	12.9	Object jumbled with other objects so search and select occur. Smaller than $1/4'' \times 1/4'' \times 1/8''$.
5	0	Contact, sliding or hook grasp.

Table V—Position*—P				
Class of Fit		**Symmetry**	**Easy to Handle**	**Difficult to Handle**
1—Loose	No pressure required.	S	5.6	11.2
		SS	9.1	14.7
		NS	10.4	16.0
2—Close	Light pressure required.	S	16.2	21.8
		SS	19.7	25.3
		NS	21.0	26.6
3—Exact	Heavy pressure required.	S	43.0	48.6
		SS	46.5	52.1
		NS	47.8	53.4

*Distance moved to engage—1″ or less.

Table VI—Release—RL		
Case	**Time (TMU)**	**Description**
1	2.0	Normal release performed by opening fingers as independent motion.
2	0	Contact release.

Table 13.1 (continued)

Table VII—Disengage—D		
Class of Fit	**Easy to Handle**	**Difficult to Handle**
1—**Loose**—Very slight effort, blends with subsequent move.	4.0	5.7
2—**Close**—Normal effort, slight recoil.	7.5	11.8
3—**Tight**—Considerable effort, hand recoils markedly.	22.9	34.7

Table VIII—Eye Travel Time and Eye Focus—ET and EF

Eye Travel Time $= 15.2 \times \dfrac{T}{D}$ TMU, with a maximum value of 20 TMU.

where T = the distance between points from and to which the eye travels.
        D = the perpendicular distance from the eye to the line of travel T.

Eye Focus Time = 7.3 TMU.

Table IX—Body, Leg, and Foot Motions			
Description	**Symbol**	**Distance**	**Time TMU**
Foot Motion—Hinged at ankle.	FM	Up to 4″	8.5
With heavy pressure.	FMP		19.1
Leg or Foreleg Motion	LM—	Up to 6″	7.1
		Each add'l. inch	1.2
Sidestep—Case 1—Complete when leading leg contacts floor.	SS-C1	Less than 12″	Use REACH or MOVE time
		12″	17.0
		Each add'l. inch	0.6
Case 2—Lagging leg must contact floor before next motion can be made.	SS-C2	12″	34.1
		Each add'l. inch	1.1
Bend, Stoop, or Kneel on One Knee.	B, S, KOK		29.0
Arise.	AB, AS, AKOK		31.9
Kneel on Floor—Both Knees.	KBK		69.4
Arise.	AKBK		76.7
Sit.	SIT		34.7
Stand from Sitting Position.	STD		43.4
Turn Body 45 to 90 degrees—			
Case 1—Complete when leading leg contacts floor.	TBC1		18.6
Case 2—Lagging leg must contact floor before next motion can be made.	TBC2		37.2
Walk.	W-FT.	Per foot	5.3
Walk.	W-P	Per pace	15.0

Table X—Simultaneous Motions

Reach			Move			Grasp			Position			Disengage			
A, E	B	C, D	A, Bm	B	C	G1A G2 G5	G1B G1C	G4	P1S	P1SS P2S	P1NS P2SS P2NS	D1E D1D	D2	Case	Motion
		w/o	w/o	w/o	w/o	w/o	w/o		E/D	E/D	E/D	E/D			
□	□	□□	□□	□□	××	□	□□	□□	□□	□×	××	□	□□	A, E	Reach
□	□	□×	□□	□×	×■	□	□×	×■	××	×■	■■	□	□×	B	
		□×	×■	×■	■■	□	×■	■■	■■	■■	■■	×	■■	C, D	
			□□	□□	□□	□	□□	□□	□□	□×	××	□	□□	A, Bm	Move
				□□	□□	□	□×	×■	××	×■	■■	□	□×	B	
					×■	□	×■	■■	■■	■■	■■	×	■■	C	
						□	□□	□□	□□	□■	■■	□	■■	G1A, G2, G5	Grasp
							■■	×■	■■	■■	■■	■	■■	G1B, G1C	
								■■	■■	■■	■■	■	■■	G4	
									×■	■■	■■	■	■■	P1S	Position
										■■	■■	■	■■	P1SS, P2S	
											■■	■	■■	P1NS, P2SS, P2NS	
												□	□□	D1E, D1D	Disengage
													□□	D2	

□ = EASY to perform simultaneously.
× = Can be performed simultaneously with PRACTICE.
■ = DIFFICULT to perform simultaneously even after long practice. Allow both times.

MOTIONS NOT INCLUDED IN ABOVE TABLE
TURN—Normally EASY with all motions except when TURN is controlled or with DISENGAGE.
APPLY PRESSURE—May be EASY, PRACTICE, or DIFFICULT. Each case must be analyzed.
POSITION—Class 3—Always DIFFICULT.
DISENGAGE—Class 3—Normally DIFFICULT.
RELEASE—Always EASY.
DISENGAGE—Any class may be DIFFICULT if care must be exercised to avoid injury or damage to object.
*W = Within the area of normal vision.
  O = Outside the area of normal vision.
**E = EASY to handle.
  D = DIFFICULT to handle.

Source: MTM Association for Standards and Research, Fair Lawn, New Jersey.

First, the analyst summarizes all left-hand and right-hand motions required to perform the job properly. Then the rated times in TMU for each motion are determined from the methods–time data tables. To determine the time required for a normal performance of the task, the nonlimiting motion values are either circled or deleted, as only the limiting motions will be summarized, provided that it is "easy" to perform the two motions simultaneously (see Table X of Table 13.1). For example, if the right hand must reach 20 in (50 cm) to pick up a nut, the classification would be R20C and the time value would be 19.8 TMU. If, at the same time, the left hand must reach 10 in (25 cm) to pick up a cap screw, a designation of R10C with a TMU value of 12.9 would be in effect. The right-hand value would be the limiting value, and the 12.9 value of the left hand would not be used in calculating the normal time.

The tabulated values do not carry any allowance for personal delays, fatigue, or unavoidable delays. When analysts use these values to establish time standards, they must add appropriate allowances to the summary of the synthetic basic motion times. Proponents of MTM-1 state that no fatigue allowance is needed in the vast majority of applications, because the MTM-1 values are based on a work rate that can be sustained at steady state for the working life of a healthy employee.

An example of an MTM-1 analysis of a clerical operation (replacing a page in a three-ring binder) is shown in Table 13.2.

Today, MTM has received worldwide recognition. In the United States, it is administered, advanced, and controlled by the MTM Association for Standards and Research. This nonprofit association is one of 12 associations comprised by the International MTM Directorate. Much of MTM systems' success is the result of an active committee structure made up of members of the association.

The MTM family of systems continues to grow. In addition to MTM-1, the association has introduced MTM-2, MTM-3, MTM-V, MTM-C, MTM-M, MTM-MEK, MTM-UAS, and the Windows-based software tool MTM-Link.

MTM-2

In an effort to further the application of MTM to work areas where the detail of MTM-1 would economically preclude its use, the International MTM Directorate initiated a research project to develop less-refined data suitable for the majority of motion sequences. The result of this effort was MTM-2. Defined by the MTM Association of the United Kingdom, it is a system of synthesized MTM data and is the second general level of MTM data. It is based exclusively on MTM and consists of

1. Single basic MTM motions
2. Combinations of basic MTM motions

The data are adapted to the operator and are independent of the workplace or equipment used. In general, MTM-2 should find application in work assignments where

1. The effort portion of the work cycle is more than 1 min.
2. The cycle is not highly repetitive.
3. The manual portion of the work cycle does not involve a large number of either complex or simultaneous hand motions.

The variability between MTM-1 and MTM-2 depends to a large extent on the length of the cycle. This is reflected in Figure 13.2, which shows the range of percentage deviation of MTM-2 from MTM-1. This range of error is considered to be the expected range for 95 percent of the time.

Table 13.2 MTM-1 A1

			MTM-1 Analysis of MTM-C			**Validation**

	ELEMENT TITLE:	Replace page in 3-ring binder				
	STARTS:	Get binder from shelf at left				ANALYST:
MTM Association	INCLUDES:	Get binder, open cover, locate correct page,				
for Standards		open rings, replace old sheet,				
and Research	ENDS:	close rings, Aside binder to shelf				DATE:

Left-Hand Description	F	LH Motion	TMU	RH Motion	F	Right-Hand Description
1. GET BINDER—OPEN COVER						
Reach to binder		R30B	25.8			
Grasp binder		G1A	2.0			
Move to desk		M30B	24.3			
Release		RL1	2.0			
Reach to cover		R7B	9.3			
Grasp edge		G1A	2.0			
Open cover		M16B	15.8			
Release		RL1	2.0			
			83.2			
2. LOCATE CORRECT PAGE						
			14.6	EF	2	Read first page data
Reach to edge	3	R3D	21.9			
Grasp	3	G1B	10.5			
Move up	3	M4B	20.7			
Regrasp		G2	—			
			43.8	EF	2 × 3	Identify pages
Move pages back		M8B	10.6			
Release		RL1	2.0			
Reach to hold		R8B	10.1	(R4B)		To edge of page
Grasp		G5	0.0	G5		Contact
			8.0	M½B	4	Slide back up
Contact	3	G5	0.0	RL2	4	Release
Move	3	(M½B)	7.5	R1B	3	To corner
			0.0	G5	3	Contact
Regrasp pages		G2	5.6			
			87.6	EF	4 × 3	Identify pages
Move pages back		M8B	10.6			
Release		RL1	2.0			
			255.5			
3. REPLACE PAGE						
To ring		R7A	7.4	R7A		To ring
Grasp		G1A	2.0	G1A		Grasp
Pull open		APB	16.2	APB		Pull open
Open		M½A	2.0	M½A		Open
Release		RL1	2.0	RL1		Release
To edge of paper		R6D	10.1			
Grasp		G1B	3.5			

(Continued)

Table 13.2   (continued)

Left-Hand Description	F	LH Motion	TMU	RH Motion	F	Right-Hand Description
To basket		M30B	24.3	(R-E)		
Release		RL1	2.0			
			10.1	R6D		To new sheet
			3.5	G1B		Grasp
			15.2	M12C		To rings
			16.2	P2SE		Align to ring
			2.0	M½C		To ring
			16.2	P2SE		Align
			2.0	M½A		Down on rings
			2.0	RL1		Release
To center ring		(R4B)	8.6	R6B		To center ring
Grasp		G1A	2.0	G1A		Grasp
Press to close		APB	16.2	APB		Press to close
Close		M½A	2.0	M½A		Close
Release		RL1	2.0	RL1		Release
			167.5			
4. CLOSE COVER AND ASIDE BINDER						
Reach to cover		R7B	9.3			
Grasp edge		G1A	2.0			
Close cover		M16B	15.8			
Release		RL1	2.0			
Reach to binder		R6B	8.6			
Grasp		G1A	2.0			
Regrasp		G2	5.6			
Move to shelf		M30B	24.3			
Release		RL1	2.0			
			71.6			
ELEMENT SUMMARY						
1. Get Binder—Open Cover			83.2			
2. Locate Correct Page			255.5			
3. Replace Page			167.5			
4. Close Cover and Aside Binder			71.6			
		TOTAL	577.8			

MTM-2 recognizes 11 classes of actions, which are referred to as categories. These 11 categories and their symbols are

GET	G
PUT	P
GET WEIGHT	GW
PUT WEIGHT	PW
REGRASP	R
APPLY PRESSURE	A
EYE ACTION	E
FOOT ACTION	F
STEP	S
BEND & ARISE	B
CRANK	C

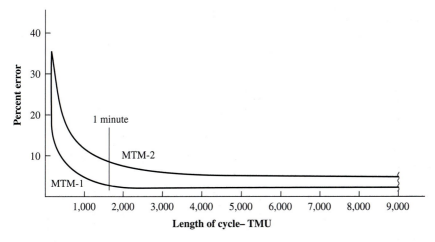

Figure 13.2 Percentage variation of MTM-1 when compared with MTM-2 at increasing cycle lengths.
Courtesy: MTM Association.

Table 13.3 Summary of MTM-2 Data

Range	Code	GA	GB	GC	PA	PB	PC
Up to 2″	−2	3	7	14	3	10	21
Over 2″–6″	−6	6	10	19	6	15	26
Over 6″–12″	−12	9	14	23	11	19	30
Over 12″–18″	−18	13	18	27	15	24	36
Over 18″	−32	17	23	32	20	30	41

	GW 1 - per 2 lb			PW 1 - per 10 lb		
	A	R	E	C	S	F	B
	14	6	7	15	18	9	61

Source: MTM Association.

   In using MTM-2, analysts estimate distances by classes; these distances affect the times of the GET (G) and PUT (P) categories. As in MTM-1, the analysts base the distance moved on the path traveled by the knuckle at the base of the index finger for hand motions, and the path traveled by the fingertips, if only the fingers move. The codes for the five tabulated distance classes correspond to the five levels of classification of motions, as discussed in Section 4.2 (see Table 13.3).

   Three variables affect the time required to perform a GET: the case involved, the distance traveled, and the weight handled. GET can be considered a composite of the therbligs *reach, grasp,* and *release,* while PUT is a combination of the therbligs *move* and *position.*

   The three cases of GET are A, B, and C. Case A implies a simple contact grasp, such as the fingers pushing an object across a desk. If an object, such as a

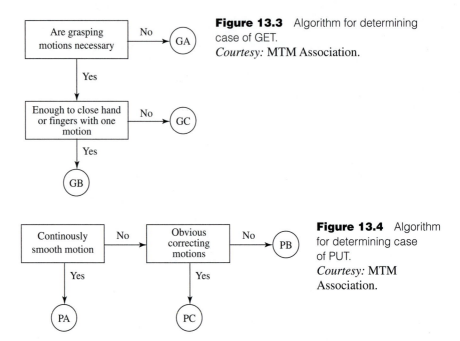

Figure 13.3   Algorithm for determining case of GET.
Courtesy: MTM Association.

Figure 13.4   Algorithm for determining case of PUT.
Courtesy: MTM Association.

pencil, is picked up by simply enclosing the fingers with a single movement, this is a B grasp. If the type of grasp is neither an A nor a B, then it is a case C GET. Analysts can resort to the decision diagram (see Figure 13.3) to assist in determining the correct case of GET. Tabular values in TMU of the three cases of GET, as applied to each of the five coded distances, appear in Table 13.3.

PUT involves moving an object to a destination with the hand or fingers. It starts with an object grasped and under control at the initial place and includes all transporting and correcting motions necessary to place the object at the destination. PUT ends with the object still under control at the intended place. The same case, distance, and weight variables affect PUT as for GET.

Just as there are three cases of GET, there are three cases of PUT. The PUT case depends on the number of correcting motions required. A correction is an unintentional stop, hesitation, or change in direction at the terminal point.

1. *PA: No correction.* This is a smooth motion from start to finish and is the action employed in laying an object aside, or placing it against a stop or in an approximate location. This is the most common PUT.
2. *PB: One correction.* This PUT occurs most often in positioning easy-to-handle objects where a loose fit occurs. It is difficult to recognize. The decision diagram shown in Figure 13.4 is designed to identify this PUT by exception.
3. *PC: More than one correction.* Multiple corrections, or several very short, unintentional motions, are usually obvious. These unintentional motions are generally caused by handling difficulty, close fits, nonsymmetry of engaging parts, or uncomfortable working positions.

Table 13.4 Comparison of Insertion and Alignment PUTs [in (mm)]

	PA	PB	PC
Insertion	Clearance > 0.4 in (10.2 mm)	Clearance < 0.4 in (10.2 mm)	Tight fit
Alignment	Tolerance > 0.25 in (6.3 mm)	0.0625 in (1.6 mm) < Tolerance < 0.25 in (6.3 mm)	Tolerance < 0.0625 in (1.6 mm)

Analysts identify cases of PUT by the decision model shown in Figure 13.4. When in doubt, analysts assign the higher class. If an engagement of parts follows a correction, and if the engagement distance exceeds 1 in (2.5 cm), an additional PUT is used. The tabular values for the three cases of PUT as applied to the five coded distances (the same as for GET) are given in Table 13.3.

A final technicality involving PUTs is that a PUT can be accomplished in one of two ways: insertion or alignment. An insertion involves placing one object into another, such as a shaft into a sleeve. With an insertion, the terminal point for a correction is the point of insertion. An alignment involves orienting a part on a surface, such as bringing a rule up to a line. Table 13.4 may be helpful in assisting the analyst in better identifying the appropriate case.

Weight in MTM-2 is determined similarly to MTM-1. The time value addition for GET WEIGHT (GW) is 1 TMU per 2 lb (1 kg). Thus, if a load of 12 lb (6 kg) is handled by both hands, the time addition due to weight would be 3 TMU, since the effective weight per hand would be 6 lb (3 kg). For PUT WEIGHT (PW), additions are 1 TMU per 10 lb (5 kg) of effective weight, up to a maximum of 40 lb (20 kg). Weights less than 4 lb (2 kg) are not considered.

The category REGRASP (R) is also similar to MTM-1. Here, however, a time of 6 TMU is assigned. The authors of MTM-2 point out that for a REGRASP to be in effect, the hand must retain control.

APPLY PRESSURE (A) has a time of 14 TMU. The authors point out that this category can be applied by any member of the body and that the maximum permissible movement for an apply pressure is ¼ in (6.4 mm).

EYE ACTION (E) is allowed under either of the following cases:

1. The eyes must move to see various aspects of the operation involving more than one specific section of the work area. This eye movement is defined as moving beyond a circle of 4-in (10-cm) diameter at a typical viewing distance of 16 in (40 cm). Note that this is equivalent to the primary visual field defined in Chapter 4.

2. The eyes must concentrate on an object to recognize a distinguishable characteristic.

The estimated value of EYE ACTION is 7 TMU. The value is only allowed when EYE ACTION is independent of hand or body motions.

CRANK (C) occurs when the hands or fingers move an object in a circular path of more than one-half revolution. For less than one-half revolution, a PUT is used instead. Under MTM-2, the category CRANK has only two variables: the number of revolutions and the weight or resistance. A time of 15 TMU is allotted for each complete revolution. Where weight or resistance is significant, PUT WEIGHT is applied to each revolution.

FOOT (F) movements are 9 TMU, and STEP (S) movements are 18 TMU. The time for STEP movement is based on a 34-in (85-cm) pace. The decision diagram (see Figure 13.5) can be helpful in ascertaining whether a given movement is a STEP or a FOOT movement.

The category BEND & ARISE (B) occurs when the body changes its vertical position. Typical movements characteristic of BEND & ARISE include sitting down, standing up, and kneeling. A time value of 61 TMU is assigned to this category. However when an operator kneels on both knees, the movement should be classed as 2 B. Table 13.3 summarizes the applicable MTM-2 values.

There are several special situations of which the analyst should be aware in performing a correct MTM-2 analysis. Motions performed with both hands simultaneously cannot always be performed in the same time as motions performed by one hand only. Figure 13.6 reflects motion patterns for which the time required for simultaneous motions is the same as that required for easy motions performed by one hand. In these instances, an open rectangle appears. An X in the rectangle indicates that, with practice, simultaneous motions can be made. A darkened rectangle indicates that it is difficult, even with practice, to perform the motions simultaneously. Figure 13.7 shows how much additional time these difficult

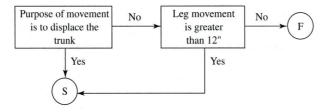

Figure 13.5 Algorithm for differentiating between STEP (S) and FOOT motion (F). *Courtesy:* MTM Association.

Motion		GET				PUT		
	Case	GA	GB	GC	PA	PB O*W	PC	
Get	GA					⊠	▓	
	GB					⊠	▓	
	GC			▓	⊠	▓	▓	
Put	PA					⊠	⊠	
	PB	⊠	⊠	▓	⊠	▓	▓	
	PC	⊠	▓	▓	⊠	▓	▓	

Easy ▭   Practice ⊠   Difficult ▓

*O = Outside; W = Within normal vision

Figure 13.6 Difficulty of simultaneous motions. *Courtesy:* MTM Association.

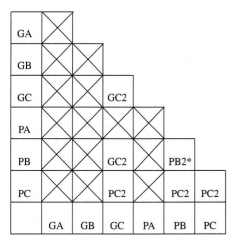

Figure 13.7 MTM-2 simultaneous hand motion allowances.

*IF PB _____ is performed simultaneously with PB _____, an addition of PB2 is made only if the actions are outside the normal area of vision.

simultaneous motions require. An example of the application of this *principle of simultaneous motions* is marked with ① in Figure 13.8. A PC2 is added to the overall time because of two simultaneous PC motions in the right and left hands.

A second situation involves the *principle of limiting motions:* for two different motions performed simultaneously by the left and right hands, the longer time predominates. This is shown as a ② in Figure 13.8, with the shorter element circled out (GB18). A GC12 is 23 TMUs, as compared to 18 TMUs for a GB18.

If one hand performs two motions simultaneously, the longer time again predominates, because of the *principle of combined motions*. This would apply for case ③; a GB18 is longer than an F. (A curve connects the two lines to indicate a combined motion.)

A similar situation appears for case ④, but it is more complicated because of the combined motion in the left hand. In this case, R is already assumed to be part of any C-type motion and is thus crossed out. Therefore, the overall time for the first two lines is 24 TMU, because of the right hands' GB18 (18 TMU) and R (6 TMU) rather than GC12 (23 TMU). However, if the R is a completely separate motion, then it would be counted. As an example, Figure 13.9 shows a complete MTM-2 analysis for a flashlight assembly.

MTM-3

The third level of methods–time measurement is MTM-3. This level was developed to supplement MTM-1 and MTM-2. MTM-3 is helpful in work situations where an interest in saving time at the expense of some accuracy makes it the best alternative. The accuracy of MTM-3 is within ±5 percent, with a 95 percent confidence level, when compared to MTM-1 analysis for cycles of approximately 4 minutes. It has been estimated that MTM-3 can be applied in about one-seventh

MTM Methods Analysis

Page   of

Operation:						
Study No.:						
Date:						
Analyst:						

Remarks:

Description	No.	LH	TMU	RH	No.	Description
	①	PC6	26	PC6		
		PC2	21	PC2		
	②	GC12	23	GB18		
	③	GA2	18	GB18		
				F1		
	④	GC12	18	GB18		
		R1	6	R		

Summary	Total TMU:	Conversion:	% Allowance:	Standard time:

Figure 13.8 Examples of MTM-2 conventions.

the time of MTM-1. However, MTM-3 cannot be used for operations that require either eye focus or eye travel time, since the data do not consider those motions. The MTM-3 system consists of only four categories of manual motions:

1. *Handle* (H). A motion sequence with the purpose of getting control of an object with the hand or fingers and placing the object in a new location.

MTM Methods Analysis

Page   of

Operation:	FLASH LIGHT ASSEMBLY	Remarks:
Study No.:	1-2	
Date:	8-22	
Analyst:	A F	

Description	No.	LH	TMU	RH	No.	Description
PICK UP HOUSING		GB12	14	GB12		GET BATTERY #1
TO WORK AREA		(PA12)	19	PB12		MOVE TO HOUSING
			6	R		REGRASP TO ORIENT
			14	GB12		GET BATTERY #2
			19	PB12		MOVE TO HOUSING
			6	R		REGRASP TO ORIENT
			14	GB12		GET END CAP
			19	PB12		MOVE END CAP
			6	R		REGRASP
			3	PA2		PUSH SPRING
			1	PW10		PUSH WITH FORCE
			21	PC2		BACK TURN-ALIGN
			15	C		FULL TURN
			14	A		TIGHTEN
SET DOWN		PA12	11			

Summary	Total TMU: 182	Conversion: 0.0006	% Allowance: 10	Standard time: 0.120

Figure 13.9 MTM-2 analysis of flashlight assembly.

2. *Transport* (T). A motion with the purpose of moving an object to a new location with the hand or fingers.
3. *Step and Foot motions* (SF). These are the same as defined in MTM-2.
4. *Bend and Arise* (B). These, too, are the same as defined in MTM-2.
5. Case A or B is determined by whether there is a correcting motion or not.

Table 13.5 Summary of MTM-3 Data

		Handle		Transport	
Inches	**Code**	**HA**	**HB**	**TA**	**TB**
6	6	18	34	7	21
6	32	34	48	16	29
		SF 18		B 61	

Table 13.5 presents MTM-3 data. Ten time standards, ranging from 7 to 61 TMU, form the basis for the development of any standard, subject to the limitations noted earlier.

MTM-V

MTM-V was developed by Svenska MTM Gruppen, the Swedish MTM Association, for use in metal cutting operations. It is of particular use with short runs in machine shops. MTM-V provides for work elements involved in (1) bringing the work to the jig, fixture, or chuck, removing the work from the machine, and placing it aside; (2) operating the machine; (3) checking the work to ensure quality of output; and (4) cleaning the nip point area of the machine, to maintain facility output and product quality. MTM-V does not cover process time involving feeds and speeds.

Analysts use this system to establish setup times for all typical machine tools. Therefore, standard times for such elements as setting up and dismounting fixtures, jigs, stops, cutting tools, and indicators can be calculated. All manual cycle times of 24 min (40,000 TMU) or more established by MTM-V are within ±5 percent of that produced by MTM-1 at the 95 percent confidence level. MTM-V is about 23 times faster than MTM-1.

MTM-C

MTM-C, which is widely used in the banking and insurance industries, is a two-level standard data system used to establish time standards for clerical-related work tasks, such as filing, data entry, and keyboarding. Both levels of MTM-C are traceable to MTM-1 data.

The system provides three distinct ranges for reach and move (Get Place). A six-place numeric coding system (similar to MTM-V) provides a detailed description of the operation being studied.

MTM-C develops standards in the same way as the other MTM systems. Analysts can combine it with existing proven standard data or standard data developed from other sources or techniques. MTM-C is available in manual or automated forms. For the latter, an MTM-C data set can be incorporated into MTM-Link.

The nine level 1 categories used in MTM-C, shown in Table 13.6, are as follows:

1. *Get Place*. This category includes the basic work divisions required to get an object, move it aside without relinquishing control, and release it. For

Table 13.6 MTM-C Level-1 Elements

Level 1 Elements	Symbol
Get Place	11 X X X X
Open Close	21 X X X X
Fasten Unfasten	31 X X X X
Organize File	4 X X X X X
Read Write	5 X X X X X
Keyboarding	6 X X X X X
Handling	7 X X X X X
Walk Body Motions	8 X X X X X
Machines	9 X X X X X

example, the coding and description of an element in this category might be 112210—Get small stack with medium move.

2. *Open/Close*. Characteristic of this category are such operations as opening or closing books, doors, drawers, binder rings, zippered objects, envelopes, and files. An example of the coding for a representative operation would be 212100—Open hinge cover, medium.

3. *Fasten/Unfasten*. This category includes attaching and removing clips, clamps, bands, and staples, all used to join materials. A representative coding for this work element would be 312130—Fasten with a large paper clip.

4. *Organize File*. This category includes the basic elements involved with filing activities and some of the organizational handling of work directly or indirectly related to filing. An example of the coding and description of this category is 410400—Arrange a pile into a stack.

5. *Read/Write*. This category includes a prose reading speed of 330 words per minute. Writing time values have been developed for letters, numerals, and symbols. Values are a weighted average based on the frequency of occurrence of each type of character in normal prose. An example of coding and a representative description would be 510600—Read average prose, per word.

6. *Keyboarding*. This category includes all the actions related to data entry and the manual keyboarding functions. An example of the coding and the description in this category follows: 613530—Insert single object in typewriter, long distance.

7. *Handling*. This category includes all the clerical activities not covered in the other categories. An example of the coding and the description of an element in this category might be 760600—Adhere envelope flap.

8. *Walk Body Motions*. This category includes walking values based on pace. Body motions include sitting, standing, and horizontal and vertical body movements while in a chair. An example of the coding and description of an element in this category follows: 860002—Move while in swivel chair.

9. *Machines*. The machine data are representative of a group of similar types of equipment. Data for keyboards are typical examples of this category.

Level 2 data are directly traceable to level 1 and to MTM-1. Level 2 elements and their symbols are shown in Table 13.7. The level 2 elements are as follows:

1. *Get/Place/Aside.* These elements are applied collectively or separately. The coding and element example of this category, with collective basic divisions, would be G5PA2—Get a pencil for use and set it aside for later.

2. *Open/Close.* Included in these data is getting the object being opened or closed. These data are applied individually or in combination, as follows: C65—Close string, tie envelope; or OC4—Open and close binder rings.

3. *Fasten/Unfasten.* For fasten (F), the element is made up of getting the objects and fastening them. For unfasten (U), the element includes getting the objects and unfastening them.

4. *Identify.* The data for this element include eye travel time values, along with eye focuses required to identify (I) single or multiple words and sets of numbers.

5. *Locate File.* The data for this element cover typical filing activities. The first position of the coding is L, for locate. The second position is also a letter that corresponds to the filing activity, such as LI (insert), LR (remove), LT (tilt and replace).

6. *Read/Write.* The reading data include the reading of words and single numbers and/or characters. Read also contains detailed Read-and-Compare, as well as Read-and-Transcribe data. The writing data include common clerical items, such as address, date, initials, and names. The coding and element description for two representative elements would be RW20—Read 20 words and RCN25—Read and compare 25 numbers.

7. *Handling.* This element includes the actual paper handling activities from level 1: organize data and handling data. In the majority of elements,

Table 13.7   MTM-C Level 2 Elements

Level 2 Elements	Symbol
Aside	A
Body Motions	B
Close	C
Fasten	F
Get	G
Handling	H
Identify	I
Locate File	L
Open	O
Place	P
Read	R
Typing	T
Unfasten	U
Write	W

objects have been obtained with a Get, as well as the action handling elements. In the coding for handling elements, H is the first coding position. The second position is the initial letter of the element activity. An example of coding for folding a sheet with two folds would be HF12.

8. *Body Motions*. These elements include walking, sitting and standing, bending, and arising, and the horizontal body motions alone or in a chair.

9. *Keyboarding*. These elements include three major sections of data: Handling, Keystroke, and Correction. An example of the coding and description is TKE17E—Type one 7-in (17.5-cm) line.

MTM-C level 1 can be calculated faster than MTM-2. Also, the speed of MTM-C level 2 is faster than MTM-3. Compare the standards for replacing a page in a three-ring binder, developed first using MTM-1 (see Table 13.2), then using MTM-C level 1 (see Table 13.8), and finally using MTM-C level 2 (see Table 13.9). Note how closely the three standards agree (see Table 13.10).

MTM-M

MTM-M is a predetermined time system for evaluating operator work using a microscope. In the development of MTM-M, basic times from MTM-1 were not used, although the beginning and endpoint definitions of motion elements were compatible with MTM-1. The data used were original data developed through the efforts of the United States/Canada MTM Association. In general, MTM-M is a higher-level system, similar to MTM-2.

This system has four major tables and one subtable. Analysts must consider four variables when selecting the appropriate data: (1) type of tool, (2) condition of tool, (3) terminating characteristic of the motion, and (4) distance/tolerance ratio. Additional factors, other than the direction of movement and these four variables, that have an impact on motion performance time include

1. Tool load state, empty or loaded
2. Microscope power
3. Distance moved
4. Positioning tolerance
5. Purpose of the motion, as determined by the manipulations involved at the motion termination (for example, workers may use tweezers to contact grasp an object, or to pick up an object)
6. Simultaneous motions

With the growing amount of microminiature manufacturing, the application of fundamental data similar to MTM-M will expand. Such data allow analysts to establish equitable standards that would be difficult to establish by stopwatch procedures. Establishing reliable elemental standards by direct observation is impossible. It is only possible to establish sound elemental and operation standards for microscope work by using standard data similar to MTM-M or by micromotion procedures.

Table 13.8 MTM-C Operation Analysis (Level 1)

	MTM-C Operation Analysis			Validation
MTM Association for Standards and Research	MTM-C Level 1 Replace page in 3-ring binder			Sheet of
DEPARTMENT: Clerical	ANALYST: CNR			DATE: 11/77
No. Description	Reference	Element (TMU)	Occurrence per Cycle	TMU per Cycle
1. OPEN BINDER				
Get binder from shelf	113 520	21	1	21
Aside to desk	123 002	22	1	22
Get cover	112 520	14	1	14
Open cover	212 100	15	1	15
2. LOCATE CORRECT PAGE				
Read on first page	510 000	7	2	14
Locate approximate	451 120	16	3	48
Identify page number	440 630	22	3	66
Locate correct page	450 130	18	4	72
Identify pages	440 630	22	3	66
3. REPLACE PAGES				
Get binder rings	112 520	14	1	14
Open rings	210 400	21	1	21
Get old sheet	111 100	10	1	10
Aside sheet to basket	123 002	22	1	22
Get new sheet	111 100	10	1	10
Insert sheet in binder	462 104	64	1	64
Get rings	112 520	14	1	14
Close rings	222 400	21	1	21
4. CLOSE COVER AND ASIDE BINDER				
Get cover	111 520	8	1	8
Close cover	222 100	13	1	13
Get binder	112 520	14	1	14
Aside binder to shelf	123 002	22	1	22
		TOTAL TMU PER CYCLE		571
		ALLOWANCES _____ %		
		STANDARD HOURS PER _____ UNIT		
		UNITS PER HOUR		

OTHER SPECIALIZED MTM SYSTEMS

Three other specialized MTM systems are MTM-TE, MTM-MEK, and MTM-UAS. The first of these, MTM-TE, was developed for electronic tests. This system has two levels of data that were developed from MTM-1 for basic test application. Level 1 includes the elements *get, move, body motions, identify, adjust,* and miscellaneous data. Level 2 includes *get* and *place, read*

Table 13.9 MTM-C Operation Analysis (Level 2)

	MTM-C Operation Analysis			Validation
	MTM-C Level 2			Sheet of
MTM Association for Standards and Research	Replace page in 3-ring binder			
DEPARTMENT: Clerical	ANALYST: CNR			DATE: 2/77
No. Description	Reference	Element (TMU)	Occurrence per Cycle	TMU per Cycle
Get and aside binder	G5A2	29	1	29
Open cover	O1	29	1	29
Read first page	RN2	14	1	14
Locate pages	LC12	129	1	129
Identify pages	130	22	6	132
Open rings	O4	35	1	35
Remove sheet	G1A2	32	1	32
New sheet on rings	HI14	84	1	84
Close rings	C4	35	1	35
Close cover	C1	27	1	27
Aside binder	G5A2	29	1	29
			TOTAL TMU PER CYCLE	575

Table 13.10 Comparison of MTM-1, MTM-C (1), and MTM-C (2)

Techniques	Number of Elements	Standard
MTM-1	57	577.8
MTM-C level 1	21	577
MTM-C level 2	11	575

and *identify, adjust, body motions,* and *writing*. A third level of data is also available in the form of synthesized level 1 data. MTM-TE data do not cover "troubleshoot" relative to electronic test operations. They do, however, provide guidelines for investigations and recommendations for work measurement for this activity.

The second specialized system, MTM-MEK, was designed to measure one-of-a-kind and small-lot production. This two-level system developed from MTM-1 can analyze all manual activities, as long as the following requirements are met:

1. The operation is not highly repetitive or organized, although it may contain similar elements that require different methods. The method used to perform a given operation typically varies from cycle to cycle.

2. The workplace, tools, and equipment used are universal in character.

3. The task is complex and necessitates employee training; yet the lack of a specific method requires a high degree of versatility by the operator.

The objectives of MTM-MEK are to

1. Provide accurate measurement of an activity connected with one-of-a-kind or small-lot production
2. Provide an easily definable description of unorganized work, thus generally identifying a procedure
3. Provide fast application
4. Provide accuracy relative to MTM-1
5. Require minimum training and application practice

The data in MTM-MEK consist of 51 time values in the following eight categories: get and place, handle tool, place, operate, motion cycles, fasten or loosen, body motions, and visual control. In addition, there are standard data for a wide range of assembly tasks in one-of-a-kind and small-lot production. These data consist of 290 time values in the following categories: fasten, clamp and unclamp, clean and/or apply lubricant/ adhesive, assemble standard parts, inspect and measure, mark and transport.

The third specialized system, MTM-UAS, is a third level-system developed to provide a process description, as well as to determine the standard times in any activity related to batch production. MTM-UAS is applicable to various activities, as long as the following characteristics of batch production are present:

1. Similar tasks
2. Workplace specifically designed for the task
3. Good levels of work organization
4. Detailed instructions
5. Well-trained operators

The MTM-UAS analyzing system consists of 77 time values in seven of the eight categories used in MTM-MEK. These are get and place, place, handle tool, operate, motion cycles, body motions, and visual control. MTM-UAS is about eight times faster than MTM-1. At cycle times of 4.6 min or more, the standard produced by MTM-UAS is within ±5 percent of that produced by MTM-1, with a 95 percent confidence level.

MTM SYSTEMS COMPARISON

Figure 13.10 illustrates the total absolute accuracy, at a 90 percent confidence level, of all the MTM systems. Table 13.11 compares the level of detail, such as the number of therbligs utilized, the time to analyze a job (expressed as a multiple of the job cycle time), and the accuracy, for the three basic MTM systems. Overall, MTM-2 may be a good compromise between the excessive time required for MTM-1 and the poor accuracy of MTM-3. A 6-min job will take approximately 600 min to analyze with MTM-2 and will be off by no more than 0.24 min.

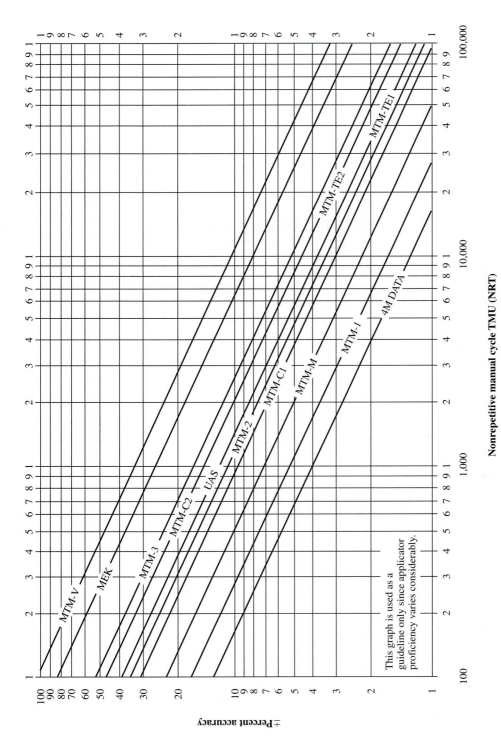

Nonrepetitive manual cycle TMU (NRT)

Figure 13.10 Total absolute accuracy at 90 percent confidence level of the various MTM systems.

Table 13.11 Comparison of MTM-1, MTM-2, and MTM-3

Therbligs utilized	MTM-1	MTM-2	MTM-3
Release			
Reach	Get	Handle	
Grasp			
Move	Put		
Position			
Time to analyze job	250 × cycle time	100 × cycle time	35 × cycle time
Relative speed	1	2.5	7
Time/accuracy - 100 TMU	15 min/± 21%	6 min/± 40%	2 min/± 70%
Time/accuracy - 10,000 TMU	1500 min/± 2.1%	600 min/± 4%	200 min/± 7%

13.2 MAYNARD OPERATION SEQUENCE TECHNIQUE (MOST)

An outgrowth of MTM, called Maynard Operation Sequence Technique (MOST), is a simplified system developed by Zandin (1980) and originally applied at Saab-Scania in Sweden in 1967. With MOST, analysts can establish standards at least five times faster than with MTM-1, with little if any sacrifice in accuracy.

Similar to MTM, there are three levels of MOST system. At the highest level, MaxiMOST is used to analyze long, infrequent operations. Such operations may range from 2 min to several hours in length, occur less than 150 times per week, and tend to have high variability. As such it is very quick but less accurate. At the lowest level, MiniMOST is used for very short and very frequent operations. Such operations are less than 1.6 min in length, are repeated more than 1500 times a week, and have little variability. Consequently, the analysis is very detailed and precise but quite time-consuming. The intermediate level of accuracy is covered by BasicMOST, which covers operations between the two ranges described above. The typical operation best suited for BasicMOST would be 0.5 to 3 min in length.

MOST identifies three basic sequence models: general move, controlled move, and tool and equipment use. The general move sequence identifies the spatial free movement of an object through the air, while the controlled move sequence describes the movement of an object when it either remains in contact with a surface or remains attached to another object during the movement. The tool and equipment use sequence is aimed at the use of common hand tools and other pieces of equipment.

To identify the exact way a general move is performed, analysts consider four subactivities: action distance (A), which is primarily horizontal movement distance; body motion (B), which is mainly vertical motion; gain control (G); and placement (P). These subactivities are grouped into three phases: get, put, and return (see Figure 13.11). Get means reaching some distance with hands, perhaps with body motion or steps, for an object and gaining manual control of the object. It uses three of the subactivities, A, B, and G, to define this phase of the

General Move

Get	Put	Return
A B G	A B P	A

Controlled Move

Get	Move/Actuate	Return
A B G	M X I	A

Tool/Equipment Use

Get	Put in place	Use	Put aside	Return
A B G	A B P	*	A B P	A

Figure 13.11 BasicMOST activities and subactivities.

general move. Put is moving the object some distance to new location (A), perhaps with body motion (B), and placing the object at a specified location (P). Return means walking back some distance to the workstation (A). This is not meant for the hands, and typically if the operator remains at the workstation, there is no return.

Each subactivity is further defined by time-related index values corresponding to the relative difficulty of the subactivity. MOST starts with the values 0, 1, 3, 6, 10, and 16, but can proceed to higher values for very specific subactivities, such as long walking distances or complicated or long controlled moves. The specific index values with their descriptions for the general move are shown in both Figures 13.12 and 13.13. These index values, when scaled by a factor of 10, yield the appropriate time values of the subactivities in TMUs.

For example, to get a washer 5 in (12.5 cm) away, place on a bolt located 5 in away, and return to the original position would yield $A_1B_0G_1A_1B_0P_1A_1$ with a total time of $(1 + 0 + 1 + 1 + 0 + 1 + 1) \times 10 = 50$ TMU. The get is defined by A_1 = reach to washer with 5-in travel, B_0 = no body motion, G_1 = grasp washer; the put is defined by A_1 = place washer with 5-in travel, B_0 = no body motion, P_1 = place washer with a loose fit. The final A_1 = return to original position with 5-in travel.

About 50 percent of manual work occurs as general moves. A typical general move may include the parameters of walking to a location, bending to pick up an object, reaching and gaining control of the object, arising after bending, and placing the object. As another example, consider walking three steps, bending down to pick up a bolt from the floor, arising, walking the three steps back, and placing the bolt in a bolthole. The analysis yields $A_6B_6G_1A_1B_0P_3A_0$ with a total time of $(6 + 6 + 1 + 1 + 3 + 0) \times 10 = 170$ TMU. The get is defined by A_6 = walk three steps to object, B_6 = bend and arise, G_1 = gain control of light object; the put is defined by A_6 = walk three steps to place object, B_0 = no body motion, P_3 = place and adjust object, and A_0 = no return.

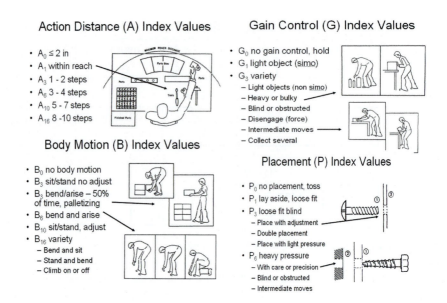

Action Distance (A) Index Values

- $A_0 \leq 2$ in
- A_1 within reach
- A_3 1 - 2 steps
- A_6 3 - 4 steps
- A_{10} 5 - 7 steps
- A_{16} 8 -10 steps

Gain Control (G) Index Values

- G_0 no gain control, hold
- G_1 light object (simo)
- G_3 variety
  - Light objects (non simo)
  - Heavy or bulky
  - Blind or obstructed
  - Disengage (force)
  - Intermediate moves
  - Collect several

Body Motion (B) Index Values

- B_0 no body motion
- B_3 sit/stand no adjust
- B_3 bend/arise – 50% of time, palletizing
- B_6 bend and arise
- B_{10} sit/stand, adjust
- B_{16} variety
  - Bend and sit
  - Stand and bend
  - Climb on or off

Placement (P) Index Values

- P_0 no placement, toss
- P_1 lay aside, loose fit
- P_3 loose fit blind
  - Place with adjustment
  - Double placement
  - Place with light pressure
- P_6 heavy pressure
  - With care or precision
  - Blind or obstructed
  - Intermediate moves

Figure 13.12   General move subactitivities.

The controlled move sequence covers such manual operations as cranking, pulling a starting lever, turning a steering wheel, or engaging a starting switch and accounts for approximately one-third of all work sequences. In controlled move sequences, the following subactivities are used: the previously defined action distance (A), body motion (B), gain control (G); and the new subactivities: move controlled (M), in which the path of the object is controlled; process time (X), controlled not by manual actions but by mechanical devices; alignment (I), the final conclusion of the controlled move process to achieve alignment of two objects. These subactivities are also grouped into three phases: get, move/actuate, and return. As for the general move, get involves reaching some distance with the hands for an object (A), perhaps with body motion or steps (B), and then gaining manual control of the object (G). The move is over a controlled path perhaps with body motion or steps (M), with a time allotment for the control process to occur or a device be actuated (X), and final alignment of the object at end of the process time (I). Finally, similar to the general move, there is a return to the workplace if needed (A). Again there are the basic 0, 1, 3, 6, 10, and 16 index values with larger ones allocated for longer process times. The subactivities are shown in Figure 13.14 and with the complete tabulation shown in Figure 13.13.

As an example of a controlled move, engaging a feed lever on a milling machine yields $A_1 B_0 G_1 M_1 X_{10} I_0 A_0$ with a total time of $(1 + 0 + 1 + 1 + 10 + 0 + 0) \times 10 = 130$ TMU. The get subactivity is defined by A_1 = reach to lever (within reach), B_0 = no

Index	Interval Mean TMU	BasicMOST Interval Limits TMU
0	0	0
1	10	1 - 17
3	30	18 - 42
6	60	43 - 77
10	100	78 - 126
16	160	127 - 196
24	240	197 - 277
32	320	278 - 366
42	420	367 - 476
54	540	477 - 601
67	670	602 - 736
81	810	737 - 881
96	960	882 - 1041
113	1130	1042 - 1216
131	1310	1217 - 1411
152	1520	1412 - 1621
173	1730	1622 - 1841
196	1960	1842 - 2076
220	2200	2077 - 2321
245	2450	2322 - 2571
270	2700	2572 - 2846
300	3000	2847 - 3146
330	3300	3147 - 3446

Time Measurement Unit

1 TMU	= .00001 hour	1 hour	= 100,000 TMU
	= .0006 minute	1 minute	= 1667 TMU
	= .036 second	1 second	= 27.8 TMU

H. B. Maynard and Company, Inc.
Seven Parkway Center, Pittsburgh, PA 15220-3880 USA
Phone: 412.921.2400 Fax: 412.921.4575
www.hbmaynard.com

© 2005 H. B. Maynard and Company, Inc.

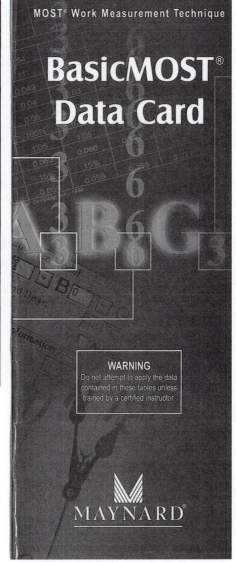

Figure 13.13 BasicMOST® Data Card.

body motion, and G_1 = gain control of light lever. The move/actuate subactivity is defined by M_1 = move lever (< 12 in) to engage the machine, with an X_{10} = process time of ~3.5 s, and I_0 = no precise alignment. The final A_0 = no return to workplace; i.e., all the activity occurred at the same workstation. In a second, more complicated example, a press operator moves a 4×8 ft sheet of thin-gauge steel a

General Move

A B G A B P A
Get Put Return

Index x 10	A — Action Distance	B — Body Motion	G — Gain Control	P — Placement	Index x 10
0	≤ 2 In. (5 cm)			Pickup / Toss	0
1	Within Reach		GRASP: Light Object / Light Objects Simo	PUT: Lay Aside / Loose Fit	1
3	1 - 2 Steps	Sit or Stand / Bend and Arise 50% occ.	GET: Light Objects Non-Simo / Heavy or Bulky / Blind or Obstructed — Disengage / Interlocked / Collect	PLACE: Loose Fit Blind or Obstructed / Adjustments / Light Pressure / Double Placement	3
6	3 - 4 Steps	Bend and Arise		POSITION: Care or Precision / Heavy Pressure / Blind or Obstructed / Intermediate Moves	6
10	5 - 7 Steps	Sit or Stand with Adjustments			10
16	8 - 10 Steps	Stand and Bend / Bend and Sit / Climb On or Off / Through Door			16

A — Action Distance (Extended Values)

Index	Steps	Feet	Meters
24	11 - 15	38	12
32	16 - 20	50	15
42	21 - 26	65	20
54	27 - 33	83	25
67	34 - 40	100	30
81	41 - 49	123	38
96	50 - 57	143	44
113	58 - 67	168	51
131	68 - 78	195	59
152	79 - 90	225	69
173	91 - 102	255	78
196	103 - 115	288	88
220	116 - 128	320	98
245	129 - 142	355	108
270	143 - 158	395	120
300	159 - 174	435	133
330	175 - 191	478	146

Controlled Move

A B G M X I A
Get Move/Actuate Return

Index x 10	M — Move Controlled (Push/Pull/Turn)	(Crank)	X — Process Time (Seconds)	(Minutes)	(Hours)	I — Alignment	Index x 10
1	≤ 12 in. (30 cm) / Button / Switch / Knob		.5 Sec.	.01 Min.	.0001 Hr.	1 Point	1
3	> 12 in. (30 cm) / Resistance / Seat or Unseat / High Control / 2 Stages ≤ 24 in. (60 cm) Total	1 Rev.	1.5 Sec.	.02 Min.	.0004 Hr.	2 Points ≤ 4 in. (10 cm)	3
6	2 Stages > 24 in. (60 cm) Total / 1 - 2 Steps	2 - 3 Rev.	2.5 Sec.	.04 Min.	.0007 Hr.	2 Points > 4 in. (10 cm)	6
10	3 - 4 Stages / 3 - 5 Steps	4 - 6 Rev.	4.5 Sec.	.07 Min.	.0012 Hr.		10
16	6 - 9 Steps	7 - 11 Rev.	7.0 Sec.	.11 Min.	.0019 Hr.	Precision	16

M — Push or Pull (Extended Values)

Index	Steps
24	10 - 13
32	14 - 17
42	18 - 22
54	23 - 28
67	29 - 34

Crank (Extended Values)

Index	Revs.
24	12 - 16
32	17 - 21
42	22 - 28
54	29 - 36

I — Alignment of Machining Tools

Index	Align To
3	Workpiece
6	Scale Mark
10	Indicator Dial

Alignment of Non-typical Objects

Index	Positioning Method
0	Against Stop(s)
3	1 Adjustment to Stop
6	2 Adjustments to Stop(s) / 1 Adjustment to 2 Stops
10	3 Adjustments to Stop(s) / 2 - 3 Adjustments to Linemark

Non-typical Object Characteristics
Flat, Large, Flimsy, Sharp, Difficult to handle

Tool Use

A B G A B P * A B P A
Get Tool Put Tool Tool Action Aside Tool Return

F L — Fasten or Loosen

Index x 10	Finger Action — Spins (Fingers, Screwdriver)	Turns (Hand, Screwdriver, Ratchet, T-Wrench)	Wrist Action — Strokes (Wrench)	Cranks (Wrench, Ratchet)	Taps (Hand, Hammer)	Arm Action — Turns (Ratchet)	Turns (T-Wrench 2-Hands)	Strokes (Wrench)	Cranks (Wrench, Ratchet)	Strikes (Hammer)	Power Tool — Screw Diam. (Power Wrench)	Index x 10
1	1	-	-	-	1	-	-	-	-	-	-	1
3	2	1	1	1	3	1	-	1	-	1	1/4 in. (6 mm)	3
6	3	3	2	3	6	2	1	-	1	3	1 in. (25 mm)	6
10	8	5	3	5	10	4	-	2	2	5		10
16	16	9	5	8	16	6	3	3	3	8		16
24	25	13	8	11	23	9	6	4	5	12		24
32	35	17	10	15	30	12	8	6	6	16		32
42	47	23	13	20	39	15	11	8	8	21		42
54	61	29	17	25	50	20	15	10	11	27		54

Figure 13.13 (continued)

A B G A B P * A B P A
Get Tool Put Tool Tool Action Aside Tool Return

Tool Use

Index x 10	C Cut — Cutoff (Pliers Wire)	Secure (Pliers)	Cut (Scissors Cuts)	Slice (Knife Slices)	S Surface Treat — Air-Clean (Nozzle sq. ft. 0.1 m²)	Brush-Clean (Brush sq. ft. 0.1 m²)	Wipe (Cloth sq. ft. 0.1 m²)	M Measure (Measuring Tool)	Index x 10
1		Grip	1	-	-	-	-		1
3	Soft		2	1	-	-	1/2		3
6	Medium	Twist Form Loop	4	-	1 Spot Cavity	1	-		6
10	Hard		7	3	-	-	1	Profile Gauge	10
16		Secure Cotter Pin	11	4	3	2	2	Fixed Scale Caliper ≤ 12 in. (30 cm)	16
24			15	6	4	3	-	Feeler Gauge	24
32			20	9	7	5	5	Steel Tape ≤ 6 ft. (2 m) Depth Micrometer	32
42			27	11	10	7	7	OD-Micrometer ≤ 4 in. (10 cm)	42
54			33					ID-Micrometer ≤ 4 in. (10 cm)	54

P Tool Placement

Tool	Index	Tool	Index
Hammer	0 (1)	Measuring Tool	1
Fingers or Hand	1 (3) (6)	Screwdriver	3
Pliers	1 (3)	Ratchet	3
Scissors	1 (3)	T-Wrench	3
Knife	1 (3)	Wrench	3
Surface Treating Tool	1	Power Tool	3
		Adjustable Wrench	6 (3)

A B G A B P * A B P A
Get Tool Put Tool Tool Action Aside Tool Return

Tool Use

Index x 10	R Record — Write (Pencil/Pen Digits)	Write Words	Copy	Mark (Marker Digits)	T Think — Inspect (Eyes/Fingers Points)	Read (Digits, Single Words)	Read (Eyes Text of Words)	Compare	Index x 10
1	1	-	-	Check Mark	1	1	3	1	1
3	2	-	1	Scribe Line	3	3 Gauge	8	2	3
6	4	1	3	2	5 Feel for Heat	6 Scale Value Date or Time	15	4	6
10	6	-	5	3	9 Feel for Defect	12 Vernier Scale	24	8	10
16	9 Signature or Date	2	8	5	14	Table Value	38	13	16
24	13	3	10	7	19		54		24
32	18	4	14	10	26		72		32
42	23	5	18	13	34		94		42
54	29	7	22	16	42		119		54

P Tool Placement

Tool	Index
Writing Tool	1
Keyboard/Electric Typewriter	1
Keypad	1
Letter/Paper Handling	1

A B G A B P * A B P A
Get Put Use Equipment Aside Return

Equipment Use

Index x 10	W Keyboard/Electric Typewriter Set	Words	K Keypad Digits	Data	H Letter/Paper Handling Operations	Jog or Tap	Staple	Stamp	Leaf Through Paper	Filing Select	Open/Close Select	File	Open/Close File	Index x 10
1	Tab	Click Mouse	2	2		1	Electric		1					1
3		1	6	6	Open Envelope	3	Hole Punch Hand Remove		4					3
6	Set Tab	2 Date	11	12	Interleaf	6		1 Ink	7	1				6
10	Set Margin	4	18	20	Seal Envelope	10		2	12	3		1		10
16		6	28	32	Fold and Crease	16		3	20	6	2	4	1	16
24	Insert and Remove	8	39	48				5	28	9	6	7	5	24
32		11	52	60				7	37	12	9	10	8	32
42		15 Address	68	79				9	47	17	12	15	11	42
54		19	85	100				11	61					54

Figure 13.13 (continued)

Move (M): Push/Pull/Pivot Index Values

- M_1 one stage $\leq$ 12 in
  - or press button/switch
- M_3 one stage > 12 in
  - or push with force
  - or seat/unseat
  - or high control
  - or two stages $\leq$ 24 in

- M_6 two stages > 24 in
  - or with 1-2 steps
- M_{10} 3-4 stages
  - or 3-4 steps
- M_{16} 6-9 steps
- Extended values

Move (M): Crank Index Values

- Crank:
  - move fingers, wrist, forearm, in circular path > ½ rev
  - if < ½ rev then it is a push/pull/pivot
- M_3 = 1 rev
- M_6 = 2-3 rev
- M_{10} = 4-6 rev
- M_{16} = 7-11 rev

Process Time (X) Index Values

Index Value	Seconds	Minutes
0	No process time	
1	0.5	0.01
3	1.5	0.02
6	2.5	0.04
10	4.5	0.07
16	7.0	0.11
330	124	2.06

Alignment (I) Index Values

- Assume within area of normal vision
- I_1 to one point in single correcting action
- I_3 to two $\leq$ 4 in apart
- I_6 to two > 4 in apart
- I_{16} with precision

Figure 13.14 Controlled move subactivities.

distance of 14 in. The sheet is aligned to two stops on opposite ends of the sheet (not necessary to reposition hands during alignment). The operator takes one step back to gain control of the sheet. The analysis yields $A_3B_0G_3M_3$ $X_0I_6A_0$ with a time of $(3 + 0 + 3 + 3 + 0 + 6 + 0) \times 10 = 150$ TMU. The get subactivity is defined by A_3 = a step back to handle the oversize sheet, B_0 = no vertical body motion, and G_3 = gain control of bulky object. The move/actuate subactivity is defined by M_3 = move sheet more than 12 in, with X_0 = zero process time, and I_6 = a two-point alignment, the points separate by more than 4 in. The final A_0 = no return to workplace; i.e., all the activity occurred at the same workstation.

The third and final activity sequence in BasicMOST is tool and equipment use. Cutting, surface treating, gaging, fastening, recording with tools, keyboarding, paper handling, and even thinking are all covered with this activity sequence which covers one-sixth of all work sequences. The tool and equipment use activity sequence embraces a combination of general move and controlled move activities with five phases of subactivities: (1) get tool, (2) put tool in place for use, (3) use the tool, (4) put the tool aside, and (5) return to workplace (if needed). Get tool means reaching some distance with the hands for the tool, perhaps with body motion or steps, and gaining control of the tool, with the same A, B, and G subactivities as for the general or controlled move. Put tool in place for use includes moving tool to where it will be used (A), perhaps with body motion or steps (B), and a final positioning for use (P). Use of the tool or the specific piece of equipment has a variety of common actions: F = fasten, i.e., assemble with fingers or tool; L = loosen, i.e., disassemble with fingers or tool, opposite of F; C = cut, separate, or divide with a sharp tool; S = surface treat, i.e., apply or remove material from the surface of an object; M = measure, i.e., compare the physical characteristics of object with a standard; R = record information with pen or pencil;

T = think, i.e., eye actions or mental activity to obtain information or to inspect an object; W = keyboard or typewriter, i.e., use a mechanical or electronic data entry device; K = keypad, i.e., use alphanumeric keypad such as on a PDA or telephone; and H = letter or paper handling, i.e., the performance of various paper filing and sorting operations. Put the tool aside (perhaps for later reuse) is similar to the put in a general move with subactivities A, B, and P. Finally, as for the general and controlled moves, there is a return to the workplace if needed (A). Again there are the basic 0, 1, 3, 6, 10, and 16 index values with larger ones for more complicated actions. The complete tabulation is shown in Table 13.12.

As an example of tool use, an operator picks up a knife from a bench two steps away, makes one cut across the top of a cardboard box, and puts the knife back on the bench. The analysis yields $A_3B_0G_1A_3B_0P_1C_3A_3B_0P_1A_0$, with a total time of $(3 + 0 + 1 + 3 + 0 + 1 + 3 + 3 + 0 + 1) \times 10 = 150$ TMU. The get phase is defined by A_3 = walk two steps, B_0 = no vertical body motion, and G_1 = gain control of light knife. The put phase consists of A_3 = walk two steps back, B_0 = no vertical body motion, and P_1 = loose fit of knife on box. The tool use phase is C_3 = slice once with knife. The put tool aside phase is defined by A_3 = walk two steps back to bench, B_0 = no vertical body motion, and P_1 = lay knife on bench. The final A_0 = no return; i.e., the operator remains at the bench.

Another example involves a testing operation, in which a technician picks up a meter lead, places it on a terminal, and reads a voltage off the meter scale. The lead is then placed aside. The analysis yields $A_1B_0G_1A_1B_0P_3T_6A_1B_0P_1A_0$ with a total time of $(1 + 0 + 1 + 1 + 0 + 3 + 6 + 1 + 0 + 1 + 0) \times 10 = 140$ TMU. The get phase is defined by A_1 = reaches for lead (within reach), B_0 = no vertical body motion, and G_1 = gain control of light lead. The put phase consists of A_1 = moves hand with lead within reach, B_0 = no vertical body motion, and P_3 = places lead with slight adjustment on terminal. The tool use phase is T_6 = read voltage value on scale. The put tool aside phase is defined by A_1 = moves lead back within reach, B_0 = no vertical body motion, and P_1 = sets lead back on bench. The final A_0 = no return; i.e., the operator remains at the bench.

The analysis is performed on the BasicMOST form (see Figure 13.15). Basic information such as job code and date ① is entered at the top right-hand corner. The area of work ② is entered below with information on the activity or the job and the conditions encountered ③. The job is broken down into appropriate activities which are numbered sequentially and entered in order in the left side of the form ④. Appropriate activity sequences—general move, controlled move, or tool and equipment use—are selected on the right side of the form ⑤ identified with the corresponding labels by the number. Next the appropriate index values are entered next to the subactivity letters. Finally, the index values are multiplied by 10 (to obtain TMU), entered in the right most column and summated in the bottom right-hand corner ⑥.

There are some general rules to be followed. Each activity sequence is fixed; i.e., no letter may be added or omitted. Index values are fixed as given; i.e., there is no interpolation; round up if necessary. Finally, there may be some adjustments for the amount of simultaneous activity (similar to MTM). In a high-level interaction

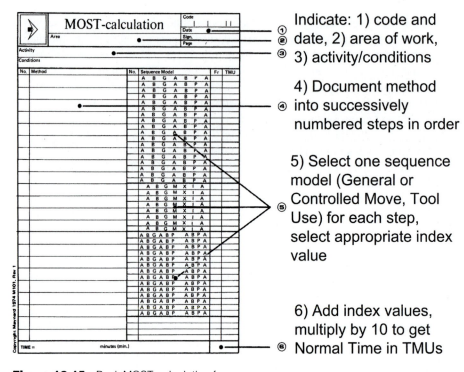

Indicate: 1) code and
② date, 2) area of work,
③ 3) activity/conditions

4) Document method
④ into successively
numbered steps in order

5) Select one sequence
model (General or
Controlled Move, Tool
⑤ Use) for each step,
select appropriate index
value

6) Add index values,
multiply by 10 to get
⑥ Normal Time in TMUs

Figure 13.15 BasicMOST calculation form.

with both hands working simultaneously, only the highest value from the two hands is used (60 TMU), and the other hand and TMU value are circled out:

| RH | $A_1 B_0 G_1$ | $A_1 B_0 P_3$ | A_0 | 60 |
| LH | $A_1 B_0 G_1$ | $A_1 B_0 P_3$ | A_0 | 60 |

In a low-level interaction with both hands working separately, the values from both hands are added to yield 120 TMU:

| RH | $A_1 B_0 G_1$ | $A_1 B_0 P_3$ | A_0 | 60 |
| LH | $A_1 B_0 G_1$ | $A_1 B_0 P_3$ | A_0 | 60 |

There can also be intermediate level of interaction with only certain phases occurring simultaneously, e.g., the GET phase below, which is circled out, yielding a total of 100 TMU.

| RH | $A_1 B_0 G_1$ | $A_1 B_0 P_3$ | A_0 | 60 |
| LH | $A_1 B_0 G_1$ | $A_1 B_0 P_3$ | A_0 | 40 |

As a final complete example, the same flashlight assembly shown in Figure 13.9, using MTM-2 analysis, is analyzed using MOST in Figure 13.16.

⏵	MOST - calculation		Code		
			Date	8-22 -	
			Sign.	AF	
			Page	1 / 1	

Activity	FLASHLIGHT ASSEMBLY
Conditions	WITH 10% ALLOWANCE

No.	Method	No.	Sequence Model	Fr.	TMU
1	GET HOUSING TO WORK AREA	1	A₁ B0 G₁ A₁ B0P0 A0		30
			A B G A B P A		
2	GET BATTERIES INTO HOUSING	2	A₁ B0 G₁ A₁ B0P₁ A0	2	80
			A B G A B P A		
3	GET END CAP, SCREW TIGHT	4	A0 B0 G0 A₁ B0 P0 A0		10
4	PUT FLASHLIGHT DOWN		A B G A B P A		
			A B G A B P A		
			A B G A B P A		
			A B G A B P A		
			A B G A B P A		
			A B G A B P A		
			A B G A B P A		
			A B G A B P A		
			A B G M X I A		
			A B G M X I A		
			A B G M X I A		
			A B G M X I A		
			A B G M X I A		
			A B G M X I A		
			A B G M X I A		
		3	A₁ B0 G₁ A₁ B0 P3 F3 A0 B0 P0 A0		90
			A B G A B P A B P A		
			A B G A B P A B P A		
			A B G A B P A B P A		
			A B G A B P A B P A		
			A B G A B P A B P A		
			A B G A B P A B P A		
			A B G A B P A B P A		
			A B G A B P A B P A		
			A B G A B P A B P A		
			A B G A B P A B P A		
			A B G A B P A B P A		

TIME =	millihours (mh)	(WITH 10% ALLOWANCE) minutes (min.)	0.119	180

Copyright Maynard 1974 M 101 -REV - 1

Figure 13.16 MOST analysis of flashlight assembly.

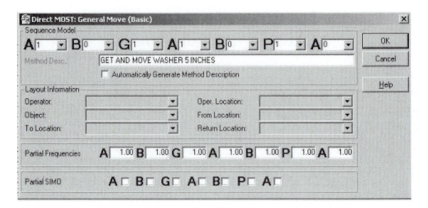

Figure 13.17   Example of a general move sequence, get and move washer 5 in, in BasicMOST.
Figure from *BasicMOST*®. Used by permission of H. B. Maynard and Co., Inc., Pittsburgh, PA.

MOST is also available in a computerized version which permits the retrieval of activity sequences, subactivities, and the indexed parameters involved in developing a standard of performance for the method under study. It has been estimated that the use of a computerized system should result in application speeds five to ten times faster than manual application speeds. Computer-developed standards are also more error-free, since the system does not accept an input that is not logical. An example of a general move sequence in Basic-MOST® is shown in Figure 13.17.

13.3 PREDETERMINED TIME APPLICATION

STANDARD DATA DEVELOPMENT

One of the most important uses of predetermined time systems is the development of standard data elements. With standard data, standard times for operations can be set much faster than by the laborious procedure of summarizing long columns of fundamental motion times. In addition, standard data usually reduce clerical errors, since less arithmetic is involved.

With sound standard data, it is economically feasible to establish standards on indirect work, such as maintenance, material handling, clerical and office, inspection, and similar expense operations. Also, with standard data, analysts can economically calculate operation times involving long cycles and consisting of many short-duration elements. For example, one company developed standard data applicable to radial drill operations in its tool room. Time study analysts developed standard data for the elements required to move the tool from one hole to the next and to present and back off the drill. They then combined these standard data elements into one multivariable chart, so that the data could be rapidly summarized.

An example illustrating the flexibility of predetermined systems is the development of a standard time for a clerical operation. This formula for sorting time slips includes the following elements:

1. Pick up pack of departmental time slips and remove elastic band.
2. Sort time slips into direct labor (incentive), indirect labor, and daywork.
3. Record the total number of time slips.
4. Get pile of time slips, put elastic band around pack, and put aside.
5. Get pile of time slips and bunch.
6. Sort incentive time slips into "parts" time slips.
7. Count piles of incentive "parts" time slips.
8. Record the number of "parts" time slips and the number of incentive time slips.
9. Sort "parts" time slips into numerical sequence.
10. Bunch piles of numerical time slips and place in one pile on desk.

Methods analysts break each element down into the fundamental motions. Once the basic values are assigned and the variables are determined, the resulting algebraic equation allows the rapid calculation of time for the clerical operation. A stopwatch can frequently be helpful in developing standard data elements. Some portions of an element may be more readily determined by predetermined times, while other portions may be better adapted to stopwatch measurement. Finally predetermined basic motion times have necessarily been developed for standard performance. Therefore, no ratings are needed, eliminating the one step which is most open to controversy and uncertainty.

METHODS ANALYSIS

An equally important use for any predetermined time system is methods analysis. Analysts who appreciate these systems look more critically at each and every workstation, thinking about how improvements may be made. Using a predetermined time system is simply developing a motion or methods analysis in greater numerical detail, identifying better ways of eliminating ineffective therbligs, and reducing the times on the remaining effective therbligs. Another checklist has therefore been developed (see Figure 13.18) to aid the analyst in better methods analysis. Key opportunities for simplifying the method (using the MTM-2 system as an example) include

1. Elimination of body motions, such as Bend and Arise, with a large time value of 61 TMU
2. Reduction of case levels, especially case C motions, resulting in a 39 percent decrease in basic motion times
3. Minimizing reach distances, with a 5-TMU decrease for each shorter distance code
4. Avoiding the lifting of heavy parts, where each 2 lb (1 kg) drops 1 TMU

GETs (G)	Yes	No
1. Can GETs be performed simultaneously with other GETs or PUTs without penalty?	❏	❏
2. Can GETs be performed during a machine cycle?	❏	❏
3. Can jigs/fixtures, gravity-feed devices, or bins be used to simplify GETs (i.e., from GC to GB or to GA)?	❏	❏
4. Can GAs be used and objects slid into position?	❏	❏
5. Can the transfer of objects from one hand to another be avoided?	❏	❏
6. Can tools be prepositioned to simplify GETs?	❏	❏
7. Can tools be palmed while performing other work (instead of being set down and later retrieved)?	❏	❏
8. Can more than one object be grasped at the same time?	❏	❏
9. Can travel distances be reduced (i.e., to lower motion classification levels)?	❏	❏
10. Are hand motions balanced in terms of case and distance?	❏	❏

PUTs (P)	Yes	No
1. Can PUTs be performed simultaneously with other GETs or PUTs without penalty?	❏	❏
2. Can tight tolerances or the accurate location of an object be avoided?	❏	❏
3. Can the delivery point of an object be chamfered or funneled?	❏	❏
4. Can fixed guides or stops be utilized?	❏	❏
5. Can the objects be made symmetrical?	❏	❏
6. Can the depth of insertion be reduced?	❏	❏
7. Can the other hand assist in complex PUTs?	❏	❏
8. Can objects be PUT together mechanically?	❏	❏
9. Can drop deliveries be utilized to simplify PUTs (i.e., from PC to PB or to PA)?	❏	❏
10. Can objects be slid to a location (i.e., use a PA)?	❏	❏
11. Are destination points in the normal area of vision?	❏	❏

Apply Pressure (A)	Yes	No
1. Can As be avoided by improved design or better processing (e.g., eliminate burrs or tight spots)?	❏	❏
2. Can unnecessary tightening from operations be avoided?	❏	❏
3. Can tight tolerances be avoided?	❏	❏
4. Can the contamination of parts due to filings, dust, dirt, etc., causing As be avoided?	❏	❏
5. Can momentum be used to eliminate As?	❏	❏
6. Are the largest muscle groups used to best advantage in applying pressure?	❏	❏
7. Can clapping devices or mechanical actions be used to eliminate As?	❏	❏

Regrasp (R)	Yes	No
1. Can Rs be avoided during PUTs?	❏	❏
2. Can tools be prepositioned in the desired orientation?	❏	❏

Figure 13.18 MTM-2 methods analysis checklist.
Adapted from Brown, 1976.

		Yes	No
3.	Can magazine feeds, stacking devices, vibratory feeders, etc., be used to present the part properly?	❑	❑
4.	Can parts be made symmetrical to avoid the need for Rs?	❑	❑
5.	Can parts be prepositioned during a machine cycle?	❑	❑

Eye Action (E)

		Yes	No
1.	Can objects and displays be placed in the normal area of vision to avoid Es?	❑	❑
2.	Is there sufficient illumination to avoid Es?	❑	❑
3.	Are bins and parts correctly identified, perhaps by use of color?	❑	❑
4.	Can parts be made symmetrical and positioned properly to avoid Es?	❑	❑
5.	Can visual checks of assembly parts be avoided (i.e., use detents and tactile feel)?	❑	❑
6.	Can visual interpretation of dial settings be avoided (i.e., use on/off or status indicators)?	❑	❑
7.	Can Es be performed during preceding manual motions without penalty?	❑	❑

Crank (C)

		Yes	No
1.	Can the wheel or crank be spun?	❑	❑
2.	Can the number of revolutions be reduced (i.e., larger thread size used)?	❑	❑
3.	Can resistance during cranking be eliminated?	❑	❑
4.	Can the crank be power-driven?	❑	❑

Step (S)

		Yes	No
1.	Is the shortest route or best layout being utilized?	❑	❑
2.	Are floor surfaces easy and clear of obstructions?	❑	❑
3.	Are the most commonly used parts located close by?	❑	❑
4.	Is any necessary information and tooling located at the workstation (i.e., avoid unnecessary Ss)?	❑	❑
5.	Can materials and parts be brought mechanically (via conveyors) to and from the workstation?	❑	❑
6.	Can vehicular transport (carts) be used?	❑	❑

Foot Motion (F)

		Yes	No
1.	Can Fs be performed simultaneously with other motions?	❑	❑
2.	Can the foot rest comfortably on the switch or pedal during the operation?	❑	❑
3.	Is the body supported by a stool (weight off the load-bearing leg)?	❑	❑
4.	Can either foot operate the pedal alternately?	❑	❑

Bend and Arise (B)

		Yes	No
1.	Can drop deliveries be utilized to avoid Bs?	❑	❑
2.	Are materials and products located between elbow and knuckle height to minimize Bs?	❑	❑
3.	Are proper lifting procedures (squat lifting, etc.) being utilized?	❑	❑
4.	Can the too frequent entry and exit of a seated workstation be avoided?	❑	❑

Figure 13.18   (continued)

5. Eliminating operations that require eye travel and eye focus, which eliminates 7 TMU each

6. Prepositioning tools, parts, and materials

In one company, $40,000 was allocated for advanced tooling to increase the rate of production on a brazing operation. Prior to retooling, analysts conducted a work measurement study of the existing method. Using predetermined time systems, they discovered that by providing a simple fixture and rearranging the loading and unloading area, the company production could increase from 750 to 1,000 pieces per hour. The total cost of the synthetic basic motion-time study was $40. As a result of the study, the company avoided the costly $40,000 retooling program.

EXAMPLE 13.1   **Methods Improvement in T-Shirt Turning**

This example considers both the productivity and health/safety aspects obtained through an MTM-2 analysis of T-shirt turning (Freivalds and Yun, 1994). Garments are sewn "inside out" so that the seams can be stitched. Once the garment is completed, it must be "turned" or inverted.

Workers on this job were highly susceptible to various cumulative trauma disorders. An MTM-2 analysis of the present method, shown in Figure 13.19a, indicated that the results were a total of 141 TMU. An obvious characteristic of this job was the high usage of case C motions. Was it possible to reduce the GETs and PUTs (questions 3 and 4 under GET in Figure 13.18)?

The proposed solution was to build a vacuum-powered device to draw the T-shirt into a pipe. Once the vacuum was shut off, the T-shirt could be removed in an inverted state. The MTM-2 analysis of the improved method (see Figure 13.19b) yielded a total of only 108 TMU. For the complete operation of turning, inspection, and folding (total of 360 TMU), this resulted in a $(141 - 108)/360 = 9.2$ percent decrease in time. Overall, the difficult and injurious case C hand motions were eliminated, with a simultaneous improvement in productivity.

MTM Methods Analysis						Page of
Operation: *T-SHIRT TURNING*			Remarks: *MANUAL HANDLING TOTAL OF 141 TMUs*			
Study No.: *(MANUAL)*						
Date: *2-12-93*						
Analyst: *A F*						
Description	No.	LH	TMU	RH	No.	Description
GET T-SHIRT		*GB18*	*18*	*GB 18*		*GET T-SHIRT*
REACH INSIDE, PINCH CLOTH		*GC12*	*23*	*GC 12*		*REACH INSIDE, PINCH CLOTH*
SIMULTANEOUS MOTION		*G C2*	*14*	*GC2*		*SIMULTANEOUS MOTION ALLOW*
PULL SLEEVE UP AND OUT		*PC32*	*41*	*PC32*		*PULL SLEEVE UP AND OUT*
SIMULTANEOUS MOTION ALLOW.		*PC2*	*21*	*PC2*		*SIMULTANEOUS MOTION*
SET T-SHIRT DOWN		*PB18*	*24*	*PB18*		*SET T-SHIRT DOWN*
			(141)			

(a) Present method

Figure 13.19 MTM-2 analysis of T-shirt turning.

MTM Methods Analysis

Page   of

| Operation: | T-SHIRT TURNING | | | | Remarks: | USING VACUUM DEVICE |
| Study No.: | (AUTOMATED) | | | | | TOTAL OF 108 TMUs |

Date: 2-12-93

Analyst: A F

Description	No.	LH	TMU	RH	No.	Description
GET T-SHIRT		GB18	18	GB18		GET T-SHIRT
PULL T-SHIRT UP PIPE		PA32	20	PA32		PULL T-SHIRT UP PIPE
			9	F		ACTIVATE PEDAL
PULL T-SHIRT DOWN		PB32	30	PB32		PULL T-SHIRT DOWN
AGAINST RESISTANCE		PW10	1	PW10		AGAINST RESISTANCE
SET T-SHIRT DOWN		PB32	30	PB32		SET T-SHIRT DOWN
			(108)			

(b) Proposed method

Figure 13.19   (continued)

SUMMARY

This chapter discusses several of the more popular predetermined time systems. There are many others, including several proprietary systems developed by industry. Many years ago, Frederick W. Taylor visualized the development of standards for basic divisions of work similar to those still in use. In his paper "Scientific Management," he predicted that the time would come when a sufficient volume of basic standards would be developed to make further time studies unnecessary. Today we have about reached this state with the vast majority of standards developed by using standard data and/or predetermined times.

However, there is still a question as to the validity of adding basic predetermined times to determine elemental times, since therblig times may vary once the sequence is changed. Thus, the time for the basic element "reach 20 in" may be affected by the preceding and succeeding elements and may not be entirely dependent on the class of reach and distance.

Therefore, the analyst should consider the main purpose of the motion pattern, as well as its complexity, characteristics, and distance. For example, when an object is palmed while the hand is moving, a simultaneous operation takes place in addition to the move. The result might be a reduction in the average speed. This would allow the hand to establish control of the object over the distance moved. The longer the distance, the more time the hand has to palm the object. Thus, the longer the combined motion, the more the motion approaches the time required for a simple reach of the same distance.

There are many compelling reasons to use predetermined time systems. They can be used to define a standard time before production begins and to estimate production costs ahead of time, when no job exists to time study. However, these systems are only as good as the people using them. The analyst must be very careful to understand the assumptions behind the systems and use them in the proper manner. They should not be installed without professional help or a complete understanding of their application. To assist in and, perhaps, even simplify this process, some of the predetermined time systems offer associated software packages, which are listed at the end of the chapter.

QUESTIONS

1. Who was originally responsible for thinking in terms of developing standards for basic work divisions? What was his contribution?
2. What are the advantages of using predetermined times?
3. What other two terms are frequently used to identify predetermined times?
4. Who pioneered the MTM system?
5. What is the time value of 1 TMU?
6. Would it be easy or difficult to perform a GB get with the left hand while simultaneously performing a PC place with the right hand? Explain.
7. Why was MTM-2 developed? Where does MTM-2 have special application?
8. Are MTM-1 and MTM-2 consistent in their handling of simultaneous motions?
9. If MTM-3 were used to study an operation of approximately 3-min duration, what could you say about the accuracy of the standard?
10. How is MTM related to the method analysis?
11. Explain the relationship of predetermined times to standard data.
12. How is MOST related to MTM?
13. Compare and contrast the three MOST systems.
14. What are the three basic motion sequences used in BasicMOST?
15. How are simultaneous activities handled in BasicMOST?
16. What are some of the advantages in using a predetermined time system as opposed to stopwatch time study?

PROBLEMS

1. Determine the time for the dynamic component of M20 B20.
2. A 30-lb bucket of sand having a coefficient of friction of 0.40 is pushed 15 in away from the operator, with both hands. What would be the normal time for the move?

3. A ¾-in-diameter coin is placed within a 1-in-diameter circle. What would be the normal time for the position element?

4. Calculate the equivalent in TMUs of 0.0075 h per piece, 0.248 min per piece, 0.0622 h per hundred, 0.421 s per piece, and 10 pieces per minute.

5. The MTM-2 (see Figure 13.20) analysis describes a simple operation in which each hand gets a part, regrasps it, and then the right hand puts it into a fixture. Pressure is applied to seat it. Next, a pin is grasped, regrasped, and inserted into the assembly. A handwheel is cranked continuously six revolutions under resistance until a pointer is aligned exactly. Identify any errors by circling them. Rewrite the analysis correctly and explain each correction.

6. Develop a BasicMOST analysis of the paper sequence of activities in Figure 13.20.

7. The Purdue Pegboard Task is a standard test for motor skills. It consists of a board with a series of holes and three types of pieces: pegs, washers, and standoffs, stored in a depression at the top of the board. The assembly, with the board rotated perpendicular to the operator's torso, is as follows:

    a. Right hand picks up a peg and inserts it into a hole with a tight clearance.

    b. As the peg is inserted, the left hand picks up a washer and mounts it on the peg (0.01-in clearance).

Completed assembly

    c. As the washer is mounted, the right hand picks up a standoff and mounts it on the peg on top of the washer (0.01-in clearance).

    d. As the standoff is mounted, the left hand picks up another washer and mounts it on the peg on top of the standoff (0.01-in clearance).

    e. As the washer is mounted, the right hand picks up another peg and starts a new assembly in the next hole down. The hands keep alternating in picking up pieces and completing assemblies.

    Develop an MTM-2 analysis for the first complete assembly. What happens as the operator proceeds down the board? Why? Why would MTM-2 not be appropriate for this task?

MTM Methods Analysis Page of

Operation:	ASSEMBLY	Remarks:
Study No.:	PROBLEM # 5	
Date:	1-27-98	
Analyst:	AF	

Description	No.	LH	TMU	RH	No.	Description
PART TO FIXTURE {		GC12	18	GB18		} PART TO FIXTURE
		R	6	R		
		PC12	30	PC6		
SEAT PART		A	14	A		SEAT PART
GET & ASSEMBLE PIN		GC12	23	GC12		} GET & ASSEMBLE PIN
		PC12	30	PC12		
		R	6	R		
			10	GB6		} CRANK AGAINST
CRANK AGAINST			5	GW10		RESISTANCE
RESISTANCE			90	C	6	
			5	PWS		
ALIGN POINTER			21	PC2		ALIGN POINTER

Summary	Total TMU: 259	Conversion: .0006	% Allowance: 10	Standard time: .171

Figure 13.20 MTM-2 analysis of simple assembly for Problem 5.

8. Develop a BasicMOST analysis for the first complete assembly in Problem 7.
9. Develop an MTM-2 analysis of the cable-clamp assembly shown in Figure 4.17.
10. Develop a BasicMOST analysis of the cable-clamp assembly shown in Figure 4.17.

REFERENCES

Antis, William, John M. Honeycutt, Jr., and Edward N. Koch. *The Basic Motions of MTM*. 3d ed. Pittsburgh, PA: The Maynard Foundation, 1971.

Baily, Gerald B., and Ralph Presgrave. *Basic Motion Time-Study*. New York: McGraw-Hill, 1958.

Birn, Serge A., Richard M. Crossan, and Ralph W. Eastwood. *Measurement and Control of Office Costs*. New York: McGraw-Hill, 1961.

Brown, A. D. "Apply Pressure." *Journal of the Methods-Time Measurement Association,* 14 (1976).

Freivalds, A., and M. H. Yun. "Productivity and Health Issues in the Automation of T-Shirt Turning." *International Journal of Industrial Engineering,* 1, no. 2 (June 1994), pp. 103–108.

Geppinger, H. C. *Dimensional Motion Times*. New York: John Wiley & Sons, 1955.

Karger, Delmar W., and Franklin H. Bayha. *Engineered Work Measurement*. New York: Industrial Press, 1957.

Karger, Delmar W., and Walton M. Hancock. *Advanced Work Measurement*. New York: Industrial Press, 1982.

Maynard, Harold B., G. J. Stegemerten, and John L. Schwab. *Methods Time Measurement*. New York: McGraw-Hill, 1948.

Quick, J. H., J. H. Duncan, and J. A. Malcolm. *Work-Factor Time Standards*. New York: McGraw-Hill, 1962.

Sellie, Clifford N. "Predetermined Motion-Time Systems and the Development and Use of Standard Data." In *Handbook of Industrial Engineering*. 2d ed. Ed. Gavriel Salvendy. New York: John Wiley & Sons, 1992.

Zandin, Kjell B. *MOST Work Measurement Systems*. New York: Marcel Dekker, 1980.

SELECTED SOFTWARE

MOD++, International MODAPTS Association, 3302 Shearwater Court, Woodbridge, VA 22192 (http://www.modapts.org/)

MOST, H. B. Maynard and Co., Eight Parkway Center, Pittsburgh, PA 15220, 2001 (http://www.hbmaynard.com/)

MTM-Link, The MTM Association, 1111 East Touhy Ave., Des Plaines, IL 60018 (http://www.mtm.org/)

TimeData. Royal J. Dosset Corp., 2795 Pheasant Rd., Excelsior, MN 55331

Work Sampling

KEY POINTS

- Work sampling is a method for analyzing work by taking a large number of observations at random times.

- Use work sampling to

  - Determine machine and operator utilization

  - Determine allowances

  - Establish time standards

- Use as many observations as practical but maintain accuracy

- Take observations over as long a period of time as feasible, preferably several days or weeks

---

Work sampling is a technique used to investigate the proportions of total time devoted to the various activities that constitute a job or work situation. The results of work sampling are effective for determining machine and personnel utilization, allowances applicable to the job, and production standards. Although the same information can be obtained by time study procedures, work sampling frequently provides the same information faster and at considerably less cost.

In conducting work sampling studies, analysts take a comparatively large number of observations at random intervals. The ratio of observations of a given activity to the total observations approximates the percentage of time that the process is in that state of activity. For example, if 1,000 observations taken at random intervals over a day showed that an automatic screw machine was turning out work in 700 instances but was idle for miscellaneous reasons in 300 instances, then the downtime of the machine would be 30 percent of the working day.

Work sampling was first applied in the British textile industry. Later, under the name ratio-delay study, the technique was brought to the United States (Morrow, 1946). The accuracy of the data determined by work sampling depends on the number of observations and the period over which the random observations are taken. Unless the sample size is sufficiently high, and the sampling period represents typical conditions, inaccurate results may occur.

The work sampling method has several advantages over the conventional time study procedure:

1. It does not require continuous observation by an analyst over a long time.
2. Clerical time is diminished.
3. The total work-hours expended by the analyst are usually much fewer.
4. The operator is not subjected to long-period stopwatch observations.
5. Crew operations can be readily studied by a single analyst.

14.1   THE THEORY OF WORK SAMPLING

The theory of work sampling is based on the fundamental law of probability: at a given instant, an event can be either present or absent. Statisticians have derived the following expression to show the probability of x occurrences of such an event in n observations:

$$P_{(x)} = \frac{n!}{x!\,(n-x)!}\,P^x q^{n-x}$$

where p = probability of a single occurrence
$q = 1 - p$ = probability of an absence of occurrence
n = number of observations

The distribution of these probabilities is known as the *binomial distribution* with mean equal to np, and variance equal to npq. As n becomes large, the binomial distribution approaches the normal distribution. Since work sampling studies involve large sample sizes, the normal distribution is a satisfactory approximation of the binomial distribution. This normal distribution of a proportion has a mean equal to p and a standard deviation equal to

$$\sqrt{\frac{pq}{n}}$$

In work sampling studies, we take a sample of size n in an attempt to estimate p. We know from elementary sampling theory that we cannot expect $\hat{p}$ ($\hat{p}$ = the proportion based on a sample) for each sample to be the true value of p. We do, however, expect the $\hat{p}$ of any sample to fall within the range of $p \pm 1.96$ standard deviations approximately 95 percent of the time. In other words, if p is the true percentage of a given condition, we can expect the $\hat{p}$ of any sample to fall outside the limits $p \pm 1.96$ standard deviations only about 5 times in 100 due to chance alone.

This theory can be used to estimate the total sample size needed to achieve a certain degree of accuracy. The expression for the standard deviation σ_p of a sample proportion is

$$\sigma_p = \sqrt{\frac{pq}{n}} = \sqrt{\frac{p(1-p)}{n}} \qquad (1)$$

where σ_p = standard deviation of a percentage
      p = true percentage occurrence of element being sought, expressed as a decimal
      n = total number of random observations upon which p is based

Based on the concept of a confidence interval consider the term $z_{\alpha/2}\sigma_p$ as the acceptable limit of error ℓ at a $(1-\alpha)100$ percent confidence error, where

$$\ell = z_{\alpha/2}\sigma_p = z_{\alpha/2}\sqrt{pq/n} \qquad (2)$$

Squaring both sides and solving for n yield

$$n = z_{\alpha/2}^2\, pq/\ell^2 = z_{\alpha/2}^2\, p\,(1-p)/\ell^2 \qquad (3)$$

For a typical application, using a 95 percent confidence interval, $z_{\alpha/2}$ is 1.96 and n becomes

$$n = 3.84\, pq/\ell^2$$

Normal Approximation of the Binomial Distribution   EXAMPLE 14.1

To clarify the fundamental theory of work sampling, it would be helpful to interpret the results of an experiment. Assume the following circumstances: One machine with random breakdowns was observed for a 100-day period. During this period, eight random observations were taken per day.

Let n = number of observations per day
      k = total number of days that observations were taken
      x_i = number of breakdown observations observed in n random observations on day i $(i = 1, 2, ..., k)$
      N = total number of random observations
      N_x = number of days that the experiment showed the number of breakdowns equal to x $(x = 0, 1, 2, ..., n)$

The probability $P(x)$ that the machine is down x times in n observations is given by the binomial distribution

$$P(x) = \frac{n!}{x!(n-x)!}p^x q^{n-x}$$

where p = probability of machine being down
      q = probability of machine running

and

$$p + q = 1$$

For our example, $n = 8$ observations per day, $k = 100$ days of observations, and $N = 800$ total observations. An all-day time study for several days revealed that $p = 0.5$. The following table shows the number of days in which x breakdowns were observed in the work sampling study ($x = 0, 1, 2, 3, ..., n$) and the expected number of breakdowns given by our binomial model, using $p = 0.5$ from the all-day time study.

X	N_x	P(x)	100P(x)
0	0	0.0039	0.39
1	4	0.0312	3.12
2	11	0.1050	10.5
3	23	0.2190	21.9
4	27	0.2730	27.3
5	22	0.2190	21.9
6	10	0.1050	10.5
7	3	0.0312	3.12
8	0	0.0039	0.39
	100	1.00*	100*

*Approximately

There is close agreement between the observed days that a specified number of breakdowns occurred N_x and the expected number computed theoretically as $kP(x)$.

$$\overline{P}_i = \frac{x_i}{n} = \text{observed proportion of downtime on day } i$$

where $i = 1, 2, 3, \ldots, k$

$$\hat{P} = \frac{\sum_{i=1}^{k} \overline{P}_i}{k} = \frac{\sum_{i=1}^{k} x_i}{n \cdot k}$$

$$= \frac{\sum_{i=1}^{k} x_i}{N} = \text{estimated proportion of machine downtime, based}$$
$$\text{on a work sampling experiment}$$

The hypothesis is that the theoretical information shows close enough agreement to the observed information for the theoretical binomial to be accepted. This may be tested using the chi-square (χ^2) distribution. The χ^2 distribution tests whether the observed distribution frequencies differ significantly from the expected frequencies.

In the example, the observed frequency is N_x and the expected frequency is $kP(x)$, and we have

$$\chi^2 = \sum_{k=0}^{k} \frac{[N_x - 100P(x)]^2}{100P(x)}$$

The quantity under the summation is distributed approximately as χ^2 for k degrees of freedom. In this example, $\chi^2 = 0.206$.

Analysts must determine whether the calculated value of χ^2 is sufficiently large to refute a null hypothesis, that is, the difference between the observed frequencies and the computed frequencies is due to chance alone. This experimental value of χ^2 is so small that it could easily have occurred through chance. Therefore, we accept the hypothesis that the experimental data "fit" the theoretical binomial distribution.

In typical industrial situations, p (which was known to have a value of 0.5) is unknown to analysts. The best estimate of p is $\hat{p}$, which may be computed as $\sum_{i=1}^{k} x_i / N$.

As the number of random observations per day n increases and/or the number of days increases, $\hat{p}$ will approach p. However, with limited observations, analysts are concerned with the accuracy of $\hat{p}$.

If a plot of $P(x)$ versus x were made from our example, it would appear as shown in Figure 14.1.

When n is sufficiently large, regardless of the actual value of p, the binomial distribution very closely approximates the normal distribution. This tendency can be seen in the example when p is approximately 0.5. When p is near 0.5, n may be small and the normal can be a good approximation to the binomial.

When using the normal approximation, set

$$\mu = p$$

and

$$\sigma_p = \sqrt{\frac{pq}{n}}$$

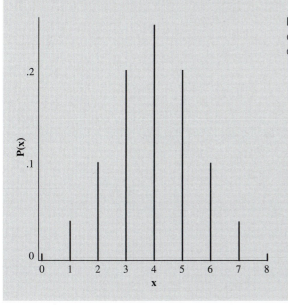

Figure 14.1 Probability distribution of breakdown observations.

To approximate the binomial distribution, the variable z is used for entry in the normal distribution (see Table A3–2, Appendix 3), and it takes the following form:

$$z = \frac{\hat{p} - p}{\sqrt{pq/n}}$$

Although p is unknown in the practical case, we can estimate p from $\hat{p}$ and determine the interval within which p lies, using confidence limits. For example, imagine that the interval defined by

$$\hat{p} - 1.96\sqrt{\frac{\hat{p}\hat{q}}{n}}$$

and

$$\hat{p} + 1.96\sqrt{\frac{\hat{p}\hat{q}}{n}}$$

contains p 95 percent of the time.

Graphically, this may be represented as

$$\hat{p} - 1.96\sqrt{\frac{\hat{p}\hat{q}}{n}} \qquad \hat{p} \qquad \hat{p} + 1.96\sqrt{\frac{\hat{p}\hat{q}}{n}}$$

We derive the expression for finding a confidence interval for p as follows: Suppose that we want an interval that contains p 95 percent of the time; that is, a 95 percent confidence interval. For n sufficiently large, the expression

$$z = \frac{\hat{p} - p}{\sqrt{\hat{p}\hat{q}/n}}$$

is approximately a standard normal variable. Therefore, we set the probability

$$P\left[z_{0.025} < \frac{\hat{p} - p}{\sqrt{\hat{p}\hat{q}/n}} < z_{0.975} \right] = 0.95$$

Rearranging the inequalities and remembering that $-z_{0.025} = z_{0.975} = 1.96$, the interval with approximately a 95 percent chance of containing p is then

$$\hat{p} - 1.96\sqrt{\frac{\hat{p}\hat{q}}{n}} < p < \hat{p} + 1.96\sqrt{\frac{\hat{p}\hat{q}}{n}}$$

These limits imply that the interval defined contains p with 95 percent confidence, since z has been selected as having a value of 1.96.

The underlying assumptions of the binomial are that p, the probability of a success (the occurrence of downtime), is constant for each random instant that we observe the process. Therefore, it is always necessary to take random observations when doing a work sampling study. This reduces any bias introduced by worker anticipation of observation times.

14.2 SELLING WORK SAMPLING

Before beginning a work sampling program, the analyst must "sell" its use and reliability to all members of the organization who will be affected by the results, including the union, the line supervisor, and company management. This can be done by conducting several short sessions with representatives of the various interested parties and explaining examples of the law of probability, thus illustrating why sampling procedures work. Even though stopwatch time study is well understood and readily accepted both unions and workers will accept work sampling techniques, once the procedure is fully explained. Factors in favor of work sampling are that it is completely impersonal and does not have the pressure of a stopwatch time study.

In the initial session, the analyst should create a simple study by tossing unbiased coins. All participants should readily recognize that a single coin toss stands a 50-50 chance of being heads. When asked how they would determine the probability of heads versus tails, they will undoubtedly propose tossing a coin a few times to find out. When asked whether two times is adequate, they will say no. Ten times may be suggested, but they may still think that is not adequate. When larger numbers are suggested, they will probably agree that 100 or more times is sufficient to achieve the desired result with some degree of assurance. This example firmly implants the principle of work sampling: adequate sample size to ensure statistical significance.

Next, the analyst should discuss the probable results of tossing four unbiased coins. Here, there is only one arrangement in which the coins can fall to show no heads, and only one arrangement that permits all heads. However, three heads or one head can result from four possible arrangements. Six possible arrangements can give two heads. With all 16 possibilities thus named, four unbiased coins tossed continually will distribute themselves as shown in Figure 14.2.

After this explanation and a demonstration of this distribution, that is, by making several tosses and recording the results, the audience should agree that 100 tosses could demonstrate a normal distribution. A thousand tosses would probably approach a normal distribution more closely, and 100,000 would give a nearly perfect distribution. However, such a distribution is not sufficiently more

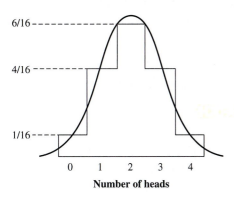

Figure 14.2 Distribution of number of heads with infinite number of tosses, using four unbiased coins.

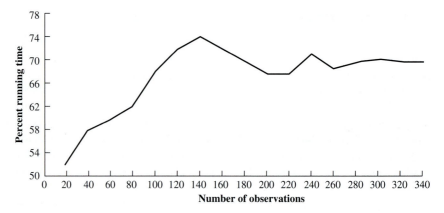

Figure 14.3 Cumulative percentage of running time.

accurate than the 1,000-toss distribution, and it is not economically worth the extra effort. This establishes the idea that significant accuracy is approached rapidly at first and then at a diminishing rate.

Next the analyst should point out that a machine or operator could figuratively be in a heads or tails state. For example, a machine could be running (heads) or idle (tails). A cumulative plot of "running" would eventually level off, giving an indication of when it would be safe to stop taking readings (see Figure 14.3). Also, "idle" machine time could be broken down into the various interruptions and delays for a more detailed understanding of such time.

14.3 PLANNING THE WORK SAMPLING STUDY

Detailed planning must be done before actual work sampling study observations are made. The plans start with a preliminary estimate of the activities on which information is sought. This estimate may involve one or more activities, and the estimate can frequently be made from historical data. If the analyst cannot make a reasonable estimate, he or she should work sample the area for two or three days and use that information as the basis for these estimates.

Once the preliminary estimates have been made, the analyst can determine the desired accuracy of the results. This can best be expressed as a tolerance, or limit of error, within a stated confidence level. Next, the analyst must estimate the number of observations to be made and determine the frequency of observations. Finally, the analyst designs the work sampling form on which to tabulate the data as well as the control charts used in conjunction with the study.

DETERMINING THE NUMBER OF OBSERVATIONS NEEDED

To determine the number of observations needed, the analyst must know the desired accuracy of the results. The more observations, the more valid the final answer. Three thousand observations give considerably more reliable results than

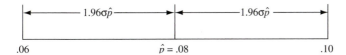

Figure 14.4 Tolerance range of the percentage of unavoidable delay allowance required within a given section of a plant.

300. However, because of the cost of obtaining so many observations and the marginal improvement in accuracy, 300 observations may be ample.

For example, suppose you want to determine the number of observations required, with 95 percent confidence, such that the true proportion of personal and unavoidable delay time is within the interval of 6 to 10 percent. The unavoidable and personal delay time is expected to be 8 percent. These assumptions are expressed graphically in Figure 14.4.

In this case, $\hat{p}$ would equal 0.08, and we assume an ℓ of 2 percent, or 0.02. Using these values, we can solve for n as follows:

$$n = \frac{3.84 \times 0.08 \times (1 - 0.08)}{0.02^2} = 707 \text{ observations}$$

If the analyst does not have the time or capability to collect 707 observations, but can only collect 500 data points, the above equation can be inverted to solve for the resulting error limit:

$$\ell = \sqrt{\frac{3.84p(1 - p)}{n}} = \sqrt{\frac{3.84(0.92)(0.08)}{500}} = 0.024$$

With 500 observations, then, the accuracy of the study would be ±2.4 percent. Thus, there is a direct trade-off between the error or accuracy of the study and the number of observations collected. Note that this 2.4 percent is an absolute accuracy. Some analysts may wish to express this as a relative accuracy of 30 percent with respect to the base proportion (0.024/0.08).

Software for determining the observations required for a work sampling study is readily available today. These programs perform all the statistical calculations required to determine sample sizes and confidence intervals. For example, they can calculate the 90 percent, 95 percent, and 99 percent confidence intervals for a sample. They can also provide the number of samples necessary to achieve 90 percent, 95 percent, and 99 percent confidence for a specified degree of accuracy.

Note that if several workers are observed simultaneously, the observations cannot be considered to be independent. This problem of correlated observations has been addressed by Richardson and Pape (1992) and results in a corrected confidence interval calculated after data collection. Instead of Equation (1), the standard deviation for the confidence interval is calculated from

$$\sigma = \left[\frac{\Sigma y(j)^2 / n(j) - np^2}{n(m - 1)} \right]^{1/2}$$

where m = number of grouped observations

$n(j)$ = number of workers at jth observation

n = total number of observations

$y(j)$ = number of workers "idle" (or other category of interest) at jth observation

Example 14.2 demonstrates the effect of correlated observations on the error in a work sampling study.

EXAMPLE 14.2   **Correlation Observations**

The owner of a strip mall would like to determine whether there are sufficient parking spots (presently 250) for her customers. Cursory observations indicate that the parking lot is roughly 80 percent full during business hours. She hires an industrial engineering analyst to conduct a more complete work sampling study. The analyst collects 10 random samples on a Wednesday from 9 A.M. to 6 P.M. with the following results.

Sample j	Empty Spots $y(j)$	$y(j)^2$
1	36	1,296
2	24	576
3	11	121
4	10	100
5	9	81
6	20	400
7	19	361
8	28	784
9	35	1,225
10	57	3,249
Total	**249**	**8,193**

The proportion of empty spots is

$$p = 249/10/250 = 0.0996$$

Since at any given sampling time the 250 observations of parking spots are going to be correlated, the limit of error must be calculated from

$$\ell = 1.96s = 1.96\left[\frac{\Sigma y(j)^2/n(j) - np^2}{n(-1)}\right]^{1/2}$$

$$= 1.96\left[\frac{8,193/250 - 2,500(0.0996)^2}{2,500(10-1)}\right] = 0.0368$$

The owner then can conclude with 95 percent confidence that 9.96 ± 3.68 percent of the parking spots will be open at any given time. This translates approximately to 16 to 34 open spots, and the owner can conclude that presently there is sufficient (but barely at certain times) parking for her customers. Note that the direct calculation of the limit of error

$$\ell = (3.8pq/n)^{1/2} = [3.84(0.0996)(0.9004)/2,500]^{1/2} = 0.0117$$

is incorrect and understates the true error. It would also be prudent for the analyst to collect samples over several days to avoid representativeness errors.

DETERMINING OBSERVATION FREQUENCY

The frequency of the observations depends, for the most part, on the number of observations required and the time available to develop the data. For example, for 3,600 observations to be completed in 20 calendar days, the analyst would need to obtain approximately $3,600/20 = 180$ observations per day.

Determination of the Required Number of Observations	EXAMPLE 14.3

An analyst wishes to determine the amount of downtime due to tool problems in an area involving 10 CNC machining centers where very fine drilling is taking place. An initial pilot study indicated that out of 25 observations, only one CNC machine was down, for a $\hat{p}$ of 0.04. The analyst wishes a more accurate study with an estimate within ± 1 percent of the true value, with a 99 percent confidence. Since $Z_{0.005}$ is 2.58, the number of observations needed is then:

$$n = \frac{2.58^2 \times 0.04 \times (1 - 0.04)}{0.01^2} = 2,557$$

Even if the analyst took 256 trips to the plant floor with 10 simultaneous observations on each trip, this is still a large number, and the analyst may wish to reconsider a lower level of confidence. Furthermore, there is also the problem of correlated observations (see Example 14.2).

Of course, the number of analysts available and the nature of the work being studied also influence the frequency of the observations. For example, if only one analyst is available to accumulate the data in Example 14.3, it may be impractical for that person to take 180 observations during one day.

After determining the number of observations per day, the analyst must select the actual time needed to record the observations. To obtain a representative sample, observations are taken randomly at all times of the day. There are many ways of randomizing the occurrence of the observations. In a manual approach, the analyst may select nine numbers daily from a statistical table of random numbers, ranging from 1 to 48 (see Appendix 3, Table A3–5). If each number carries a value, in minutes, equivalent to 10 times its size, the numbers selected can then set the time, in minutes, from the beginning of the day to the time for taking the observations. For example, the random number 20 would mean that the analyst should make a series of observations 200 min after the beginning of the shift.

Another approach considers four adjacent digits in the random number table. Digit 1 is the day identifier, with numbers 1 to 5 identifying the workday Monday through Friday. Digit 2 is the hour identifier, with numbers 0 to 8 added to the starting time of work (e.g., 7:00 A.M.). Digits 3 and 4 are the minutes identifiers, with numbers between 0 and 60 acceptable. Obviously, the easiest approach is to write a small program using the random number generator on any of the commercial spreadsheets or to use the feature in DesignTools or QuikSamp.

The study should be long enough to include normal fluctuations in production. The longer the overall study, the better the chance of observing average conditions. Usually, work sampling studies are made over a block of time ranging from two to four weeks.

Another alternative to help analysts decide when to make daily observations is a random reminder. This pocket-sized instrument beeps at random times, letting analysts know when to make the next observation. The user preselects an average sampling rate (observations per hour, or observations per day) and responds with a trip to the data collection area upon hearing the beep. Typically, the instrument can be preset at any of the following average beeps per hour: 0.64, 0.80, 1.0, 1.3, 1.6, 2.0, 2.5, 3.2, 4.0, 5.0, 6.4, and 8.0. This instrument is especially useful for self-observation, discussed later in this chapter. A table with times prepared in advance can require too much of the analyst's time when he or she is attempting to record data conscientiously at the listed times.

DESIGNING THE WORK SAMPLING FORM

The analyst should design an observation form to record the data to be gathered during the work sampling study. A standard form is usually not acceptable, since each work sampling study is unique from the standpoint of the total observations needed, the random times that observations are made, and the information being sought. The best form is tailored to the study objectives.

Figure 14.5 is an example of a work sampling study form. An analyst designed this form to determine the time utilized for various productive and nonproductive states in a maintenance repair shop. The form accommodates 20 random observations during the workday. Some analysts prefer to use a specially designed card that allows observations to be made without the attention caused by a clipboard. The card can be sized so that it can be carried conveniently in the shirt or coat pocket. For instance, the form shown in Figure 14.5 could easily be split into two sections, with one section on each side of a 3-in by 5-in card that could be carried in the shirt pocket.

USING CONTROL CHARTS

The control chart techniques used in statistical quality control work can readily be applied to work sampling studies to identify problem areas. Since these studies deal exclusively with percentages or proportions, analysts use the p chart most frequently.

The first problem in setting up a control chart is the choice of limits. In general, a balance must be sought between the cost of looking for an assignable cause when none is present and the cost of not looking for an assignable cause when one is present. As an arbitrary choice, the analyst should use the $\pm 3\sigma$ limits as control limits on the p chart. (More extreme limits, such as the six sigma process developed by Motorola, can also be utilized.) Substituting 3σ for 1.96σ in Equation (1) yields

$$\ell = 3\sigma = 3\sqrt{p(1-p)/n}$$

Suppose that p for a given condition is 0.10 and that 180 observations are taken each day. Solving for ℓ yields

$$\ell = 3 \times [0.1 \times 0.9/180]^{1/2} = 0.067 \approx 0.07$$

A control chart similar to Figure 14.6 could then be constructed, and the p' values for each day would be plotted on that chart.

Work Sampling Study

Main repair shop

Remarks _____

Number working this study _____ Date _____ By _____

Obs. nos.	Random time	Productive occurrences							Nonproductive occurrences							Total observations	Percentage productive	Percentage nonproductive
		Mch	Weld	Pipe fit	Gen. labor	Elect.	Carpen.	Janitor	Get tools	Grind tools	Wait job	Wait crane	Confer foreman	Personal	Idle			
1																		
2																		
3																		
4																		
5																		
6																		
7																		
8																		
9																		
10																		
11																		
12																		
13																		
14																		
15																		
16																		
17																		
18																		
19																		
20																		
Total																		

Figure 14.5 Work sampling study form.

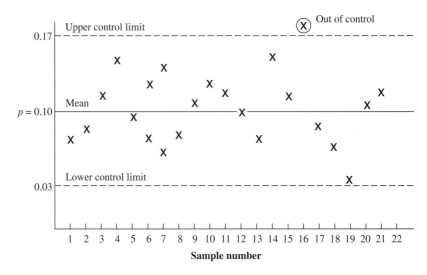

Figure 14.6   Sample control chart.

In quality control work, the control chart indicates whether the process is in control. In a similar manner, in work sampling, the analyst considers points beyond $\pm 3\sigma$ limits of p as being out of control. Thus, a sample that yields a value of p' is assumed to have been drawn from a population with an expected value of p if p' falls within the $\pm 3\sigma$ limits of p. Expressed another way, if a sample has a value p' that falls outside the $\pm 3\sigma$ limits, the sample is assumed to be from some different population, or the original population is assumed to be changed.

As in quality control work, points other than those out of control may be of some statistical significance. For example, it is more likely that a point will fall outside the $\pm 3\sigma$ limits than that two successive points will fall between the $\pm 2\sigma$ and $\pm 3\sigma$ limits. Hence, two successive points between these limits would indicate that the population has changed. A series of significant sets of points has been derived. This idea is discussed in most statistical quality control texts under the heading "runs."

| EXAMPLE 14.4 | Use of Control Charts in Work Sampling |

The Dorben Company wishes to measure the percentage of machine downtime in the lathe department. An original estimate showed the downtime to be approximately 0.20. The desired results were to be within ± 5 percent of p, with a confidence level of 0.95. Analysts took 6,400 readings over 16 days at the rate of 400 readings per day. They computed a p' value for each daily sample of 400 and set up a p chart for $p = 0.20$ and subsample size $N = 400$ (see Figure 14.7).

Each day, they took readings and plotted p'. On the third day, the point for p' went above the upper control limit. An investigation revealed that following an accident in the plant, several workers left their machines to assist the injured employee. Since an assignable cause of error was discovered, they discarded this point from the study. If they had not used a control chart, these observations would have been included in the final estimate of p.

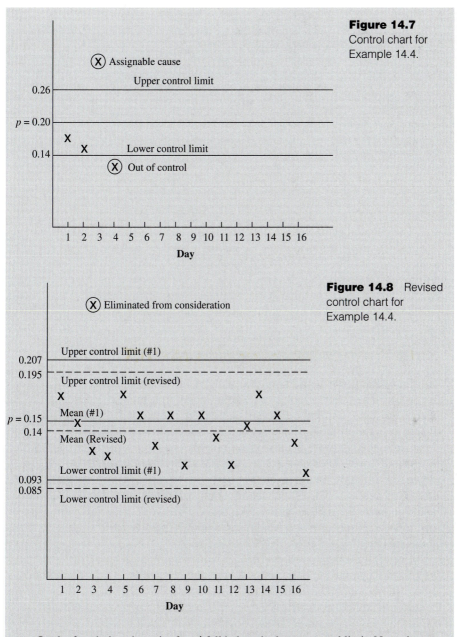

Figure 14.7
Control chart for
Example 14.4.

Figure 14.8 Revised
control chart for
Example 14.4.

On the fourth day, the point for p' fell below the lower control limit. No assignable cause could be found for this occurrence. The industrial engineer in charge of the project also noted that the p' values for the first two days were below the mean p and decided to compute a new value for p, using the values from days 1, 2, and 4. The new estimate of p turned out to be 0.15. To obtain the desired accuracy, n was then 8,830 observations. The control limits also changed (see Figure 14.8).

The analysts took observations for 12 more days and plotted the individual p' values on the new chart. As can be seen, all the points fell within the control limits. They then calculated a more accurate value of p, using all 6,000 observations, and determined that the new estimate of p was 0.14. A recalculation of achieved accuracy showed it to be slightly better than the desired accuracy. As a final check, the analysts computed new control limits, using p equal to 0.14. The dashed lines superimposed on Figure 14.8 show that all points were still in control using the new limits. If a point had fallen out of control, the analysts would have eliminated it and computed a new value of p. They would have repeated this process until the desired accuracy was achieved and all p' values were in control.

The percentage downtime in Example 14.4 will not remain the same forever. Methods improvement should be a continuing process, and the percentage downtime should diminish. Control charts can also be used to show the progressive improvement of work areas. This idea is especially important if work sampling studies are used to establish standard times, since such standards must change whenever conditions change if they are to remain realistic.

14.4 RECORDING OBSERVATIONS

In approaching the work area, the analyst must not anticipate the expected recording. The analyst should walk to a fixed point, make the observation, and record the facts. It might be helpful to make an actual mark on the floor to show where to stand for the repeat observations. If the operator or machine being studied is idle, the analyst must also determine the reason for the idleness and mark this on the form. The cause for a worker's idleness, whether a machine breakdown, lack of raw material, etc., is very important in redesigning the job for improved productivity. The analyst must learn to take the visual observations, and then make the written entries after leaving the work area. This minimizes the operators' feelings of being watched and allows them to perform in their accustomed manner.

Even if the analyst observes proper work sampling protocol, the data tend to be biased when the technique is used for studying people. The arrival of an analyst at the work center immediately influences the activity of the operator. The operator becomes productively engaged as soon as he or she sees the analyst approaching the work center. Then, too, there is a natural tendency for the analyst to record what has just happened or what will be happening, rather than what is actually happening at the exact moment of the observation.

A camcorder can be very useful in performing unbiased work sampling studies involving only people. A work sampling study was made by the authors over a 10-day period on data processing workers with the elements "working" and "not working." The 2,520 observations collected indicated a statistically significant ($p <$ 0.001) difference of a 12.3 percent greater "not working" average with the camcorder than with the personal observation method. Whereas the workers hesitated to indicate that they were idle, the camcorder accurately recorded the ongoing activity.

Similarly some of the work measurement software described later comes with dedicated or personal digital assistants (PDAs) to facilitate data recording and uploading of the data to desktop PCs for later data analysis. These devices can also be set to provide audible beeps to signal the appropriate random time for data collection. Otherwise, stand-alone random reminder devices can be utilized.

14.5 MACHINE AND OPERATOR UTILIZATION

Analysts can use work sampling to determine machine and operator utilization. As an example, consider machine utilization in a heavy machine shop. Management estimated that the actual cutting time in this section should be 60 percent of the workday, to comply with the quotations being submitted. There were 14 machines involved, and the analysts had to take approximately 3,000 observations to get the accuracy desired.

Analysts designed a work sampling form (see Figure 14.9, but with cells blank) to accommodate the 16 possible states that each of the 14 machines might be in at the time of an observation. Then they set up a random pattern of 6 observations of the 14 operations per each of 36 separate shifts.

Since the principal purpose of the study was to learn the status of the actual cutting time in this section, an analyst kept a cumulative percentage machine cutting chart (see Figure 14.10). At the beginning of each day's study, the analysts took a ratio of all previous cutting observations to the total observations to date. By the end of the 10th day of the study, the percentage of machine cutting time began to level off at 50.5 percent.

After all 3024 observations were collected, the analysts divided the sum of all observations in each category by the total number of observations, resulting in percentages that represented the distributions of cutting time, setup time, and the various operations listed. Figure 14.9 illustrates the summary sheet for this study with cutting time amounting to 50.7 percent. The percentage of time required by the various delays such as 9.6 percent for setup or 10.8 percent for tool handling, may indicate potential areas for methods improvement that would help increase the cutting time. Example 14.5 shows a similar approach for determining operator utilization.

DATE ___7-15___
OBSERVER ___R. Guild___

MACHINE	DWG.	CUTTING	SETUP	MACHINE IDLE	CRANE WAIT	WAIT-INSPECTION	AID INSPECTION	WAIT-TOOLS NOT AVAILABLE	WAIT-TOOL TROUBLE	CONFER WITH OTHER SHIFT	TOOL HANDLING	GET OR GRIND TOOLS	CONFER WITH FOREMAN, INSP.	WAIT FOR JOB	REMOVE CHIPS CLEAN TABLE	MISCELLANEOUS	NO OPERATOR
20″ VBM		101	7	14	2	3	1			2	37	5	3			6	35	216
16′ VBM		102	34	14	15	3	1		1	1	28	5	1	7	4			216
28′ VBM		119	34	10	5	5	2				18	2	1	2			18	216
12′ VBM		109	24	12	13	6	1			3	26	6	2	3	3	2	6	216
16′ PLANER		127	17	6	9	2	2				22		15			4	12	216
8′ IMM		64	18	17	16	3	1			2	30	2	3	3	1	28	28	216
16′ VBM		147	19	10	14	3					15	3		1		1	3	216
14′ PLANER		140	8	5	7	2				2	17	3		3		11	18	216
72″ E. LATHE		99	13	12	7	3			1	1	32	8		1		3	36	216
96″ E. LATHE		89	9	29	18	11		3		2	29	8	3	2		3	10	216
96″ E. LATHE		109	14	12	8	10		3			32	9	2	10	1	1	5	216
160″ E. LATHE		72	34	13	14	6				4	21	4	3	3	1	4	37	216
11-1/2′ PLNR		106	35	11	10	4				1	11	6	1	5	2	8	16	216
32′ VBM		151	23	8	7	1			1	1	10	3		5	1	5		216
		1535	289	173	145	62	8	6	3	19	328	64	34	45	13	76	224	3024 =
%		50.7	9.6	5.9	4.8	2.1	.3	.2	.1	.6	10.8	2.1	1.1	1.5	.4	2.5	7.4	100%

Figure 14.9 Work sampling summary sheet.

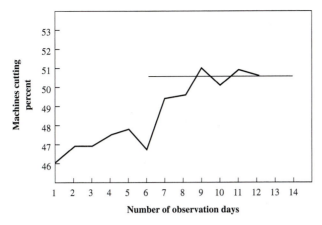

Figure 14.10 Cumulative percentage machine cutting.

EXAMPLE 14.5

Work Sampling to Determine Operator Utilization

Management in a semiconductor plant was considering increasing the number of taping machines monitored by one operator from 10 to 12. Typically the machines run automatically other than requiring regular resupply of raw components and having irregular stoppages due to lot changes, loading errors, and electrical arc problems, which need to be attended to by the operator. Since there are long machine cycles with irregular breakdowns, work sampling is the best approach to determine operator utilization.

A total of 185 observations were taken over a one-week period, resulting in 125 observations in which the operator was busy with a variety of activities, shown in

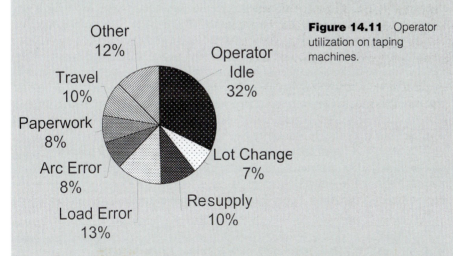

Figure 14.11 Operator utilization on taping machines.

Figure 14.11. Correspondingly 60, or 32.4 percent, of the observations indicated that the operator was idle. The resulting limit of error is

$$\ell^2 = 3.84pq/n = 3.84(0.324)(0.676)/185 = 0.00455 \quad \text{or} \quad \ell = 0.067$$

Thus management concluded with 95 percent confidence that the operator will be idle 32.4 ± 6.7 percent of the time, and thus the operator could easily monitor an additonal 2 machines.

Dept._____		ACME Electric Products, Inc.			Date _____
Operation	Personal	Other Idle	Interference	Working	Total Observations	Percent Allowance
Bench	80	39	26	2,750	2,895	.95
Machine	20	9	27	1,172	1,228	2.30
Inspection	61	8	7	984	1,060	0.71
Spray	63	199	43	1,407	1,712	3.06

Figure 14.12 Work sampling summary of idle, interference, and working activities for determining unavoidable allowance.

14.6 DETERMINING ALLOWANCES

The determination of allowances must be correct, if fair standards are to be developed. Prior to the introduction of work sampling, analysts frequently determined allowances for personal reasons and unavoidable delays by taking a series of all-day studies on several operations and then averaging the results. Thus, they recorded, timed, and analyzed trips to the rest room, trips to the drinking fountain, interruptions, and so forth. Although this method provided a fair allowance, it was costly and time-consuming and was fatiguing to both the analyst and the operator.

Through a work sampling study, analysts take a great number of observations (usually, over 2,000) at different times of the day and of different operators. They can then divide the total number of idle observations that involve normal operators by the total number of working observations. The result equals the percentage allowance that should be given to the operator for the class of work being studied. The different elements that enter into personal and unavoidable delays can be kept separate, and an equitable allowance can be determined for each class or category.

Figure 14.12 illustrates a summary of a work sampling study for determining unavoidable delay allowances on bench, machine, inspection, and spray operations. There were interferences in 26 cases out of 2,895 observations made on the bench operations. This indicated an unavoidable delay allowance of 0.95 (26/2,750) percent for this class of work.

14.7 DETERMINING STANDARD TIME

Work sampling can be very useful for establishing time standards on both direct and indirect labor operations. The technique is the same as that used for determining allowances. The analyst must take a large number of random observations. The percentage of the total observations that the operator is working approximates the percentage of total time for that state.

Calculation of Standard Time for a Single Operation

EXAMPLE 14.6

Table 14.1 lists the information necessary for the calculations, the sources for the information, and the specific data used in this example of a drill press operator.

$$\text{OT} = \frac{T}{P} \times \frac{n_i}{n} = \frac{480}{420} \times 0.85 = 0.971 \text{ min}$$

The normal time (NT) is found by scaling the observed time by the average rating $\overline{R}$:

$$\text{NT} = \text{OT} \times \overline{R}/100 = 0.971 \times 110/100 = 1.069 \text{ min}$$

Table 14.1 Information on Drill Press Operator

Information	Source	Data
Total working day (working + idle)	Time card	480 min
Number of units drilled	Inspection department	420 units
Working fraction	Work sampling	85 percent
Average rating	Work sampling	110 percent
Allowances	Work sampling	15 percent

Finally, the standard time is found by adding allowances (using the multiplier approach) to the normal time:

$$\text{ST} = \text{NT} \times (1 + \text{Allowance}) = 1.069 \times 1.15 = 1.229 \text{ min}$$

More specifically, the observed time OT (see Chapter 10) for a given element is calculated from the working time divided by the number of units produced during that time

$$\text{OT} = \frac{T}{P} \times \frac{n_i}{n}$$

where T = total time
   n_i = number of occurrences for element i
   n = total number of observations
   P = total production for period studied

The normal time NT is found by multiplying the observed time by the average rating:

$$\text{NT} = \text{OT} \times \overline{R}/100$$

where $\overline{R}$ = average performance rating = $\Sigma R/n$. Finally, the standard time is found by adding allowances to the normal time.

Example 14.6 shows this procedure for a single operation while Example 14.7 shows this for multiple elements.

EXAMPLE 14.7	Calculation of Standard Time for Multiple Elements

An analyst made 30 observations over 15 min on a work assignment involving three elements, during which time 12 units were produced. The resulting data appear in Table 14.2. The observed times were, respectively,

$$OT_1 = \frac{15}{12} \times \frac{9}{30} = 0.375 \text{ min}$$

$$OT_2 = \frac{15}{12} \times \frac{7}{30} = 0.292 \text{ min}$$

$$OT_3 = \frac{15}{12} \times \frac{12}{30} = 0.500 \text{ min}$$

and the respective normal times were

$$NT_1 = 0.375 \times \frac{860}{9 \times 100} = 0.358 \text{ min}$$

$$NT_2 = 0.292 \times \frac{705}{7 \times 100} = 0.294 \text{ min}$$

$$NT_3 = 0.500 \times \frac{1180}{11 \times 100} = 0.492 \text{ min}$$

Table 14.2 Tabular Data of a Three-Element Work Sampling Study

Observation number	Performance rating observed			
	Element 1	**Element 2**	**Element 3**	**Idle**
1	90			
2				100
3		110		
4	95			
5	100			
6		100		
7			105	
8	90			
9			110	
10	85			
11			95	
12		90		
13			100	
14			95	
15	80			
16			110	
17		105		
18			90	
19	100			
20			85	
21			90	
22			90	

| Observation | Performance rating observed | | | |
number	Element 1	Element 2	Element 3	Idle
23	110			
24			100	
25		95		
26				100
27		105		
28		100		
29			110	
30	110			
$\sum$ Rating	860	705	1,180	100

Assuming a constant 10 percent allowance for all elements, the final standard time was

$$ST = (0.358 + 0.294 + 0.492)(1 + 0.10) = 1.258 \text{ min}$$

14.8 SELF-OBSERVATION

Managers and salaried workers should periodically take work samples of their own work to evaluate the effectiveness of their time usage. In many cases, managers spend less time on the important aspects than they think they are spending. They also spend more time on unimportant aspects, such as personal and avoidable delays, than they believe they are spending. Once managers learn how much time is being taken by functions that could be readily handled by subordinates and clerical personnel, they can take positive action.

For example, a university professor may decide to conduct a personal work sampling study to determine how much time is spent on various activities over an 8-week period during the academic year. This period should be representative and not subject to seasonal variation. The professor sets a random reminder to provide 2 samples per hour on average. Thus, over the 8-week study period, the professor would have approximately 640 observations (8 weeks $\times$ 40 h/wk $\times$ 2 observations/h). For a more accurate study the professor could have chosen to take samples at a higher rate within the study interval.

To record the data, a form similar to that shown in Figure 14.13 was used to record one week of daily random observations. Each time the random reminder beeped (e.g., Diviliss Electronics), the professor recorded the code letter for the applicable category and the time.

At the end of the 8-week study, 80 of the total 640 observations were coded I (committee participation), which meant that about 12.5 percent of work time was spent in committee participation. The 95 percent confidence interval would be

$$\pm 1.96 \sqrt{\frac{0.125(1 - 0.125)}{640}} = \pm 0.026$$

Name   A. B. Jones   Week of   3/25   Study No.   B-47

T Teaching	**C** Continuing Ed.	**D** Personal
R Research	**A** Student Advising	**I** Committee
P Preparation	**S** Professional Development	

MON.		TUES.		WED.		THUR.		FRI.		SAT.		SUN.		NOTES
TIME	CO	TIME	CO	TIME	CO	TIME	CO	TIME	CO	TIME	CO	TIME	CO	
8:17	T	8:06	C	8:58	P	8:02	I	8:49	T					I - Exec . Com.
8:52	T	8:32	C	9:08	R	8:31	A	9:07	C					
9:04	D	8:58	A	9:25	R	8:45	S	9:17	C					
9:27	R	9:32	P	10:01	I	9:32	T	9:51	I					I - Res. Com. + Personnel Policy
9:50	R	10:11	T	10:50	S	10:17	T	10:11	R					
10:11	I	11:00	S	10:57	S	10:40	S	10:32	R					I - Personnel Policy
10:18	A	11:05	S	11:26	A	11:35	P	10:53	A					
11:01	P	11:55	I	11:40	P	11:59	D	11:17	P					I - Dept. Curriculum
11:25	P	1:42	P	1:17	D	1:04	R	11:42	P					
1:05	P	1:59	P	2:05	T	1:27	R	1:11	I					I - Dept Curriculum
2:01	T	2:11	R	2:35	T	1:47	I	1:47	I					I - Dept. Curriculum
2:35	T	2:37	R	3:00	I	2:17	R	2:15	T					I - Entertainment Committee
2:55	S	3:25	R	3:24	S	2:46	A	2:45	T					
3:45	S	3:40	S	4:14	S	3:40	S	3:00	T					
4:11	P	3:57	R	4:38	P	4:11	S	4:02	S					
4:42	R	4:15	A	5:00	P	4:37	S	4:25	D					

WEEK SUMMARY

T 13	**P** 15	**A** 7	**D** 4				
R 14	**C** 4	**S** 14	**I** 9	TOTAL 80			

Figure 14.13 A work sampling form specially designed for self-observation.

Therefore, the professor is 95 percent confident that committee work is occupying 12.5 ± 2.6 percent of the time. Based on this and the other activity percentages, the professor altered the calendar to utilize time and energy in a more positive manner.

14.9 WORK SAMPLING SOFTWARE

Using a computer can save an estimated 35 percent of the total work sampling study cost, because of the high percentage of clerical effort relative to actual observation time. The majority of the effort involved in summarizing work sampling data is clerical: calculating percentages and accuracies, plotting data on control charts, determining the number of observations required, determining the daily observations required, determining the number of trips to the area being studied per day, determining the time of day for each trip, and so on.

There are several work sampling software packages available for the analyst with a variety of features. WorkSamp by Royal J. Dossett Corp. provides a built-in beeper to signal for data entry at random intervals, several different summary reports, and an RS233C connection to upload data into a desktop PC. A slight disadvantage is the use of a custom-built Datawriter to collect the data electronically. Other packages use more versatile pocket PCs or Palm PDAs to collect data, allowing them to be used for other tasks by the analyst. For example, CAWS/e by C-Four Consulting uses the pocket PC, while SamplePro by Applied Computer Services, WorkStudy+™3.0 by Quetech, Ltd., UmtPlus by Laubrass, Inc., or QuikSamp (available with the textbook, see Figure 14.14) use the Palm

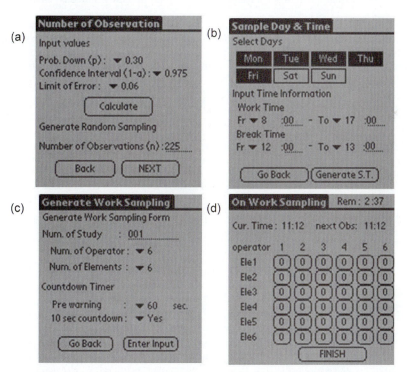

Figure 14.14 The QuickSamp work sampling program on Palm PDAs.
(a) Generating the number of observations required; (b) selecting time and days for sampling; (c) selecting the operators and work elements to be sampled; (d) maximum-size data entry screen.

PDA. One advantage of the pocket PC programs is the ability to perform more detailed data analysis by linking directly into Excel. However, any of these software products will provide the following benefits to the analyst:

1. The amount of industrial engineering time is increased through the reduction of clerical routines.
2. Results of the study are achieved more rapidly, and the data are presented in a professional manner.

EXAMPLE 14.8   ## Workload Analysis in the Service Industry

A large case study of community service operations was undertaken by Sterling Associates (1999) for the Department of Social and Health Services (DSHS) of Washington state. Since community service caseloads had decreased over the past few years, DSHS was interested in knowing whether the workload for social workers, financial specialists, and office assistants had also decreased. A total of 304 staff members were given pagerlike random reminders to "buzz" at specified random times, at which time they were to record the specific task they were performing on an electronic Web-based form. This format allowed for a quick access, regular updating, and easy analysis of the data.

The study encompassed a two-month study period (specifically 90,385 worker-hours) collecting 91,371 observations for 17 different tasks, for 15 different programs in six different district offices. Assuming a 99 percentile confidence interval, a worst-case scenario of $p = 0.5$ for the task of interest, and using Equation (3) in reverse, yielded

$$\ell = \sqrt{\frac{(2.58^2)(0.5)(0.5)}{91,371}} = 0.0042$$

Thus, for a task occurring roughly 50 percent of the time, the resulting accuracy is ± 0.41 percent. For any task occurring a smaller percentage of the time, the accuracy would be even better.

Time spent for a given task (e.g., case audits) was calculated as follows. Out of the 91,371 observations, 2,224 were for case audits, during which time 2,217 case audits were completed. Thus, the time per audit was

$$\text{Time/audit} = \frac{90,385}{2,217} \times \frac{2,224}{91,371} = 0.99 \text{ h/audit}$$

The primary results included the following. There was a large disparity (see Figure 14.15) between the amount of time spent on serving various programs and the actual number of cases handled by that program (as measured in assistance units, AU). This may mean that a redistribution of effort or workforce may be needed to better handle the various programs. To assist in this, a staffing needs model was developed. Although there were some differences, the overall results were similar between the various district offices. Finally, although the setting of standards was not emphasized in this study, the overall idle time (roughly 4 percent) was quite low. However, this result should be tempered by the fact that, as in all self-observation studies, workers may be hesitant to accurately indicate that they are not working.

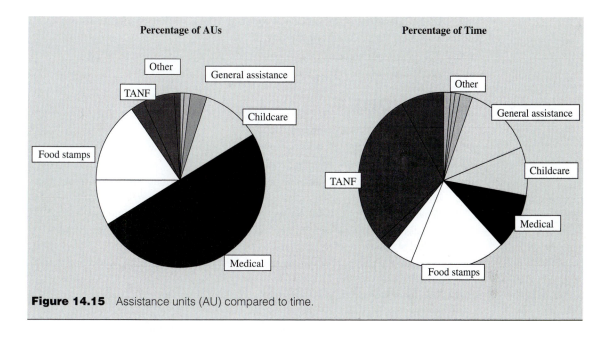

Figure 14.15 Assistance units (AU) compared to time.

3. The cost of conducting work sampling studies is significantly reduced.
4. Accuracy of the computations is improved.
5. Fewer errors are committed by analysts.
6. The automated system provides an incentive to make greater use of the work sampling technique.

SUMMARY

Work sampling is another tool that allows time and methods analysts to obtain information on machine and operator utilization as well as to set time standards. Performance-rated work sampling is especially useful in determining the amount of time that should be allocated for unavoidable delays, work stoppages, and the like. The extent of these interruptions is a suitable area for study to improve productivity. Work sampling is also being used more heavily for establishing standards on production support labor, maintenance, and service labor.

Everyone in the field of methods-time study and wage payment should become familiar with the advantages, limitations, uses, and applications of this technique. In summary, the following considerations should be kept in mind:

1. Explain and "sell" the work sampling method before using it.
2. Confine individual studies to similar groups of machines or operations.
3. Use as large a sample size as is practicable.

4. Take individual observations at random times, so that observations are recorded for all hours of the day.

5. Take the observations over an extended period of time so as to be representative of the actual conditions.

QUESTIONS

1. Where was work sampling first used?
2. What are advantages of work sampling over stopwatch time study?
3. In what areas is work sampling applicable?
4. How can you determine the time of day to make the various observations, such that biased results do not occur?
5. What considerations should be kept in mind when doing work sampling studies?
6. Discuss the statistical basis for the trade-off between the number of observations collected and the accuracy of work sampling.
7. What are the principal advantages of using a random reminder in connection with a self-observation study?
8. Over how long a period is it desirable to collect observations?
9. Discuss the trade-offs of observing simultaneously 10 clerks in a large bank.
10. How can the validity of work sampling be "sold" to the employee not familiar with probability and statistical procedures?
11. What are the pros and cons for using work sampling to establish performance standards?

PROBLEMS

1. The analyst in the Dorben Reference Library decides to use the work sampling technique to establish standards. Twenty employees are involved. The operations include cataloging, charging books out, returning books to their proper location, cleaning books, record keeping, packing books for shipment, and handling correspondence. A preliminary investigation resulted in the estimate that 30 percent of the group's time was spent in cataloging. How many work sampling observations would be made if it were desirable to be 95 percent confident that the observed data were within a tolerance of ± 10 percent of the population data? Describe how the random observations should be made.

   The following table illustrates some of the data gathered from 6 of the 20 employees. The number of volumes cataloged equals 14,612. From these data, determine a standard, in hours per hundred, for cataloging. Then design a control chart based on $\pm 3\sigma$ limits for the daily observations.

Item	Operators					
	Smith	Apple	Brown	Green	Baird	Thomas
Total hours worked	78	80	80	65	72	75
Total observations (all elements)	152	170	181	114	143	158
Observations involving cataloging	50	55	48	29	40	45
Average rating	90	95	105	85	90	100

2. The work measurement analyst in the Dorben Company is planning to establish standards for indirect labor, using the work sampling technique. This study will provide the following information:

   T = total operator time represented by the study

   N = total number of observations involved in the study

   n = total observations of element under study

   P = production for period under study

   R = average performance rating factor during study

   Derive the equation for estimating the normal elemental time for an operation.

3. The analyst in the Dorben Company wishes to measure the percentage of downtime in the drop hammer section of the forge shop. The superintendent estimates the downtime to be about 30 percent. The desired results, using a work sampling study, are to be within ± 5 percent of p, with a confidence level of 0.95. The analyst decides to take 300 random readings a day for three weeks. Develop a p chart for $p = 0.30$ and subsample size $N = 300$. Explain the use of this p chart.

4. The Dorben Company is using the work sampling technique to establish standards for its typing pool. This pool has varied responsibilities, including typing from tape recordings, filing, Kardex posting, and copying. The pool has six typists who each works a 40-h week. Over a 4-week period, 1,700 random simultaneous observations of all 6 typists were made. During the period, the typists produced 9,001 pages of routine typing. Of the random observations, 1,225 showed that typing was taking place. Assuming a 20 percent allowance and an adjusted performance rating factor of 0.85, calculate the hourly standard per page of typing.

5. How many observations should be recorded in determining the allowance for personal delays in a forge shop, if it is expected that a 5 percent personal allowance will suffice, and if this value is to remain between 4 and 6 percent across 95 percent of the time?

6. To get ± 5 percent precision on work that is estimated to take 80 percent of the workers' time, how many random observations are required at the 95 percent confidence level?

7. If the average handling activity during a 10-day study is 82 percent and the number of daily observations is 48, what is the confidence interval for each day's percentage activity?

8. The following data were collected by an intern working at the Mole Hill Ski Resort regarding the efficiency of the V-8 Ford engine that was used to power the rope tow for the bunny slope. Managers would like to know how many hours of service they can expect to achieve out of their obviously failing engine out of a typical 16-h skiing day. After studying the data, you can state with 95 percent confidence that the engine will be able to run for _____ h $\pm$ _____ h of a 16-h day.

Running on 8 cylinders	|||||
Running on 7 cylinders	||||| ||||| ||
Running on 6 cylinders	||||| ||||| ||||| ||
Not running, engine clogged	||||| ||||| ||||| ||||| ||||| ||

9. Data from an 8-h work sampling study are shown below.
   a. What percentage of the time is the machine running?
   b. What percentage of the time is the machine broken?
   c. What is the limit of accuracy of this study for $\alpha = 0.05$?
   d. What is the 95 percent confidence interval of the machine running?
   e. Your manager says that the range in part d is unacceptably large. She would like to narrow the range down to ± 1 min of running time. How many observations should you collect to achieve this?

Running	IIIII IIIII IIIII IIIII IIIII II	
Idle	Broken	IIIII
	Out of stock	IIIII IIIII I
	Other	IIIII II

10. Dorben Co. provides a 10% allowance for its operators. A work sampling study of one operator over an 8-h shift yields the following data. Average rated performance is 110%.

   a. What is the observed time (in minutes) for loading?
   b. What is the normal time for unloading?
   c. What is the overall standard time?
   d. What is the accuracy of the study?

Load	IIIII IIIII IIIII I
Unload	IIIII IIIII IIII
Process	IIIII II
Idle	IIIII IIIII III

REFERENCES

Barnes, R. M. *Work Sampling*. 2d ed. New York: John Wiley & Sons, 1957.

Morrow, R. L. *Time Study and Motion Economy*. New York: Ronald Press, 1946.

Richardson, W. J. *Cost Improvement, Work Sampling and Short Interval Scheduling*. Reston, VA: Reston Publishing, 1976.

Richardson, W. J., and E. S. Pape. "Work Sampling." In *Handbook of Industrial Engineering*. 2d ed. Ed. Gavriel Salvendy. New York: John Wiley & Sons, 1992.

Sterling Associates Ltd. *Community Service Office Workload Study*, 711 Capitol Way South, Suite 610, Olympia, WA 98501, 1999.

SELECTED SOFTWARE

CAWS/E. C-Four, P.O. Box 808, Pendleton, SC 29670 (http://www.c-four.com/)

JD-7/JD-8 Random Reminder. Divilbiss Electronics, RR #2, Box 243, Chanute, KS 66720 (http://www.divilbiss.com)

DesignTools and QuikSamp (available from the McGraw Hill text website at www.mhhe.com/niebel-freivalds), New York: McGraw-Hill, 2002.

Workstudy+™ 3.0, Quetech, Ltd., 866 222-1022 (www.quetech.com)

TimerPalm. Applied Computer Services, Inc., 7900 E. Union Ave., Suite 1100, Denver CO 80237.

UMT Plus, Laubrass, Inc., 3685 44e Ave., Montreal, QC H1A 5B9, Canada (www.laubrass.com)

WorkSamp. Royal J. Dosset Corp., 2795 Pheasant Rd., Excelsior, MN 55331.

SELECTED VIDEOTAPES

Work Sampling Fundamentals. ½″ VHS, C-Four, P.O. Box 808, Pendleton, SC 29670

Indirect and Expense Labor Standards

CHAPTER
15

KEY POINTS

- Use both time studies and predetermined time systems to develop standards for relatively predictable indirect labor.

- Use slots or similar job categories to establish standards for relatively unpredictable indirect and expense labor.

- Utilize work sampling and historical records to establish standards for professional expense labor.

- Use queuing theory to calculate waiting times for jobs.

- Use Monte Carlo simulation to predict delays or downtimes on jobs.

Since 1900, the percentage increase of indirect and expense workers has more than doubled that of direct labor workers. This is especially true with the recent growth of service industries such as health care, insurance and banking, retailing, information technology, and even arts, entertainment, and the leisure industry. Groups traditionally classified as *indirect labor* include shipping and receiving, trucking, stores, inspection, material handling, toolroom, janitorial, and maintenance. *Expense labor* includes all positions not coming under direct or indirect: office clerical, accounting, sales, management, engineering, and so on.

The rapid growth in office workers, maintenance workers, and other indirect and expense employees is due to several factors. First, the increased mechanization of industry and the complete automation of many processes, including the use of robots, have decreased the need for craftspeople and operators. This trend toward increased automation has resulted in a huge demand for electronics specialists, instrumentation specialists, computer hardware and software technicians, and other service people.

Second, the tremendous increase in paperwork brought on by federal, state, and local legislation is responsible, to a large extent, for an ever-increasing number of clerical employees. Third, office and maintenance work has not been subjected to the methods study and technical advances that have been applied so effectively to direct labor in industrial processes. With a large share of most payrolls being earmarked for indirect and expense labor, management is beginning to realize the opportunities for the application of methods and standards in this area.

15.1 INDIRECT AND EXPENSE LABOR STANDARDS

The systematic approach to methods, standards, and wage payment is just as applicable to the indirect and expense areas as it is to direct labor. Careful fact-finding, analysis, proposed method development, presentation, installation, and job analysis development should precede a program for establishing standards on indirect and expense labor. The methods analysis procedure in itself can introduce economies.

Work sampling is a good technique for determining the severity of a problem and the savings potential in the indirect and expense areas. It is not unusual to find that the workforce is productively engaged only 40 to 50 percent of the time, or even less. For example, in maintenance work, which represents a large share of the total indirect cost, analysts may find the following reasons for much of the time lost during the workday:

1. *Inadequate communication.* It is quite common to find incomplete and even incorrect job instructions on work orders. This necessitates additional trips to the toolroom and supply room to obtain parts and tools that should have been available when the work was started. As an example, a work order that merely states "Repair leak in oil system" does not provide sufficient detail on whether a new valve, pipe, or gasket is needed, or whether the valve needs to be repacked.

2. *Unavailability of parts, tools, or equipment.* If maintenance workers do not have the facilities and parts to do the job, they are obliged to improvise, which usually wastes time and frequently results in inferior work.

3. *Interference of production employees.* With improper scheduling, maintenance employees may find that they are unable to begin a repair, service, or overhaul operation, because the machine is still being used by production employees. This can result in the technician idly waiting until the production department is ready to turn over the equipment.

4. *Overstaffing of the maintenance job.* Too often, a crew of three or four is supplied, when only two are needed, resulting in wasted time.

5. *Unsatisfactory work that must be redone.* Poor planning frequently results in an attitude of "this will get by" on the part of the mechanic. This results in the repair work having to be redone.

6. *Improper planning.* Good planning ensures that there is sufficient work scheduled for maintenance so as to minimize idle waiting time.

INDIRECT LABOR STANDARDS

Standards for indirect labor departments, such as clerical, maintenance, and toolmaking, should be developed on any operation or group of operations that can be quantified and measured. These operations must be first broken down into direct, transportation, and indirect elements. The tools used for establishing standards for direct work are the same as discussed previously: time study, predetermined time systems, standard data, formulas, and work sampling.

Analysts can therefore establish standards for such tasks as hanging a door, rewinding a 1-hp motor, painting a centerless grinder, sweeping the chips from a department, or delivering a skid of 200 forgings. For each of these operations, analysts establish standard times by measuring the time required for the operator to perform the job. They then performance-rate the study and apply an appropriate allowance.

Careful study and analysis often reveals that crew balance and interference cause more unavoidable delays in indirect work than in direct work. Crew balance is the wasted time encountered by one member of a crew while watching other crew members perform job elements. Interference time is the time that a worker waits for others to do necessary work before that worker can start. Both crew balance and interference delays are unavoidable delays; however, they are usually characteristic of indirect labor operations only, such as those performed by maintenance workers. Queuing theory, explained later in this chapter, is a useful tool for estimating the magnitude of waiting time.

Because of the high degree of variability characteristic of most maintenance and material handling operations, it is necessary to conduct sufficient independent time studies of each operation to ensure that the resulting standard is representative of the time needed for the normal operator to do the job under average conditions. For example, if a study indicates that it takes 47 min to sweep a machine floor 60 ft wide by 80 ft long, the analyst must ensure that average conditions prevail during the study. The work of a sweeper will be considerably more time-consuming if the shop is machining cast iron rather than an alloy steel, because alloy steel is much cleaner and alloy steel chips are easier to handle. Also, the use of slower speeds and feeds results in fewer chips. Therefore, a 47-min standard established when the shop is working with alloy steel will be inadequate for producing cast-iron parts. Additional time studies would ensure that average conditions prevailed and that the resulting standards represented those conditions. Similarly, analysts can establish standards based on square footage for painting.

Toolroom work is similar to the work done in job shops. Analysts can therefore predetermine the method by which tools such as a drill jig, milling fixture, form tool, or die can be made. Analysts use time study and/or predetermined elemental times to establish a sequence of elements and to measure the normal time required for each element. Work sampling provides an adequate tool for determining the allowances that must be added for fatigue, personal, unavoidable, and special delays. The standard elemental times thus developed can be tabularized in the form of standard data and then used to design time formulas for pricing future work.

FACTORS AFFECTING INDIRECT AND EXPENSE STANDARDS

All indirect and expense work is a combination of four divisions: (1) direct work, (2) transportation, (3) indirect work, and (4) unnecessary work and delays.

Direct work is that segment of the operation that discernibly advances the progress of the work. For example, in the installation of a door, the direct work elements may include these actions: cut door to rough size, plane to finish size, locate and mark hinge areas, chisel out hinge areas, mark for screws, install screws, mark for lock, drill out for lock, and install lock. Such direct work can be easily measured using conventional techniques, such as a stopwatch time study, standard data, or fundamental motion data.

Transportation is the work performed in movements during the course of the job or from job to job. Transportation may be horizontal or vertical or both. Typical transportation elements include walk up and down stairs, ride elevator, walk, carry load, push truck, and ride on motor truck. Transportation elements are also easy to measure and to establish as standard data. For example, one company uses 0.50 min per 100-ft zone as its standard for horizontal travel time and 0.30 min as its standard time for 10 ft of vertical travel.

As a general rule, analysts cannot evaluate the indirect portion of indirect or expense labor by physical evidence in the completed job, or at any stage during the work, except by deductive inferences from certain features of the job. Indirect work elements may be separated into three divisions: (1) tooling, (2) material, and (3) planning.

Tooling work elements include the acquisition, disposition, and maintenance of all tools needed to perform an operation. Typical elements under this category would include getting and checking tools and equipment; returning tools and equipment at the completion of the job; cleaning tools; and repairing, adjusting, and sharpening tools. The tooling work elements are easy to measure by conventional means; statistical records provide data on the frequency of their occurrence. Waiting line or queuing theory provides information on the expected waiting time at supply centers.

Material work elements involve acquiring and checking the material used in an operation and disposing of scrap. Making minor repairs to materials, picking up and disposing of scrap, and getting and checking materials are characteristic of material work elements. As with tooling elements, material elements are readily measured, and their frequency can be accurately determined through historical records. Queuing theory provides the best estimate of waiting time for acquiring material from storerooms.

The planning elements represent the most difficult area in which to establish standards. These elements include consulting with the supervisor, planning work procedures, inspecting, checking, and testing. Work sampling techniques, especially self-observation, provide a basis for determining the time required to perform the planning elements.

Planning and methods improvement can eliminate unnecessary work and delays, which may represent as much as 40 percent of the indirect and expense payroll. Much of this wasted time is management-oriented; for this reason, analysts should follow the systematic procedure recommended in Chapter 1 prior to establishing standards. In this work, analysts should get and analyze the facts, and develop and install the method, before establishing the standard.

Much of the delay time encountered in indirect and expense work is due to queues. Workers are obliged to stand in line at the tool crib, the storeroom, or the stockroom, waiting for a forklift truck, a desk calculator, or some other piece of equipment. Through the application of queuing theory, analysts can frequently determine the optimum number of servicing units required in the given circumstances.

BASIC QUEUING THEORY

Queuing system problems may occur when the flow of arriving traffic (people, facilities, and so on) establishes a random demand for service at facilities with a limited service capacity. The time interval between arrival and service varies inversely with the level of the service capacity. The greater the number of service stations and the faster the rate of servicing, the smaller the time interval between arrival and service.

Methods and work measurement analysts should select an operating procedure that minimizes the total cost of operation. There must be an economic balance between waiting times and service capacity. The following four characteristics define queuing problems:

1. *The pattern of arrival rates.* The arrival rate (for example, a machine breaking down for repair) may be either constant or random. If random, the pattern is a probability distribution of the values of the intervals between successive arrivals. Also, probability distribution of the random pattern may be definable or undefinable.

2. *The pattern of the servicing rate.* The servicing time may also be either constant or random. If random, analysts should define the probability distribution that fits the random pattern.

3. *The number of servicing units.* In general, multiple-service queuing problems are more complex than those of single-service systems. However, most problems are the multiple-service type, such as the number of mechanics required to keep a battery of machines in operation.

4. *The pattern of selection for service.* Service is usually on a first-come, first-served basis; however, in some cases, the selection may be completely random, or may follow some set of priorities.

The solutions to queuing problems fall into two broad categories: analytic and simulation. The analytic category covers a wide range of problems for which mathematical probability and analytic techniques have provided equations representing systems with various assumptions about queuing characteristics. One of the most common assumptions about the arrival pattern, or arrivals per unit time interval, is that they follow Poisson's probability distribution:

$$p(k) = \frac{a^k e^{-a}}{k!}$$

where a = mean arrival rate

  k = number of arrivals per time interval

$$P(c, a) = \sum_{x=c}^{\infty} \frac{e^{-a}a^x}{x!}$$

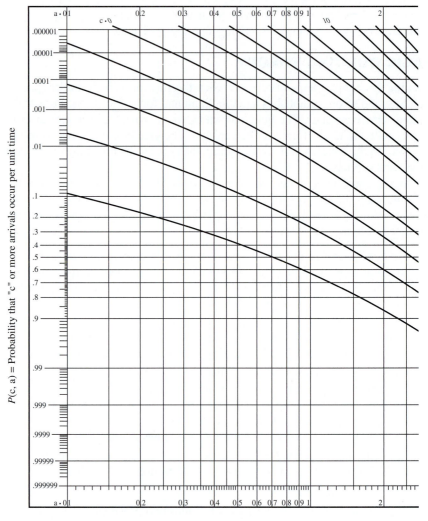

a = Average number of arrivals per unit time

Figure 15.1 Poisson distribution of arrivals.

A helpful graphic presentation of the cumulative Poisson probabilities appears in Figure 15.1.

A further consequence of a Poisson-type arrival pattern is that the random variable time between arrivals obeys an exponential distribution with the same

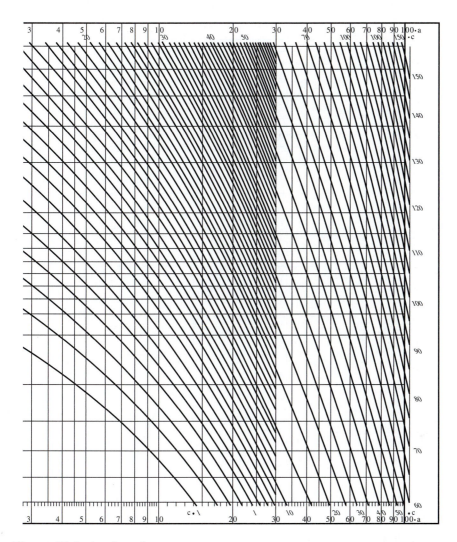

Figure 15.1 (continued)

parameter a. Being a continuous distribution, the exponential distribution has a density function

$$f(x) = ae^{-ax}$$

The exponential distribution has a mean $\mu = 1/a$ and a variance $= 1/a^2$, and μ can be recognized as the mean interval between arrivals. In some queuing systems, the number of services per unit time may follow the Poisson pattern; the service times subsequently follow the exponential distribution. Figure 15.2 illustrates the exponential curve $F(x) = e^{-x}$, which shows the probabilities of exceeding various multiples of any specified service time.

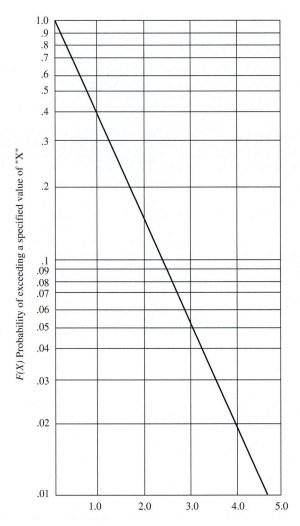

Figure 15.2 The exponential distribution.

The basic equations governing the queue applicable to Poisson arrivals and service, in arrival order, fall into these five categories:

1. Any service time distribution and a single server
2. Exponential service time and a single server
3. Exponential service time and finite servers
4. Constant service time and a single server
5. Constant service time and finite servers

Equations have been developed for each of these categories. These equations provide quantitative answers to such problems as mean delay time in the waiting line and mean number of arrivals in the waiting line.

Two examples demonstrate the application of queuing theory. The first (Example 15.1) is the determination of a standard for an inspection process. This fits into the first of the five categories, a single server with any service time distribution. The second example (Example 15.2) applies to toolroom delays and falls into the second category, a single server with exponential service time. Many industrial problems fall into this category.

Use of Queuing Theory to Establish a Standard Time for Inspection

EXAMPLE 15.1

An analyst wishes to determine a standard for inspecting the hardness of large motor armatures. The time is composed of two distinct quantities: the time the inspector takes to make his Rockwell observations, and the time the operator must wait until the next armature shaft is made available for inspection. The following assumptions apply: (1) single server; (2) Poisson arrivals; (3) arbitrary servicing time; and (4) first-come, first-served discipline. This is a situation that fits into the first of the five categories, and the following equations apply:

$$(a)\ P > 0 = u = \frac{ah}{s}$$

$$(b)\ w = \left[\frac{uh}{2(1-u)}\right]\left[1 + \left(\frac{\sigma}{h}\right)^2\right]$$

$$(c)\ m = \frac{w}{P > 0} = \frac{w}{u}$$

where a = average number of arrivals per unit of time
 h = mean servicing time
 w = mean waiting time of all arrivals
 m = mean waiting time of delayed arrivals
 n = number of arrivals present (both waiting and being served) at any given time
 s = number of servers
 u = servers' occupancy ratio = $\dfrac{ah}{s}$
 σ = standard deviation of servicing time
 $P(n)$ = probability of n arrivals being present at any random time
 $P(\geq n)$ = probability of at least n arrivals being present at any random time
 t = unit of time
 $P > t/h$ = probability of a delay greater than t/h multiples of the mean holding time
 $P(d > 0)$ = probability of any delay (delay greater than 0)
 L = mean number of waiting individuals among all individuals

A stopwatch time study establishes a normal time of 4.58 min per piece for the actual hardness testing. The standard deviation of the service time is 0.82 min, and 75 tests are made in every 8-h workday. From these data, we have

$$s = 1$$

$$a = \frac{75}{480} = 0.156$$

$$h = 4.58$$

$$\sigma = 0.82$$

$$u = (0.156)(4.58) = 0.714$$

$$w = \left[\frac{(0.714)(4.58)}{2(1 - 0.714)} \right]\left[1 + \left(\frac{0.82}{4.58}\right)^2 \right]$$

$$= 5.95 \text{ min mean waiting time of an arrival}$$

Thus, the analyst is able to determine a total time of $4.58 + 5.95 = 10.53$ min per shift.

EXAMPLE 15.2

Use of Queuing Theory to Establish a Standard Time for Toolroom Service

Toolroom service time can be modeled as a single server with Poisson arrivals with exponential service. The equations that apply are

$$(a)\ P > 0 = u$$

$$(b)\ P > t/h = ue^{(u-1)(t/h)}$$

$$(c)\ P(n) = (1 - u)u^n$$

$$(d)\ P(\geqq n)u^n$$

$$(e)\ w = \frac{h(P > 0)}{1 - u} = \frac{uh}{1 - u}$$

$$(f)\ m = \frac{w}{P > 0} = \frac{h}{1 - u}$$

$$(g)\ L = \frac{m}{h} = \frac{1}{1 - u}$$

The arrivals at a tool crib are considered to be Poisson, with an average time of 7 min between one arrival and the next. The length of the service time at the crib window is distributed exponentially, with a mean of 2.52 min determined through a stopwatch time study. The analyst wishes to determine the probability that a person arriving at the crib will have to wait, and the average length of the queues that form from time to time. Using this information, the analyst can evaluate the practicality of opening a second tool-dispensing window.

$a = 0.14$ average number of arrivals per minute

$h = 2.52$ min mean service time

$s = 1$ server

$P > 0 = u$

$u = \dfrac{ah}{s} = 0.35 =$ probability of a person arriving at crib having to wait

$L = \dfrac{1}{1 - u} = 1.52 =$ average length of queue formed

MONTE CARLO SIMULATION

Simulation can be used to solve queuing problems for which neither standard nor empirical formulas are obtainable. Instead the analyst introduces a sample set of input values drawn from specified arrival and service time distributions. These input data generate a sample output distribution of waiting line results for the period. This approach, termed *Monte Carlo simulation*, develops an optimum solution through a proper balance of service stations, servicing rates, and arrival rates and is most helpful for analyzing the waiting line problem involved in the centralized–decentralized storage location of tools, supplies, and service facilities.

Example 15.3 demonstrates the use of the technique to determine the minimum number of operators needed to set up machines in an automatic screw machine section.

Use of Monte Carlo Simulation to Determine the Optimum Number of Operators
EXAMPLE 15.3

Fifteen machines are presently being serviced by three operators. The labor rate is $12.00 per hour, while the machine rate is $48.00 per hour. An analysis of past records reveals the following probability distributions of work stoppages per hour and the time required to service a machine:

Work stoppages per hour	Probability	Hours to get machine into operation	Probability
0	0.108	0.100	0.111
1	0.193	0.200	0.254
2	0.361	0.300	0.009
3	0.186	0.400	0.007
4	0.082	0.500	0.005
5	0.040	0.600	0.008
6	0.018	0.700	0.105
7	0.009	0.800	0.122
8	0.003	0.900	0.170
	1.000	1.000	0.131
		1.100	0.075
		1.200	0.003
			1.000

The time to get a machine into operation results in a bimodal distribution and does not conform to any standard distribution. The analyst assigns random blocks of three-digit numbers (000 to 999), in direct proportion to the probabilities associated with the data for both the arrival and the service rates, to simulate the expected behavior in the screw machine section over a period of time. The analyst takes a set of random observations, to simulate the work stoppages occurring during one day of activity (8 working hours) on the floor. These random numbers yield the following:

Hour	Random number	Work stoppages
1	221	1
2	193	1
3	167	1
4	784	3
5	032	0
6	932	5
7	787	3
8	236	1
9	153	1
10	587	2
11	573	2

To estimate the time required to get a machine into operation after it has stopped, the analyst selects a different set of random numbers as input for each work stoppage.

Hour	Random number	Hours to get machine into operation
1	341	0.200
2	112	0.200
3	273	0.200
4	106	0.100
5	597	0.800
6	337	0.200
7	871	1.000
8	728	0.900
9	739	0.900
10	799	1.000
11	202	0.200
12	854	1.000
13	599	0.800
14	726	0.900
15	880	1.000

Table 15.1 shows the predicted machine downtime because of insufficient operators, based on the indicated outcomes for 8 h of operation in the automatic screw machine department.

The simulated model indicates 2.8 h of machine downtime every day because of the lack of an operator. At a machine rate of $48 per hour, this amounts to a daily cost of 2.8 × 48 = $134.40. Since the cost of a fourth operator only results in an added daily direct labor cost of 8 × 12 = $96, it appears that three workers are not economically optimum to service the operation.

Note, however, that ten random numbers make a very small simulation and may lead to incorrect results. With larger numbers, there is a greater chance of converging to an optimum solution. In fact, with 80 random numbers, the present setup of three workers to service the 15 automatic screw machines does appear to be the optimum solution, as only 19.6 h (instead of the extrapolated 28) is lost over the course of 80 h of operation.

Table 15.1 Results of Monte Carlo Simulation of Machine Downtime

Hour	Random number	Work stoppages	Random number	Hours to get machine into operation	Operators available for next work stoppage	Downtime hours because of lack of operators for servicing
1	221	1	341	0.200	2	
2	193	1	112	0.200	2	
3	167	1	273	0.200	2	
4	784	3	106	0.100	2	
			597	0.800	1	
			337	0.200	0	
5	032	0	—	—	3	
6	932	5	871	1.000	2	
			728	0.900	1	
			739	0.900	0	
			799	1.000	0	0.9
			202	0.200	0	0.9
7	787	3	854	1.000	0	
			599	0.800	0	0.9
			726	0.900	0	0.1
8	236	1	880	1.000	1	
9	153	1	495	0.700	2	
10	587	2	128	0.200	2	
			794	1.000	1	—
					Total	2.8

EXPENSE LABOR STANDARDS

More and more, management is recognizing the need to determine accurately the appropriate office force for a given volume of work. To control office payroll, management must develop time standards, since they are the only reliable yardsticks for evaluating the size of any task.

As for direct labor work, methods analysis should precede work measurement in all expense operations. The flow process chart is the ideal tool for displaying the facts of the present method, allowing that method to be critically reviewed. Using the primary approaches to operation analysis, the analyst then considers such factors as the purpose of the operation, design of forms, office layout, elimination of delays resulting from poor planning and scheduling, and the adequacy of existing equipment.

After the completion of a thorough methods program, standards development can commence. Many office jobs are repetitive; consequently, it is not particularly difficult to set fair standards. Word processors; call centers, billing groups, and data input operators are representative groups that readily lend themselves to work measurement by stopwatch, standard data, and basic motion data techniques.

In studying office work, analysts should carefully identify element endpoints, so that standard data may be established for pricing future work. For example, in inputting physician orders to the computer system the following elements of work normally occur for each physician order.

1. Pick up physician order from pile.
2. Read physician order instructions.
3. Input headings on order, using keyboard:
   a. Date
   b. Patient name
   c. Lab test or drug required
   d. Ordering physician/department

Once analysts have developed standard data for most of the common elements used in an office, they can calculate time standards rapidly and economically. Of course, many clerical positions comprise a series of diversified activities that do not readily lend themselves to measurement. Such work is not made up of a series of standard cycles that continually repeat themselves; consequently, the work is more difficult to measure than direct labor operations. Because of this characteristic of some office routines, it is necessary to take many time studies, each of which may only be one cycle in duration. Then, by calculating all the studies taken, analysts can develop a standard for typical or average conditions. For example, they can calculate a time standard based on a page of physician orders. Granted, some orders require symbols, Latin terminology, and other special characters or spacing and therefore take considerably longer than routine orders. Yet, if the technical keyboarding is not representative of average conditions, it does not result in an unfair influence on the operator's performance over a period of time; simple keyboarding and shorter orders balance out the extra time needed to input complex orders.

It is usually not practical to establish standards on service positions that require creative thinking. One should carefully consider before trying to establish time standards for the physician placing the order. If such standards are established, they should be used for scheduling, control, or labor budgeting, not for incentive wage payment. Putting pressure on employees such as physicians, may retard creative thinking and patient care and may even result in outright rebellion. Also, physicians should be evaluated under professional performance standards (see next section.) The result may be more costly to the health maintenance organization than the amount saved through the greater productivity of the physician.

SUPERVISORY STANDARDS

It is possible to establish standards for supervisory work (see Example 15.4). The work sampling technique is an especially good tool for determining equitable supervisor loads and for maintaining a proper balance between supervisors, facilities, clerical employees, and direct labor. Analysts could obtain the same information through all-day time studies, but the cost of reliable data would usually be prohibitive. Supervisory standards can be expressed in effective machine running hours or some other benchmark.

EXAMPLE 15.4

The Setting of Supervisory Standards

A study of a panel manufacturer revealed that 0.223 supervisory hour was required per machine running hour in a given department (see Figure 15.3). The work sampling study showed that out of 616 observations, the supervisor was working with the grid machines, inspecting grids, doing desk work, supplying material, walking, or engaging in activities classified as miscellaneous allowances for a total of 519 times. Converted to prorated hours, this figure revealed that 518 indirect hours were required for 2,461 machine running hours. Adding a 6 percent personal allowance, analysts computed a standard of 0.223 supervisory hour per machine running hour.

$$\frac{518}{2,461} \times 1.06 = 0.223$$

Thus, in a department operating 192 machine running hours a week, a supervisor's efficiency would be

$$\frac{192 \times 0.223}{40 \text{ h/week}} = 107 \text{ percent}$$

INDIRECT LABOR STANDARD

JOB- SUPERVISION   DEPT.- GRID   DATE- 4-16

Cost Center	Number of Observations	Percent of Observations	Prorated Hours	Base Indirect Hours	Effective Machine Running Hours (EMRH)	Direct Labor Hours	Base Indirect Hours per Machine Running Hour (includes 6% personal)
Grid Machines	129	21	130	130	2,461		0.223
Inspect Grids	161	26	160	160			
Desk Work	54	9	56	56			
Supply Material	18	3	18	18			
Misc. Allow.	150	24	148	148			
Walking	7	1	6	6			
Out of Dept.	11	2	12				
Idle	86	14	86				
Total	616	100	616	518	2,461		0.223

Figure 15.3 Summary of supervisory work.

15.2 STANDARD INDIRECT AND EXPENSE LABOR DATA

FUNDAMENTALS

Developing standard data to establish standards on indirect and expense labor operations is quite feasible. In view of the diversification of indirect labor operations, standard data are probably more appropriate for office, maintenance, and other indirect work than for standardized production operations.

As individual time standards are calculated, the analyst should tabularize the elemental times, for future reference. As the inventory of standard data is built up, the cost of developing new time standards declines proportionally. For example, tabularizing standard data for forklift truck operations can be based on six different elements: travel, brake, raise fork, lower fork, tilt fork, and the manual elements required to operate the truck. Once standard data have been accumulated for each of these elements, through the required range, analysts can determine the standard time required for any fork truck operation by summarizing the applicable elements. In a similar manner, standard data can readily be established on janitorial work elements, such as sweep floor, wax and buff floor, dry mop, wet mop, vacuum rugs, or clean, dust, and mop lounge.

The maintenance job of "inspecting seven fire doors in a plant and making minor adjustments" can be estimated from standard data. For example, the Department of the Navy developed the following standard:

Operation	Unit time(h)	No. units	Total time(h)
Inspection of fire door, roll-up type (manual chain, crankshaft, or electrically operated) fusible link. Includes minor adjustment	0.170	7	1.190
Walk 100 ft between each door, obstructed walking	0.00009	600	0.054
			1.24

The standard data time of 0.00009 h/ft of obstructed walking was established from predetermined time systems, and the inspection time of 0.170 h per fire door was established from a stopwatch time study.

UNIVERSAL INDIRECT LABOR STANDARDS (UILS)

Where maintenance and other indirect operations are numerous and diversified, efforts have been made to reduce the number of different time standards for indirect operations, through universal indirect labor standards (UILS). The principle behind universal standards is the assignment of the major proportion of indirect operations (perhaps as much as 90 percent) to appropriate groups. Each group has its own standard, which is the average time for all indirect operations assigned to the group. For example, group A may include the following indirect operations: replace defective union, repair door (replace two hinges), replace limit switch, and replace two sections (14 ft of 1-in pipe). The standard time for any indirect operation performed in group A may be 48 min. This time represents the mean $\bar{x}$ of all jobs within the group, and the dispersion of the jobs within the group for $\pm 2s$ is some predetermined percentage of $\bar{x}$ (perhaps 610 percent).

The three principal steps in introducing a universal indirect labor system, expressed as *slotting*, are to

1. Determine the number of standards (groups or slots) to do a satisfactory job. (Twenty slots should be used when the range is up to 40 h.)
2. Determine the numerical standard representative of each group of operations contained in each slot.
3. Assign the standard to the appropriate slot of indirect labor work as it occurs.

The initial step is to determine good benchmark standards, based on measurements of an adequate sample of the indirect labor for which the UILS system is being developed. This is the most time-consuming and costly step in the process. The analyst must establish a relatively large number of standards (200 or more) that are representative of the entire population of indirect work. Competent analysts can develop these measured benchmark standards by using proven industrial engineering tools, including stopwatch time study, standard data, formulas, fundamental motion data, and work sampling.

Once established, the benchmark standards are arranged in numerical sequence. Thus, assuming 200 benchmark standards, the shortest would be listed first, the next-shortest second, and so on, ending with the longest. If there are 20 slots, and if a uniform distribution is used, the time standard for the first slot is computed by calculating the mean of the first 10 benchmark standards. Similarly, the value of the second slot is calculated by taking the mean of benchmark standards 11 through 20. The last slot would be equal to the average of the benchmark standards 191 through 200. Engineers have used this procedure extensively in the development of UILS.

More reliable UILS result from using the normal, rather than the uniform, distribution. For 20 slots, the 200 standards would not be assigned as 10 per slot. Instead, the standard normal variable would be divided into 20 equal intervals (truncation of the two tails allows this). For example, the standard normal variable may have a truncated range of

$$- 3.0 \leq z \leq + 3.0$$

which accounts for 99.74 percent of the area under the curve. The range of each interval would be 0.3. The benchmark standards used in the compilation of the mean of each of the 20 slots (intervals) would equal

$$P(z \in \text{interval})(200)/0.9987$$

Slot numbers 1 and 20 (because of symmetry) would have

$$\frac{P(-3.0 \geq z \leq -2.7)(200)}{0.9987} = \frac{P(2.7 \leq z \leq 3.0)(200)}{0.9987}$$

$$= \frac{(0.9987 - 0.9965)(200)}{0.9987} = 0.4406 \approx 0 \text{ standards}$$

and slot numbers 10 and 11 would have

$$\frac{P(-0.3 \geq z \leq 0.0)(200)}{0.9987} = \frac{P(0.0 \leq z \leq 0.3)(200)}{0.9987}$$

$$= \frac{(0.6179 - 0.5000)(200)}{0.9987} = 23.61 \approx 24 \text{ standards}$$

Rounding off the fractions is dictated by the fact that all 200 jobs must be assigned to a slot. The universal standard time for each slot is the average of the benchmark standards assigned to the slot. When studying a new part to be made, analysts can fit jobs to categories where similar jobs have been studied and standards established.

To determine the relative accuracy of the uniform and normal techniques, analysts used 270 maintenance standards developed by the Department of the Navy as benchmark standards. They divided these benchmark standards into 20 slots, using the two approaches. To compare the results of the uniform and normal techniques, a simulation was done. For each of 25 weeks, analysts selected jobs at random until the sum of the actual standard times exceeded or equaled 40 h. They then determined the universal maintenance standard for each job and calculated the weekly sum. They assumed that each job was properly slotted.

For each week, an error was calculated as

$$\left| \frac{\text{Actual standard time} - \text{universal standard time}}{\text{Actual standard time}} \right| \times 100 \text{ percent}$$

The results of the simulations for the uniform and normal distributions are given in Table 15.2. This study confirmed that the normal gives better results than the uniform distribution.

Increasing the pay period from one week (40 h) to two weeks (80 h) would markedly reduce the cumulative error per pay period. The magnitude of the error would also decrease as the number of groups (slots) increased.

UILS offers an opportunity to introduce standards for a majority of indirect operations at a moderate cost, and it minimizes the cost of maintaining the indirect standards system.

15.3 PROFESSIONAL PERFORMANCE STANDARDS

The cost of salaried workers is a sizable proportion of the total expense budget, especially in service organizations, In most manufacturing and business operations, the salaries of professional employees in engineering, accounting,

Table 15.2 Results of 25-Week Simulation

	Absolute percent error	
Week number	**Uniform**	**Normal**
1	5.97	7.18
2	16.01	6.93
3	8.49	6.42
4	10.94	4.03
5	25.78	1.67
6	2.61	0.47
7	4.79	6.08
8	0.88	3.37
9	4.51	5.34
10	0.05	6.45
11	30.78	0.32
12	21.93	1.75
13	8.23	4.24
14	6.67	7.55
15	2.37	2.37
16	0.06	0.87
17	12.53	2.88
18	3.73	5.21
19	6.85	1.52
20	11.50	2.29
21	20.18	2.48
22	6.44	8.31
23	3.46	6.72
24	2.96	0.45
25	11.74	1.01
Mean	9.18	3.84
Variance	151.78	21.62
Std. dev.	12.32	4.65

purchasing, sales, and general management represent a significant proportion of total cost. If the productivity of these employees can be improved by as little as a few percent, the overall impact on the firm's business is consequential. Establishing standards for salaried employees and utilizing these standards as achievable goals inevitably enhance productivity.

The difficulties in developing professional standards lie, first, in determining what to count and, second, in determining the method for counting these outputs. In determining what to count, the analyst can begin by stating the objectives of the professional employees' positions. For example, buyers in the purchasing department might have the following objective: "to procure quality components and raw materials at the lowest price, in time to meet company production and delivery schedules." To be effective, a count of buyers' outputs must consider five things: (1) proportion of deliveries made on schedule; (2) proportion of deliveries that meet or exceed quality requirements; (3) proportion of shipments that represent the lowest available price; (4) number of orders placed during some interval of time, such as one month; and (5) total dollar value of the purchases made during a period of time.

The next problem is to establish achievable goals. In such instances, using historical records, supplemented with work sampling analyses, to determine how time is utilized can serve as the basis for the development of professional standards.

Returning to the example of establishing buyers' standards, it would not be difficult to identify the purchases made by the various buyers, and to review what proportion of these orders was delivered on schedule over a 6-month period. The historical data study may reveal something analogous to the following:

Buyer	Proportion of orders delivered on or before schedule (percent)
A	70
B	82
C	75
D	50
E	80

Based on this record, skilled buyers should be able to procure at least 72 percent of their purchases on schedule (the mean of their performance).

Similarly, a quality review of the purchases made by the five buyers may disclose the following:

Buyer	Proportion of orders delivered with less than 5 percent rejects (percent)
A	85
B	90
C	80
D	95
E	80

Here, the quality standard could be that 86 percent of orders received have less than 5 percent rejects (the mean of the past performance of the five buyers).

For the proportion of procurements purchased at the lowest available price, historical records can once again provide a comparison of the performance of the five buyers. Assume the following performance record applied:

Buyer	Proportion of orders procured at the lowest available price (percent)
A	45
B	50
C	60
D	47
E	40

The average of these values, 48.4 percent of the orders placed at the lowest available price, could be used as the normal performance.

Historical records might also indicate that, on average, a buyer placed 120 orders per month, with a total monetary worth of $120,840.

These five criteria could then be used to develop an overall performance standard: delivery, quality, price, number of orders, and order value. For example, one method would add the means of the first three criteria (0.72 + 0.86 + 0.484) plus 0.002 times the mean of the order placed plus 0.000001 times the average monetary worth of purchases. The buyer's standard in this fictitious operation would then be

$$0.72 + 0.86 + 0.484 + 0.24 + 0.12 = 2.424$$

Another example may help clarify how performance standards may be developed for managerial personnel. Consider the position of director of personnel administration. An analysis may suggest four specific objectives of this position:

1. Establish a methodology for identifying both the quantity and quality of the company's human resources.
2. Establish a procedure for attracting, employing, and retaining the kinds and numbers of employees required for the successful operation of the company.
3. Establish policies, programs, and practices that facilitate the achievement of departmental objectives and maintain employee morale.
4. Administer and maintain the company's benefit program.

Now that the objectives have been stated, it is relatively easy to develop a performance standard in terms of time. For example, the standard for objective 1 may be: "Within the next 3 months, train staff representatives to conduct an audit of the company's personnel, to determine projected needs from the standpoint of both numbers and type."

The performance standard for objective 2 might be: "Within the next 12 months, employ (a) 2 Ph.D. chemists; (b) 7 M.S. degrees in industrial and/or mechanical engineering; and (c) 35 B.S. degrees, with a distribution of 10 in business, 20 in engineering, and 5 in liberal arts. Employ (based on anticipated turnover and expansion) 75 hourly employees. Investigate the turnover rate of professional employees in the past year, and prepare a report showing how turnover may be reduced."

For objective 3, the performance standard might be as follows: "Within the next 3 months, update the current management handbook, bringing the salary administration program up to date. Within the next 6 months, develop and distribute a booklet for all hourly employees, describing the new grievance procedure established in the new labor contract. The booklet should explain not only the importance of reducing the number of grievances, but also how this can be done."

The performance standard for objective 4 could be: "Within the next 12 months, review the company's entire fringe benefit program and compare our benefits to those of similar-sized companies in this area. Make any appropriate recommendations to management."

These objectives identify performance standards for finite time periods. The standards may change as time passes, since each standard is result-based. The standards established for succeeding periods may include different work assignments to meet the stated objectives.

In the establishment of professional performance standards, the professionals themselves should assist in identifying the objectives of each position, gathering the historical performance records, and developing the standards. Performance standards developed without the complete involvement of the professionals are seldom realistic.

When gathering historical data to facilitate the development of professional standards, the analyst should take a work sampling study during the period that serves as the basis of the historical record data. This work sampling study can reveal how much working time was spent on the various necessary work routines, or on work assignments that could better be handled by clerical or semiprofessional employees. It could also reveal how much time was literally wasted. After reviewing the work sampling study, the analyst can performance-rate the average data gathered over the historical period, to obtain a standard that is more representative of normal professional experience.

In the development of professional standards, the following guidelines should be observed:

1. Each manager should be involved with setting the standards for his or her professional subordinates. Professional standards should be developed jointly by employees and their supervisors.
2. Standards should be result-based and should be worded to include measurement references.
3. Standards must be realistically attainable by at least one-half of the group concerned.
4. Standards should be periodically audited and revised if necessary.
5. It is helpful to work sample managers to ensure that they have adequate clerical and administrative aid support and are using their time judiciously.

SUMMARY

It is more difficult to study and determine representative standard times for non-repetitive tasks, which are characteristic of most indirect labor operations, than for repetitive tasks. Since indirect labor operations are difficult to standardize and study, they are only infrequently subjected to methods analysis. Consequently, this area usually offers a greater potential for reducing costs and increasing profits through methods and time study than does any other.

The usual procedure is to take a sufficiently large sample of stopwatch time studies to ensure the representation of average conditions, and then to tabularize the allowed elemental times in the form of standard data. Predetermined time systems data also have wide application in establishing standards on indirect

Table 15.3 Guide for Establishing Indirect Labor and Expense Standard

Indirect and expense type work	Recommended method of establishing standards
Routine maintenance. Work standards 0.5 to 3 hrs	Standard data, MTM-2, MTM-3, MOST
Complicated maintenance, standards 3 hrs to 40 hrs	Slotting based on universal indirect labor standards
Shipping and receiving	Standard data MTM-2, MTM-3, MOST
Toolroom	Slotting based on universal indirect labor standards
Inspection	Standard data, MTM-2, MOST
Tool design	Slotting based on universal indirect labor standards
Buying	Standards based on historical records, analysis, and work sampling
Accounting	Standards based on historical records, analysis, and work sampling
Plant engineering	Standards based on historical records, analysis, and work sampling
Clerical	Standards based on standard data, MTM-2, MOST
Janitorial	Standard data, slotting based on universal indirect labor standards
General management	Standards based on historical records, analysis, and work sampling

work. This is especially true of those systems that utilize larger blocks of fundamental motions, such as MTM-2 and MOST.

Analysts can accurately estimate indirect elements involving waiting time by using waiting line or queuing theory. Where the problem does not fit established waiting line equations, analysts can use Monte Carlo simulation as a tool for determining the extent of the waiting line problem in the work area. To assist in establishing standards on indirect and expense work, Table 15.3 can serve as a guide for choosing the appropriate method.

Standards on indirect work offer distinct advantages to both the employer and the employee. Some of these advantages are as follows:

1. The installation of standards leads to many operating improvements.
2. The efficiency of various indirect labor departments can be determined.
3. Labor loads can be better scheduled and budgeted.
4. System improvements can be evaluated prior to installation, avoiding costly mistakes.
5. Wage incentive plans can be installed, allowing employees to increase their earnings.
6. Employees require less supervision and perform better, as a program of work standards tends to enforce itself.

QUESTIONS

1. Differentiate between indirect labor and expense labor.
2. Explain queuing theory.
3. Which four divisions constitute indirect and expense work?
4. Are standards established on the unnecessary delays portion of indirect and expense work? Why or why not?
5. Why has there been a marked increase in indirect workers?
6. Why do more unavoidable delays occur in maintenance operations than on production work?
7. What is meant by crew balance? By interference time?
8. Explain how time standards could be established on janitorial operations.
9. Which office operations can be readily time studied?
10. Why are standard data especially applicable to indirect labor operations?
11. Summarize the advantages of standards established on indirect work.
12. Explain the application of slotting for indirect or expense labor.
13. Why will a universal standards system involving as few as 20 benchmark standards work in a large maintenance department where thousands of different jobs are performed each year?

PROBLEMS

1. Work measurement procedures establish an average time of 6.24 min per piece on the inspection of a complex forging. The standard deviation of the inspection time is 0.71 min. Usually, 60 forgings are delivered to the inspection station on the line every 8-h turn. One operator performs this inspection. Assuming that the castings arrive in Poisson fashion and that the service time is exponential, what would be the mean waiting time of a casting at the inspection station? What would be the average length of the casting queue?
2. In the tool and die room of the Dorben Company, the work measurement analyst wishes to determine a standard for the jig boring of holes on a variety of molds. The standard will be used to estimate mold costs only. It will be based on operator wait time for molds coming from a surface grinding section and on operator machining time. The wait time is based on a single server, Poisson arrivals, exponential service time, and first-come, first-served discipline. A study revealed the average time between arrivals was 58 min. The average jig-boring time was 46 min. What is the possibility of a delay of a mold at the jig borer? What is the average number of molds in back of the jig borer?
3. What would be the expected waiting time per shipment if a stopwatch time study established that the normal time to prepare a shipment was 15.6 min? Twenty-one shipments are made every shift (8 h). The standard deviation of the service time has been estimated as 1.75 min. It is assumed that the arrivals are Poisson distributed and that the servicing time is arbitrary.

4. Using Monte Carlo methods, what would be the expected downtime hours due to the lack of an operator for servicing if four operators were assigned to the work situation described in Example 15.3?

5. The arrivals at the company cafeteria are Poisson, with an average time of 1.75 min between arrivals during the lunch period. The average time for a customer to obtain lunch is 2.81 min, and this service time is distributed exponentially. What is the probability that a person arriving at the cafeteria will have to wait? How long?

REFERENCES

Knott, Kenneth. "Indirect Operations: Measurement and Control." In *Handbook of Industrial Engineering*. 2d ed. Ed. Gavriel Salvendy. New York: John Wiley & Sons, 1992.

Lewis, Bernard T. *Developing Maintenance Time Standards*. Boston, MA: Industrial Education Institute 1967.

Nance, Harold W. *Office Work Measurement*. Malabar, FL: Krieger, 1983.

Newbrough, E. T. *Effective Maintenance Management*. New York: McGraw-Hill, 1967.

Pappas, Frank G., and Robert A. Dimberg. *Practical Work Standards*. New York: McGraw-Hill, 1962.

Raghavachari, M. "Queuing Theory." In *Handbook of Industrial Engineering*. 2d ed. Ed. Gavriel Salvendy. New York: John Wiley & Sons, 1992.

Standards Follow-Up and Uses

KEY POINTS

- Follow-up of methods and standards is necessary for both fairness to workers and profitability.

- Use the appropriate approach for setting and revising standards.

- Use standards to

  - Set wage incentives.

  - Compare methods.

  - Determine plant capacity.

  - Determine labor costs and budgets.

  - Enforce quality standards.

  - Improve customer service.

Follow-up is the eighth and last of the systematic steps in installing a methods improvement program. Although follow-up is as important as any of the other steps, it is frequently the most neglected. Analysts have a natural tendency to consider the methods improvement program complete after developing the time standards. However, a methods installation and resultant standard should never be considered complete.

Follow-up is necessary to ensure that the proposed method is being followed, that the established standards are being realized, and that the new method is being supported by labor, supervision, the union, and management. Follow-up usually results in additional benefits accruing from new ideas and new approaches, which eventually stimulate the desire to improve a methods engineering program for an existing design or process. The procedure is to repeat the methods improvement cycle shortly after it is completed, so that each process and each design are continually being scrutinized for possible further improvement. This is a necessary component of any sound continuous improvement program.

Without follow-up, it is easy for the proposed methods to revert to the original procedures. We have made innumerable methods studies where follow-up revealed that the method under study was slowly reverting to, or had reverted to, the original method. Humans are creatures of habit, and a workforce must develop the habit of the proposed method, if it is to be preserved. Continual follow-up is the only way to ensure that the new method is maintained long enough for all those associated with its details to become completely familiar with its routines.

16.1 MAINTAINING STANDARD TIMES

Both labor and management have emphasized the necessity of establishing fair standard times, and once fair standards are introduced, it is equally important that they be maintained. Although it is the normal function of production supervision to spot-check and monitor standards, the extensiveness of this job seldom permits adequate time for completely effective follow-up. Consequently, the methods and standards department should schedule regular follow-ups.

The initial follow-up or audit for production jobs should take place approximately one month after the development of time standards. A second audit should be made two months later, and a third three to nine months after that. The frequency of the audits should be based on the expected hours of application per year. As an example, one large company uses the data shown in Table 16.1 to determine the frequency of the audit of methods and standards.

On each follow-up, analysts should review the original method report as well as the development of the standard, to be certain that all aspects of the new method are being followed. At times, they may find that portions of the new method are being neglected and that workers have reverted to the old ways. Workers sometimes conceal methods changes they have personally implemented, so they can increase their earnings or diminish their effort while achieving the same production. Often methods changes that increase the time required to perform the task may have also developed. These changes may have been initiated by the supervisor or an inspector and, in their opinion, may be of insufficient consequence to adjust the standard. When this happens, the supervisor should be contacted immediately and the analyst should attempt to determine

Table 16.1   Frequency of Audits

Hours of application per standard per year	Frequency of audit
0–10	Once per 3 years
10–50	Once per 2 years
50–600	Once per year
Over 600	Twice per year

Source: Courtesy Industrial Engineers Division, Procter & Gamble, Co.

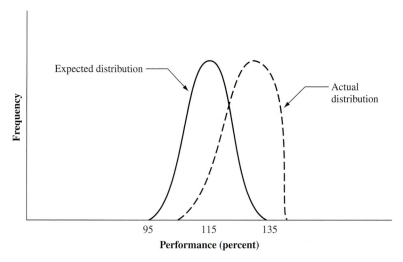

Figure 16.1 Distribution of performance in plant where standards are loose compared to expected distribution of average performance of 115 percent of normal.

why the unauthorized changes have taken place. If no satisfactory reasons can be given for the change, the analyst should insist that the correct procedure be followed.

Both the method and the performance of the operator should be followed up. The worker's performance should be equal to or greater than standard. Performance should be compared to typical learning curves (see chapter 18) for the class of work. If the operator is not making sufficient progress, a careful study, including a meeting with the operator, should be conducted to determine if any unforeseen difficulties have been encountered.

Usually, workers' performance approximates the normal curve, as described in Chapter 11. However, several common variations from the normal curve indicate problems and the need for an audit. Figure 16.1 illustrates that the standards are loose and that workers are holding back so that they do not earn at a rate above 140 percent, feeling that if they perform beyond this point, the time standards will be adjusted downward.

Figure 16.2 illustrates the output in an environment where the method has not been standardized. Variation in material is another cause of this flat distribution. In both of these instances, an audit can ensure that the best method is being used. Thus, the developed standard reflects the time required for average experienced operators, working with good skill and effort, to perform an operation at a pace that can be maintained for 8 h, allowing for personal and unavoidable delays and fatigue.

An audit should typically be performed if the overall cycle time and the existing time standards vary by more than ± 5 percent. In the majority of cases, a detailed time study should reveal that, a change in method is the cause. This can

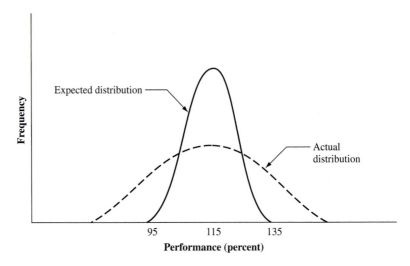

Figure 16.2   Distribution of performance in plant where methods and/or material has not been standardized.

be ascertained by referring to the original time study, which includes a complete description of the method.

In addition, the analyst should review all factory layouts to ensure that the ideal flow of materials and product is taking place and examine any new equipment acquired in conjunction with the method, to ensure that the anticipated productivity is being realized.

Also the analyst should audit the job evaluation after the worker has performed the new method for 6 months. This review should ensure that the compensation of all employees associated with the developed method is competitive with equivalent jobs in the area. Absentee rates should be audited, to obtain an additional measure of operator acceptance. Although auditing methods and standards require time and expense, a thorough and regular follow-up system will ensure the success of the program.

16.2   USING STANDARDS

STANDARDS REVIEW

Time standards are fundamental to the operation of any manufacturing enterprise or business, providing the one common denominator from which all elements of cost evolve. In fact, everyone uses time standards for practically everything she or he does or wants anyone else to do. Examples in everyday life include a worker who allocates one hour to wash, shave, dress, eat breakfast, and get to work; the student who schedules two hours to complete an assignment; and the bus driver who follows a specific schedule of arrivals and departures.

We are particularly interested in time standards used in the effective operation of a manufacturing company, a service enterprise, or a business. Such time standards may be determined in one or more of the following ways:

1. By estimate or performance records
2. By stopwatch time study (see Chapter 10)
3. By standard data (see Chapter 12)
4. By time study formulas (see Chapter 12)
5. By predetermined time systems (see Chapter 13)
6. By work sampling studies (see Chapter 14)
7. By queuing theory (see Chapter 15)

Methods 2, 3, 4, 5, and 6 give considerably more reliable results than methods 1 or 7. If standards are used for wage payment, they should be determined as accurately as possible. Consequently, standards determined by estimate or by performance records are better than no standards at all. All these methods are applicable under certain conditions, and all have limitations on accuracy and installation cost. The summaries presented in Tables 16.2 and 16.3 may prove helpful in selecting the appropriate approach.

WAGE INCENTIVE PLANS BASIS

While standards have many other uses in the operation of an enterprise, the need for reliable and consistent standards is most pronounced in the wage payment area (see Chapter 17). Without equitable standards, no incentive plan that compensates in proportion to output can possibly succeed, and without this yardstick, it is not possible to measure individual performance.

Similarly, any type of supervisory bonus keyed to productivity depends directly on equitable time standards. Since workers receive more and better supervisory attention under a plan where the supervisory bonus is related to output, the majority of supervisory plans consider worker productivity as the principal criterion for the supervisory bonus. Other factors usually are indirect labor costs, scrap cost, product quality, and methods improvements.

METHODS COMPARISONS

Since time is a common measure for all jobs, time standards are a basis for comparing various methods of doing the same work. For example, suppose an operator thought it might be advantageous to install broaching on a close-tolerance inside diameter, rather than ream the part to size, as is currently being done. To make a sound decision on the practicality of the change, analysts would develop time standards for each procedure and then compare the results.

EFFECTIVE SPACE UTILIZATION

Time is the basis for determining how much of each kind of equipment is needed. Management can only make the best possible utilization of space by knowing the

Table 16.2 Comparison of Different Methods for Establishing a Time Standard

Advantages	Disadvantages
Stopwatch Time Study	
1. Only method directly measuring operator times	1. Requires rating worker performance
2. Allows detailed observation of complete cycle and method	2. Does not force a detailed record of the method, motions, tools, etc., used
3. May cover relatively infrequent elements	3. May not properly evaluate noncyclic elements
4. Provides quick and accurate values for machine-controlled elements	4. Bases standard on the bias of one analyst studying one worker using one method
5. Is relatively simple to learn and explain	5. Requires ongoing production work
Predetermined Time Systems	
1. Forces a detailed record of the method, motions, tools, etc., used	1. Depends on complete description of method, motions, tools, etc., for accurate standard
2. Encourages work simplification	2. Requires more training of analysts
3. Eliminates performance rating	3. Is more difficult to explain to workers
4. Permits establishment of standard before actual production	4. Requires more time to establish standards
5. Permits easy adjustment of standard due to methods changes	5. Requires other data sources for process- or machine-controlled elements
6. Establishes consistent standards	
Standard Data, Formulas, and Queuing Methods	
1. Eliminates performance rating	1. May require more training of analysts
2. Establishes consistent standards	2. Is more difficult to explain to workers
3. Permits establishment of standard before actual production	3. May not accommodate small variations in the method
4. Permits easy adjustment of standard due to methods changes	4. May be inaccurate if extended beyond the scope of the data used in their development
Work Sampling	
1. Reduces tension caused by constant observation of the worker	1. Impairs accuracy of performance rating
2. Establishes an average standard over varying conditions	2. Requires relatively large number of random observations for accuracy
3. Permits simultaneous development of standards for a variety of operations	3. Requires accurate records of working hours and units produced
4. Best for analysis of machine utilization, work activities, and delays	4. Assumes that worker uses a standard method

Table 16.3 Guide for Selecting Appropriate Time Standard Method

Best method:	For work:
Stopwatch time study	1. With repetitive cycles of any duration 2. With wide variety of dissimilar elements 3. With process- or machine-controlled elements
Predetermined time systems	1. With operator-controlled elements 2. With repetitive cycles of short to medium duration 3. Not yet in production 4. With controversy over rating and consistency of standards
Standard data, formulas, and queuing methods	1. With similar elements of any duration 2. With controversy over rating and consistency of standards
Work sampling	1. With large variations from cycle to cycle 2. With controversy over stopwatch use 3. With controversy over constant observation of worker 4. Where machine utilization, activity levels, and delay allowances are needed

exact requirements for facilities. If a company requires 10 milling machines, 20 drill presses, 30 turret lathes, and 6 grinders in one machining department, the manager can plan for the best layout of this equipment. Without time standards, the company could overprovide for one facility and underprovide for another, inefficiently utilizing the space available.

As another example, to determine the size of storage and inventory areas, managers consider the length of time a part will be in storage as well as the demand for the part. Here again, time standards are the basis for determining the size of such areas.

PLANT CAPACITY DETERMINATION

Through time standards, the machine, department, and plant capacity can all be determined. Once the available facility hours and the time required to produce a unit of product are known, it is a matter of simple arithmetic to estimate product potential. For example, if the bottleneck operation requires 15 min per piece, and if 10 facilities for this operation exist, then based on a 40-h/week operation, the plant capacity for this product would be

$$\frac{40 \text{ h} \times 10}{0.25 \text{ h}} = 1,600 \text{ pieces per week}$$

NEW EQUIPMENT PURCHASE BASIS

Since time standards allow analysts to determine machine, department, and plant capacity, they also provide the information necessary to determine how many of which facilities are needed for a given production volume. Accurate comparative time standards also highlight the advantages of one facility over its competitors. For example, a plant may find it necessary to purchase three additional single-spindle, bench-type drill presses. By reviewing available standards, plant managers can procure the style and design of drill press that produce the most favorable output per unit of time.

WORKFORCE VERSUS AVAILABLE WORK

Having concrete information on the required production volume, as well as the time needed to produce a unit, enables analysts to determine the required labor force. For example, if the production load for a given 40-h week is 4,420 h, the company needs 4,420/40, or 111, operators. This use of standards is especially important in a retrenching market, in which production volume is going down. Without a yardstick to determine the actual number of people needed to perform the reduced load, when overall volume diminishes, the entire workforce may slow down to make the available work last. Unless the workforce balances with the available work volume, unit costs progressively rise. Under these circumstances, production operations start producing a loss, necessitating increased selling prices with potential reductions in sales volume. The cycle repeats with greater losses.

In an expanding market, it is equally important to be able to budget labor. Rising customer demands necessitate a greater volume of personnel. Companies must determine the exact number and type of personnel to add to the payroll, so that these workers can be recruited in sufficient time to meet customer schedules. If accurate time standards exist, it is a matter of simple arithmetic to convert product requirements to departmental work-hours.

Figure 16.3 illustrates how overall plant capacity may be increased in an expanding market. Here, the plant anticipates doubling its work-hour capacity between January and November. This budget projects the scheduled contracts in terms of work-hours and allows a reasonable cushion (crosshatched section) for receiving additional orders. This process can be extended into budgeting the labor needs of specific departments as well.

IMPROVING PRODUCTION CONTROL

Production control is the phase of an operation that schedules, routes, expedites, and follows up production orders, in an effort to achieve operating economies and satisfy customer requirements. The whole function of production control is based on determining where and when the work can be done. This cannot be achieved without a concrete idea of "how long."

Scheduling, one of the major functions of production control, is usually handled in three degrees of refinement: (1) long-range, or master, scheduling; (2) firm order scheduling; and (3) detailed operation scheduling, or machine loading.

Long-range scheduling is based on the existing and anticipated production volume. In this case, specific orders are not given any particular sequence, but are merely lumped together and scheduled in appropriate time periods.

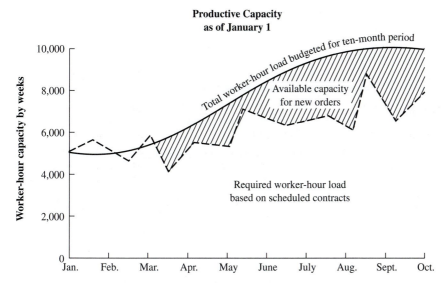

Figure 16.3 Chart illustrating actual projected work-hour load and budgeted work-hour load.

Firm order scheduling involves scheduling existing orders to meet customer demands, while still operating in an economical fashion. Here, workers assign degrees of priority to specific orders, and anticipated shipping promises evolve from this schedule.

Detailed operation scheduling, or machine loading, involves assigning specific operations to individual machines day by day. The scheduling is planned to minimize setup time and machine downtime, while meeting firm order schedules. Figure 16.4 illustrates the machine loading of a specific department for one week. It shows that considerable capacity exists on milling machines, drill presses, and internal thread grinders.

Regardless of the degree of refinement of the scheduling procedure, scheduling would be impossible without time standards. Time standards help predetermine the flow of materials and work in progress, thus forming a basis for accurate scheduling. The success of any schedule is directly related to the accuracy of the time values used in determining the schedule. If time standards do not exist, schedules formulated on judgment only cannot be expected to be reliable.

DETERMINING LABOR EFFICIENCY

With reliable time standards, a plant does not have to have an incentive wage payment system to determine and control its labor costs. The ratio of departmental earned production hours to departmental clock hours reveals the efficiency of that department. The reciprocal of the efficiency multiplied by the average hourly rate gives the hourly cost of standard production. For example, the finishing

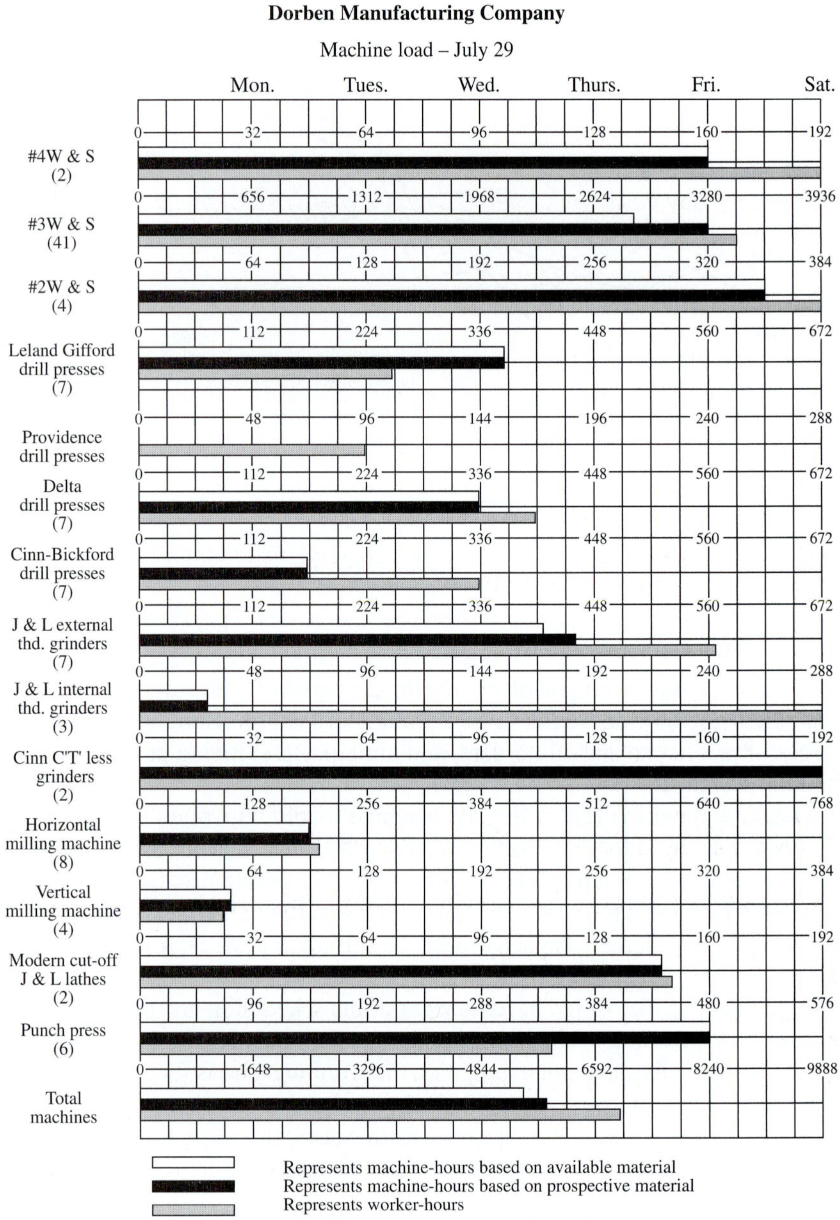

Dorben Manufacturing Company

Machine load – July 29

Represents machine-hours based on available material
Represents machine-hours based on prospective material
Represents worker-hours

Figure 16.4   Machine loading of a machining department for one week.
(Notice that several schedules depend on receiving additional raw material.)

department in a plant using straight daywork may have 812 clock hours H_c of labor time for 876 earned hours H_e of production. The departmental efficiency will then be

$$E = \frac{H_e}{H_c} = \frac{876}{812} = 108 \text{ percent}$$

If the average daywork (see Chapter 17) hourly rate in the department is $16.80, then the hourly direct labor cost, based on standard production, is

$$\frac{1}{1.08} \times \$16.80 = \$15.56$$

In a second example, assume that in another department, the clock hours are 2,840 and the earned hours of production for the period are only 2,760. In this case, the efficiency would be

$$\frac{2,760}{2,840} = 97 \text{ percent}$$

and the hourly direct labor cost, based on standard production, with an average daywork rate of $16.80, would equal

$$\frac{1}{0.97} \times \$16.80 = \$17.32$$

In the latter case, management would realize that its labor costs were running $0.52 per hour more than base rates, and it could increase supervision to bring total labor costs into line. In the first example, labor costs were running less than standard, which would allow a downward price revision, increasing the production volume, or making some other adjustment suitable to both management and labor.

QUALITY STANDARDS ENFORCEMENT

Time standards force quality requirements to be maintained. Since production standards are based on the quantity of acceptable pieces produced in a unit of time, and since no credit is given for defective work, there is a constant intense effort by all workers to produce only good parts. If an incentive wage payment plan is in effect, operators are only compensated for good parts; to keep their earnings up, they keep their scrap down. Sampling inspection is invariably more effective under incentive conditions. The operator has already ensured that the quality of each piece turned out is satisfactory before the piece is released. When some of the pieces are defective, either the operator who produced the parts is held responsible for the salvage, or the worker's earnings are adjusted so that compensation is received only for satisfactory parts.

MANAGEMENT PROBLEMS

Time standards are accompanied by many control measures, such as scheduling, routing, material control, budgeting, forecasting, planning, and standard costs.

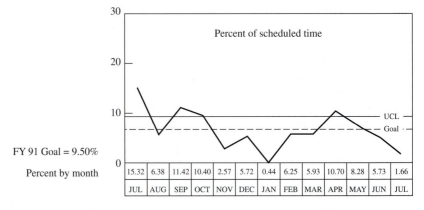

Figure 16.5 ETF monthly lost-time analysis. July 14–20, 1991.
(*Courtesy:* Ramesh C. Gulati, Sverdrup Technology, Inc., AEDC Group)

Having controls on practically every phase of an enterprise, including production, engineering, sales, and cost, minimizes management problems. By exercising the "exception principle," in which attention is given only to those items deviating from the planned course of events, management can confine its efforts to only a small segment of the total activity of the enterprise.

For example, Figure 16.5 illustrates a Weekly Lost Time Analysis developed so that management can take positive action when scheduled hours are not attained as planned. Note that a goal of 9.50 percent hours lost of scheduled time was established and that the months of July, September, October, and April exceeded that goal.

CUSTOMER SERVICE

Experience has proved that companies that have developed sound standards based on measurement are more likely to meet scheduled delivery dates for their products. The use of time standards allows the introduction of up-to-date production control procedures, with the resulting advantage to customers who get their merchandise when they want and need it. Also, time standards tend to make any company more time- and cost-conscious; this usually results in lower selling prices. As has been explained, quality is maintained under a work standards plan, thus assuring the customers of more parts made to required specifications.

16.3 COSTING

Costing refers to the procedure of accurately determining costs in advance of production. The advantage of being able to predetermine cost is obvious. Most contracts today are let on a "firm cost" basis, which means that the producer must predetermine production costs in order to set the firm price high enough to make a margin of profit. By having time standards on direct labor operations, producers can preprice those elements using the *prime cost* of the product. Prime cost is usually thought of as the sum of the direct material and direct labor costs.

Costs are the basis of actions within an organization. When the costs of processing a part become too high compared to competitive production methods, consideration will be given to making a change. Invariably, several alternatives exist for producing a given functional design, alternatives that compete on a cost basis. For example, casting will compete with forging, reaming with broaching, die casting with plastic molding, powdered metal with automatic screw machine, and so on.

Manufacturing costs can be classified into four groups: direct material costs, direct labor costs, factory expense, and general expense. The first two are directly involved with production, while the latter two are expenses beyond production costs, sometimes termed *overhead. Direct material* costs include the costs of raw materials, purchased subcomponents, standard commercial items (fasteners, wires, connectors, etc.), and subcontracted items. The industrial engineer begins by calculating the basic quantity required for the design. To this value are added losses for scrap, from either manufacturing or process errors; waste, from design errors; and shrinkage due to theft or environmental effects. The resulting increased quantity, multiplied by unit price, yields final material costs, with a factor subtracted for any anticipated salvage:

$$\text{Cost}_{\text{materials}} = Q \times (1 + L_{sc} + L_w + L_{sh}) \times C - S$$

where Q = base quantity in weight, volume, area, length, etc.
L_{sc} = loss factor due to scrap (same units)
L_w = loss factor due to waste (same units)
L_{sh} = loss factor due to shrinkage (same units)
C = unit cost of materials
S = value of salvaged materials

Direct labor refers to workers that are involved in direct production of the product. Direct labor costs are calculated from the time required to produce the product (the standard time, as discussed in previous chapters) multiplied by the wage rate.

Estimating the Prime Cost of a Component	EXAMPLE 16.1

In estimating the prime cost to produce an ABS component that is injection-molded, the analyst first multiplies the cost per pound of the ABS resin by the weight in pounds of the product, with due allowance for the sprue, runners, and normal shrinkage (typically, 3 to 7 percent on complex thermoplastic parts). To this figure, the analyst adds the direct labor cost. For example, if the injection press operator services five machines and has an hourly rate (including the cost of fringe benefits) of $18, the direct labor cost per injection press is $3.60 per hour, or $0.06 per minute. If the cycle time for molding the part is 0.5 min, the direct labor cost per piece is $0.03.

Let us assume the cost of the resin to be $1.20 per pound and the weight of each piece to be 1 oz. Let us also assume that the weight of the runners and sprue is 0.1 oz, and that 5 percent shrinkage is characteristic of this particular molding room. Then the estimated material cost is

$$\$1.20 \times 1.1 \times 1.05/16 = \$0.087$$

In this example, the prime cost per piece is

$$\$0.03 + 0.087 = \$0.117$$

Factory expense includes such costs as indirect labor, tooling, machine, and power costs. *Indirect labor* typically includes shipping and receiving, trucking, warehousing, maintenance, and janitorial services. Indirect labor, machine, and tool costs may have greater influence on the selection of a particular process than material and direct labor costs. For example, in the previous illustration, the single-cavity mold may cost $30,000, and the machine rate (cost of operating the injection press, exclusive of the press operator) may be $20.00 per hour. (In a complex machining center, machine rates are often as low as a few dollars per hour and range as high as $50 or more per hour.)

The allocation of tool costs also has a significant relationship with the quantity to be produced. Going back to our example, we assume that 10,000 pieces are to be produced. This gives a unit tool cost of $30,000/10,000, or $3.00 per piece. This is much greater than the combined material and direct labor cost (more than 10 times). If 1,000,000 pieces are to be produced, the unit cost for the mold will be only $0.03 (about one-third the cost of direct labor and material). Let us assume that the machine rate of this equipment (not including the mold cost) is $20.00 per hour, or $0.333 per minute and that 1 million pieces are desired. Then the total factory cost (direct material + direct labor + factory expense) is

$$\$0.087 + \$0.03 + \$0.333/2 = \$0.2835$$

General expense includes such costs as *expense labor* (accounting, administration, clerical, engineering, sales, etc.), rent, insurance, and utilities. The industrial engineer is primarily concerned with factory cost, since this is the cost that impacts the choice of alternative ways of producing a given design.

Figure 16.6 illustrates the various cost and profit elements in the development of selling price. An understanding of the basis of cost will enable the engineer to select the materials, processes, and functions to create the best product with cost usually being the deciding factor. The relationship between cost, sales, profit or loss, and volume is best revealed in a break-even or crossover chart (from Chapter 9). Figure 16.7 illustrates a typical break-even chart in costing.

The distribution of cost factors will vary dramatically with the number of units to be produced. When quantities are low, the proportion of development costs is high compared to the cost of manufacturing expense, direct labor, raw material, and purchased parts. The development cost includes design, drawings preparation, manufacturing information compilation, tool design and construction, testing, inspection, and many other items incidental to placing the first parts into production. As the number of units increases, the emphasis centers on a

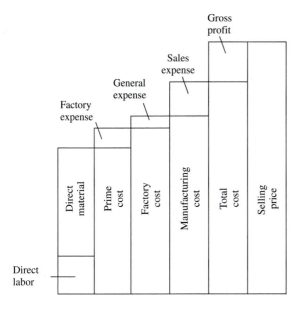

Figure 16.6 Elements of cost and gross profit (profit before taxes) entering into the development of selling price.

Note that, in this particular product, material cost is approximately 53 percent of total cost, and 17.5 percent of total cost is anticipated as gross profit.

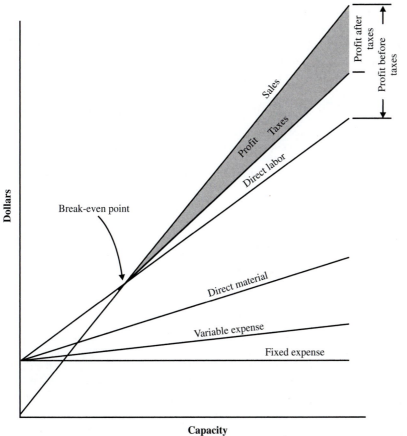

Figure 16.7 Break-even chart indicating relationship between cost, sales, profit or loss, and volume.

Note that for each category the previous category is added to yield a cumulative sum.

reduction in factory overhead, direct labor, and material costs, through advanced process engineering and manufacturing methods. For example, if an automobile manufacturer produces 2 million cars per year, of which 1 million have four cylinders per engine and the other 1 million have six cylinders per engine, and if there are four piston rings per cylinder, then 40 million piston rings must be produced per year. A savings of $0.01 on each ring would equal $400,000 per year. Therefore, to obtain the minimum cost, considerable engineering effort can be profitably applied to the production effort, beginning with raw material and continuing up to the installation of the finished product.

There is constant competition between materials and processes, based on costs that are influenced by the number of pieces made during a period of time. The parts activity affects the amount of time the activity is operated, compared with the hours available. The ratio of hours operated to hours available has considerable effect upon cost. Consider the following representative example.

A large hydraulic extrusion press, including the hydraulic pumps and the building to house the press, costs $3,000,000. Depreciation, maintenance, and interest on investment amount to 20 percent ($600,000) per year. Normally there are 2,000 working hours per year in one shift (8 h/day × 5 days/week = 40 h/week × 50 weeks/y = 2000 h/yr). Three shifts would represent 6,000 h/yr available. The minimum cost of the facility during a 24-h day would therefore be 600,000/6,000 = $100 per hour. However if the sales department can sell only enough to keep the equipment busy 8 h/day, the machine costs increase to 600,000/2,000 = $300 per hour. If sales decrease further, the machine cost will increase even more, making it difficult to operate the business at a profit.

EXAMPLE 16.2	Costing a Picture Window

As an example, consider new window construction. Hess Manufacturing Co. produces a large variety of vinyl replacement windows and storm doors. A standard data system and costing are a necessity for Hess to consider any changes in the styles, shapes, and features of windows and doors. Consider the simplest element in window construction, the picture window, or the fixed pane of glass in an aluminum frame. One of the smaller picture windows is nominally 2 ft × 3 ft, but is actually 24 in × 35.5 in.

Direct materials for a picture window include the vinyl-clad 6069 aluminum extrusions for the head, sill, two jambs, and four glazing stops; two glazing blocks and two weep-hole covers; a double pane of glass separated by a butyl spacer for insulation purposes; and packaging materials. By separate count (see Table 16.4), these total $18.42 in direct material costs. Note that several different scrap factors are utilized in calculating the material costs. For the aluminum extrusions, an 8 percent scrap factor increases the cut length and cost by 8 percent:

$$24.25 \text{ in} \times (1 + 0.08) \times \$0.08275/\text{in} = \$2.167$$

A flow process chart of the manufacturing operation indicates that the following basic operations are needed: cutting the aluminum extrusions and glazing stops to size, punching weep holes, welding the frame, cleaning the corners, assembling the final product, and packaging. From the standard data of Table 16.5 and required components, a total standard assembly time of 19.193 min for a 2-ft × 3-ft picture window can be calculated (see Table 16.6). Of the 19.193 min, 14.522 min is due to manual

Table 16.4 Bill of Materials—New Construction Picture (2-ft × 3-ft) Window

Part-Materials	Units	Length (in)	Scrap factor	Total length (in)	Unit cost ($/in)	Cost ($)
Extrusions						
Head, 6069	1	24.250	0.08	26.190	0.08275	2.167
Sill, 6069	1	24.250	0.08	26.190	0.08275	2.167
Jamb, 6069	2	35.697	0.08	38.552	0.08275	6.380
Glazing stop, top	2	15.290	0.08	16.513	0.00733	0.242
Glazing stop, sides	2	19.750	0.08	21.330	0.00733	0.313
Hardware						
Glazing block	2	—	0.01	—	0.019	0.0388
Weep hole cover	2	—	0.01	—	0.085	0.1717
Glass						
Clear glass*	2	5.92	0.10	6.51	0.258	3.360
Swiggle†	1	119.00	0.10	130.90	0.0246	3.220
Packaging						
Corner boots	4	—	0.03	—	0.056	0.231
Stretch wrap‡	1	—	—	—	0.131	0.131
				Total direct material costs		18.42

*Glass size is given as an area in feet and unit cost as $/foot.
†A swiggle is the butyl spacer between the double panes of glass.
‡Not specifically measured; an average value is used.

Table 16.5 Standard Data (Minutes) for New-Construction (2-ft ×3-ft) Windows

Operation	Oper. Code	Window type				Observed time		Operator rating	Normal time	
		sh	sci	dh	pic	Operator	Mach.		Operator	Mach.
Cut	CT	•	•	•		0.125	—	115	0.144	—
	CT	•	•	•	•	0.232	—	102	0.236	—
	CT	•	•	•	•	0.432	—	122.5	0.529	—
Mill	ML	•	•			0.305	—	125	0.381	—
Drill	DL	•	•	•		0.275	—	115	0.316	—
	DL	•		•		0.242	—	117	0.283	—
Punch	PC	•		•		0.145	—	115	0.167	—
	PC	•	•	•	•	0.208	—	122.5	0.255	—
Balance assembly	BA	•		•		0.757	—	120	0.908	—
Reinforcing bar	RB	•	•	•		1.233	—	115	1.418	—
Wool pile	WP	•	•	•		0.163	—	115	0.187	—
Weld	WD	•	•	•	•	0.767	0.717	107.5	0.825	0.717
Corner clean	CC	•	•	•		1.133	—	122.5	1.388	—
	CC	•	•	•	•	0.220	2.942	100	0.220	2.942
Hardware	HW	•	•	•		1.673	—	112.5	1.882	—
Drop-in glazing	DG	•	•	•	•	3.210	—	107.5	3.451	—
Final assembly	FA	•	•	•	•	3.390	—	115	3.899	—
Packaging	PK	•	•	•	•	0.373	0.790	105	0.392	0.790

Note: sh = single-hung window; sci = slider single-hung window; dh = double-hung window; pic = picture window.

Table 16.6   Assembly Time (Minutes) Analysis—New-Construction Picture (2 ft × 3 ft) Window

Process	Head	Sill	Jamb	Glazing stop	Machine elements
	Manual elements				
Cut	0.529	0.529	0.529	0.236	—
Punch weep holes	—	0.255	—	—	—
# Parts/frame	1	1	2	4	—
Subtotal time		3.315			—
Weld		0.825			0.717
Corner clean		0.220			2.942
Drop-in glazing		3.451			—
Final assembly		3.899			—
Packaging		0.392			0.790
Total assembly time		12.102			4.449
Allowances		20%[1]			5%[2]
Standard assembly time		14.522			4.671
			Total standard assembly time		19.193

[1]Includes 5 percent personal allowance, 5 percent basic fatigue allowance, 5 percent delay allowance, 5 percent material handling allowance
[2]Considers malfunctioning and maintenance of machine

Table 16.7   Costing of New Construction Picture (2 ft × 3 ft) Window

Type of cost	Minutes	Hours	Rate ($/h)	Cost ($)
Direct materials	—	—	—	18.42
Direct labor	19.193	0.320	7.21	2.31
Factory expense	19.193	0.320	9.81	3.14
			Total factory cost	23.87

elements of the operator (with a 20 percent allowance, 12.102 × 1.2 = 14.522 min), and 4.671 min is due to machine elements (with a 5 percent allowance for machine malfunction and maintenance, 4.449 × 1.05 = 4.671 min).

The 19.193 min corresponds to 0.32 h (19.193/60) of direct labor, which, at an average rate of $7.21 hour, yields $2.31 in direct labor costs. An overhead rate of 136 percent of direct labor costs yields $3.14 in factory expense (0.32 × 1.36 × 7.21). Adding direct material costs of $18.42 to the two previous costs yields a total factory cost of $23.87 (see Table 16.7). Based on this cost, Hess can determine a suggested retail cost that will maintain its budget and desired profitability.

Note that these *standard costs* are carefully predetermined target costs that should be attained. They are used to determine product costs, evaluate performance, and, in general, form budgets. As work is done, however, actual costs are incurred, which may reveal *variances* when compared to the standard costs. These variances are considered favorable when actual costs are less than the budgeted or standard costs, or unfavorable when actual costs exceed standard costs. The variances provide feedback on what should be modified to run a productive line.

SUMMARY

Thorough and regular follow-up ensures the expected benefits from the new method. This calls for maintaining time standards to ensure a satisfactory rate structure. Periodically, all standards should be checked to verify that the methods being employed are identical to those in use at the time the standards were established. A continuing methods analysis is a must.

There are many uses of time standards in all areas of any enterprise. Probably the most significant result of time standards is the maintenance of overall plant efficiency. If efficiency cannot be measured, it cannot be controlled, and without control, it will markedly diminish. Once efficiency goes down, labor costs rapidly rise, and the result is eventual loss of competitive position in the market. By establishing and maintaining effective standards, a business can standardize direct labor costs and control overall costs.

QUESTIONS

1. Compare and contrast the different ways of determining a standard time.
2. How can valid time standards help develop an ideal plant layout?
3. Explain the relationship between time standards and plant capacity.
4. In what way are time standards used for effective production control?
5. How do time standards allow the accurate determination of labor costs?
6. How does developing time standards help maintain the product quality?
7. In what way is customer service improved through valid time standards?
8. What is the relationship between labor cost and efficiency?
9. Explain how inventory and storage areas can be accurately predicted.
10. An audit revealed that a standard as originally established was 20 percent loose. Explain the methodology for rectifying the rate.
11. What is the relationship between the accuracy of time standards and production control? Does the law of diminishing returns apply?
12. Explain the different costs comprising factory cost. Which ones would a methods analyst have greatest control over?
13. Explain the benefits of increasing production in relation to manufacturing cost.
14. When is it no longer necessary to follow up the installed method?

PROBLEMS

1. In the XYZ Company, management is considering going from two 8-h shifts per day to three 8-h shifts per day, or two 10-h shifts per day, to increase capacity. Management realizes that shift start-up results in a loss of productivity that averages 0.5 h per employee. The premium for the third shift is 15 percent per hour. Time over 8 h worked per day gives the operator 50 percent more pay. To meet projected demands, it is necessary to increase the work-hours of production by 25 percent. In view of insufficient space and capital equipment, this increase cannot

be accommodated by increasing the employees on either the first or the second shift. How should management proceed?

2. A time standard is established allowing the operator 11.28 min per piece. The sales department expects to sell at least 2,000 of these parts in the next year. How many audits of this standard would you recommend be scheduled during the next 12 months?

3. If a daywork shop were paying an average rate of $12.75 per hour and had 250 direct labor employees working, what would be the true direct labor cost per hour if, during a normal month, 40,000 h of work were produced? Assume 21 working days per month.

4. The Dorben Co. is deciding whether it is worth releasing a new "improved" model of its widget. The following details have been collected:

Standard time (min)	= 1.00
Direct material costs per widget	= $0.50
Direct labor costs per widget	= $1.00
Indirect labor costs per widget	= $0.50
Expense labor costs per widget	= $0.50
Fixed overhead costs including sales costs/year	= $50,000

   a. Plot the prime cost as a function of production.

   b. Plot total costs as a function of production.

   c. Given a sales price of $3.00 per widget, how many widgets must Dorben sell to make a profit? Show this point on the plot.

5. Operators assemble pumps at an hourly rate of $10.00. The standard time for the assembly is 20 min. Direct material costs are $19.50 per pump. Indirect labor and other overhead are calculated simply at a rate of $5.00 per hour, while general office expenses are calculated at a rate of $2.00 per hour. What is the factory cost for one pump?

REFERENCES

Buffa, Elwood S. *Modern Production Operations Management*. 8th ed. New York: John Wiley & Sons, 1987.

Carter, W. K., and M. F. Usry. *Cost Accounting*. 12th ed. Houston: Dane Publications, 1999.

Lucey, T. *Costing*. New York: Continuum, 2002.

Narasimhan, S. *Production Planning and Inventory Control*. New York: Prentice-Hall, 1995.

Sims, E. R. *Precision Manufacturing Costing*. New York: Marcel Dekker, 1995.

Wage Payment

- Set simple but fair incentives based on proven standards.

- Guarantee basic hourly rates.

- Provide individual incentives above the base rates.

- Tie incentives directly to increased production of a quality product.

- Avoid offsetting productivity gains with increased injury costs.

To generate highly productive and satisfied workers, a company must reward and recognize them for effective performance. The reward must be meaningful to the employees, whether it is financial, psychological, or both. Experience has proved that workers do not give extra or sustained effort unless some incentive, either direct or indirect, is provided. Incentives, in one form or another, have been used for many years. Today, with the increasing need for U.S. business and industry to improve productivity to retard inflation and maintain or improve their position in the world market, management should not overlook the advantages of wage incentives. Only about 25 percent of manufacturing employees are now on incentives. If this figure were doubled in the next decade, large increases in productivity could be obtained.

With fringe benefits becoming increasingly significant (today, they average about 40 percent of direct labor), these costs must be spread over higher production output which can be achieved through wage incentives. At present, fringe benefits include not only basic features such as pensions, vacation time, and medical benefits (the costs of which have been skyrocketing), but also added features such as disability insurance and educational benefits, as shown in Table 17.1.

Typical nonincentive, fixed wage payment is known as *day work*, while any incentive plan that increases the employee's production may be referred to as a *flexible compensation plan*. Four types of flexible plans are discussed briefly in this chapter: piecework and standard labor hour plans, the gain-sharing plans—Scanlon, Rucker, IMPROSHARE; ESOPs; and profit sharing. Before analysts design wage payment plans for specific plants, they should review the strengths and weaknesses of past plans, including day work and all nonfinancial plans.

Table 17.1   Typical Fringe Benefits Provided by a Company

Benefit	Approximate percentage*
Health insurance	13–18
Vision insurance	½–1
Dental insurance	2–4
Vacations (up to 4 weeks per year)	20–25
Personal leave (up to 5 days per year)	2–5
Holidays (up to 10 per year)	10–12
Term life insurance	2–5
Long-term disability	1–3
Pension	25–30
Educational expense reimbursement	1–2

*As percentage of total fringe benefits.

17.1   DAY WORK PLANS

Day work plans compensate the employee on the basis of number of hours worked times an established hourly base rate. Company policies that are fair, with relatively high base rates (based upon job evaluation and merit rating), a guaranteed annual wage, and relatively high fringe benefits, build healthy employee attitudes, stimulate employee morale, and tend to indirectly increase productivity.

From the company's perspective, it would seem that a day work plan is ideal. Unit labor costs (employee wages divided by productivity or worker performance) decrease as worker productivity increases (see Figure 17.1). Mathematically, the normalized unit labor costs y_c can be expressed as

$$y_c = y_w/x$$

where y_w = normalized hourly base rate (= 1)
   x = normalized productivity or performance

Unfortunately, all day work plans have one weakness: they allow too broad a gap between employee benefits and productivity. After a period of time, employees take the benefits for granted, and the company never realizes the expected lower unit labor costs. The theories, philosophies, and techniques of day rate plans are beyond the scope of this text; for more information in this area, refer to books on personnel administration.

17.2   FLEXIBLE COMPENSATION PLANS

Flexible compensation plans include all plans in which the worker's compensation is related to output. This category includes both simple, individual incentive plans and group incentive plans. In simple individual plans, each employee's performance for the period governs that employee's compensation. Group plans are applicable to two or more persons who are dependent on one another working as a team. In these plans, each employee's compensation within the group is based on his or her base rate and on the performance of the entire group for the period.

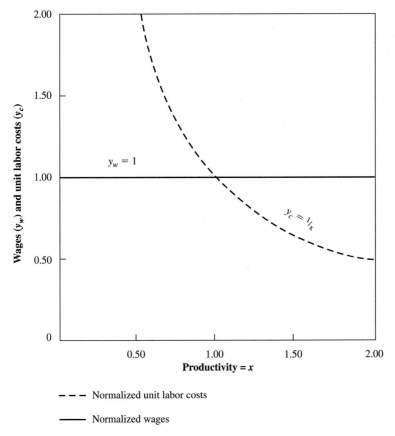

Figure 17.1 Relationship between costs, wages, and productivity in day work plans. (Adapted from Fein, 1982; Reprinted by permission of John Wiley & Sons, Inc.)

The incentive for high or prolonged individual effort is not nearly as great in group plans as it is in individual plans. Hence, industry favors individual incentive methods. In addition to lower overall productivity, group plans have other drawbacks: (1) personnel problems brought about by nonuniformity of production, coupled with uniformity of pay; and (2) difficulties in justifying base rate differentials for the various opportunities within the group.

However, group plans do offer some decided advantages over individual incentives: (1) ease of installation, due to ease of measuring group, rather than individual, output; and (2) reduction of administration cost due to reduced paperwork, less verification of inventory in process, and less in-process inspection.

In general, individual incentive plans foster higher production rates and lower product unit costs. If it is practical to install, the individual incentive plan should be given preference over group systems. On the other hand, the group approach works well where it is difficult to measure individual output and where individual work is variable and frequently performed in cooperation with another employee. For example, if four operators are working together in the operation of an extrusion press, it would be virtually impossible to install an individual incentive system; rather, a group plan would be applicable. Similarly, any job rotation scheme, to minimize repetitive-motion injuries, would necessarily have to use a group incentive plan.

PIECEWORK PLAN

Under the *piecework* plan, all standards are expressed in money, and the operators are rewarded in direct proportion to output. Under piecework, the day rate is not guaranteed. Since federal law requires a minimum guaranteed hourly rate, piecework is no longer used in the United States. Prior to World War II, piecework was used more extensively than any other incentive plan. The reasons for the popularity of piecework were that it was easily understood by the worker, was easily applied, and was one of the oldest wage incentive plans. Figure 17.2 illustrates graphically the relationship between an operator's wages and unit direct labor costs under a piecework plan.

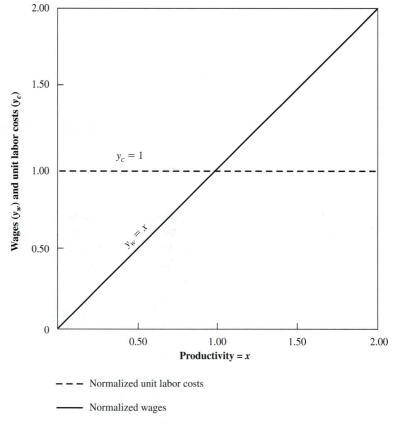

Figure 17.2 Relationship between costs, wages, and productivity in piecework.

Since the unit labor costs remain constant regardless of worker productivity, the company does not appear to benefit from a piecework plan. However, that is not true if the reader remembers the different costs that enter into factory expenses (see Chapter 16). Relatively constant overhead costs would decrease if considered on a unit cost basis.

STANDARD HOUR PLAN

The *standard hour plan* with a guaranteed base rate, established by job evaluation, is by far the most popular incentive plan in use today. It offers all the advantages of piecework and eliminates the major legal problem. Graphically, the relationship between operator wages and unit direct labor cost, plotted against productivity, is a combination of Figures 17.1 and 17.2 (see Figure 17.3). The worker operates under a day work plan up to 100 percent productivity, and under piecework, beyond 100 percent productivity. For example, a standard may be expressed as 0.02142 h

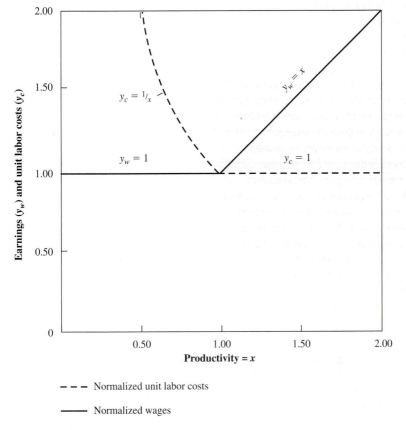

- - - Normalized unit labor costs

——— Normalized wages

Figure 17.3 Relationship between costs, wages, and productivity in the standard hour plan.
(Adapted from Fein, 1982) (Reprinted by permission of John Wiley & Sons, Inc.)

per piece or 373 pieces per 8-h shift. Once the base rate is known, it is easy to calculate either the money rate or the operator's wages. If the operator has a base rate of $12, then the money rate of this job is $12.00 \times 0.02142 = \$0.257$ per piece. If the operator produces 412 pieces in an 8-h workday, wages for the day are $412 \times 0.257 = \$105.88$, and hourly wages are $\$105.88/8 = \$13.24$. The operator's efficiency for the day, in this case, is then 412/373, or 110 percent.

A variation of the standard hour plan is a scheme whereby the incentives are applied to each worker based on group output, creating a group incentive scheme. This is especially useful for work cells, as part of job enlargement or job rotation, or in situations where individual performance cannot easily be measured (e.g., shipbuilding, aircraft manufacturing). These schemes have some advantages in allowing greater flexibility for workers, reducing competition, and encouraging group spirit and teamwork. On the other hand, individual incentive is reduced, and better workers may become discouraged.

MEASURED DAY WORK

During the early 1930s, shortly after the era of efficiency experts, organized labor tried to get away from time study practice and, in particular, piecework. At that time, *measured day work* became popular as an incentive system that lessened the direct relationship between a standard and worker's earnings. Many modifications of measured day work installations are in operation today, and the majority follow a specific pattern. First, job evaluations establish base rates for all jobs falling under the plan. Second, some form of work measurement determines standards for all operations. Third, analysts keep a progressive record of each employee's efficiency for one to three months. This efficiency, multiplied by the base rate, forms the basis of a guaranteed base rate for the next period. For example, the base rate of a given operator may be $12.00 per hour. Assume that the controlling performance period is one month, or 173 working hours. If, during the month, the operator earned 190 standard hours, his or her efficiency for the period would be 190/173, or 110 percent. Then, in view of the performance, the operator would receive a base rate of $1.10 \times 12.00 = \$13.20$, for every hour worked during the next period, regardless of performance. However, achievement during this period would govern the base rate for the succeeding period.

In all measured day work plans today, the base rate is guaranteed; thus, an operator falling below standard (100 percent) for any given period would receive the base rate for the following period. The length of time used in determining performance usually runs three months, to diminish the clerical work of calculating and installing new guaranteed base rates. Of course, the longer the period, the less incentive effort can be expected. When the spread between performance and reward is too great, the effect of incentive performance diminishes.

The principal advantage of measured day work is that it takes the immediate pressure off workers. They know what their base rates are, and they realize that, regardless of performance, they will receive that amount for the period.

The limitations of measured day work are apparent. First, because of the length of the performance period, the incentive feature is not particularly strong. Second, to be effective, the plan places a heavy responsibility on supervisors for maintaining production above standard. Otherwise, the employee's performance drops, thus lowering the base rate for the following period and causing employee dissatisfaction. Third, keeping detailed rate records and making periodic adjustments are costly in all base rates. In fact just as much clerical work is involved under measured day work as under any straight incentive plan in which employees are rewarded according to output.

GAIN-SHARING PLANS

Gain-sharing plans, also known as *productivity sharing* plans, are characterized by sharing the benefits of improved productivity, cost reduction, and/or quality improvement. Many firms throughout the United States today have some form of gain-sharing plan as a supplement, rather than replacing existing compensation systems. The principle is to reward employees for improvements in productivity and/or cost, whether or not the improvements are solely due to performance above normal or improvements in work methods.

Under plans of this type, management computes incentives on a monthly basis. Customarily, only two-thirds of the incentive earned in a given pay period is distributed. The remaining one-third is placed in a reserve fund to be used for any month in which performance falls below standard. The three productivity sharing plans discussed here are Scanlon, Rucker, and IMPROSHARE. They differ in the formula used to compute productivity savings and in the implementation method. The Scanlon and Rucker plans measure the payroll of the firm against total dollar sales and compare the result to the average of the past several years. The IMPROSHARE plan measures output against total hours worked. Thus, the Scanlon and Rucker plans use dollars as the measurement unit, while IMPROSHARE uses hours. All three of these productivity plans are flexible regarding the personnel included in the plan. Direct and indirect workers as well as all levels of management may be included.

Scanlon Plan During the Great Depression, Joseph Scanlon developed the *Scanlon plan* to save a failing company. Three fundamental principles form the basis of this plan: bonus payment, identity with the company or firm, and employee involvement. Scanlon plans recognize the value and contribution of each member of the firm, encourage decentralized decision making, and seek to get each employee to identify with the organization's objectives through financial participation.

Before the bonus is calculated, a base ratio must be computed. This traditionally is

$$\text{Base ratio} = \frac{\text{payroll costs to be included}}{\text{value of production}}$$

Analysts make a historical study of approximately one year to gather data prior to calculating a proper base ratio. For example, if the base ratio is 15 percent and if during the past month, the value of production (sales plus or minus inventory) equals $2 million, then allowed labor equals $300,000 ($0.15 \times 2,000,000$). An actual labor cost of $270,000 generates a bonus pool of $30,000. Typically, the company keeps a portion of this pool to provide for capital expenditures. The remainder is distributed to the employees as a monthly bonus, based on a percentage of their wages.

To stimulate identity with the company, the Scanlon plan recommends a continuing program of management development in which all employees, through effective communication, learn about the goals, objectives, opportunities, and problem areas characteristic of the plan. The Scanlon plan incorporates most "quality of work life" variables, including job enlargement, job enrichment, feeling of achievement, and recognition.

Employee involvement is typically accomplished through a formalized suggestion system and two overlapping committee systems. Elected employee representatives meet at least monthly with their departmental supervisors to review productivity, cost reductions, and quality improvement suggestions. These committees frequently have certain decision-making authority for less costly suggestions. More costly suggestions, or those affecting another department, are referred to a higher-level committee.

Rucker Plan This plan came into being during the early 1940s. It was conceived by Allen W. Rucker, who noted the relationship between payroll costs and actual net sales plus or minus inventory changes minus purchased materials and services.

Like the Scanlon plan, the *Rucker plan* emphasizes identity with the company and employee involvement, through the establishment of a suggestion system, Rucker committees, and good labor–management communications. The Rucker plan provides a bonus in which everyone, excluding top administration, shares a percentage of the gains. In the evaluation of the bonus, a historical relationship between labor and value added must be established. For example:

Net sales (for period of 1 year)	$1,500,000
Inventory change (decrease)	200,000
	$1,300,000
Less materials and supplies used	700,000
Production value added	$600,000

$$\text{Rucker standard} = \frac{\text{payroll costs included in group}}{\text{production value}}$$

Assuming that the labor costs in the base 1-year period are $350,000, the Rucker standard becomes

$$\frac{\$350,000}{\$600,000} = 0.583$$

Thus, in any future period (usually one month) that the actual labor costs are less than 0.583 of production value, employees earn bonuses. Typically, 30 percent of this bonus is reserved for deficit months, a portion is kept by the company for future improvements, and the remainder (often 50 percent) is distributed to employees. With 50 percent of the bonus distributed to employees and 30 percent retained for deficit months, gain-sharing additives for rework production and delivery performance can often utilize the remaining 20 percent of the bonus, rather than that amount being kept by the company to provide improvements.

Since materials and supplies used are deducted from net sales, the Rucker plan calculation partially accounts for variables, such as product mix. This plan also encourages employees to conserve supplies and materials, since the employees would benefit from these savings.

IMPROSHARE   The IMproved PROductivity through SHARing plan was developed by Mitchell Fein in 1974. Its goal is to produce more products in fewer hours of direct and indirect labor. Unlike the Scanlon and Rucker plans, IMPROSHARE does not emphasize employee involvement, but rather measures performance and encourages workers to improve productivity.

IMPROSHARE compares the work-hours saved for a given number of units produced to the hours required to produce the same number of units during a base period. The savings are shared by the company and the direct and indirect employees involved with the production of the product. Base productivity is measured by comparing the labor hour value of completed production to the total labor input for this production. Only acceptable products are counted. Thus,

$$\text{Work-hour standard} = \frac{\text{total production work-hours}}{\text{units produced}}$$

IMPROSHARE Incentive Plan	EXAMPLE 17.1

Assume that in a single-product plant, 122 employees produced 65,500 units over a 50-week period. If the total hours worked were 244,000, the work-hour standard would be

$$\frac{244,000}{65,500} = 3.725 \text{ h/unit}$$

If, in a week, 125 employees worked a total of 4,908 h and produced 1,650 units, the value of the output would be $1,650 \times 3.725 = 6,146.25$ h. The gain would be $6,146.25 - 4,908 = 1,238.25$ h. Typically, one-half of this amount, or 619.125 h, goes to the employees. This would be a 12.6 percent (619.125/4,908) bonus or additional pay to each employee.

The company also benefits, since labor costs are reduced. The unit labor cost of 3.725 h established for the base period is reduced to $(4,908 + 619.25)/1,650 = 3.350$ h per unit.

Graphically, the IMPROSHARE plan can be considered a variation of the standard hour plan of Figure 17.3, except that the slope of the piecework segment is not 1, but a fraction less than 1 (see Figure 17.4). This fraction, or slope p, is the participation fraction and could vary between companies. If the split between the company and employees is 50/50 (as discussed previously), then p is equal to 0.5. In the standard hour plan, the participation is 100 percent and $p = 1$.

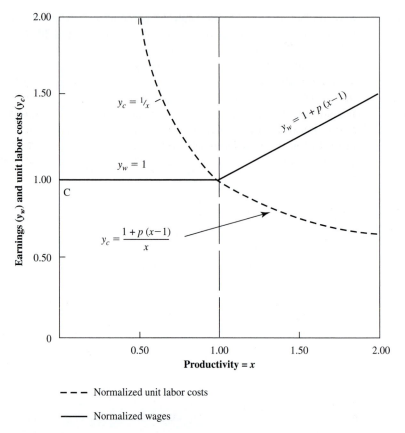

Figure 17.4 Relationship between costs, wages, and productivity in incentive plans with variable participation (p = participation fraction).
(Adapted from Fein, 1982; Reprinted by permission of John Wiley & Sons, Inc.)

Employee Stock-Ownership Plans (ESOPs)   Employee stock-ownership plans have become increasingly popular in the past decade. The Bureau of National Affairs 1984 survey of 195 employers to determine type of productivity improvement programs administered indicated that 37, or 19 percent, had installed employee stock-ownership plans. These plans involved the creation of a trust that holds company stock for its employees. Although 100 percent worker ownership plants are rare, an ESOP can be used to develop such an organization.

Profit Sharing   *Profit sharing* can be defined as a procedure under which, in addition to regular pay, an employer pays all employees special current or deferred sums based on the prosperity of the company. No one specific type of profit sharing has received general industrial acceptance. In fact, just about every installation has certain "tailor-made" features that distinguish it from others. However, the majority of profit-sharing systems fall into one of the following broad categories: (1) cash plans, (2) deferred plans, and (3) combined plans.

As the name implies, the straight cash plan involves the periodic distribution of money from the profits of the business to the employees. The payment is not included with the regular pay envelope, but is made separately, to identify it as an extra reward, brought about by the individual and combined efforts of the entire operating force. The amount of the cash distribution is based on the degree of financial success of the enterprise for the bonus period. The shorter the period, the closer the connection between effort and financial reward to the employees. Longer periods are selected because they average out the vagaries of business cycles.

Deferred profit-sharing plans feature the periodic investment of portions of the profits for employees. Upon retirement or separation from the company, employees have a source of income other than wages. Deferred profit-sharing plans obviously do not provide the incentive stimulus that cash plans do. However, deferred profit-sharing plans do offer the advantage of being easier to install and administer. Also, plans of this type offer more security than the cash reward plans. This makes them especially appealing to stable workers.

Combined plans arrange to have some of the profits invested for retirement and similar benefits, and some distributed as cash rewards. This class of plans can realize the advantages of both the deferred plans and those employing the straight cash system. A representative installation might provide for sharing one-half the profits with the employees. Of this amount, one-third may be distributed to employees as extra bonus checks, one-third may be held in reserve to be given out during a less successful financial period, and the remaining one-third may be placed with a trustee for deferred distribution.

There are three methods for determining the amount of money to be given to individual employees from the company's profits. The first and least used is the "share and share alike" plan. Here, each employee, regardless of job class, receives an equal amount of the profits, after attaining the prescribed period of company service. Proponents of this method believe that individual base rates already take care of the relative importance of the different workers to the company. The "share and share alike" plan supplies a feeling of teamwork and importance to each employee, no matter what his or her position in the plant.

The most commonly used method of distribution under profit sharing is based on the regular compensation paid to workers. The theory is that the employee who was paid the most during the period contributed the greatest to the company's profits and consequently should share in them to the greatest extent. For example, a toolmaker earning $15,000 during a six-month period would receive a greater share of the company's profits than a chip hauler paid $7,000 during the same period.

Another popular means of profit distribution involves the allocation of points. Points are given for each year of seniority, each $100 of pay, and other factors such as attendance and cooperation. The number of points accumulated for the period determines each employee's share of the profits. Perhaps the principal disadvantage of the point method is the difficulty of maintaining and administering the complex and detailed records.

For a profit-sharing plan to succeed, worker representation and union cooperation are essential. The emphasis should be placed on partnership and not on

| EXA.MPLE 17.2 | Comparison of Two Incentive Schemes |

A company would like to evaluate two incentive schemes. The first is similar to the IMPROSHARE gain-sharing plan, with a 50/50 split above 100 percent productivity. The second is constant day work up to 100 percent productivity, a *kicker* up to 120 percent (a step increase in wages to induce workers to reach a certain level of productivity), and then a participative gain-sharing plan but with the workers only receiving 20 percent and the company 80 percent. The plans and unit labor costs are shown in Figure 17.5. The company would like to know the point at which the two plans break even above 100 percent productivity.

The first plan's normalized unit labor costs can be expressed as

$$y_{c_1} = 0.5 + 0.5/x$$

while the second plan's are

$$y_{c_2} = 0.2 + 1/x$$

Equating the two equations and solving for x yield $x = 1.67$. Therefore, with plan 1, the company benefits up to 167 percent productivity, and with plan 2, with higher productivity levels. The company must decide whether it is reasonable to expect workers to reach such high levels of productivity.

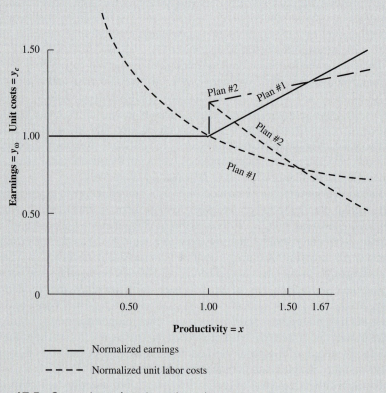

— — — Normalized earnings

- - - - - Normalized unit labor costs

Figure 17.5 Comparison of two incentive schemes

management benevolence. Management should recognize that the plan should be dynamic and is not a panacea for all problems. Profit sharing should not be used as an excuse for paying lower than prevailing wages.

Perhaps the greatest concern about profit sharing is the employee taking for granted that an extra check will be received at the end of the year. Employees will come to expect these checks, and they may even feel cheated if the company has a lean year and cannot pay them. For these reasons, any employer should be very cautious before embarking on a profit-sharing program. On the other hand, many companies today are experiencing high worker efficiency, decreased costs, scrap reduction, and better worker morale as a result of profit sharing.

INDIRECT FINANCIAL PLANS

Company policies that stimulate employee morale and increase productivity, without directly relating compensation to production, fall in the indirect financial plans classification. Overall company policies such as fair and relatively high base rates, equitable promotion practices, sound suggestion systems, a guaranteed annual wage, and relatively high fringe benefits build healthy employee attitudes, which stimulate and increase productivity. Thus, they are classified as indirect financial plans.

The weakness of all indirect incentive methods is that they allow too broad a gap between employee benefits and productivity. After a period of time, employees take the benefits for granted and fail to realize that continuance of those benefits must result entirely from employee productivity. The theories, philosophies, and techniques of indirect incentives are beyond the scope of this text; for more information in this area, refer to books on personnel administration.

17.3 IMPLEMENTATING WAGE INCENTIVES

PREREQUISITES

The majority of companies that have incentive plans installed favor their continuance and believe that their plans are (1) increasing the production rate, (2) lowering overall unit costs, (3) reducing supervision costs, and (4) promoting increased employee earnings. Before installing a wage incentive program, management should survey its plant to be sure that the plant is ready for an incentive plan. Initially, a policy of methods standardization must be introduced so that valid work measurement can be accomplished. If different operators follow different patterns while performing their work, and if the sequence of elements has not been standardized, the organization is not ready for the installation of wage incentives.

Work schedules should create a backlog of orders for each operator, so that the chances of running out of work are minimized. This implies that adequate inventories of material are available, and that machines and tools are properly maintained. Also, established base rates should be fair, and should provide for a sufficient spread between job classes, to recognize the positions that demand more skill, effort, and responsibility. Preferably, management will have established base rates through a sound job evaluation program.

Finally, fair performance standards must be developed before wage incentives are installed. These rates should never be set by judgment or past performance records. To be sure that the rates are correct, some form of work measurement, based on time study, predetermined time systems, standard data, formulas, or work sampling should be used.

Once these prerequisites are complete and management is fully sold on incentive wage payment, the company is in a position to design the system.

DESIGN

To be successful, an incentive plan must be fair to both the company and its operators. The plan should give operators the opportunity to earn approximately 20 to 35 percent above base rate. Next, a good incentive plan must be simple. The simpler the plan, the more easily it is understood by all parties, and understanding enhances the chances for approval. Individual incentive plans are more easily understood, and they work the best, if individual output can be measured.

The plan should guarantee the basic hourly rate set by job evaluation, and the rate should be a good living wage comparable to the prevailing wage rate of the area for each job in question. There should be a range of pay rates for each job, and these should be related to total performance, which encompasses quality, reliability, safety, and attendance as well as output. For performance greater than standard, operators should be compensated in direct proportion to output, thus discouraging any restriction of production.

To help employees associate effort with compensation, paycheck stubs should clearly show both the regular and the incentive earnings. It is also advisable to indicate on a separate form the efficiency of the operator for the past pay period. This is calculated as the ratio of the standard hours produced during the period to the hours worked during the period.

ADMINISTRATION

Once the plan has been installed, management must accept responsibility for maintaining it. Management must exercise its right to change the standards when the methods and/or equipment is changed. Employees must be guaranteed an opportunity to present their suggestions, and the advisability of their requests must be proved before any changes are made. Compromising on standards must be avoided, or it will lead to a failure of the plan.

Management must keep all employees aware of how the plan works and of any changes to the plan. One technique frequently used is to distribute to all employees an "operating instructions" manual detailing both company policy relative to the plan and all working details, with examples. The manual should thoroughly explain the basis of job classifications, time standards, performance-rating procedure, allowances, and grievance procedures. It should also describe the technique of handling any unusual situation. Finally, it should present the objectives of the organization and the role of each employee in the fulfillment of those objectives.

Next a motivational climate must be established to accompany the formal incentive plan. First a management style must emphasize a supporting role, rather than a directive role supervisors assisting workers as best they can.

Second, realistic goals of the plan should be clearly established and should be broken down into division, department, work center, and individual goals. They should emphasize both quality and quantity, as well as reliability and any other characteristic essential to the success of the business.

Third, there should be regular feedback to all employees, about the results of their efforts and the impact of these efforts on the established goals. Fourth, every work situation should be designed so that operators are in a position to control, to a large extent, the assignments they are given. A sense of responsibility is an important source of motivation, as is recognition for achievement. More details on motivational theories and approaches are presented in Chapter 18.

Administrators of the plan should make a daily check of low and excessively high performance, in an effort to determine their causes. Low performance not only is costly to management in view of the guaranteed hourly rate, but also leads to employee unrest and dissatisfaction. Unduly high performance is a symptom of loose standards, or the introduction of a methods change for which no standard revision has been made. In addition, a loose rate would lead to the dissatisfaction among neighboring workers, who see the operator with the low standard as having a "soft job." A sufficient number of such poor standards can cause the whole incentive plan to fail. Also, operators who have the loose rate may restrict their daily production, in fear that management will adjust the standard. This restriction of output is costly both to the operators and to the company.

Management should make a continuous effort to include a greater share of the employees in the incentive plan. When only a portion of the plant is on standard, there will be a lack of harmony among operating personnel because of significant differentials in take-home pay. However, work generally should not be put on incentive unless

1. It can readily be measured.
2. The volume of available work is sufficient to justify economically an incentive installation.
3. The cost of measuring the output is not excessive.

Fundamental to the administration of any wage incentive plan keyed to production is the constant adjustment of standards in response to changes in the work. No matter how insignificant a methods change may be, the standard should be reviewed for possible adjustment. Several minor methods improvements, in aggregate, can amount to a sufficient time differential, causing a loose rate, if the standard is not changed. When revising time standards due to methods changes, it is only necessary to study those elements affected by the changes.

Effective administration of the plan requires a continuing effort to minimize the nonproductive hours of direct labor. This nonproductive time, for which allowance must be given, represents lost time due to machine breakdowns, material shortages, tool difficulties, and long interruptions of any sort not covered in the allowances applied to the individual time standards. Managers must watch this time—frequently referred to as "blue ticket time" or "extra allowance time"—carefully, or it will destroy the purpose of the entire plan.

Under incentive effort, production performance is considerably higher than under day work operation. With the shorter accompanying in-process time of materials, very careful inventory control is needed to prevent material shortages. Likewise, management should introduce a program of preventive maintenance to ensure the continuous operation of all machine tools. Equally important to material control is the control of all nondurable tools, so that shortages, with resulting operator delays, do not develop.

It is essential that exact piece counts which determines the operator's earnings be recorded at each workstation. This is usually done by the operator; however, controls must be established to prevent operators from falsifying their production

---

EXAMPLE 17.3   Administration of a Wage Incentive Plan

Assume that on a certain job, the production rate is 10 pieces per hour, and that an hourly rate of $12 is in effect under a straight day work operation. Thus, the unit direct labor cost is $1.20. Now, this shop changes over to incentive wage payment for which the day rate of $12 per hour is guaranteed, and, above task, the operator is compensated in direct proportion to his or her output. Let us assume that the standard developed through time study is 12 pieces per hour, and that for the first 5 h of the workday, a certain operator averages 14 pieces per hour. His earnings for this period would then be

$$(\$12.00)(5)\left(\frac{14}{12}\right) = \$70.00$$

Now, assume that for the remainder of the workday, the operator could not be productively engaged in work, due to a material shortage. The worker would then expect at least the base rate, or

$$(3)(\$12.00) = \$36.00$$

which would give earnings for the day of

$$\$70.00 + \$36.00 = \$106.00$$

This would result in a unit direct labor cost of

$$\frac{\$106.00}{70} = \$1.514$$

Under day work, even with the low performance, the operator would have produced the 70 pieces in less than the working day. Here, the earnings would be 8 × $12.00, or $96.00, and the unit direct labor cost would be $96.00/70, or $1.371. Therefore, any nonproductive time should be carefully controlled.

output. When the number of pieces produced is small, the operators' count at the end of the shift is verified by the immediate supervisor, who initials the production report. When the number of pieces is large, a large box holding round numbers of the work, such as 10, 20, or 50 is used. Thus, at the end of the shift, it is a simple matter for the operators' supervisors to authenticate production reports by counting the boxes and multiplying by the number each box holds.

Basically, management establishes wage incentive plans to increase productivity. In a sound and properly maintained installation, the percentage of incentive earnings of those workers on incentive would remain relatively constant over time. If an analysis shows that incentive earnings continue to rise over a period of years, the installation probably has problems that will ultimately erode the effectiveness of the plan. If, for example, the average incentive earnings increases from 17 percent to 40 percent over a period of 10 years, the 23 percent rise is most likely not due to a proportionate increase in productivity, but to a creeping looseness in standards

A checklist for fundamental principles to be applied in a sound wage incentive plan is given in Table 17.2.

INCENTIVE PLAN FAILURE

An incentive plan may be classified as a failure when it costs more for its maintenance than the plan actually saves; it thus must be discontinued. Usually, it is not

Table 17.2 Checklist for Sound Wage Incentive Plans

	Yes	No
1. Is there agreement between management and labor on general principles?	☐	☐
2. Is there a sound foundation of job evaluations and wage rate structures?	☐	☐
3. Are there individual, group, or plantwide incentives?	☐	☐
a. Is the most weight applied to individual incentives?	☐	☐
4. Are incentives in direct proportion to increased production?	☐	☐
5. Is the plan as simple as possible?	☐	☐
6. Is quality tied to incentives?	☐	☐
7. Is the establishment of incentives preceded by methods improvements?	☐	☐
8. Are the incentives based on proven techniques?		
a. From detailed time studies	☐	☐
b. From basic motion data or predetermined time systems	☐	☐
c. From standard data or formulas	☐	☐
9. Are the standards based on standard performance under normal conditions?	☐	☐
10. Are standards changed when methods change?	☐	☐
a. By mutual agreement between management and labor representatives	☐	☐
11. Are temporary standards kept to a minimum?	☐	☐
12. Are basic hourly rates guaranteed?	☐	☐
13. Are incentives established for indirect workers?	☐	☐
14. Are accurate records kept for piece counts, unmeasured work, setup, and downtime?	☐	☐
15. Are good human relations maintained?	☐	☐

Table 17.3 Most Common Reasons for Incentive Plan Failures

	Percent
Fundamental deficiencies	**41.5**
Poor standards	11.0
Low incentive coverage of direct productive work	8.6
Ceiling on earnings	7.0
No indirect incentives	6.8
No supervisory incentives	6.1
Complicated pay formula	2.0
Inept human relations	**32.5**
Insufficient supervisor training	6.9
No guarantee of standards	5.7
A fair day's work not required	5.0
Standards negotiated with the union	4.8
Plan not understood	4.1
Lack of top-management support	3.6
Poorly trained operators	2.4
Poor technical administration	**26.0**
Method changes not coordinated with standards	7.8
Faulty base rates	5.1
Poor administration, i.e., poor grievance procedure	4.9
Poor production planning	3.2
Large group on incentive	2.8
Poor quality control	2.2

possible to put a finger on the precise cause of failure; there may be numerous reasons for a plan's lack of success. One survey (Britton, 1953; see Table 17.3) listed the principal causes of plan failure as poor fundamentals, inept human relations, and poor administration, resulting in too costly a program. Most of these are due to incompetent management—management that permits the installation of a plan with poor scheduling, unsatisfactory methods, a lack of standardization or loose standards, and the compromising of standards.

Also, without the complete cooperation of the employees, the union, and management to foster a team spirit, ultimate success of the incentive plan cannot be achieved.

17.4 NONFINANCIAL PERFORMANCE MOTIVATION PLANS

Nonfinancial incentives include any rewards that have no relation to pay, and yet they improve employees' spirits to such an extent that added effort is evident. Elements or company policies that fall under this category include periodic shop conferences, quality control circles, frequent talks between supervisors and employees, proper employee placement, job enrichment, job enlargement (see Chapter 18), nonfinancial suggestion plans, ideal working conditions, and the posting of individual production records. Effective supervisors and capable, conscientious managers also use many other techniques, such as treating the employee and spouse to dinner, providing tickets to sporting events or the theater, or arranging special trips to trade shows or other companies for exposure to state-of-the-art technology. All these approaches seek to motivate by improving the work environment. They are frequently referred to as "quality of work life" plans.

The management team also needs to set an example of high performance and the pursuit of excellence. Thus, employees will understand that the culture of their company is top performance in the manufacture of products of the best quality. The results of this philosophy by all workers will be a feeling of great pride in their work. In concert with this philosophy should be individual and group programs that recognize teamwork and team results.

SUMMARY

Incentive principles have been applied in job shops and production shops, in the manufacture of both hard goods and soft goods, in manufacturing and service industries, and in direct and indirect labor operations. Incentives have been used to increase productivity, improve product quality and reliability, reduce waste, improve safety, and stimulate good working habits, such as punctuality and regular attendance.

Probably the best wage incentive plan applied to individual workers today is the standard hour plan with a guaranteed day rate. However, profit sharing, employee stock ownership, and other related cost improvement savings plans have also met with success in many cases. In general, they tend to be more effective when they are installed in addition to, rather than instead of, a simple incentive plan. Group plans must guarantee their respective day rates to all members of the group, and should reward group members in direct proportion to their productivity once standard performance is achieved.

Table 17.4 illustrates the thinking of 508 personnel/industrial relations managers regarding the various plans. The majority feel that simple incentives—piecework, standard hour plans, and measured day work—are the best from the standpoint of raising productivity and being easy to explain.

Table 17.4 Characterization of Flexible Compensation Plans by Survey Respondents

	Profit sharing (%)	Employee stock-ownership plans (%)	Gain sharing Scanlon, Rucker, IMPROSHARE (%)	Simple std. hr. pc. rate (%)
Best for:				
Raising productivity	28	5	26	41
Increasing loyalty	48	17	19	14
Providing for retirement	80	13	n.a.	n.a.
Linking labor cost to performance	53	n.a.	28	19
Easiest to:				
Explain to employees	32	9	4	49
Administer	40	7	4	38

n. a. = not asked.
Source: Adapted from Broderick and Mitchell (1977).

Soundly administered incentive systems possess important advantages, for both workers and management. The chief benefit to employees is that these plans make it possible for the employees to increase their total wages, not at some indefinite time in the future, but immediately—in their next paycheck. Management obtains a greater output and, assuming that some profit is being made on each unit produced, a greater volume of profits. Normally, profits increase not in proportion to production, but when a higher rate of production occurs, so that overhead costs per unit decrease. Also, the higher wages that result from incentive plans improve employee morale and tend to reduce labor turnover, absenteeism, and tardiness.

In general, the harder the work is to measure, the more difficult it will be to install a successful wage incentive plan. Usually, it is not advantageous to install incentives unless the work can reasonably be accurately measured. Furthermore, it is usually not advantageous to introduce incentives if the availability of work is limited to less than 120 percent of normal. Also wage incentive plans can increase production costs and decrease total unit costs. Usually, however, they more than compensate for the increased costs of industrial engineering, quality control, and timekeeping that may result from their use.

One final caveat is that there is a definite trade-off between increasing the work pace with an incentive plan and increasing the risk of injury from repetitive motion, especially if the job or workplace has not been ergonomically designed. The authors have seen many instances, especially in the garment industry, in which jobs with low base rates but high incentives (so that to achieve a decent wage, the sewers must perform at very high rates—well above 150 percent) have experienced high injury rates. Undoubtedly a better designed job may decrease the injury rate. However, even with the best of conditions, high rates (over 20,000 hand motions per 8-h shift) may still lead to some injuries. Therefore, even neglecting worker health and safety issues, the standards engineer must decide whether the added medical costs, under today's escalating conditions, offset the gains obtained with a given incentive plan.

QUESTIONS

1. What are the three general categories under which the majority of wage incentive plans may be classified?
2. Differentiate between individual wage payment plans and group-type plans.
3. What is meant by fringe benefits?
4. Which company policies are included under nonfinancial incentives?
5. What are the characteristics of piecework? Plot the unit cost curve and operator earning curve for day work and piecework on the same set of coordinates.
6. Why did measured day work become popular in the 1930s?
7. How does IMPROSHARE differ from the Rucker and the Scanlon plans?
8. Define profit sharing.
9. Which three broad categories cover the majority of profit-sharing installations?
10. Under the cash plan, what does the amount of money distributed depend on?

11. What determines the length of the period between bonus payments under the cash plan? Why is it poor practice to have the period too long? What disadvantages are there to the short period?

12. What are the characteristic features of the deferred profit-sharing plan?

13. What is the basis for the "share and share alike" method of distribution?

14. What are the fundamental prerequisites of a successful wage incentive plan?

15. Why is it fundamental to keep time standards up to date if a wage incentive plan is to succeed?

16. What does unduly high performance indicate?

17. How would you go about establishing a climate to increase worker motivation?

18. What are some of the reasons for the failure of a wage incentive scheme?

PROBLEMS

1. In a single-product plant where IMPROSHARE was installed, 411 employees produced 14,762 product units over a 1-year period and recorded 802,000 clock hours. In a given week, 425 employees worked a total of 16,150 h and produced 348 units. What would be the hourly value of this output? What percentage bonus would each of these 425 workers receive? What would be the unit labor cost in hours for this week's production?

2. Analysts established a standard time of 0.0125 h per piece for machining a small component. A setup time of 0.32 h was also established, as the operator performed the necessary setup work on incentive. Compute the following:
   a. Total time allowed to complete an order of 860 pieces
   b. Operator efficiency, if job is completed in an 8-h day
   c. Efficiency of the operator who requires 12 h to complete the job

3. A 100 percent participation plan for incentive payment is in operation. The operator base rate for this class of work is $10.40. The base rate is guaranteed. Compute:
   a. Total earnings for the job at the efficiency determined in Problem 2(b)
   b. Hourly earnings
   c. Total earnings for the job, at the efficiency determined in Problem 2(c)
   d. Direct labor cost per piece from (a), excluding setup
   e. Direct labor cost per piece from (c), excluding setup

4. A rate of 0.42 min per piece is set for a forging operation. The operator works on the job for a full 8-h day and produces 1,500 pieces. Use a standard hour plan.
   a. How many standard hours does the operator earn?
   b. What is the operator's efficiency for the day?
   c. If the base rate is $9.80 per hour, compute the earnings for the day.
   d. What is the direct labor cost per piece at this efficiency?
   e. What would be the proper piece rate (expressed in dollars) for this job, assuming that the time standard is correct?

5. A 60/40 participation plan is used in a plant (base rate is guaranteed and operator receives 60 percent of proportional gain after exceeding 100 percent). The established time value on a certain job is 0.75 min, and the base rate is $8.80. What is the direct labor cost per piece when operator efficiency is as follows?

   a.   50 percent of standard
   b.   80 percent of standard
   c.   100 percent of standard
   d.   120 percent of standard
   e.   160 percent of standard

6.   In a plant where all the rates are set on a money basis (piece rates), a worker is regularly employed at a job for which the guaranteed base rate is $8.80. This worker's regular earnings are in excess of $88 per day. Due to the pressure of work, the operator is asked to help out on another job, classified to pay $10 per hour. This employee works 3 days on this job and earns $80 each day.
   a.   How much should the operator be paid for each day's work on this new job? Why?
   b.   Would it make any difference if the operator had worked on a new job for which the base rate was $8 per hour and had earned $72? Explain.

7.   An incentive plan employing a kicker is in use. Below standard performance the worker is guaranteed a rate of $6 per hour, and above standard performance the worker is paid $9.20 per hour. A job is studied and a rate of 0.036 h per piece is set. What is the direct labor cost per piece at the following efficiencies?
   a.   50 percent
   b.   80 percent
   c.   98 percent
   d.   105 percent
   e.   150 percent

8.   A company would like to evaluate two incentive schemes that take effect once the worker exceeds standard performance. In the first case, the benefits are split 50/50 between the worker and the company. In the second case, the worker receives a kicker up to 120 percent earnings and then maintains level performance up to 150 percent, after which all the earnings go to the worker.
   a.   Plot the normalized unit labor costs for each scheme.
   b.   Derive the equations for worker earnings and unit labor costs for each scheme.
   c.   Find the point at which the two plans break even.
   d.   Which do you think the company would prefer?

9.   A company would like to evaluate two incentive schemes that take effect once the worker exceeds standard performance. In the first case, the benefits are split 30 percent to the worker and 70 percent to the company up to 120 percent performance. If the worker exceeds 120 percent performance, all the earnings go to the worker. In the second case, all earnings beyond standard performance are split 50/50 between the worker and the company.
   a.   Plot the earnings for each scheme.
   b.   Derive the equations for worker earnings and normalized unit labor costs for each scheme.
   c.   Find the point at which the two plans break even.
   d.   Which do you think the company would prefer?

REFERENCES

Britton, Charles E. *Incentives in Industry*. New York: Esso Standard Oil Co., 1953.

Broderick, R., and D. J. B. Mitchell. "Who Has Flexible Wage Plans and Why Aren't There More of Them?" *IRRA 29th Annual Proceedings*, 1977, pp. 163–64.

Fay, Charles H., and Richard W. Beatty. *The Compensation Source Book*. Amherst, MA: Human Resource Development Press, 1988.

Fein, M. "Financial Motivation." *In Handbook of Industrial Engineering*. Ed. Gavriel Salvendy. New York: John Wiley & Sons, 1982.

Globerson, S. *Performance Criteria and Incentive Systems*. *Amsterdam*: Elsevier, 1985.

Lazear, E. P. *Performance Pay and Productivity*. Cambridge, MA: National Bureau of Economic Research, 1996.

Lokiec, Mitchell. *Productivity and Incentives*. Columbia, SC: Bobbin Publications, 1977.

U.S. General Accounting Office. *Productivity Sharing Programs: Can They Contribute to Productivity Improvement?* Gaithersburg, MD: U.S. Printing Office, 1981.

Von Kaas, H. K. *Making Wage Incentives Work*. New York: American Management Associations, 1971.

Zollitsch, Herbert G., and Adolph Langsner. *Wage and Salary Administration*. 2d ed. Cincinnati, OH: South-Western Publishing, 1970.

SELECTED SOFTWARE

DesignTools, (available from the McGraw-Hill text website at www.mhhe.com/niebel-freivalds), New York: McGraw-Hill, 2002.

Training and Other Management Practices

CHAPTER 18

KEY POINTS

- Train workers to minimize injuries and to reach the standard time more quickly.
- Use learning curves to adjust standards for new workers and small batches.
- Recognize and understand worker needs.
- Use job enlargement and job rotation to minimize repetitive injuries and increase worker self-esteem.

I n a 1954 survey on subjects found in industrial engineering curriculums, educators ranked motion and time study first in a list of 41 subject areas (Balyeat, 1954). A similar survey 10 years later of more than 8,700 nonclerical employees in industrial engineering in 250 large U.S. manufacturing companies found that industrial engineers still spent most of their time on work measurement (Anonymous, 1964). A more recent survey (Freivalds et al., 2000) of 61 practicing industrial engineers indicated that traditional work measurement topics (time study, standard data, work sampling) were no longer at the top of the list (although methods tools were). On the other hand, several nontraditional work organizational items (teamwork, job evaluations, and training) jumped to the top ten. Because of this demand, a greater emphasis has been placed on these topics in this book.

Another trend is the spread of industrial engineering techniques to all areas of modern business, including marketing, finance, sales, and top management as well as into the service sector and health management areas. To meet the demand, and to reap the benefits of training in this field more quickly, many industries have embarked on education programs conducted in their own plants on company time. For example, an extensive survey of over 5,300 U.S. companies revealed that 80 percent were providing formal training programs for first-line supervisors, and 42 percent of these training programs dealt with work simplification, methods, and lean manufacturing. Therefore, these topics are also emphasized in this chapter.

18.1 OPERATOR TRAINING

TRAINING APPROACHES

A company's labor force is one of its main resources. Without skilled workers, production rates would be slower, product quality would be poorer, and overall productivity would be lower. Therefore, once the new method is installed and the proper standard is set, the operators must be trained appropriately to follow the prescribed method and attain the desired standard. If this is done, the operators will have little difficulty in meeting or exceeding the standard. Many excellent sources of training material, programs, and consultants are readily available and are not discussed in detail here. However, it is important to be aware of some of the major options in training programs, such as those that follow.

On-the-Job Learning   Putting the operators directly on a new job without any training is a sink-or-swim approach. Although the company may think it is saving money, it definitely is not. Some of the operators will muddle along and eventually adapt to the new technique, theoretically "learning." However, they may learn the incorrect method and may not attain the desired standard. Or they may take a longer time to reach the proper standard. This means a longer learning curve. Other operators may watch or ask questions of their coworkers and eventually learn the new method. However, during this period, they will have slowed down both the other operators and overall production. Worse yet, the coworkers may be using an incorrect method, which would then be passed on to the new operator. In addition, the new operator may experience considerable anxiety during the whole learning process, which may hinder that process.

Written Instructions   Simple written descriptions of the correct method are an improvement over on-the-job learning, but only for relatively simple operations, or in situations where the operator is relatively knowledgeable of the process and only needs to adjust for minor variations. This assumes that the operator understands the language in which the instructions are written or has sufficient education to read well. In these days, with greater diversity in the workplace, neither can be assumed.

Pictorial Instructions   Still pictures or photographs used with written instructions have proved to be very effective in training operators. This also allows less-educated workers and non-native English speakers to acquire the new method more easily. Line drawings generally have an advantage over photographs in emphasizing specific details, omitting extraneous details, and allowing exploded views. On the other hand, photographs are easier to produce and store and are more true to life (Konz and Johnson, 2000), if properly exposed and focused.

Videotapes and DVDs   Movies can show the dynamics of the process, such as the interrelationships of motions, parts, and tools, much better than still pictures. Whether in the form of videotapes or DVD, movies are inexpensive and easy to produce and show. Furthermore, they allow the operator the freedom to control the time and rate of viewing, backing up if necessary and reviewing procedures. Also, both modalities can be easily stored, erased, and rerecorded.

Physical Training   Training involving physical models, simulators, or real equipment is best for complex tasks. This allows the trainee to perform the job activities under valid real-life conditions, experience emergency conditions under safe controls, and have performance monitored for feedback. Such physical training is best exemplified by the high-fidelity flight simulators for pilot training used by several airlines, and the simulated coal mine for roof bolting or continuous miner operator training at the Bureau of Mines Bruceton research facility near Pittsburgh, Pennsylvania.

One advantage of physical training is that the operators also undergo *work hardening* in the process; that is, they perform muscle exertions or wrist motions under controlled conditions and reduced frequencies, so that the body can gradually build up to the more extreme conditions found on the job. This procedure has been quite successful, for example, in reducing work-related musculoskeletal disorders for meat packers, mentioned by OSHA in its meatpacking guidelines (OSHA, 1990) and recommended by the American Meat Institute in its Ergonomics and Safety Guidelines.

18.2   THE LEARNING CURVE

Industrial engineers, ergonomists, and other professionals interested in the study of human behavior recognize that learning is time-dependent. Even the simplest operation may take hours to master. Complicated work may take days and even weeks before the operator achieves the coordinated mental and physical qualities enabling him or her to proceed from one element to another without hesitation or delay. This time period and the related level of learning form the learning curve, a typical one of which appears in Figure 18.1.

Once the operator reaches the flatter section of the learning curve, the problem of performance rating is simplified. However, it may not always be convenient to wait this long to develop a standard. Analysts may need to establish the standard early in the learning process, where the slope of the curve is greatest. In such cases, it is helpful to have available learning curves representative of the various types of work being performed. This information can be useful, both for determining the point in time at which it would be desirable to establish the standard, and for providing a guide as to the expected productivity level of the average operator.

By plotting learning curve data on logarithmic paper, analysts may linearize those data, making them easier to use. For example, plotting both the dependent variable (cycle time) and the independent variable (number of cycles) shown in Figure 18.1 on log-log paper results in a straight line, as seen in Figure 18.2.

The theory of the learning curve proposes that as the total quantity of units produced doubles, the time per unit declines at some constant percentage. For example, if analysts expect an 80 percent rate of learning, then as production

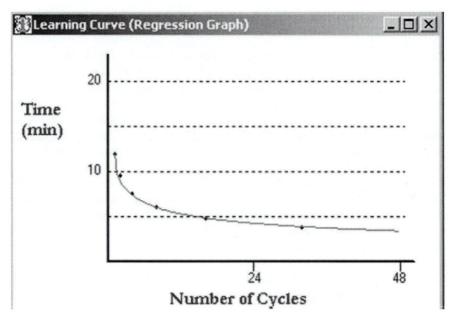

Figure 18.1   Typical productivity increase graph.

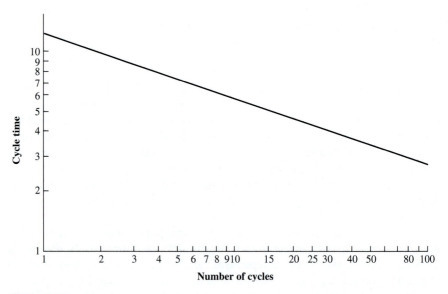

Figure 18.2   Estimated cycle times based on a 20 percent reduction each time the quantity doubles.

Table 18.1 Learning Data in a Tabular Form

Number of cycles	Cycle time (min)	Ratio to previous time
1	12.00	—
2	9.60	80
4	7.68	80
8	6.14	80
16	4.92	80
32	3.93	80

doubles, the average time per unit declines 20 percent. Table 18.1 illustrates the decline in the cycle time as the number of cycles increases; with successive doubling of the cycles, an 80 percent rate of improvement is realized. These are the same data as plotted in Figures 18.1 and 18.2. Typical rates of learning are as follows: large or fine assembly work (such as aircraft), 70 to 80 percent; welding, 80 to 90 percent; machining, 90 to 95 percent. Counterintuitively, 70 percent is the highest rate of learning, characteristic of every manual operation, while 100 percent would be no learning, for fully automated operations.

When linear graph paper is used, the learning curve is a power curve of the form $y = kx^n$. On log-log paper, the curve is represented by

$$\log_{10} y = \log_{10} k + n \times \log_{10} x$$

where y = cycle time
x = number of cycles or of units produced
n = exponent representing the slope
k = value of first cycle time

By definition, the learning in percent is then equal to

$$\frac{k(2x)^n}{kx^n} = 2^n$$

Taking the log of both sides gives

$$n = \frac{\log_{10} (\text{learning percent})}{\log_{10} 2}$$

For 80 percent learning,

$$n = \frac{\log_{10} 0.80}{\log_{10} 2} = \frac{-0.0969}{0.301} = -0.322$$

Also, n can be found directly from the slope:

$$n = \frac{\Delta y}{\Delta x} = \frac{\log_{10} y_1 - \log_{10} y_2}{\log_{10} x_1 - \log_{10} x_2} = \frac{\log_{10} 12 - \log_{10} 4.92}{\log_{10} 1 - \log_{10} 16} = -0.322$$

Note that k is 12, and the final equation for the learning curve is

$$y = 12x^{-0.322}$$

Table 18.2 presents the slopes of the common learning curves as a function of the learning percentages. Example 18.1 should help clarify the use of the learning curve.

Table 18.2 Relationship of the Slope of the Learning Curve to the Learning Curve Percentage

Learning curve percentage	Slope
70	−0.514
75	−0.415
80	−0.322
85	−0.234
90	−0.152
95	−0.074

EXAMPLE 18.1

Calculation of Learning Curve

Assume that it takes 20 min to produce the 50th unit and 15 min to produce the 100th unit. What is the learning curve?

$$n = \frac{\Delta y}{\Delta x} = \frac{\log_{10} 20 - \log_{10} 15}{\log_{10} 50 - \log_{10} 100} = \frac{1.301 - 1.176}{1.699 - 2.000} = -0.4152$$

The learning curve percentage is

$$2^{-0.4152} = 75\%$$

To complete the learning curve equation, we substitute one of the data points, such as (20, 50), into the equation and solve for k:

$$k = y/x^n = 20/50^{-0.4152} = 101.5$$

Thus, the analyst's costs for the first units produced would be based upon 101.5 min of time to produce one assembly, not the 10 min developed from standard data or predetermined time systems.

The analyst may next wish to determine how many cycles are needed to reach a specific time, for example, a standard time of 10 min. Substitute $y = 10$ min into the learning equation, take the logs of both sides, and solve for x, and we get

$$10 = 101.5\, x^{-0.4152}$$
$$\log_{10}(10/101.5) = -0.4152 \log_{10} x$$
$$\log_{10} x = (-1.006)/(-0.4152) = 2.423$$
$$x = 10^{2.423} = 264.8 \approx 265 \text{ cycles (always round up)}$$

Thus, it would take the worker 265 cycles to reach the standard time.

Next, the analyst may desire to know how long it takes in actual time to reach a standard time of 10 min. This is the area under the learning curve, which may be found by integrating under the curve:

$$\text{Total time} = \int_{x_1 - 1/2}^{x_2 + 1/2} kx^n dx = k[\,(x_2 + 1/2)^{n+1} - (x_1 - 1/2)^{n+1}\,]/(n+1)$$

$$= 101.5(265.5^{0.5848} - 0.5^{0.5848})/0.5848 = 4.424 \text{ min}$$

Thus, for Example 18.1, it takes a total of 4,424 min, or approximately 73.7 h, to reach a cycle time of 10 min. The average cycle time would be 4,424/265 = 16.7 min.

Note that there are two learning curve models. The one presented is the *Crawford model*, also known as the *unit model*, because in this model the improvement gained is for a particular unit (Crawford, 1944). The other model is the original *Wright model* developed by Wright (1936) for the aircraft industry and also known as the *cumulative average model*, because in this model the improvement gained is for the cumulative average unit rather than for a specific unit. The cumulative average value is necessarily greater than the unit cost for the nth production unit; however, as the number of units increases, the models will converge. Some analysts consider the Crawford model to be more practical because it does not mask the individual variability as much as the Wright model does.

An interesting question involves what happens when the operator takes a vacation. Does the operator forget some of the learning? This does in fact happen and is known as *remission* (Hancock and Bayha, 1982). The amount of remission is a function of the operator's position on the learning curve when the break occurs. The amount of remission is approximated by extrapolating a straight line from the time of the first cycle to the standard time (see Figure 18.3). The equation for this remission line is

$$y = k + \frac{(k - s)(x - 1)}{1 - x_s}$$

where s = standard time
x_s = number of cycles to standard time

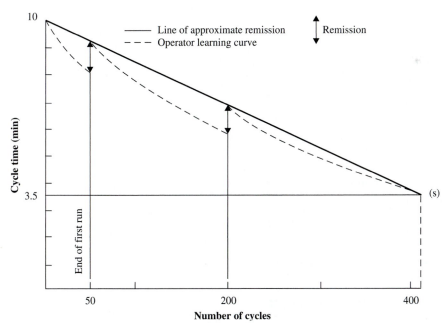

Figure 18.3 The effect of breaks on operator learning
(*From:* Hancock and Bayha, 1982) (Reprinted by permission of John Wiley & Sons, Inc.)

Being able to estimate the time for the first unit produced and the time for successive units can be extremely helpful in estimating relatively low quantities, if the analyst has the standard data and learning curve information. Since standard data are usually based on worker performance when learning has leveled off or reached the flat portion of the learning curve, the data need to be adjusted upward to ensure that adequate time is allowed per unit under low-quantity conditions. For example, let us assume that the analyst wants to know the time needed to produce the first unit of a complex assembly. The standard data analysis suggests a time of 1.47 h, which is the cycle time for the nth unit, or the point where the learning curve begins to flatten. The nth unit, in this case, is estimated to be 300 assemblies. Based on other similar jobs, the analyst expects a 95 percent learning rate. From Table 18.2, the exponent n, representing the slope, is -0.074. Then k, the value for the first cycle time, is

$$k = 1.47/300^{-0.074} = 2.24 \text{ h}$$

EXAMPLE 18.2	Calculation of Learning Curve with Remission

In Example 18.1, the operator stops after 50 cycles for a two-week vacation. His cycle time for the 51st cycle will be determined from the remission function

$$y = 101.5 + \frac{(101.5 - 10)(51 - 1)}{1 - 265} = 84.17$$

The operator's cycle time without a break would have been

$$y = 101.5x^{-0.4152} = 101.5 \times 51^{-0.4152} = 19.84$$

Therefore, there has been a remission of $84.17 - 19.84 = 64.33$ min, and a new learning curve with a new $k = 84.17$ starts. The 51st cycle now becomes the first cycle of the new learning curve of $y = 84.17x^{-0.4152}$.

Thus, the analyst's costs would be based on 2.24 h of time to produce one assembly, not the 1.47 h developed from the standard data.

Many factors affect human learning. Job complexity is very important. The longer the cycle length, the greater the uncertainty in movements, and the more C-type or simultaneous motions (see Chapter 13)—the more training will be required. Similarly, individual capabilities, such as age (rate of learning declines with age), prior training, and physical capabilities all affect the ability to learn.

18.3   EMPLOYEES AND MOTIVATION

EMPLOYEE REACTIONS

Industrial engineers must have a clear understanding of the psychological and sociological reactions of employees' attitudes toward the methods, standards, and wage payment. Three points should always be recognized:

1. Most people do not respond favorably to change.
2. Job security is uppermost in most workers' minds.
3. People have a need for affiliation and are consequently influenced by the group to which they belong.

Most people, regardless of their positions, have an inherent resistance to changing anything associated with their work patterns or work centers. This is due to several psychological factors. First, change indicates dissatisfaction with the present situation. But the natural tendency is to defend the present way, since it is intimately associated with the individual. No one likes others to be dissatisfied with his or her work; if a change is even suggested, the immediate reaction is to oppose the proposed changes.

Second, people tend to be creatures of habit. Once a habit is acquired, it is difficult to give up, and there is resentment if someone endeavors to alter the habit. For example, anyone in the habit of eating at a certain place is reluctant to change to another restaurant, even though the food may be better and less expensive.

Third, people naturally desire security in their position, which is just as basic as the instinct for self-preservation. In fact, security and self-preservation are related and form the second level of Maslow's hierarchy of human needs (see next section). Most workers prefer job security over high wages when choosing a place to work.

Fourth, to the worker, all methods and standards changes appear to be an effort to increase productivity. The immediate and understandable reaction is to believe that if production goes up, the demand will be filled in a shorter period; and without demand, there will be fewer jobs.

The solution to the need for job security lies principally in the sincerity of management. When methods improvement results in job displacement, management is responsible for making an honest effort to relocate those who have been displaced. This may include providing for retraining. Some companies have gone as far as to guarantee that no one will lose employment as a result of methods improvement. Since the labor turnover rate is usually greater than the improvement rate, the natural attrition through resignation and retirement can usually absorb any people displaced as a result of improvement.

Fifth, the sociological need for affiliation and the resulting impact of "behaving as the group wants everyone to behave" also influence change. Frequently, the worker, as a union member, feels that he or she is expected to resist any change that has been instituted by management; consequently, the worker is reluctant to cooperate with any contemplated changes resulting from methods and standards work. Another factor is the resistance toward anybody who is not part of one's own group. A company represents a "group" that has several groups within its major boundaries. These individual groups respond to basic sociological laws. Change proposed by someone outside one's own group is often received with open hostility. The worker is associated with a different group than that of the methods analyst and tends to resist any effort from analysts that might interfere with the usual performance within the individual's group.

MASLOW'S HIERARCHY OF HUMAN NEEDS

Psychosocial factors such as stress, needs, or rewards can be very important aspects of worker productivity. Workers naturally want to work with the least amount of stress and the greatest amount of rewards. Maslow (1970) quantified these wants into a hierarchy comparable to a set of steps leading to the top of a pyramid, or the ultimate goal (see Figure 18.4). Each lower want or need must be satisfied before a worker will seek rewards at the next-higher level. The lowest level includes the *physiological needs* corresponding to survival, food, water, and health. Job-related factors at this level could be sufficient pay or other monetary rewards.

Once these physiological wants are satisfied, the second level, *safety needs*, becomes important. Safety needs include the need for security, in both the physical and psychological sense. These could be as simple as trying to avoid physical injury on the job, or as complex as seeking a "nice" supervisor who doesn't threaten or demean the worker. With the prevalence of company downsizing starting in the late 1990s, safety needs could include job security and seniority rights.

The third level, *social needs*, includes the need for attention, friendship, social belonging, and meaningful relationships with coworkers. In the fourth

Figure 18.4   Maslow's Hierarchy of Human Needs.

level, *self-esteem needs*, workers strive for competence and achievement, express a desire for self-respect, or seek to satisfy their egos.

At the top of the pyramid, the final or fifth level is *self-fulfillment*. The workers have finally achieved all their needs, they are personally fulfilled, and their egos are satisfied. This level can vary considerably from individual to individual. Where some people may be satisfied making widgets day in and day out, others may only be satisfied running their own businesses.

The industrial engineer may wonder what purpose Maslow's hierarchy serves on the plant floor, or how these wants can be satisfied for the production worker. Consider the first level of physiological needs. One tactic, though very negative in terms of labor–management relationships, is the threat of termination for failure to meet production quotas, or for violation of safety rules. Other scare procedures or hard-sell approaches fall in the same category. A more positive approach is the implementation of wage incentives (Chapter 17). This is classic conditioning, or positive reinforcement, at its simplest. Many workers are willing to work at relatively tedious jobs, or at higher production rates, given sufficient monetary incentive. Thus, workers trade increased satisfaction with the job, provided by the extra pay, for decreased satisfaction on the job. Unfortunately, with increasing wealth and progressive income taxes, additional income becomes less meaningful, and the industrial engineer may need to proceed to higher levels of Maslow's hierarchy.

At the second level, safety or security needs, overall job security is the concern, especially with the increasing trend toward downsizing or right-sizing. Traditionally, in other cultures, especially in Japan, a job was a lifetime guarantee with that company. In the United States, it is not unusual for a worker to change jobs every 5 to 6 years and work for a half dozen employers during his or her lifetime. Employment can perhaps be guaranteed for a fixed number of years. At the worksite level, specific regulations regarding work practices, physical guarding of unsafe machines, or safety contests can improve the overall safety or the working climate.

At the third level, social needs, workers seek "belonging" within a social system. In terms of work, this could imply having friendly coworkers, comfortable interaction with management, participation on ergonomics or safety committees, and so on. Such formal organizations are much more common in Japan with its quality circles; in Germany, where workers elect a work council (*Betriebsrat*) to handle grievances and negotiate with management; and in Sweden, where auto plants have work groups (*arbetsgrupper*).

At the fourth level, the workers seek an increase in self-esteem. This could be provided by making the work more challenging, adding more responsibility, and providing greater variety. The latter can be done through *job enlargement*, a horizontal expansion of work. Instead of just tightening one set of bolts all day, the worker could perform the complete assembly. This not only increases the worker's sense of responsibility, but also utilizes a variety of his or her muscles and joints, dividing the work stress over a larger part of the body and thus reducing the risk of cumulative trauma disorders. Tied into job enlargement is also *job enrichment*, a vertical expansion of work, which allows workers both to start and to complete a given task, diversifying duties so that no one person has all of a boring task, delegating decision making, and rotating job assignments. *Job rotation* is similar to job enlargement in that any worker gets the opportunity to do a variety of tasks, while adhering to a more rigid schedule. Job rotation has effects similar to job enrichment in varying job stressors and allowing fatigued muscles and body parts to recover.

VOLVO APPROACH

All these concepts (job enlargement, job enrichment, job rotation, and work groups) were pioneered in the 1960s in Sweden. The impetus was increasing absenteeism, wildcat strikes, worker unrest, and general employee dissatisfaction. Drastic changes had to be made. Therefore, under the direction of its president Pehr Gyllenhammer, Volvo devised a revolutionary plan and built a completely new auto assembly plant at Kalmar in 1974. The traditional conveyor line was replaced by an automated guided vehicle (AGV) system on which the car assembly took place. The AGV was guided by an electronic system of cables embedded in the floor. A central computer controlled the movement of the AGVs throughout the plant, but could be overridden by the employees at any time. In addition, there was a drastic change in work organization: employees were fully involved and formed work groups that received and examined factory orders, decided exactly which group member would do which task for the given day, inspected their own work, completed paperwork after assembly, and, at the end of the day, had a brief discussion of the day's happenings and problems. Job enlargement was carried out to the highest degree in that one group of workers assembled over 25 percent of one car.

The Kalmar design was successful from the beginning, as the work became more meaningful and workers assumed more responsibility. Absenteeism and employee turnover were greatly reduced, while cost and production targets were met. Because of the success at Kalmar, similar new plants were opened at Uddevalla and Gteborg (Torslunda). Unfortunately, due to a shifting market and radically lower sales figures, Volvo eventually closed the Uddevalla and Kalmar plants. In 1997, the Uddevalla plant reopened with the production of a new sports car.

Note that all three forms of job reorganization—job enlargement, job enrichment, and job rotation—were in place in the Volvo plants. Cycle times increased to many hours, decreasing the repetitiveness of motion for any one limb or set of muscles.

At the fifth and highest level of Maslow's hierarchy, the worker would be expected to devote himself or herself completely to the company. Other than in Japan, this is probably not feasible in any large-scale company. On the other hand, in small, start-up companies, not only the owner, but also some of the closest colleagues may put in most of the waking hours in keeping the company afloat. Then, the company and work truly become one's self-actualization.

MOTIVATION

An interesting *motivation maintenance theory* was developed by Herzberg (1966), based on a survey of factors leading to satisfaction or dissatisfaction for 1,500 employees in 12 different organizations. Similar to Maslow, Herzberg found two basic but different needs in individuals. If workers were dissatisfied with their jobs, their main concern was the working environment. However, if they were satisfied with their jobs, their satisfaction dealt with the actual work itself.

Herzberg classified the environmental factors as *extrinsic* and potential dissatisfiers. These included such factors as the administration, supervision, working conditions, salary, and interpersonal relations. The potential satisfiers or motivators, which included achievement, recognition, responsibility, and advancement, he termed *intrinsic factors*. The extrinsic factors had little positive effect, but could be strong dissatisfiers, leading to large negative feeling. The intrinsic factors encouraged workers to be more productive and satisfied. Therefore, it is in the manager's interest to maximize the intrinsic factors and minimize the negative effects of the extrinsic factors.

One of the most effective intrinsic motivation techniques is job enrichment, which is the opposite of job simplification. With work methods and the principles of motion economy, the typical goal of an industrial engineer is job simplification. If the job is simple and repetitive, little learning is required and workers can be easily interchanged. This approach was developed for the machinelike consistency required on an assembly line. However, workers are not machines, and when subjected to such conditions, they may become bored and dissatisfied, leading to increased absences and job changes. Even worse, as recent statistics show, are increased stress levels leading to increases in cumulative trauma disorders. It is not worth saving pennies on more repetitive jobs when thousands of dollars are lost in the resulting injuries.

Herzberg also found some interesting deviations in the survey results, depending on the populations examined. These could be used to a company's advantage, depending on the composition of the worker population. For example, younger workers were less concerned about job security than older workers and were generally more satisfied with the whole organizational reward system. More highly educated and more highly paid workers favored the intrinsic rewards. Extrinsic rewards ranked higher overall than intrinsic rewards, but were most prized by less-educated, lower-paid, and older workers.

18.4 HUMAN INTERACTIONS

Interactions between employees at the workplace are an important component of morale and productivity. Several approaches can be used to deal and communicate with people, two of which are discussed here: transactional analysis and the Dale Carnegie approach.

TRANSACTIONAL ANALYSIS

Transactional analysis, as developed by Berne (1964), consists of several components: (1) ego states, (2) transactions, (3) stroking and stamps, and (4) more complex games and lifestyles. There are three *ego states,* which are found to some degree in all people at all times. The parent *ego state* reflects attitudes and values absorbed from parents as authority figures and produces a statement such as "That's really a dumb mistake." Bill Cosby of television's *Cosby Show* would be a good example of the parent ego state. The *adult ego state* logically analyzes facts, makes rational decisions and conclusions, and operates with phrases such as "Let's examine that problem carefully." Mr. Spock of the *Star Trek* series would be a perfect example of the adult ego state. The *child ego state* is more complex and may take up to three different forms. A naive state produces responses such as "Oh, I didn't know that." The adaptive state establishes internal rules based on social conditioning, such as "Respect your elders." The manipulative state may fake injuries to get out of something unpleasant, such as pretending to have a cold to get out of school.

Interactions between the ego states occur in the form of *transactions*. Participants can both send and receive messages from any of the three ego states. If the messages are sent and received at the same ego state level, such as adult to adult, the transactions are termed *complementary* and are considered to result in a positive and successful exchange (Figure 18.5). A parent-to-child transaction (Figure 18.6), if occurring at a parallel level, is still considered complementary, but may not be as effective as a transaction occurring at the same level.

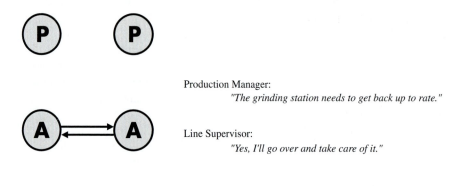

Production Manager:
   "The grinding station needs to get back up to rate."

Line Supervisor:
   "Yes, I'll go over and take care of it."

Figure 18.5   Complementary transaction: adult to adult—the message is sent and received appropriately.
(Adapted from Berne, 1964)

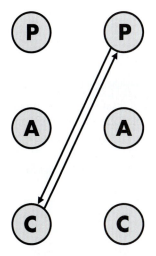

Production Manager (with a troubled facial expression):
"I was wondering if you could get those grinders back up to rate."

Line Supervisor (in a patronizing tone):
"Now don't you worry, I'll take care of it."

Figure 18.6 Complementary transaction: parent to child—this is not as effective as the adult-to-adult complementary transaction, but it is still useful. (Adapted from Berne, 1964)

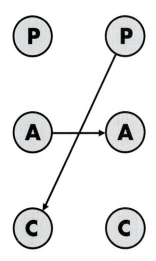

Production Manager (as adult to adult):
"What has been done to correct the material handling problem?"

Line Supervisor (as parent to child):
"Don't you have anything better to do but bug me all the time?"

Figure 18.7 Crossed transaction—this can result in anger and hostile feelings. (Adapted from Berne, 1964)

A *crossed transaction* occurs when each party assumes a different transaction level, and this often results in anger or hostile feelings (Figure 18.7). *Ulterior transactions*, although appearing logical on the surface, always have a hidden meaning, and they form the basis for games (Figure 18.8). As an example, a line supervisor conducts an adult-to-adult transaction on the surface, but in reality only goes through the motions and produces a parent-to-child transaction. This may be

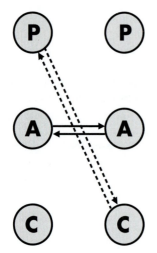

Production Manager (in apparent adult to adult transaction):
> *"I'll find out what's causing that machine to jam up and get back to you."*

But, is really thinking:
> *"Oh, all right, I'll check on it and get back to you."*

Operator hears a condescending tone and sees it as a parent to child transaction.

Figure 18.8   Ulterior transaction—although this approach appears logical, it can have a hidden meaning and form the basis for games. (Adapted from Berne, 1964)

how an operator wants to be treated; if not, the line supervisor should not be surprised when the employees complain that nobody ever listens to them.

Transactional analysis stresses that all people feel the need to be recognized in some manner. This need (the fourth step in Maslow's hierarchy) probably starts in childhood and continues well into adulthood. Recognition can come as either *positive* or *negative strokes*, positive being based on good attributes, such as recognition for intelligence, helpfulness, and compassion, while negative is based on bad attributes, such as deceit and selfishness. Only positive strokes ("you're OK") keep a person mentally healthy. Negative strokes may leave a person with a chip on the shoulder and a bad view of the world. Excessive negative stroking (criticism) in childhood may carry into adulthood, with the person seeking transactions leading to sympathy or dependency. Some individuals may become extreme positive stroke seekers.

As the transactions become more complex, they take the form of rituals, pastimes, and games. Rituals are the simplest cultural ties, such as simple morning greetings, "Hi, how's it going?" Pastimes are more complex interactions such as conversations regarding work, sports, or friends in social functions. Games are the most complex transactional interactions, which may replace intimacy in private life or may produce accident-prone behavior (a child seeking forgiveness) at work.

In general, the industrial engineer or manager should try to have a basic understanding of transactional analysis, to interact better with production workers and other personnel. Complex games should be avoided by switching from a parent or child ego to the adult ego. This works in "yes, but" situations in which, because of an ulterior motive for one of the participants, the problem-solving effectiveness of the situation is reduced. The manager should sense that when all suggestions for improving the workstation design are rejected with "yes, but can't" comments, the transaction has proceeded into a parent-to-child mode. Switching to an adult-to-adult transaction with "Yes, that is indeed difficult—what can you do about it?" will short-circuit the game and get directly to the problem. In other words, in crossed transactions, it is better to change egos, even if they are at a parent-to-child level and are thus less effective than adult-to-adult transactions. Finally, it may even be necessary to participate at a lower level of games, such as giving or receiving strokes. In many companies, there is a "Calamity Jane or Joe" who is always involved in one sort of problem or another, whether jamming a machine or damaging a tool. These individuals may have carried a quest for negative strokes and forgiveness through from childhood into adulthood. Switching to an adult ego with "I accept responsibility for assigning you to that job" would eventually stop the game, but may also create an enemy. Another approach may be to counteract the negative strokes by providing more positive strokes in the form of recognition of the good things done by that operator (e.g., above average performance, high quality, etc.) (Denton, 1982).

Above all, the industrial engineer should take time to talk to the operators and get their ideas and reactions. The work progresses much more smoothly and effectively if operators become part of the team. However, they must be asked, not instructed, to "join the team." The operators are closer to their job situations than anyone else and usually have more specific knowledge of details than anyone else. This knowledge should be realized, respected, and utilized. Accept operators' suggestions gratefully; if they are practical and worthwhile, put them into effect as soon as possible. If they are used, be sure that the operators are appropriately rewarded. If they cannot be used at present, give a complete explanation as to why they cannot be used. At all times, analysts should imagine themselves in the workers' place and then use the approach they would like used toward themselves. Friendliness, courtesy, cheerfulness, and respect, tempered with firmness, are the human characteristics that must be practiced to be successful in this work. In short, the golden rule must be applied.

DALE CARNEGIE APPROACH

The human approach to handling people, making people like you, influencing the thinking of people, and changing people, has been developed to a fine art by Dale Carnegie (1936) in his series of courses, still popular today. Carnegie's principles and thoughts are summarized in Table 18.3.

18.5 COMMUNICATIONS

Industrial engineers as middle-level managers spend a considerable amount of their time in interpersonal communications. Therefore, mastery of the ability to communicate effectively goes a long way toward selling an argument or a design,

Table 18.3 The Dale Carnegie Approach

Fundamental Techniques in Handling People

1. Instead of criticizing people, try to understand them.
2. Remember that all people need to feel important; therefore, try to figure out the other person's good points. Forget flattery; give honest, sincere appreciation.
3. Remember that all people are interested in their own needs; therefore, talk about what they want and show them how to get it.

Six Ways to Make People Like You

1. Become genuinely interested in other people.
2. Smile.
3. Remember that a person's name is to him or her the sweetest and most important sound in the English language.
4. Be a good listener. Encourage others to talk about themselves.
5. Talk in terms of the other person's interest.
6. Make the other person feel important—and do it sincerely.

Twelve Ways to Win People to Your Way of Thinking

1. The only way to get the best of an argument is to avoid it.
2. Show respect for the other person's opinions. Never tell anyone they are wrong.
3. If you are wrong, admit it quickly and emphatically.
4. Begin in a friendly way.
5. Get the other person saying yes, immediately.
6. Let the other person feel that the idea is his or hers.
7. Let the other person do a great deal of the talking.
8. Try honestly to see things from the other person's point of view.
9. Be sympathetic with the other person's ideas and desires.
10. Appeal to the nobler motives.
11. Dramatize your ideas.
12. Throw down a challenge.

Nine Ways to Change People Without Giving Offense or Arousing Resentment

1. Begin with praise and honest appreciation.
2. Call attention to people's mistakes indirectly.
3. Talk about your own mistakes before criticizing the other person.
4. Ask questions instead of giving direct orders.
5. Let the other person save face.
6. Praise the slightest improvement and praise every improvement. Be hearty in your approbation and lavish in your praise.
7. Give the other person a fine reputation to live up to.
8. Use encouragement. Make the fault seem easy to correct.
9. Make the other person happy about doing the thing you suggest.

even though it may be worthy on its own merit. Communications can be divided into five major types: verbal, nonverbal, one-to-one, small group, and large audience (Denton, 1982).

VERBAL COMMUNICATIONS

In verbal communications, words are very powerful and their meanings become very important. Thus, the word *production* is all-powerful, while others terms such as *safety* or *human factors* may carry a negative connotation, because they imply, whether true or not, coddling the workers or slowing down the production. A person's name (and family member names) is very important to that person. Therefore, managers should know the workers' names (and a little about their background), to stimulate the interest of the opposing party and to make the conversation more rewarding.

One problem in any language is the specific meaning of a given word. With greater diversity in the workplace, there is a greater chance that the other individual may assign a slightly different meaning, make a different inference, or perhaps not even understand the meaning of some words.

Managers must also be careful not to dichotomize the world. Categorizing things as either good or bad, safe or unsafe, polarizes events and causes individuals to concentrate on differences rather than similarities.

NONVERBAL COMMUNICATIONS

Some data indicate that more than 50 percent of a message, especially related to feelings, is presented through nonverbal channels, including voice characteristics, facial expressions, and body language. In voice characteristics, a rapid speech pattern indicates excitement, while a slower rate with pauses indicates passive emotion. Facial expressions and body language involve such nonverbal behavior as head nodding to indicate attention to the other person's discourse, raising of the eyebrows to indicate surprise, maintaining eye contact to indicate trust, crossing of arms or clenching of fists to indicate a defensive attitude, crossing of the legs to indicate superiority or a lack of involvement.

Other factors, such as the amount of space around the individual, can also affect communications. For example, people try to maintain a certain amount of open space around them; closing this space forces a greater amount of discomfort even though it may increase interaction.

ONE-TO-ONE

One-to-one, or *dyadic*, communications occur frequently between a manager and a worker in a face-to-face situation. The purpose of such communication is generally to bring about an understanding of the goals between the two individuals. One of the two may then seek to obtain approval for a proposed idea, and available solutions may need to be presented. To obtain the expected solution, it may be necessary to use motivational techniques, such as guided questioning, which can be either leading questions that guide the answers in a certain direction, or closed-end (yes/no or limited choice) questions to elicit commitment, or open-ended questions to elicit discussion.

Unfortunately, conflicts can arise during the conversation. Simple conflicts arise when each side knows the other's goals, but neither can win without the other failing. In such cases, delaying further discussion until both sides can cool down and find a rational solution may be appropriate. Pseudoconflicts arise because of ineffective communication and can only be diffused when accurate data are provided and distortions eliminated. The worst conflicts are ego conflicts that relate to the previously discussed Berne's (1964) transactional analyses.

SMALL GROUPS

Typically, small-group communications are centered on problem solving. Problems can be quite complex, and no one individual may have all the solutions. Therefore, the concept of having a group of individuals working on a problem seems logical. Additional benefits are that extreme individual judgments tend to be moderated, overall judgments tend to improve in accuracy, and a wider range of information or opinions is included in the discussions. There are also trade-offs. By their very nature, small groups are time-consuming. Also, lack of coordination, low motivation, and personality conflicts among the group members can result in the failure of the group to meet its objectives. Therefore, it is important to organize and administer small groups effectively.

The basic problem-solving procedures (e.g., fish diagrams) must be followed in small-group communications. The group must identify the problem, analyze the details, develop a variety of ideas, select specific ideas for further development, evaluate the different alternatives, and then specify and sell the solution. To improve this process, the group's facilitator must allow easy access to information to encourage the building of trust. High standards and proper planning are also very important, as are specific interactional techniques that can increase the effectiveness of the process.

Facilitating Agreement   Facilitating agreement among all members, such as obtaining a consensus, can be improved by positively involving all group members, reinforcing their self-esteem, using open-ended questions, summarizing each individual's comments before proceeding to the next person, and summarizing the pros and cons of a discussion before proceeding to the next topic.

Role Playing   Role playing can help strengthen a group's problem-solving ability by presenting appropriate situations or events. This can be followed by further participation and discussion through *buzz groups* (smaller subgroups). One group member may act as recorder to write down the group's ideas quickly. This can easily lead to *brainstorming* sessions, for which these are the basic guidelines: (1) ideas are encouraged, regardless of how wild; (2) the more ideas, the better; (3) no ideas are criticized (sometimes the contributors are not identified); and (4) participants are encouraged to build on or combine previous ideas. Usually, a time limit of 10 min is set, after which the ideas are ranked, with possible solutions being included. The pros and cons of each idea are discussed, and the potential solutions are voted upon. The top vote getters are further reviewed and voted upon again, until the process of elimination leaves only the best solutions (Denton, 1982).

Quality Circles *Quality circles* are a small-group format developed in Japan in 1963 to assist in solving quality control problems. The essence is participative problem solving in groups of 8 to 10 people, including workers, engineers, and managers. It is important to have participants from the different departments involved with this product. These volunteers are given special training in statistical quality control techniques and typically hold meetings once or twice a month. With the help of a facilitator, the group selects a problem that is a cause of product defects and could potentially have a solution. Typically, exploratory operational tools, such as Pareto distributions and fish diagrams (Chapter 2), are used to help in identifying the problem and the factors involved. The group then recommends potential solutions, such as improved procedures or design changes, and it attempts to implement the solution. All this is done with the cooperation of management (Konz and Johnson, 2000).

Ergonomics Teams A logical extension of quality circles, to combat the high rates of musculoskeletal disorders in U.S. companies, is the ergonomics team. These are typically interdisciplinary teams consisting of an ergonomist (if there is one on staff), an industrial engineer, a safety specialist, a medical person (typically, the plant nurse), several interested production workers, a labor union member, and perhaps a representative from higher management. This committee typically meets once or twice a month and follows a procedure similar to that used by quality circles in seeking solutions to problem-causing jobs. In the authors' experience, many companies, including such large companies as auto manufacturers and smaller companies of fewer than 500 employees, have experienced considerable success in using such teams.

LARGE AUDIENCES

Industrial engineers or middle managers rarely present information to large groups, and this topic will not be considered here. Considerable information on producing presentations and using effective delivery techniques for large audiences is available in other sources.

LABOR RELATIONS

Every business owner recognizes the importance of harmonious labor–management relations. The lack of consideration of the human element in work measurement procedures may cause turmoil and reduce the profitability of operations. Management should identify and implement conditions most likely to enable employees to achieve the organization's objectives.

To understand the relationship between work measurement and labor relations, analysts must understand the principal objectives of the typical labor union, which one to secure for its members higher wage levels, decreased working hours per workweek, increased social and fringe benefits, improved working conditions, and job security. The philosophy underlying the union movement has, in the past, had much to do with opposition to incentive systems. Unions attempted to unite workers by seeking ends common to all members, and was not to the advantage of the early unions to emphasize differences in workers' abilities and interests. Consequently, organized labor usually sought percentage wage increases for all members of a group, rather than means by which pay could be adjusted to the worth of the individual worker. Work measurement analysts came to be looked upon as means by which management sought to destroy the solidarity of workers by stressing the differences in their capabilities.

With the recent changes to a more competitive nature of manufacturing, with more foreign labor and outsourcing, and with a more fluid movement of capital and production facilities, labor unions have lost much of their previous power. Unions, therefore, are less combative units than bodies concerned with the orderly negotiation of wage contracts for their members. To satisfy most of their members, unions must obtain equitable wages (recognizing workers' different skills and qualities), as well as high wages, for all. In fact, unions have already done this in many instances.

Many unions today train their own work measurement personnel. In most instances, however, these time study people are employed to check standard times and to explain them to workers, rather than to take part in their initial establishment. In many instances, company training of union time study representatives has been quite successful as a means of promoting a more cooperative atmosphere in the installation and maintenance of methods, standards, and wage payment systems. This procedure provides for the joint training of both company and union personnel. Having received this training, union representatives are much more qualified to evaluate the fairness and accuracy of the technique and to discuss any technical points relative to a specific case.

18.6 MODERN MANAGEMENT PRACTICES

LEAN MANUFACTURING

Because of its emphasis on a tight and efficient manufacturing process using methods similar to those described in previous chapters while enlisting the active participation of the worker, the Toyota Production System (TPS) deserves special mention. The Toyota Production System was developed by Toyota Motor Corporation as a means of eliminating waste in the aftermath of the 1973 oil embargo. Its primary purpose is the improvement of productivity and the reduction of costs following the footsteps of the Taylor system of scientific management and the Ford mass assembly line. Yet, it is much broader in concept, targeting not only manufacturing costs, but also sales and administrative and capital costs. Toyota felt that it would be dangerous to follow the Ford mass production system blindly, which worked fine in times of high growth. In times of lower growth, it was important to give greater attention to cutting waste, decreasing costs, and increasing efficiency. In the United States this TPS approach has been termed *lean manufacturing*.

TPS highlights seven types of *muda* or waste (Shingo, 1981): (1) overproduction, (2) waiting for the next step, (3) unnecessary transportation, (4) overprocessing,

(5) excess inventory, (6) unnecessary motion, and (7) defective products. These are very similar to the operations analysis techniques and methods study approaches presented in Chapters 2 and 3. For example, waiting and transportation are elements examined directly with flow process charts for potential elimination or improvement through more efficient layouts and better material handling. Overprocessing means that energy is being expended by operators and machines with little value added to the product and goes back to the basic questions asked in operations analysis in Chapter 3. Wasted motion summarizes the Gilbreths' lifelong work in motion study, culminating in the principles of work design and motion economy. It also includes gross movements of the operators, which can be minimized through more efficient layout of the workstation or facilities. The wastes of overproduction and inventory are based on common sense in the additional storage requirements and material handling requirements to move items in and out of storage in addition to lighting, heating, and maintenance costs. Finally, the waste of defective products is obvious, requiring rework. Rather than focusing on an acceptable quality level found in mass production, lean manufacturing emphasizes just-in-time production which necessitates a zero-defects policy in parts quality.

Other key elements of TPS include (1) *keiretsu*, a favored vendor system, which supplies quality parts in a timely manner; (2) *poke-yoke*, a quality control error prevention system; (3) *just-in-time* (JIT) production with its associated *jidoka*, autonomous error control or stopping machines to prevent defective units from disrupting a subsequent process; (4) the *kanban* system, a taglike card with product information that follows the product completely through the production cycle, to maintain JIT; (5) flexible workforce, such as varying the number of workers in response to demand changes; (6) *kaizen*, or continuous improvement activities (Imai, 1986); and (7) respect for the worker and a "creative thinking"worker suggestion system.

JIT is based on a *pull system* of production control in which the demand for parts comes from the downstream station, rather than a *push system*, in which parts are produced irrespective of the needs of the system, resulting in large queues and bottlenecks. A necessary component of JIT is the single-minute exchange of die (SMED). SMED is a series of techniques pioneered by Shingo (1981) for changeovers of production machinery in less than 10 min. Obviously, the long-term objective is zero setup, in which changeovers are instantaneous and do not interfere in any way with continuous work flow.

A corollary to the seven mudas is the *5S* system to reduce waste and optimize productivity by maintaining an orderly workplace and consistent methods. The 5S pillars are (1) *sort* (seiri), (2) *set in order* (seiton), (3) *shine* (seiso), (4) *standardize* (seiketsu), and (5) *sustain* (shitsuke). *Sort* focuses on removing all unnecessary items from the workplace and leaving only the bare essentials. *Set in order* arranges needed items so that they are easy to find and use. Once the clutter is removed, *shine* ensures further cleanliness and tidiness. Once the first three pillars have been implemented, *standardize* serves to maintain the order and consistent approach to housekeeping and the methods. Finally, *sustain* maintains the full 5S process on a regular basis.

Remarkable successes have been shown with the implementation of TPS, ranging from Toyota itself to tiny suppliers, such as Showa (Womack and Jones, 1996). For more details on the Toyota Production System, consult Shingo (1981), Imai (1986), and Ohno (1988) for the original sources, and Monden (1993), and Womack and Jones (1996) for easier reading and understanding.

TOTAL QUALITY

Quality is a concept that everybody intuitively understands, but it is difficult to define. Everybody can relate to eating out in a restaurant and judging its quality by the taste of the food, the promptness and courtesy of the service, cost, and ambience. Two aspects that cross all these factors are results and customer satisfaction. In other words, does the product or service meet or exceed customer satisfaction? Furthermore quality is an ever-changing state that must be continually maintained through a *continuous improvement* program. *Total quality* is a much broader concept that encompasses not just the results aspect, but also the quality of the process, materials, environment, and people.

The total quality movement, like work measurement, could be considered to have evolved from F. W. Taylor's *Principles of Scientific Management*. Later development came about because of the impact of World War II on U.S. and Japanese industries. As U.S. companies were focused more on meeting delivery dates than on quality, which continued well past the war, Japanese companies were forced to compete with established companies in the rest of the world. This could only be done by emphasizing the quality of its products over the next 20 years.

The Japanese effort on continuous quality improvement and quality circles was initiated primarily by the philosophies and works of three individuals: W. E. Deming, J. M. Juran, and A. V. Feigenbaum. Following his work in the United States during World War II, Deming became a consultant to Japanese industries and convinced their top management of the power of statistical methods and the importance of quality as a competitive weapon. He is best known for his 14 points (see Table 18.4) and the Deming prize for quality established by the Japanese Union of Scientists and Engineers.

Juran is one of the founders of statistical quality control and is best known for his *Quality Control Handbook* (Juran, 1951), a standard reference in the area. The Juran philosophy is based on the organization and implementation of improvements through "managerial breakthroughs" highlighted in the 10 steps to quality improvement (see Table 18.5).

Feigenbaum was the first to introduce the concept of a companywide quality control program, in his book *Total Quality Control* (1991, 3d ed.), which was widely used in Japan in the 1950s. Only in the late 1980s and early 1990s did the total quality concept start gaining wide acceptance in the United States under the names of *total quality management* (TQM), *total quality assurance* (TQA), or more specialized company-specific programs, such as Motorola's Six Sigma.

Table 18.4 Deming's Fourteen Points

1. Create constancy of purpose toward the improvement of products and services in order to become competitive, stay in business, and provide jobs.
2. Adopt the new philosophy. Management must learn that it is a new economic age and awaken to the challenge, learn its responsibilities, and take on leadership for change.
3. Stop depending on inspection to achieve quality. Build in quality from the start.
4. Stop awarding contracts on the basis of low bids.
5. Improve continuously and forever the system of production and service, to improve quality and productivity, and thus constantly reduce costs.
6. Institute training on the job.
7. Institute leadership. The purpose of leadership should be to help people and technology work better.
8. Drive out fear so that everyone may work effectively.
9. Break down barriers between departments so that people can work as a team.
10. Eliminate slogans, exhortations, and targets for the workforce. They create adversarial relationships.
11. Eliminate quotas and management by objectives. Substitute leadership.
12. Remove barriers that rob employees of their pride of workmanship.
13. Institute a vigorous program of education and self-improvement.
14. Make the transformation everyone's job and put everyone to work on it.

Table 18.5 Juran's Ten Steps to Quality Improvement

1. Build awareness of both the need for improvement and opportunities for improvement.
2. Set goals for improvement.
3. Organize to meet the goals that have been set.
4. Provide training.
5. Implement projects aimed at solving problems.
6. Report progress.
7. Give recognition.
8. Communicate results.
9. Keep score.
10. Maintain momentum by building improvement into the company's regular systems.

In general, total quality (TQ) is a way of doing business that maximizes the competitiveness of a company through the continuous improvement of its products, service, people, process, and environment. The key elements of TQ include a companywide strategic focus, even obsession, on quality, with the customer as its driver. TQ utilizes a scientific approach, employee involvement (especially teamwork), education and training, a long-term commitment, and unity of purpose. The process is not always easy to achieve and must be continually worked on to achieve improvements. Also, cost reductions through better awareness of life-cycle costs, improved product/process designs, and better process controls through the whole manufacturing process are important factors in the success of total quality. More details on total quality and specific program components can be found in Goetsch and Davis (1997).

ISO 9000

Related to total quality is ISO 9000 certification. ISO 9000 is a standard for quality control developed by the International Standards Organization (ISO 9000, 1993). (It now actually comprises a set of five standards, ISO 9000–9004.) By definition, ISO 9000 is concerned only with quality management procedures for contract review, design, development, production, installation, and servicing of products and/or services.

Certification for ISO 9000 ensures that a company's products and/or services are consistently up to a certain level of quality. In the United States, such certification is done by a Registration Accreditation Board staffed jointly by the American National Standards Institute and the American Society for Quality Control. However, this is a private volunteer group, and it does not carry the weight of government authorization, as in other countries. Where ISO 9000 is primarily limited to the processes used in a company, total quality encompasses every aspect of that company, including the workforce and the environment. Thus, ISO 9000 is compatible with and typically a subset of total quality, serving to ensure that a company is competitive in a global marketplace (Goetsch and Davis, 1998).

SUMMARY

To a large extent, the work of industrial engineers influences labor relations within an enterprise. At the least, they should know the nature of training that is being provided to the operators and the effect it will have on learning curves and the setting of production standards. In addition, since they are affecting the wage payment of the operators, they also need to understand the attitudes, concerns, and problems of the workers as well as those of the unions representing the workers. In setting new methods and standards, whether using traditional tools or more current lean manufacturing approaches, they must always act in a reasonable and fair manner, to both the company and the workers. At all times, they must be cognizant of the necessity of using the human approach.

QUESTIONS

1. Why is training necessary for operators?
2. How is learning quantified?
3. What is the difference between the Crawford and Wright models of learning?
4. What is remission and how does it affect learning?
5. How should the standards analyst utilize learning curves?
6. Which five states related to the psychological and sociological reactions of the operator should be recognized by the analyst?
7. What do we mean by the human approach?
8. Name 12 ways you can get people to agree with your ideas.

9. Why is plantwide training in the areas of work measurement a healthy management step?

10. Why should experienced analysts be continually checked on their ability to performance-rate?

11. Why do unions often train their own time study analysts?

12. What are the ego states in transactional analysis?

13. What is a crossed transaction?

14. What levels of transactions work best in dealing with workers?

15. What is an ulterior transaction?

16. What is a quality circle?

17. Compare and contrast intrinsic and extrinsic motivators.

18. How does role-playing enter into ergonomic teams?

19. How does total quality management enter into modern management practices?

20. How does job enrichment differ from job enlargement?

21. What are the seven types of waste?

22. What is continuous improvement and why is it important?

23. Define lean manufacturing.

24. What is meant by just in-time production?

25. What is the 5S systems?

26. What is the difference between a push and a pull system?

27. What is a Kanban?

28. What does SMED mean and how does it relate to JIT?

PROBLEMS

1. A new employee at the Dorben Co. took 186 and 140 min to assemble the fourth and eighth assemblies, respectively. The standard time for assembling this product is 100 min.
   a. Calculate this worker's learning curve.
   b. How many assemblies does it take the worker to achieve the standard time? How long is this?

2. A training expert suggests that one should allocate a minimum of 40 h of learning. What time would the worker in Problem I have achieved at this point?

3. Workers new to carburetor assembly take 15 min to complete their first assembly. Assuming a 95 percent learning curve, how long would it take them to reach a standard time of 10 min?

4. From an MTM synthesis, the standard time for one C assembly is calculated to be 1.0 min. A new worker typically requires about 2 min for the first assembly, and this drops to about 1.7 min by the fifth assembly.
   a. Calculate and plot the learning curve.
   b. What is the percent learning?
   c. How long will it take the new worker to reach standard performance?

5. The Dorben Co. uses a standard hour plan with a base wage of $9.00 per hour. It hires an IE consultant to plan a batch production run of 300 units. This consultant has a worker on the line produce 2 units. The first takes 10 min and the second 9.7 min.

    a. Find the equation of the learning curve.

    b. What is the percent learning?

    c. How long would it take the worker to reach the standard time of 8.0 min?

    d. Assume that the worker tries to improve and earn the incentive pay. What would the worker be paid for the first 20 h under the standard hour plan?

    e. Assuming that the worker continues to improve her performance, what will she be paid for the full batch of 300 units under the standard hour plan?

    f. What is the unit cost for the first unit made? What is the cost for 300th unit?

6. An analyst has estimated that there will be an 84 percent learning curve for an assembly operation. The first assembly takes 48 min, and the standard time is set at 6 min.

    a. How long will it take the operator to reach the standard time?

    b. Unfortunately, the operator falls ill after the first week on the new assembly operation and returns after one week. What will be estimated time for the operator to reach standard time now?

REFERENCES

Anonymous. "Just What Do You Do? Mr. Industrial Engineer." *Factory*, 122 (January 1964), pp. 83–84.

Balyeat, R. E. "A Survey: Concepts and Practices in Industrial Engineering." *Journal of Industrial Engineering*, 5 (May 1954), pp. 19–21.

Berne, E. *Games People Play*. New York: Grove Press, 1964.

Carnegie, Dale. *How to Win Friends and Influence People*, New York: Simon & Schuster, 1936. (www.dalecarnegie.com)

Crawford, J. R. "Statistical Accounting Procedures in Aircraft Production." *Aero Digest*, (March 15, 1944), pp. 78–81, 222, 224, and 226.

Deming, W. Edwards. *Out of the Crisis*. Cambridge, MA: MIT Center for Advanced Engineering Study, 1986.

Denton, K. *Safety Management, Improving Performance*. New York: McGraw-Hill, 1982.

Feigenbaum, A. V. *Total Quality Control*. 3d ed. New York: McGraw-Hill, 1991.

Freivalds, A., S. Konz, A. Yurgec, and J. H. Goldberg. "Methods, Work Measurement and Work Design: Are We Satisfying Customer Needs?" *The International Journal of Industrial Engineering*, 7, no. 2 (June 2000), pp. 108–114.

Goetsch, D. L., and S. B. Davis. *Introduction to Total Quality*. Upper Saddle River, NJ: Prentice-Hall, 1997.

Goetsch, D. L., and S. B. Davis. *Understanding and Implementing ISO 9000 and ISO Standards*. Upper Saddle River, NJ: Prentice-Hall, 1998.

Hancock, W. M., and F. H. Bayha. "The Learning Curve." In *The Handbook of Industrial Engineering*. Ed. G. Salvendy. New York: John Wiley & Sons, 1982.

Herzberg, F. *Work and the Nature of Man*. Cleveland, OH: World Publishing, 1966.

Imai, M. *Kaizen*. New York: Random House, 1986.

ISO 9000: International Standards for Quality Management. 3d ed. Geneva, Switzerland: International Standards Organization, 1993.

Juran, J. M. *Quality Control Handbook*. New York: McGraw-Hill, 1951.

Konz, S., and S. Johnson. *Work Design*. 5th ed. Scottsdale, AZ: Holcomb Hathaway, Inc., 2000.

Majchrzak, A. "Management of Technological and Organizational Change." In *Handbook of Industrial Engineering*. 2d ed. Ed. Gavriel Salvendy. New York: John Wiley & Sons, 1992.

Maslow, A. *Motivation and Personality*. 2d ed. New York: Harper & Row, 1970.

Monden, Y. *Toyota Production System*. Norcross, GA: Industrial Engineering and Management Press, 1993.

Ohno, T. *Toyota Production System: Beyond Large-Scale Production*. Cambridge, MA: Productivity Press, 1988.

OSHA. *Ergonomics Program Management Guidelines for Meatpacking Plants*. OSHA 3123. Washington, DC: The Bureau of National Affairs, Inc., 1990.

Santos, J., R. Wysk, and J. M. Torres. *Improved Production with Lean Thinking*. New York: John Wiley & Sons, 2006.

Shingo, S. *Study of Toyota Production System from Industrial Engineering Viewpoint*. Tokyo, Japan: Japan Management Association, 1981.

Taylor, F. W. *The Principles of Scientific Management*. New York: Harper, 1911.

Womack, J. P., and D. T. Jones. *Lean Thinking*. New York: Simon & Schuster, 1996.

Wright, T. P. "Factors Affecting the Cost of Airplanes." *Journal of the Aeronautical Sciences*, 3 (February, 1936), pp. 122–128.

SELECTED SOFTWARE

DesignTools, (available from the McGraw-Hill text website at www.mhhe.com/niebels-freivalds).

WEBSITE:

www.dalecarnegie.com

3 E's   The preferred approach for corrective action: **engineering** redesign to ensure strict safety without relying on operator compliance; **education** that relies on operator compliance in a positive way; and **enforcement** of strict rules which may have negative connotations.

5S   System to reduce waste and optimize productivity by maintaining an orderly workplace and consistent methods.

ABC model   A behavior-based safety model in which **antecedent** events lead to certain worker **behavior** patterns that may result in unpleasant **consequences** such as accidents. To correct such worker behavior, these antecedents must first be modified.

abnormal time   Elemental time values that are either considerably higher or lower than the mean of the majority of observations taken during a time study.

absolute judgment   Differentiation between two stimuli if no direct comparison can be made.

accident prevention   A tactical approach to controlling workers, materials, tools and equipment, and the workplace for the purpose of reducing or preventing the occurrence of accidents.

accident-ratio triangle   Theory that claims for each major injury there occurred at least 29 minor injuries and 300 no-injury accidents, with untold hundreds or thousands of unsafe acts leading up to the base of the triangle. Coast-effective accident prevention efforts should focus on the large base rather than just on the one major injury.

actual time   The average elemental time actually taken by the operator during a time study.

administrative law   Law established by the executive branch or government agencies.

aerobic   Muscular work for which the oxygen intake is adequate.

affordance   A perceived property that results in the desired action, for example, a door with a handle that pulls open.

agonist   The primary muscle involved in the desired motion.

algorithm   Step-by-step specifications of the solution to a problem, usually represented by a flowchart, which eventually is translated into a program.

alignment chart   *See* **nomogram**.

allowance   The time added to normal time to provide for personal delays, unavoidable delays, and fatigue.

alphanumeric   The set of alphabetic letters (a to z) and numeric digits (0 to 9).

anaerobic   Muscular work for which the oxygen intake is inadequate.

AND gate   Type of Boolean logic that requires that all the inputs occur for the output to occur.

antagonist   The muscle that opposes the agonist and the desired motion.

anthropometry   The science that deals with measuring the physical size of the human.

Ashcroft's method   Tables of machine interference times for various servicing times as developed by H. Ashcroft.

assemble   The act of bringing two mating parts together.

assignable cause   A source of variation that can be isolated in a process or operation.

associations   Concrete relationships of a name or word with previous knowledge utilizing the user's expectations and stereotypes; useful in maintaining long-term memory.

assumption of risk   Legal concept in which a worker, who was aware of the hazards of the job, but continued working there, assumed the risks, and could not recover damages in case of injury even though it occurred through no fault of his or her own.

ATP   Adenosine triphosphate, the immediate energy unit for muscle contraction.

attention resources   The amount of cognitive capacity devoted to a particular task or processing stage, or simply, attention.

attention time   Portion of the cycle time when the operator is observing a process to maintain the efficient progress of the operation.

automation   Increased mechanization to produce goods and services.

available machine time   That portion of a time cycle during which a machine could be performing useful work.

average cycle time   The sum of all average elemental times divided by the number of cycle observations.

average elemental time   The mean elemental time taken by the operator to perform the task during a time study.

average hourly earnings   The mean dollar-and-cent moneys paid to an operator on an hourly basis, determined by dividing the hours worked per period into the total wages paid for the period.

avoidable delay   A cessation of productive work due entirely to the operator and not occurring in the regular work cycle.

B

balanced motion pattern   A sequence of motions made simultaneously by both the right and left hands in directions that facilitate rhythm and coordination.

ballistic movement   The motion of arms (usually) or legs with smooth, flowing, rapid muscle action from the start to the termination of the action.

bandwidth   The maximum information processing speed of a given communication channel.

base wage rate   The hourly money rate paid for a given work assignment performed at a standard pace by a qualified operator.

basic event   Events in a fault tree identified by circles, at the bottom of the fault tree that can not be developed any further.

basic fatigue allowance   Constant allowance given to account for the energy expended while carrying out typical work and the alleviation of monotony.

basic motion   A fundamental motion related to primary physiological and/or biomechanical performance capabilities of body members.

basic times   See **predetermined times**.

basilar membrane.   Membrane that splits the cochlea lengthwise, containing hair cells.

benchmark   A standard that is identified with characteristics in sufficient detail that other classifications can be compared as being above, below, or comparable to the identified standard.

beta   The ratio of the height of the signal to the noise curves of the response criterion.

binomial distribution   A discrete probability distribution with mean $= np$ and variance $= np(1 - p)$ having a probability function

$$P(k) = C_{n,k}\, p^k\, (1 - p)^{n-k} \text{ for } k = 0, 1, 2 \ldots$$

biomechanics   The application of mechanical principles, such as levers, mechanical advantage, and forces, to the analysis of body part structure and movement.

bit   The amount of information obtained with two equally likely alternatives.

body discomfort chart   A method of assessing a worker's health status by checking the level of discomfort for various body parts.

bonus earnings   Those moneys paid in addition to the regular wage or salary.

Borg RPE scale   Rating of perceived exertion scale ranging from 6 through 20, corresponding roughly to the heart rate (divided by 10); used for assessing the perceived exertion during dynamic whole-body activities.

bottom-up processing   Data-driven information processing guided by sensory features.

brainstorming   Discussion sessions in which ideas are encouraged, regardless of how wild they are.

break-even chart   See **crossover chart.**

break-point   A readily distinguishable point in the work cycle selected as the boundary between the completion of one element and the beginning of another element.

buttons   Isolated picture-in-picture windows within a display that can be selected by the user to invoke specific actions, an essential part of a graphical user interface.

buzz groups   Small discussion subgroups.

C

CAD   Computer-aided design.

candela   A measure of the luminous intensity of a light source.

carpal tunnel syndrome   Median nerve compression due to inflammation and swelling within the carpal tunnel of the wrist, causing pain and loss of sensation and motor control.

cause–effect diagrams   See **fish diagrams.**

cervical   The part of the vertebral column located in the neck.

changeover time   The time required to modify or replace an existing workplace. Includes both the teardown time for the existing condition and the setup of the new condition.

channel capacity   See **bandwidth.**

check time   Sum of the time elapsed before the study and the time elapsed after the study.

Chi-square analysis   A statistical tool used in identifying whether one department is significantly more hazardous than another based on the Chi-square goodness of fit test between a sample and a population distribution in the form of categorical data in contingency table.

choice-reaction time   Time for an operator to respond to one of several stimuli each with an appropriate response.

chord keyboard   Type of keyboard requiring the simultaneous activation of two or more keys for an individual character, as opposed to the sequential activation in standard keyboards.

chunking   The grouping of similar items to facilitate recall.

circadian rhythms   The roughly 24-h variation in bodily functions in humans.

citation   Statement of fault and potential penalties issued by OSHA with regard to violations of OSHA standards.

classification method   A method of job evaluation based on a series of definitions to differentiate between jobs.

clo   The amount of thermal insulation in clothing needed to maintain comfort for a person sitting in a normally ventilated room at 70°F and 50 percent relative humidity; roughly equal to a light business suit.

cochlea   The coiled fluid-filled structure of the middle ear split lengthwise by the basilar membrane.

color rendering   The closeness with which the perceived colors of an object being observed match the perceived colors of the same object when illuminated by standard light sources.

common law   Law derived from unwritten customs and typical usage but adjusted and interpreted by the courts through judicial decisions.

compatibility   The relationship between a stimulus and a response that is consistent with human expectations and minimizes conflict; for example, a red light is associated with danger or stopping.

compensatory damages   Court-awarded payments to the plaintiff for medical costs, lost wages, and other direct losses.

complementary transaction   Transaction sent and received at the same level of ego states.

cones   Photoreceptors of the eye that are sensitive to colors, especially in daylight, and have good visual acuity.

conservative   Observer behavior as the criterion shifts to the right, causing a decrease of both hits and false alarms.

consistency   The absence of noticeable or significant variation in behavioral or numerical data.

constant element   An element whose performance time does not vary significantly when changes in the process or dimensional changes in the product occur.

constant fatigue allowance   The combination of personal needs and basic fatigue allowances, which typically are constant for all workers within a company.

construction program   A facilities layout program generating the best solution from scratch.

continuous improvement   An ongoing process ensuring total quality in a company.

continuous-timing method   An operation study method in which the stopwatch is kept running continuously during the course of the study and is not snapped back at elemental termination.

contrast   The ability of a target to stand out from its background; typically measured as the difference in luminances between target and background.

control–response ratio   Ratio of the amount of movement in a control to the amount of movement in the response; used to define system responsiveness.

control system   A system that has as its primary function the collection and analysis of feedback from a given set of functions, for the purpose of controlling the functions.

cornea Outer protective covering of the pupil of the eye, also assists in focusing the light rays on the retina.

correct rejection Correct rejection of a signal when no signal is present.

cost-benefit analysis Type of analysis in which the total expected costs are weighed against the total expected benefits for one or more interventions in order to choose the best or most profitable option.

costing Procedure for accurately determining costs in advance of production.

counter A digital display showing precise numeric values.

coverage The number of jobs that have been assigned a standard during the reporting period, or the number of personnel whose jobs have been assigned a standard during the reporting period.

CP Creatine phosphate, the immediate precursor to ATP.

CR-10 (category ratio) scale Rating scale for the level of pain or body discomfort on a logarithmic scale, ranging from 0 to 10; used for assessing muscular exertions.

Crawford model Learning curve model in which the improvement gained is defined for a particular unit. Also known as the unit model.

criteria of pessimism A decision-making strategy in which the outcome with minimum negative consequence is selected.

criticality Measure of the severity associated with the head event in a fault tree.

crossed transaction Transaction between ego states that are not parallel.

crossover chart A method for plotting the increase in cost as a function of some variable. The point at which the two lines cross is known as the crossover or break-even point, and the cost for each method is the same.

cues Pieces of information used in decision making.

Cumulative average model *See* **Wright model**.

cumulative trauma disorder (CTD) Work-related musculoskeletal injuries due to highly repetitive motions involving excessive joint motions with high force; also termed **repetitive-motion injuries**.

curve A graphic representation of the relation between two factors, one of which is usually time.

cycle A series of elements that occur in regular order and make an operation possible. These elements repeat themselves as the operation is repeated.

D

d' A measure of sensitivity in signal detection theory; the distance between the mean of the noise distribution to the mean of the signal plus noise distribution.

danger The relative exposure to or potential consequences of a hazard.

database A collection of data items that can be processed by a variety of applications.

day work Any work for which the operator is compensated on the basis of time rather than output.

dBA A measure of sound intensity on a weighting scale approximating the response characteristics of the human ear; most commonly used to assess the noise exposure of workers.

de minimis violation Type of violation that has no immediate relationship to safety or health and typically has no penalty.

deadman's control A control requiring the continual application of force. Once released, it returns to the zero (or safe) position.

deadspace The amount of control movement resulting in no system response.

decibel (dB) The unit for sound intensity; a logarithmic ratio of the measured intensity to a reference intensity.

decimal hour stopwatch A stopwatch used for work measurement, the dial of which is graduated in 0.0001 h.

decimal minute stopwatch A stopwatch used for work measurement, the dial of which is graduated in 0.01 min.

decision making Information processing involving the evaluation of alternatives and selection of appropriate responses.

decision tables A structured approach for evaluating information and selecting the best among several alternative methods changes.

defendant   The person or entity, typically an employer or the manufacturer of a product, defending a suit in court.

delay   Any cessation in the work routine that does not occur in the typical work cycle.

design for adjustability   Anthropometric design principle typically used for equipment or facilities that can be adjusted to fit a wider range of individuals.

design for averages   "One size fits all" anthropometric design principle.

design for extremes   Anthropometric design principle in which a specific feature is a limiting factor in determining either the maximum or the minimum value of a population variable to be accommodated, for example, stature for doorways.

detection   The determination of whether a stimulus is actually present.

digitizing tablet   A flat pad placed on the desktop that is activated by the movement of a stylus across it.

direct labor   Labor performed on each piece that advances the piece toward its ultimate specifications.

direct lighting   Type of lighting that places more of the light on the work surfaces and the floor.

direct material costs   Cost of raw materials and components.

disassemble   The basic motion that takes place when two mating parts are separated.

discounted cash flow   Economic tool computing the ratio of the present worth of cash flow to the original investment.

disk (intervertebral disk)   Soft tissue, composed of a gellike center and surrounded by onionlike layers of fibers, separating the vertebral bones.

disk herniation   Bulging of the intervertebral disk, causing pressure on spinal nerves with resulting pain.

display modality   Information format corresponding to one of the five senses.

dissociability   The property of a desired signal being as different as possible from other signals (or noise) in terms of its characteristics.

distributive laws   Expressions of Boolean logic that allow for the simplification of more complicated expressions.

divided attention   Attention resources applied in a diffuse manner to various parts of or even all the human information processing system.

domino theory   Theory of accident causation in which the accident is the fifth domino in a sequence of falling dominoes.

downtime   The time represented by operation cessation due to machine or tool breakdown, lack of material, and so on.

drop delivery   The disposal of a part by dropping it on a conveyor or a gravity chute, thus minimizing move and position therbligs.

dry-bulb temperature   Basic ambient temperature, with the thermometer shielded from radiation.

Dvorak keyboard   Alternate keyboard design that optimizes the finger loading.

dyadic communications   One-to-one communications, typically face to face.

E

eardrum   Membrane separating the outer ear canal from the middle ear, also known as the tympanic membrane.

earned hours   The standard hours credited to a worker or a workforce as a result of the completion of a job or group of jobs.

effective time   Total of all observed time.

efficiency   The ratio of actual output to standard output. Also, light output per unit energy.

effort   The will to perform either mental or manual productive work.

effort time   The portion of the cycle time that depends on the skill and effort of the operator.

ego states   The three psychic stages an individual can achieve: adult, parent, or child.

elapsed time   The actual time that has transpired during the course of a study or an operation.

element   A division of work that can be measured with stopwatch equipment and that has readily identified terminal points or break points.

EMG (electromyogram)   The electrical activity in a muscle.

equivalent wind chill temperature   The ambient temperature that in calm conditions would produce the equivalent wind chill index as the actual combination of air temperature and wind velocity.

erector spinae   Primary muscles of the back that provide the force for lifting loads.

ergonomics   The science of fitting the task or workplace to the abilities and limitations of the human operator; sometimes termed **human factors**.

expense labor   Labor not involved in the manufacture of a product, typically engineering, research, sales, clerical, accounting, and other administrative functions.

exponential distribution   A continuous probability distribution with mean $= 1/a$ and variance $= 1/a^2$, and having a density function $= ae^{-ax}$.

extension   Joint motion in which the included angle becomes larger.

extra allowance   An allowance to compensate for required work in addition to that which is specified in the standard method.

extrinsic factors   Environmental factors such as administration, supervision, and working conditions in Herzberg's motivation maintenance theory acting as dissatisfiers.

F

facilitating agreement   Process for obtaining a consensus by positively involving all group members.

factor comparison   A method of job evaluation based on comparing various job factors.

factory cost   Direct material costs plus direct labor costs plus factory expense.

factory expense   Costs such as indirect labor, tooling, machine, and power costs.

fail-safe design   Type of system design such that, in the case of failure, it goes to the lowest energy level.

fair day's work   The amount of work performed by an operator that is fair to both the company and the operator, considering the wages paid. It is the "amount of work that can be produced by a qualified employee when working at a standard pace and effectively utilizing his time where work is not restricted by process limitations."

false alarm   Incorrect identification of a signal when no signal is present

fatigue   A lessening in the capacity to work.

fatigue allowance   Type of allowance providing time for the worker to recover from fatigue incurred as a result of the job or work environment.

fault event   Events in a fault tree identified by rectangles that are to be expanded further.

fault tree analysis   A probabilistic deductive process using a graphical model of parallel and serial combinations of events, or faults, leading to the overall undesired event, e.g., an accident.

feature analysis   The breaking down of objects into component geometric shapes or text into words and character strings.

feed   The speed at which the cutting tool is moved into the work, as in drilling and turning, or the rate at which the work is moved past the cutting tool.

feedback   The return of meaningful information to the operator such that performance can be appropriately modified.

film analysis   The frame-by-frame observation and study of a film of an operation or process, with the objective of improving that operation or process.

finishing time   The clock time at which a time study ends.

fish (cause–effect) diagrams   A method of defining an occurrence of a typically undesirable event or problem; that is, the effect, as the fishhead, and then identifying contributing factors, that is, the causes, as fish bones attached to a backbone and the fishhead.

Fitts' law   Extension of the Hick-Hyman law with respect to movement time; i.e., the longer the distance and the smaller the target, the longer the movement will take.

Fitts' Tapping Task   A series of positioning movements to and from targets demonstrating Fitts' law.

fixture   A device that is usually clamped to the workstation and that holds the material being worked on.

flexible compensation plans   Any incentive plan that increases employee wages or benefits as a function of increased production.

flexion   Joint motion in which the included angle becomes smaller.

flextime   Shift system in which the starting and stopping times are established by the workers, within limits set up by management.

float   The amount of material not directly employed or worked on in a system or process at a given time.

flow diagram   A pictorial representation of the layout of a process, showing the location of all activities appearing on the flow process chart and the travel paths of the work.

flow process chart   A graphic representation of all operations, transportations, inspections, delays, and storages occurring during a process or procedure. The chart includes information considered desirable for analysis, such as the time required and the distance moved.

focused attention   Attention resources applied in a very directed manner, such as a spotlight on a particular part of the human information processing system.

footcandle   The measure of light falling on a surface. One footcandle equals 10.8 lumens per square meter.

foot-lambert   A unit of luminance (emitted or reflected light). One foot-lambert is equal to 3.43 candelas per square meter.

force–length relationship   The inverted-U relationship in which muscle force is greatest at its resting length.

force–velocity relationship   The trade-off between slower movements providing greater force and faster movements being weaker.

foreign element   An interruption in the regular work cycle.

fovea   Part of retina with greatest sensitivity of cones.

frame   The space occupied by a single picture on a motion picture film or videotape.

frame counter   A device that automatically tabulates how many frames have passed the lens of the projector.

frequency of use   Principle used in laying out controls or displays based on how often each is used.

fringe benefits   The portion of tangible compensation that is not paid in wages, salaries, or bonuses given by the employer to employees. These include insurance, retirement funds, and other employee services.

from–to chart   *See* **travel chart.**

full model   The more complex model in a general linear test.

functional layout   *See* **process layout.**

functionality   The principle used in laying out controls or displays by similar function.

G

gain sharing   Any method of wage payment in which the worker participates in all or a portion of the added earnings resulting from above-standard production.

gang process chart   A chart of the simultaneous activities of one or more machines and/or one or more workers.

Gantt chart   A series of graphs consisting of horizontal lines or bars in positions and lengths that show schedules or quotas and progress plotted on a common time scale.

general duty clause   Introductory paragraph of the OSHAct stating that each employer "must furnish a place of employment which is free from recognized hazards that cause or are likely to cause death or serious physical harm to employees." This phrase is sometimes used by OSHA to issue citations for situations not specifically covered in the standards.

general expense   Cost for expense labor, rent, insurance, and so on.

general linear test   A formalized procedure for finding the best model using statistically significant decreases in variances between alternative models.

get   The act of picking up and gaining control of an object. It consists of the therbligs reach and grasp, and move; it also sometimes includes search and select.

glare   Excessive brightness in the field of vision, impairing visibility.

globe temperature   Measure of radiative load, using a thermometer in a 6-in-diameter black copper sphere.

glucose   The primary carbohydrates component that enters the biochemical pathways for energy production.

grade description plan   *See* **classification method.**

graphical user interface (GUI)   A type of screen interface identified by four main elements: windows, icons, menus, and pointers.

grasp   The elemental hand motion of closing the fingers around a part.

gravity feed   Conveyance of materials either to or away from the workstation by using the force of gravity.

gross negligence   Higher form of negligence with failure to show the slightest care.

H

Hawthorne effect   A phenomenon in which increased employee productivity may be attributed due to perceived employer interest in improving methods or the workplace; however, it also applies to the confounded results from uncontrolled or unconsidered variables in a methods change.

hazard   A condition with the potential of causing injury or damage.

hazard action table   Decision table for specifying certain actions for given hazards.

heart rate creep   The slow increase in heart rate during heavy work, indicating fatigue.

heart rate recovery   The return of the heart rate to resting levels after work.

hertz   The unit of frequency, in cycles per second. One hertz equals one cycle per second.

Hick-Hyman law   The linear relationship between response time and the amount of stimulus information conveyed in bits.

hit   Correct identification of a signal when the signal is present.

human factors   See **ergonomics**.

I

icons   Small representations of objects that act as windows in a graphical user interface.

idle time   The time an operator or a machine is idle or not working.

illumination   The amount of light striking a surface, measured in footcandles.

imminent of danger   Type of OSHA citation in which there is reasonable certainty that a danger exists that can be expected to cause death or serious physical harm either immediately or before the danger can be eliminated through normal enforcement procedures. This may result in a cessation of the operation or even a complete shutdown of the plant.

importance   The principle used in laying out controls or displays based on the importance of each.

IMPROSHARE   A gain-sharing plan based on employee productivity as measured in working hours.

improvement program   Facilities layout program that improves upon an initial layout.

incentive   Reward, financial or other, that compensates the worker for high and/or continued performance above standard.

incentive pace   A performance that is above standard level.

independent   Descriptive of two events if the occurrence of one event doesn't affect the occurrence of another event.

index of difficulty   Relationship defining the distance of movement D and target width W as a form of information in Fitts' law:

$$\text{Index of difficulty} = \log_2 (2D/W)$$

indirect labor   Labor that does not directly enter into transforming the material used in making the product, but is necessary to support the manufacture of the product.

indirect lighting   Type of lighting in which the ceiling is illuminated, which in turn reflects the light downward.

ineffective time   The sum of all foreign element times.

information   Knowledge received regarding a particular fact or the reduction of uncertainty about that fact.

information theory   The science of measuring and understanding information.

inspect   Elemental motion of comparing object with standard.

interference time   Idle machine time due to insufficient operator time to service one or more machines that need servicing, because the operator is engaged in other assigned work.

interlock   Type of safety measure involving a sequence of steps or mechanisms to ensure that a given event doesn't occur at the same time as another event.

intrinsic factors   Potential satisfiers such as achievement, recognition, and advancement in Herzberg's motivation maintenance theory.

isokinetic strength   Type of muscle contraction in which the muscle contracts at a constant velocity.

isometric strength   Type of muscle contraction in which the muscle contracts in a fixed static position and produces the maximum force; also known as static strength.

isotonic strength   Type of muscle contraction in which the muscle contracts against a constant force; sometimes termed dynamic strength.

J

Jidoka   Autonomous error control within the Toyota Production System, stopping machines to prevent defective units from disrupting a subsequent process.

jig   A device that may or may not be clamped to the workstation and is used both to hold the work and to guide a tool.

job analysis   A procedure for making a careful appraisal of each job and then recording the details of the work so that it can be equitably evaluated.

job enlargement   A horizontal expansion or diversification of work, to avoid repetitive work.

job enrichment   A vertical expansion of work, allowing workers to both start and complete a given task, providing greater diversification and fulfillment.

job evaluation   A procedure for determining the relative worth of various work assignments.

job rotation   Similar to job enlargement in providing a worker the opportunity to do a variety of tasks to avoid repetitive work, but on a more rigid schedule.

job safety analysis   A procedure for identifying the hazards associated with each step of a job for the purpose of improving the safety of the overall job.

job/worksite analysis guide   Analysis tool identifying potential problems within a particular job, area, or worksite.

just-in-time (JIT)   Refers to a lean manufacturing technique of streamlining production flow by decreasing setup times and requiring suppliers to deliver parts only as needed rather than maintaining large inventories.

just noticeable difference (JND)   The smallest difference detectable between two stimuli.

K

kaizen   System of continuous improvement activities.

kanban   A taglike card with product information that follows the product completely through the production cycle to maintain JIT.

Karnaugh maps   A graphical tool to simplify Boolean algebraic expressions.

keiretsu   Interlocking relationship between a Japanese manufacturer and its suppliers.

key job   A job representative of similar jobs or classes of work in the same plant or industry.

kicker   A step increase in earnings to induce workers to reach a certain level of productivity.

L

lactic acid   The by-product of anaerobic metabolism, causing sensations of fatigue.

lean manufacturing   Manufacturing management theory in which production engineers work together to eliminate waste, cut costs, and increase efficiency.

learning curve   A graphic presentation of the decrease in time to perform a task; improvement of performance as a result of learning.

lens   Part of eye that focuses light rays on the retina.

level of aspiration   A decision-making strategy based on an outcome value which represents the consequence of what one is willing to settle for if one is reasonably sure to get at least this consequence most of the time.

liability   The obligation to provide compensation for damages or injury.

life-change unit theory   Theory in which situational factors tax a person's capacity to cope with stress in the workplace, leaving the person more likely to suffer an accident as the amount of stress increases.

lighting efficiency   Light output per unit energy, typically lumens per watt.

light pen   A special stylus linked to the computer by an electrical cable that senses the electron scanning beam at the particular location on the screen.

line balancing   The problem of determining the ideal number of workers to be assigned to a production line.

lock in   Type of safety measure to maintain an event or a device in an energized status so that it can't be accidentally turned off.

lock out   Type of safety measure to prevent unauthorized individuals from entering a dangerous area.

long-term memory   Long-term storage of information for later use.

loose rate   An established allowed time permitting the qualified operator to achieve standard performance with less than standard effort.

lordosis   The natural inward curvature of the lumbar portion of the spine.

lumbar   The area of the back most prone to injuries; approximately at the belt line.

lumbar support   Support for the lower back (lumbar area) in the form of an outward bulge in the seat back or a pad placed at the belt level.

luminaire   A lighting source, such as a lamp.

luminance   The amount of light reflected from a surface, measured in foot-lamberts.

luminous intensity   Light intensity of a source, measured in candelas.

lux   The unit of illuminance equal to 1 lumen per square meter, or 0.093 footcandle.

M

machine coupling   The practice of having one employee operate more than one machine.

machine interference   Situation in which two or more machines have stopped and require operator attention; since the operator can only service one of the machines at a time, the remaining machines will have to wait to be repaired and consequently will be nonproductive.

machine pacing   The machine or mechanical control over the rate at which the work progresses.

mapping   A clear representation of compatibility between controls and responses.

maximum working area   The area readily reached by the operator when the arms are fully extended, while in a normal working position.

measured day work   An incentive system in which hourly rates are periodically adjusted on the basis of operator performance during the previous period.

mental workload   Cognitive demands placed on the human information processor.

menu   An ordered list of operations, services, or information that is available to the user in a graphical user interface.

merit rating   A method of evaluating an employee's worth to a company in terms of quantity and quality of work, dependability, and general contribution to the company.

method   The technique employed to perform an operation.

methods study (methods engineering)   Analysis of an operation to increase the production per unit of time and consequently reduce the unit cost.

micromotion study   The division of a work assignment into therbligs, accomplished by analyzing motion pictures frame by frame and then improving the operation by eliminating unnecessary movements and simplifying the necessary movements.

Miller's rule   The upper limit for the capacity of working memory at approximately 7 ± 2 items.

minimax regret criterion   A decision-making strategy in which a matrix of regret values (differences between actual and projected payoffs) is calculated. The analyst selects the minimum of maximum regrets.

miss   Incorrect rejection of a signal when the signal is present.

mnemonics   An acronym or phrase, the letters of which represent a series of items, to facilitate recall.

mod ratio   The ratio of actual losses to losses expected of similar employers with 1.00 being average; used to set workers compensation insurance premiums.

modulation   Variation of signal level in a periodic cycle so as to increase attentional demands.

motion study   The analysis and study of the motions constituting an operation, to improve the motion pattern by eliminating ineffective motions and shortening the effective motions.

motivation maintenance theory   A motivational theory by F. Herzberg in which extrinsic factors (administration, working conditions) act as potential dissatisfiers and intrinsic factors (achievement, recognition) act as satisfiers.

motivation-reward-satisfaction model   Behavior-based safety model using a positive feedback cycle in which the better the worker performs, the better the worker is rewarded, the more the worker is satisfied and the greater the worker's motivation to perform better.

motor unit   The functional unit of muscles comprised of a nerve fiber and all the muscle fibers that it innervates.

mouse   A handheld device with a roller ball in the base for controlling cursor position.

move   Hand movement with a load.

MTM (methods-time measurement)   A procedure for analyzing any manual operation or method, to determine the basic motions required to perform the operation, and to assign a predetermined time standard to each motion, based on the nature of the motion and the conditions under which it is made.

muda   In Japanese industry, wastes to be eliminated.

multiple causation   Extension of the domino theory; behind each accident or injury there may be numerous contributing factors, causes, and conditions.

multiple-criteria decision making   A quantitative decision-making procedure in the presence of conflicting information.

multitasking Performing several tasks simultaneously.

musculoskeletal system The system of muscle and bones in the body allowing for movement.

mutually exclusive Boolean state in which two events can not intersect.

myofibrils Subdivision of muscle fiber containing thick and thin protein filaments.

N

natural frequency The internal frequency of vibration within a system, determined by its mass, spring, and dashpot characteristics.

negligence The failure to exercise a reasonable amount of care in preventing injury.

negligence per se Higher form of negligence, with no proof needed.

network analysis A planning technique used to analyze the sequence of activities and their interrelationships within a project.

NIOSH lifting guidelines A set of guidelines for manual lifting developed by NIOSH to provide a measure of safety in preventing overexertion injuries resulting from job demands exceeding a worker's capacity.

noise Sensory stimulation such as unwanted sound that interferes with the detection of a signal.

noise dose Total daily noise exposure, consisting of exposures to several different noise levels, each resulting in partial doses.

noise reduction rating Measure of earplug effectiveness, in terms of decibels of noise level attenuation.

nomogram A graph that usually contains three parallel scales graduated for different variables, so that when a straight line connects values of any two, the related value may be read directly from the third at the point intersected by the line.

nonschedule injury A permanent partial disability that is of less specific nature; such as disfigurement, with payments prorated to a schedule injury.

nonserious violation Type of OSHA violation that probably would not cause death or serious physical harm; incurs a penalty of up to $7,000 for each violation.

normal distribution A continuous probability distribution with mean $=\mu$ and variance $=\sigma^2$ and having a density function equal to

$$f(x) = \frac{1}{\sigma\sqrt{2\Pi}}e^{-\left[\frac{(x-\mu)^2}{2\sigma^2}\right]}$$

normal time The time required for the standard operator to perform the operation when working at a standard pace, without delay for personal reasons or unavoidable circumstances.

normal working area The space at the work area that can be reached by either the left or right hand when both elbows are pivoted on the edge of the workstation.

NOT Type of Boolean logic indicating the negation of an event.

O

objective rating Method of rating which establishes a single work assignment to which the pace of all other jobs is compared.

observation The gathering and recording of the time required to perform an element, or one watch reading.

observed time The elemental time for one cycle, obtained either directly or by subtracting successive watch times.

observer The analyst taking a time study of a given operation.

octave-band analysis Noise analysis with a special filter attachment to the sound-level meter that decomposes the noise into component frequencies.

operation The intentional changing of a part toward its ultimate desired shape, size, form, and characteristics.

operation analysis An investigative process dealing with operations in factory or office work. Usually, the process leading to operation standardization, including motion and time study.

operation card A form outlining the sequence of operations, the time allowed, and the special tools required in manufacturing a part.

operation process chart A graphic representation of an operation, showing all methods, inspections, time allowances, and materials used in a manufacturing process.

operator attention time *See* **attention time.**

operator process chart *See* **two-hand process chart.**

OR gate Type of Boolean logic that needs at least one of the inputs to occur for the output to occur.

OSHA 300 log Form for maintaining records of occupational injuries and illnesses required by the OSHact.

outline process chart A simple graphic representation of all the operations used in a manufacturing process.

output The total production of a machine, process, or worker for a specified unit of time.

overhead Any costs of a business above prime costs.

oxygen debt The increased metabolic activity after work to repay the oxygen deficit.

oxygen deficit The deficit of oxygen incurred during the initial or heavy stages of work; supplied by anaerobic metabolism.

P

pace rating Method of speed rating using a series of benchmarks to assist defining performance.

pallet A load carrier, usually with a rectangular standardized load carrier.

parallel Arrangement of components in a system, such that the total system succeeds if any one component succeeds.

Pareto analysis An exploratory tool in which items of interest are identified and measured on a common scale and then are ordered in ascending order, creating a cumulative distribution; typically 20 percent of the ranked items account for 80 percent or more of the total activity, leading to the alternative term *80-20 rule*.

payback method Economical analysis tool that uses the time to return the cost of the original investment.

perception The comparison of incoming stimulus information with stored knowledge to categorize the information.

performance The ratio of the operator's actual production to the standard production.

performance rating The assignment of a percentage to the operator's average observed time, based on the actual performance of the operator as compared to the observer's conception of standard performance.

permanent partial disability Type of disability in which the worker will not fully recover from injuries but can still perform some work. It is further subdivided into schedule and nonschedule injuries.

permanent total disability Type of disability which is so serious that the employee will be prevented from ever working in regular employment.

personal digital assistants (PDAs) Pocket-sized PCs operated with a stylus.

personal needs allowance A percentage added to the normal time to accommodate the personal needs of the operator.

PERT (Program Evaluation and Review Technique) chart A planning and control method that graphically portrays the optimum way to attain some predetermined objective, generally in terms of time.

phototropism Tendency for the eyes to be drawn directly to the brightest light source.

physiological needs First step in Maslow's hierarchy of human needs, basic concerns regarding survival, food, water, and health.

piecework A standard of performance expressed in money per unit of production.

pinch grip A type of grip in which not all fingers are involved, as opposed to a power grip; consequently, less power but more precision is produced.

plaintiff The person originating a suit in court.

plan A basic motion involving the mental process of determining the next action.

plunger criterion An optimistic decision-making strategy in which the best outcome is expected, regardless of the alternative chosen.

point A unit of output identified as the production of one standard operator in one minute. Used as a basis for establishing standards under the point system.

pointer A device for directly manipulating an icon or cursor on a graphical user interface.

point size Size of type; one point equals 1/72 in (0.035 mm).

point system A method of job evaluation in which the relative worth of different jobs is determined by totaling the points assigned to the various factors applicable to the different jobs.

Poisson distribution A discrete probability distribution, with mean $=\lambda$ and variance $=\lambda$, and having a probability function equal to

$$P(k) = \frac{\lambda^k e - \lambda}{k!} \qquad \text{for } k = 0,1,2\ldots$$

poke-yoke The quality control error prevention system within the Toyota Production System.

policy allowance Allowance to provide a satisfactory level of earnings for a specified level of performance for exceptional circumstances.

position   An element of work that consists of locating an object such that it will be properly oriented in a specific location.

positive reinforcement   A Skinnerian concept in which response strength increases as the stimulus strength is increased.

postlunch dip   Dip in performance and circadian rhythms after midday.

power grip   Optimal cylindrical grip for power that utilizes all digits and in which the thumb barely overlaps the index finger.

predetermined time system   System based on basic motion times used to calculate a standard time.

predetermined times   Times assigned to fundamental motions that cannot be precisely evaluated with ordinary stopwatch time study procedures but are the result of detailed videotape studies of many operations; they are predetermined because they are used to predict standard times for new work as a result of methods changes; they are sometimes termed *synthetic* in that they are often the result of logical combinations of therbligs; they are also *basic* in that further refinement is typically not possible.

preposition   A basic motion that consists of positioning an object in a predetermined place so that it may be grasped in the position in which it is to be held when needed.

primary task   The task to which an operator devotes the greatest amount of attention resources.

prime cost   Direct material costs plus direct labor costs.

principle of combined motions   If one hand performs two motions simultaneously, the longer time is counted in certain predetermined time systems.

principle of insufficient reason   A decision-making strategy in which the conditions or states are considered as being equally likely.

principle of limiting motions   For two different motions performed simultaneously by the left and right hands, the longer time is used in certain predetermined time systems.

principle of simultaneous motions   Motions performed simultaneously with both hands cannot always be performed in the same time as the same motions performed by each hand separately; thus a time penalty may be added in some predetermined time systems.

privity   Legal doctrine requiring a direct relationship, as in the form of a contract, between the two contesting parties.

process   A series of operations that advances the product toward its ultimate size, shape, and specifications.

process chart   A graphic representation of a manufacturing process.

process layout   Type of layout in which workstations or machines are grouped by the similarity of the process or function; also known as functional layout.

product layout   Type of layout in which workstations are located along a continuous line such that the flow from one operation to the next is minimized for any product class; also known as a straight-line layout.

productive time   Any time spent in advancing the progress of a product toward its ultimate specifications.

profit sharing   Any procedure in which an employer pays to employees special current or deferred sums, in addition to good rates of regular pay, based not only on individual or group performance, but also on the prosperity of the business as a whole.

pronation   Rotation of the forearm such that the palm faces down or supination with the palm up.

pull system   Type of just-in-time production control system in which the demand for parts comes from the downstream station.

punitive damage   Additional court-awarded payments to the plaintiff, but specifically used to punish the defendant.

pupil   The opening in the eye that lets in light.

put   The act of moving an object under control to a new position. It consist of the therbligs move and position.

psychophysical strength   Type of strength in which the operator subjectively determines the acceptable load to be lifted.

push system   Type of just-in-time production control system in which parts are produced irrespective of the needs of the downstream stations.

Q

qualified operator   An operator who can achieve the established standard of performance when following the prescribed method and working at a standard pace.

quality circles Small groups for participative problem solving.

queuing theory Mathematical analysis of the laws governing arrivals, service times, and the order in which arriving units are taken into service.

QWERTY keyboard Popular type of keyboard with the sequence of the first six leftmost keys in the third row being QWERTY.

R

rad Unit of radiation dose equivalent to the absorption of 0.01 joule per kilogram (J/kg).

radial deviation Bending of the wrist such that the thumb moves toward the arm.

random servicing The interaction between the operator and machine that occurs on a random basis.

random variable A chance number resulting from a trial from among the set of numbers x_1, x_2, and so on.

range effect Tendency of overshooting close targets and undershooting far targets, typically resulting from fatigue.

ranking method A method of job evaluation that arranges jobs in the order of their importance or according to their relative worth.

rapid rotation Type of shiftwork in which the worker changes shifts every 2 or 3 days.

rate A standard expressed in dollars and cents.

rate setting The act of establishing money rates or time values on any operation.

rating *See* **performance rating**.

rating by the watch An incorrect rating procedure whereby the analyst uses previously observed times to rate the operator.

rating method Method of job evaluation based on arranging jobs in order of importance.

rating of perceived exertion Means for assessing exertion during dynamic whole-body activities.

Raynaud's syndrome Cold-induced occlusion of blood flow to the hands, reducing dexterity.

reciprocal inhibition Type of reflex in which the agonist muscle is activated and the antagonist is inhibited so as to reduce counterproductive muscle contractions.

recommended weight limit (RWL) From the NIOSH lifting guidelines, an optimum weight that be handled by almost everyone, with adjustments to it for various factors related to variables in a lifting task.

recording error Percent error for unaccounted time during a time study.

red circle rates Employee rates higher than that called for by the job evaluation plan.

red flagging Signal for a potential safety problem indicated by injuries going above an upper control limit.

reduced model The simpler model in a general linear test.

redundancy The reduction of information from the maximum amount possible due to unequal probabilities of occurrence of the alternatives.

reflectance Percentage of light reflected from a surface.

regression to the mean Tendency of a novice analyst to rate closer to standard performance than the true performance.

regret matrix *See* **minimax regret criterion**.

rehearsal Mental repetition of items to facilitate recall.

relationship chart Chart expressing the relative degrees of closeness among different activities, areas, departments, rooms, and so on, for facilities layout purposes.

relative judgment Differentiation between two stimuli based on a direct comparison.

reliability The probability of the success of a system, which necessarily depends on the reliability or the success of its individual components.

rem Roentgen equivalent man or the equivalent dose from any type of ionizing radiation which produces a biological effect of 1 roentgen of X-rays in humans.

remission Increase in cycle time on the learning curve due to forgetfulness.

repeated violation Type of OSHA violation issued upon reinspection and finding the recurrence of a previously cited violation. Each repeated violation can bring a fine of up to $70,000.

repetitive-motion injuries *See* **cumulative trauma disorders**.

resonance Situation in which forced vibrations induce larger-amplitude vibrations in a system.

response criterion   Level of sensory stimulation used to decide if a signal is present.

resting length   Length of the muscle while it is in a neutral, uncontracted state.

retina   Layer of photosensitive receptors at the back of the eyeball.

return on investment   Method of economic analysis using the ratio of the yearly profit to the life of the product.

return on sales   Method of economic analysis using the ratio of the yearly profit to the yearly sales.

risk analysis   A probabilistic process for determining what risk factors are present, their characteristics, and the overall value for success or failure.

risky   Observer behavior as the criterion shifts to the left, causing an increase of hits but at the cost of a corresponding increase of false alarms.

rods   Photoreceptors of the eye that are sensitive to black and white, especially at night, and have poor visual acuity.

roentgen   A unit of radiation exposure that measures the amount of ionization produced in air by X or gamma radiation.

Rucker plan   A gain-sharing plan based on employee productivity as measured by one or more of the following: gross production, net sales, and inventory changes.

S

safety factor   The ratio of the strength of a structure to the maximum stress applied; for safety, it should be well above 1.

safety management   A strategic approach for the overall planning, education, and training of safety activities.

safety needs   Second step in Maslow's hierarchy of human needs, the need for security on the job.

Scanlon plan   A gain-sharing plan based on employee productivity as measured by one or more of the following: gross production, net sales, and inventory changes.

schedule injury   A permanent partial disability that is allocated a specific payment for a specified time period according to a specific schedule.

secondary task   A task that must be performed in addition to the primary task.

self-esteem needs   Fourth step in Maslow's hierarchy of human needs, a desire for competence, achievement, and self-respect.

self-fulfillment   Final achievement of all the desired needs in Maslow's hierarchy. The worker is personally fulfilled and the ego is satisfied.

sensitivity   Resolution of the sensory system, measured as the separation between the signal and noise distributions.

sensory store   The transient memory located at the input stage of a sensory channel.

sequence of use   Design principle in which controls and displays are placed in physical locations corresponding to the order in which they are used.

series   Arrangement of components in a system, such that every component must succeed for the total system to succeed.

serious violation   Type of OSHA violation in which there is substantial probability that death or serious harm could result, stemming from a hazard about which the employer knew or should have known.

servicing time   Total time an operator interacts directly with a machine; could include loading, unloading, maintenance, etc.

setup   The preparation of a workstation or a work center to accomplish an operation or a series of operations.

shiftwork   Working at times other than daytime hours.

short-term memory   *See* **working memory**.

signal   Sensory stimulation that yields information.

signal detection theory   Theory used to explain a situation in which an observer needs to identify a signal from confounding noise.

simo chart   A two-hand process chart with times measured with a microchronometer as part of a micromotion study.

simple reaction time   Time for an operator to respond to a single stimulus with a predetermined response.

simultaneous motions   Two or more elemental motions performed simultaneously by different body members. In MTM, for difficult motions, a penalty is added.

situational awareness   An evaluation of all cues received from the surrounding environment.

size principle   The orderly recruitment of motor units, from small to large.

skeletal muscle   The muscles attached to the bones that provide the driving force for motion.

skill   Proficiency at following a prescribed method.

sliding filament theory   Theory of muscle contraction in which the component filaments slide over one another.

slipped disk   *See* **disk herniation**.

slotting   The placement of jobs, standards, etc., in specific categories.

SMED (single minute exchange of die)   A series of techniques for changing over production machinery in less than 10 min.

snapback timing   Time study technique in which, after the watch is read at the break point of each element, the time is returned to zero.

social needs   Third step in Maslow's hierarchy of human needs: need for attention, friendship, and social belonging.

special allowance   A variety of allowances that are related to the process, equipment, materials, etc.; further subdivided into unavoidable delays, avoidable delays, extra, and policy allowances.

speed-accuracy trade-off   The inverse relationship between speed and accuracy, that is, as accuracy increases, the response time becomes slower.

speed rating   Method of rating in which the analyst measures the effectiveness of the operator against the concept of a qualified operator doing the same work at standard performance.

stamps   Strokes collected as a form of debt to be repaid.

standard costs   Prepriced or budget costs on which production runs and sales decisions are made.

standard data   A structured collection of normal time values for work elements, codified in tabular or graphic form.

standard hour plan   A wage incentive plan using day work up to 100 percent performance and piecework beyond 100 percent performance.

standard performance   The performance expected from a qualified operator when following the prescribed method and working at a standard pace for an 8-h shift without undue fatigue, for example, walking at 3 mi/h.

standard time   A unit time value for a work task, as determined by the proper application of appropriate work measurement techniques by qualified personnel.

starting time   The clock time at which a time study starts.

statute law   Written law enacted by legislators and enforced by the executive branch.

storage   Handling for the purpose of positioning and/or securing goods in the space intended.

straight-line layout   *See* **product layout**.

strict liability   Higher level of liability, in which the plaintiff need not prove negligence or fault.

strokes   Positive recognition of an individual and the individual's accomplishments.

Stroop color-word task   The presentation of conflicting color and word stimuli showing effects of redundancy and interference.

supination   Rotation of the forearm such that the palm faces up.

sustained attention   The ability of an operator to maintain attention resources and remain alert over prolonged periods of time; also termed **vigilance**.

synchronous servicing   An ideal case in which both the worker and the machine interact on a fixed, repetitive cycle.

synthetic rating   A system of performance rating in which elemental observed times are compared to times developed through predetermined time systems.

synthetic times   *See* **predetermined times**.

T

temporary partial disability   Type of disability in which the worker is incapable of performing any work for a limited time, but full recovery is expected. The worker can still perform most duties but may suffer some lost time and/or wages.

temporary standard   A standard established for a limited number of pieces or a limited time, to account for the newness of the work or some unusual job condition.

temporary total disability   Type of disability which is sufficiently serious to prevent an employee from ever working in regular employment.

tendinitis   Inflammation of a tendon, caused by repetitive work.

tenosynovitis   Inflammation of tendon sheaths, caused by repetitive work.

therblig   One of 17 basic motions or work elements defined by the Gilbreths.

thick filaments   Microscopic filaments in muscle fibers comprised of long proteins with molecular heads called myosin; thicker in size than the mating thin filaments; both needed for muscular contraction to occur.

thin filaments   Microscopic filaments in muscle fibers comprised of globular proteins called actin; thinner in size than the mating thick filaments; both needed for muscular contraction to occur.

thoracic vertebrae   The 12 vertebral bones of the upper back.

tight rate   A time standard that allows less time than that required by a qualified operator to do the work while working at a standard pace.

time check   Error checking procedure in a time study by accounting for all the times recorded during the study.

time elapsed after study (TEAF)   The stopwatch time readout when the analyst snaps the watch at very end of a time study at the master clock.

time elapsed before study (TEBS)   The stopwatch time readout when an analyst snaps the watch at the start of the first element of a time study.

time study   The procedure using stopwatch timing to establish standards.

time study board   A convenient board used to support the stopwatch and hold the time study form during a time study.

time study form   A form designed to accommodate the elements of a given time study, with spaces for recording their durations.

time value of money   Economic concept of a dollar today being worth more than a dollar in the future.

time-weighted average   The sound level that would produce a given noise dose if a worker were continuously exposed to that sound level over an 8-h workday.

timesharing   *See* **multitasking**.

toolbars   A collection of buttons or icons on a graphical user interface.

top-down processing   Conceptually driven information processing using high-level concepts to process low-level perceptual features.

total recorded time   The sum of check time, effective time, and ineffective time during a time study.

total quality   A Japanese management approach that encompasses quality in all aspects of a business (processes, materials, people, environment) through a continuous improvement process.

touchpad   A form of digitizing tablet integrated into the keyboard for controlling cursor position.

touch screen   Screen with a touch-sensitive overlay on the screen that is activated as the finger approaches the screen.

track point   A force joystick in the middle of a keyboard used to control the cursor.

trackball   An upside-down mouse for controlling cursor position.

transactional analysis   An approach to dealing and communicating with people using the concepts of (1) ego states, (2) transactions, and (3) stroking and stamps.

transactions   Interactions between ego states in Berne's transaction analysis.

travel chart   A table that provides distances traveled between points in a manufacturing or business facility.

trigger finger   Tendinitis in the index finger, caused by repetitive triggering of a power tool.

two-hand process chart   A chart showing the motions made by one hand in relation to those made by the other hand, and using standard therblig abbreviations or symbols.

tympanic membrane   Membrane separating the outer ear canal from the middle ear, also known as the eardrum.

U

ulnar deviation   Bending of the wrist such that the little finger moves toward the arm.

ulterior transactions   Transactions with a hidden meaning.

unaccounted time   Time miscounted during a time study; the difference between elapsed time and recorded time.

unavoidable delay   An interruption in the continuity of an operation that is beyond the control of an operator.

unit labor costs   Employee wages divided by the productivity or worker performance.

unit load   A material in a packed state. Frequently, a standardized-size transport unit.

unit model   *See* **Crawford model.**

use   A basic motion that occurs when either hand or both hands have control of an object during that part of the cycle when productive work is being performed.

V

value engineering   A method for evaluating alternatives using values and weights for alternatives in a payoff matrix.

variable element   An element whose time is affected by one or more characteristics, such as size, shape, hardness, or tolerance, such that as these conditions change, the time required to perform the element changes.

variable fatigue allowance   Fatigue allowances which typically are adjusted for individual workers within a company depending on job or working conditions.

variance   The difference between actual and standard or budgeted costs.

vasoconstriction   The occlusion of peripheral blood vessels due to cold conditions.

vasodilation   Increased peripheral blood flow due to hot conditions.

vertebrae   The bones that form the structural support of the back.

vigilance   *See* **sustained attention**.

visibility   Ability to see fine detail.

visual angle   The angle subtended at the eye by the target.

W

wage incentive   A financial inducement for effort above standard performance.

wage rate   The money rate expressed in dollars and cents per hour, paid to the employee.

waiting time   The time when the operator is unable to do useful work because of the nature of the process, or because of the immediate lack of material.

warehouse   An installation for storing products during long gaps between production stages, or for storing finished products.

Warrick's principle   Principle of display design in which points closest on the display and control move in the same direction, providing the best compatibility.

watch time   Time recorded from a watch reading.

WBGT (wet-bulb globe temperature)   Heat stress index based on a weighted average of wet-bulb, globe, and dry-bulb temperatures.

Weber's law   A rule that states that the ratio of the just noticeable difference to the stimulus level remains constant over the normal sensory range.

Westinghouse rating system   Method of rating developed at Westinghouse Corp., based on four factors, skill, effort, conditions, and consistency.

wet-bulb temperature   Measure of evaporative cooling, using a thermometer with a wet wick and natural air movement.

white finger   The occlusion of blood flow to the hand due to the effects of vibration. Results in loss of dexterity and feeling.

wild value   *See* **abnormal time**.

willful violation   Type of OSHA violation intentionally or knowingly committed by the employer i.e., the employer is aware that a hazardous condition exists and has made no reasonable effort to eliminate it.

wind chill index   Cold stress index describing the rate of heat loss by radiation and convection as a function of ambient temperature and wind velocity.

windows   Areas of the screen, in a graphical user interface, that behave as if they were an independent part of the screen.

work cycle   The total sequence of motions and events that comprise a single operation.

work design   The design process that uses ergonomics to fit the task and workstation to the human operator.

work hardening   Physical training on a simulated job to acclimatize the worker to production line conditions.

work measurement   One of several procedures (time study, work sampling, and predetermined time systems) for establishing standards.

work pace   The rate at which an operation or activity is done.

work sampling   A method of analyzing work by taking a large number of observations at random intervals, to establish standards and improve methods.

workstation   The area where the worker performs the elements of work in a specific operation.

worker–machine process chart   A chart showing the exact relationship in time between the working cycle of the operator and the operating cycle of the machine or machines.

workers compensation   A form of insurance that provides medical care and compensation for employees and their families in the event of injury in the course of employment regardless of fault or negligence.

working memory   Temporary storage of information while it is being processed for a response; also termed **short-term memory**.

Wright model   Learning curve model in which the improvement gained is defined for the cumulative average unit rather than for a particular unit. Also known as the cumulative average model.

Wright's formula   Machine interference time expressed as a percentage of mean servicing time for seven or more machines.

Y

Yerkes-Dodson inverted-U curve   Theory that explains the change in performance as a function of arousal, which has an inverted-U shape, that is, low performance at very low and high levels of arousal, and best performance at some intermediate level of arousal.

Helpful Formulas

(1) *Quadratic*

$$Ax^2 + Bx + C = 0$$

$$x = \frac{-B \pm \sqrt{B^2 - 4AC}}{2A}$$

(2) *Logarithms*

$$\log ab = \log a + \log b$$

$$\log \frac{a}{b} = \log a - \log b$$

$$\log a^n = n \log a$$

$$\log \sqrt[n]{a} = \frac{1}{n} \log a$$

$$\log 1 = 0$$

$$\log_2 x = 1.4427 \ln x$$

(3) *Binomial theorem*

$$(a + b)^n = a^n + na^{n-1}b + \frac{n(n-1)}{2!}a^{n-2}b^2$$
$$+ \frac{n(n-1)(n-2)}{3!}a^{n-3}b^3 + \cdots$$

(4) *Circle*

$$\text{Circumference} = 2\pi r$$
$$\text{Area} = \pi r^2$$

(5) *Prism*

$$\text{Volume} = Ba$$

(6) *Pyramid*

$$\text{Volume} = \tfrac{1}{3} Ba$$

(7) *Right circular cylinder*

$$\text{Volume} = \pi r^2 a$$
$$\text{Lateral surface} = 2\pi ra$$
$$\text{Total surface} = 2\pi r(r + a)$$

(8) *Right circular cone*

$$\text{Volume} = \tfrac{1}{3}\pi r^2 a$$

$$\text{Lateral surface} = \pi rs$$

$$\text{Total surface} = \pi r(r + s)$$

(9) *Sphere*

$$\text{Volume} = \tfrac{4}{3}\pi r^3$$

$$\text{Surface} = 4\pi r^2$$

(10) *Frustum of a right circular cone*

$$\text{Volume} = \tfrac{1}{3}\pi a(R^2 + r^2 + Rr)$$

$$\text{Lateral surface} = \pi s(R + r)$$

(11) *Measurement of angles*

$$1 \text{ degree} = \frac{\pi}{180} = 0.0174 \text{ rad}$$

$$1 \text{ radian (rad)} = 57.29 \text{ degrees}$$

(12) *Trigonometric functions*

   a. Right triangles:

- The sine of the angle A is the quotient of the opposite side divided by the hypotenuse: $\sin A = \dfrac{a}{c}$.
- The tangent of the angle A is the quotient of the opposite side divided by the adjacent side: $\tan A = \dfrac{a}{b}$.
- The secant of the angle A is the quotient of the hypotenuse divided by the adjacent side: $\sec A = \dfrac{c}{b}$.
- The cosine, cotangent, and cosecant of an angle are, respectively, the sine, tangent, and secant of the complement of that angle.

   b. Law of sines:

$$\frac{a}{\sin A} = \frac{b}{\sin B} = \frac{c}{\sin C}$$

   c. Law of cosines:

$$a^2 = b^2 + c^2 - 2bc \cos A$$

(13) *Equations of straight lines*

   a. Slope-intercept form

$$y = a + bx$$

   b. Intercept form

$$\frac{x}{a} + \frac{y}{b} = 1$$

Special Tables

Table A3.1 Natural Sines and Tangents

Angle	Sin	Tan	Cot	Cos	
0	0.0000	0.0000	∞	1.0000	**90**
1	0.0175	0.0175	57.2900	0.9998	**89**
2	0.0349	0.0349	28.6363	0.9994	**88**
3	0.0523	0.0524	19.0811	0.9986	**87**
4	0.0698	0.0699	14.3007	0.9976	**86**
5	0.0872	0.0875	11.4301	0.9962	**85**
6	0.1045	0.1051	9.5144	0.9945	**84**
7	0.1219	0.1228	8.1443	0.9925	**83**
8	0.1392	0.1405	7.1154	0.9903	**82**
9	0.1564	0.1584	6.3138	0.9877	**81**
10	0.1736	0.1763	5.6713	0.9848	**80**
11	0.1908	0.1944	5.1446	0.9816	**79**
12	0.2079	0.2126	4.7046	0.9781	**78**
13	0.2250	0.2309	4.3315	0.9744	**77**
14	0.2419	0.2493	4.0108	0.9703	**76**
15	0.2588	0.2679	3.7321	0.9659	**75**
16	0.2756	0.2867	3.4874	0.9613	**74**
17	0.2924	0.3057	3.2709	0.9563	**73**
18	0.3090	0.3249	3.0777	0.9511	**72**
19	0.3256	0.3443	2.9042	0.9455	**71**
20	0.3420	0.3640	2.7475	0.9397	**70**
21	0.3584	0.3839	2.6051	0.9336	**69**
22	0.3746	0.4040	2.4751	0.9272	**68**
23	0.3907	0.4245	2.3559	0.9205	**67**
24	0.4067	0.4452	2.2460	0.9135	**66**
	Cos	**Cot**	**Tan**	**Sin**	**Angle**

(*continued*)

Table A3.1 (continued)

Angle	Sin	Tan	Cot	Cos	
25	0.4226	0.4663	2.1445	0.9063	**65**
26	0.4384	0.4877	2.0503	0.8988	**64**
27	0.4540	0.5095	1.9626	0.8910	**63**
28	0.4695	0.5317	1.8807	0.8829	**62**
29	0.4848	0.5543	1.8040	0.8746	**61**
30	0.5000	0.5774	1.7321	0.8660	**60**
31	0.5150	0.6009	1.6643	0.8572	**59**
32	0.5299	0.6249	1.6003	0.8480	**58**
33	0.5446	0.6494	1.5399	0.8387	**57**
34	0.5592	0.6745	1.4826	0.8290	**56**
35	0.5736	0.7002	1.4281	0.8192	**55**
36	0.5878	0.7265	1.3764	0.8090	**54**
37	0.6018	0.7536	1.3270	0.7986	**53**
38	0.6157	0.7813	1.2799	0.7880	**52**
39	0.6293	0.8098	1.2349	0.7771	**51**
40	0.6428	0.8391	1.1918	0.7660	**50**
41	0.6561	0.8693	1.1504	0.7547	**49**
42	0.6691	0.9004	1.1106	0.7431	**48**
43	0.6820	0.9325	1.0724	0.7314	**47**
44	0.6947	0.9657	1.0355	0.7193	**46**
45	0.7071	1.0000	1.0000	0.7071	**45**
	Cos	**Cot**	**Tan**	**Sin**	**Angle**

Table A3.2 Cumulative Probabilities of the Standard Normal Distribution

Entry is area A under the standard normal curve from $-\infty$ to $z(A)$.

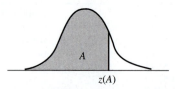

z	0.00	0.01	0.02	0.03	0.04	0.05	0.06	0.07	0.08	0.09
0.0	0.5000	0.5040	0.5080	0.5120	0.5160	0.5199	0.5239	0.5279	0.5319	0.5359
0.1	0.5398	0.5438	0.5478	0.5517	0.5557	0.5596	0.5636	0.5675	0.5714	0.5753
0.2	0.5793	0.5832	0.5871	0.5910	0.5948	0.5987	0.6026	0.6064	0.6103	0.6141
0.3	0.6179	0.6217	0.6255	0.6293	0.6331	0.6368	0.6406	0.6443	0.6480	0.6517
0.4	0.6554	0.6591	0.6628	0.6664	0.6700	0.6736	0.6772	0.6808	0.6844	0.6879
0.5	0.6915	0.6950	0.6985	0.7019	0.7054	0.7088	0.7123	0.7157	0.7190	0.7224
0.6	0.7257	0.7291	0.7324	0.7357	0.7389	0.7422	0.7454	0.7486	0.7517	0.7549
0.7	0.7580	0.7611	0.7642	0.7673	0.7704	0.7734	0.7764	0.7794	0.7823	0.7852
0.8	0.7881	0.7910	0.7939	0.7967	0.7995	0.8023	0.8051	0.8078	0.8106	0.8133
0.9	0.8159	0.8186	0.8212	0.8238	0.8264	0.8289	0.8315	0.8340	0.8365	0.8389
1.0	0.8413	0.8438	0.8461	0.8485	0.8508	0.8531	0.8554	0.8577	0.8599	0.8621
1.1	0.8643	0.8665	0.8686	0.8708	0.8729	0.8749	0.8770	0.8790	0.8810	0.8830
1.2	0.8849	0.8869	0.8888	0.8907	0.8925	0.8944	0.8962	0.8980	0.8997	0.9015
1.3	0.9032	0.9049	0.9066	0.9082	0.9099	0.9115	0.9131	0.9147	0.9162	0.9177
1.4	0.9192	0.9207	0.9222	0.9236	0.9251	0.9265	0.9279	0.9292	0.9306	0.9319
1.5	0.9332	0.9345	0.9357	0.9370	0.9382	0.9394	0.9406	0.9418	0.9429	0.9441
1.6	0.9452	0.9463	0.9474	0.9484	0.9495	0.9505	0.9515	0.9525	0.9535	0.9545
1.7	0.9554	0.9564	0.9573	0.9582	0.9591	0.9599	0.9608	0.9616	0.9625	0.9633
1.8	0.9641	0.9649	0.9656	0.9664	0.9671	0.9678	0.9686	0.9693	0.9699	0.9706
1.9	0.9713	0.9719	0.9726	0.9732	0.9738	0.9744	0.9750	0.9756	0.9761	0.9767
2.0	0.9772	0.9778	0.9783	0.9788	0.9793	0.9798	0.9803	0.9808	0.9812	0.9817
2.1	0.9821	0.9826	0.9830	0.9834	0.9838	0.9842	0.9846	0.9850	0.9854	0.9857
2.2	0.9861	0.9864	0.9868	0.9871	0.9875	0.9878	0.9881	0.9884	0.9887	0.9890
2.3	0.9893	0.9896	0.9898	0.9901	0.9904	0.9906	0.9909	0.9911	0.9913	0.9916
2.4	0.9918	0.9920	0.9922	0.9925	0.9927	0.9929	0.9931	0.9932	0.9934	0.9936
2.5	0.9938	0.9940	0.9941	0.9943	0.9945	0.9946	0.9948	0.9949	0.9951	0.9952
2.6	0.9953	0.9955	0.9956	0.9957	0.9959	0.9960	0.9961	0.9962	0.9963	0.9964
2.7	0.9965	0.9966	0.9967	0.9968	0.9969	0.9970	0.9971	0.9972	0.9973	0.9974
2.8	0.9974	0.9975	0.9976	0.9977	0.9977	0.9978	0.9979	0.9979	0.9980	0.9981
2.9	0.9981	0.9982	0.9982	0.9983	0.9984	0.9984	0.9985	0.9985	0.9986	0.9986
3.0	0.9987	0.9987	0.9987	0.9988	0.9988	0.9989	0.9989	0.9989	0.9990	0.9990
3.1	0.9990	0.9991	0.9991	0.9991	0.9992	0.9992	0.9992	0.9992	0.9993	0.9993
3.2	0.9993	0.9993	0.9994	0.9994	0.9994	0.9994	0.9994	0.9995	0.9995	0.9995
3.3	0.9995	0.9995	0.9995	0.9996	0.9996	0.9996	0.9996	0.9996	0.9996	0.9997
3.4	0.9997	0.9997	0.9997	0.9997	0.9997	0.9997	0.9997	0.9997	0.9997	0.9998

Selected Percentiles							
Cumulative probability A:	0.90	0.95	0.975	0.98	0.99	0.995	0.999
$z(A)$:	1.282	1.645	1.960	2.054	2.326	2.576	3.090

Source: J. Neter, W. Wasserman, and M. H. Kutner, *Applied Linear Statistical Models.* 2d ed. Homewood, IL: Richard D. Irwin, 1985. Reproduced with permission of the McGraw-Hill Companies.

Table A3.3 Percentage Points of the *t* Distribution

						Probability *P*							
n	0.9	0.8	0.7	0.6	0.5	0.4	0.3	0.2	0.1	0.05	0.02	0.01	0.001
1	0.158	0.325	0.510	0.727	1.000	1.376	1.963	3.078	6.314	12.706	31.821	63.657	636.619
2	0.142	0.289	0.445	0.617	0.816	1.061	1.386	1.886	2.920	4.303	6.965	9.925	31.598
3	0.137	0.277	0.424	0.584	0.765	0.978	1.250	1.638	2.353	3.182	4.541	5.841	12.941
4	0.134	0.271	0.414	0.569	0.741	0.941	1.190	1.533	2.132	2.776	3.747	4.604	8.610
5	0.132	0.267	0.408	0.559	0.727	0.920	1.156	1.476	2.015	2.571	3.365	4.032	6.859
6	0.131	0.265	0.404	0.553	0.718	0.906	1.134	1.440	1.943	2.447	3.143	3.707	5.959
7	0.130	0.263	0.402	0.549	0.711	0.896	1.119	1.415	1.895	2.365	2.998	3.499	5.405
8	0.130	0.262	0.399	0.546	0.706	0.889	1.108	1.397	1.860	2.306	2.896	3.355	5.041
9	0.129	0.261	0.398	0.543	0.703	0.883	1.100	1.383	1.833	2.262	2.821	3.250	4.781
10	0.129	0.260	0.397	0.542	0.700	0.879	1.093	1.372	1.812	2.228	2.764	3.169	4.587
11	0.129	0.260	0.396	0.540	0.697	0.876	1.088	1.363	1.796	2.201	2.718	3.106	4.437
12	0.128	0.259	0.395	0.539	0.695	0.873	1.083	1.356	1.782	2.179	2.681	3.055	4.318
13	0.128	0.259	0.394	0.538	0.694	0.870	1.079	1.350	1.771	2.160	2.650	3.012	4.221
14	0.128	0.258	0.393	0.537	0.692	0.868	1.076	1.345	1.761	2.145	2.624	2.977	4.140
15	0.128	0.258	0.393	0.536	0.691	0.866	1.074	1.341	1.753	2.131	2.602	2.947	4.073
16	0.128	0.258	0.392	0.535	0.690	0.865	1.071	1.337	1.746	2.120	2.583	2.921	4.015
17	0.128	0.257	0.392	0.534	0.689	0.863	1.069	1.333	1.740	2.110	2.567	2.898	3.965
18	0.127	0.257	0.392	0.534	0.688	0.862	1.067	1.330	1.734	2.101	2.552	2.878	3.922
19	0.127	0.257	0.391	0.533	0.688	0.861	1.066	1.328	1.729	2.093	2.539	2.861	3.883
20	0.127	0.257	0.391	0.533	0.687	0.860	1.064	1.325	1.725	2.086	2.528	2.845	3.850
21	0.127	0.257	0.391	0.532	0.686	0.859	1.063	1.323	1.721	2.080	2.518	2.831	3.819
22	0.127	0.256	0.390	0.532	0.686	0.858	1.061	1.321	1.717	2.074	2.508	2.819	3.792
23	0.127	0.256	0.390	0.532	0.685	0.858	1.060	1.319	1.714	2.069	2.500	2.807	3.767
24	0.127	0.256	0.390	0.531	0.685	0.857	1.059	1.318	1.711	2.064	2.492	2.797	3.745
25	0.127	0.256	0.390	0.531	0.684	0.856	1.058	1.316	1.708	2.060	2.485	2.787	3.725
26	0.127	0.256	0.390	0.531	0.684	0.856	1.058	1.315	1.706	2.056	2.479	2.779	3.707
27	0.127	0.256	0.389	0.531	0.684	0.855	1.057	1.314	1.703	2.052	2.473	2.771	3.690
28	0.127	0.256	0.389	0.530	0.683	0.855	1.056	1.313	1.701	2.048	2.467	2.763	3.674
29	0.127	0.256	0.389	0.530	0.683	0.854	1.055	1.311	1.699	2.045	2.462	2.756	3.659
30	0.127	0.256	0.389	0.530	0.683	0.854	1.055	1.310	1.697	2.042	2.457	2.750	3.646
40	0.126	0.255	0.388	0.529	0.681	0.851	1.050	1.303	1.684	2.021	2.423	2.704	3.551
60	0.126	0.254	0.387	0.527	0.679	0.848	1.046	1.296	1.671	2.000	2.390	2.660	3.460
120	0.126	0.254	0.386	0.526	0.677	0.845	1.041	1.289	1.658	1.980	2.358	2.617	3.373
∞	0.126	0.253	0.385	0.524	0.674	0.842	1.036	1.282	1.645	1.960	2.326	2.576	3.291

Source: Reprinted from Table III of R. A. Fisher and F. Yates, *Statistical Tables for Biological, Agricultural, and Medical Research* (Edinburgh: Oliver & Boyd, Ltd.), by permission of the authors and publishers.

Note: Probabilities refer to the sum of the two tail areas; for a single tail, divide the probability by 2.

Table A3.4   Percentiles of the χ^2 distribution

Entry is $\chi^2(A; \nu)$ where $P\{\chi^2(\nu) \leq \chi^2(A; \nu)\} = A$

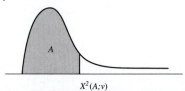

ν	\multicolumn{10}{c}{A}

ν	0.005	0.010	0.025	0.050	0.100	0.900	0.950	0.975	0.990	0.995
1	0.0^4393	0.0^3157	0.0^3982	0.0^2393	0.0158	2.71	3.84	5.02	6.63	7.88
2	0.0100	0.0201	0.0506	0.103	0.211	4.61	5.99	7.38	9.21	10.60
3	0.072	0.115	0.216	0.352	0.584	6.25	7.81	9.35	11.34	12.84
4	0.207	0.297	0.484	0.711	1.064	7.78	9.49	11.14	13.28	14.86
5	0.412	0.554	0.831	1.145	1.61	9.24	11.07	12.83	15.09	16.75
6	0.676	0.872	1.24	1.64	2.20	10.64	12.59	14.45	16.81	18.55
7	0.989	1.24	1.69	2.17	2.83	12.02	14.07	16.01	18.48	20.28
8	1.34	1.65	2.18	2.73	3.49	13.36	15.51	17.53	20.09	21.96
9	1.73	2.09	2.70	3.33	4.17	14.68	16.92	19.02	21.67	23.59
10	2.16	2.56	3.25	3.94	4.87	15.99	18.31	20.48	23.21	25.19
11	2.60	3.05	3.82	4.57	5.58	17.28	19.68	21.92	24.73	26.76
12	3.07	3.57	4.40	5.23	6.30	18.55	21.03	23.34	26.22	28.30
13	3.57	4.11	5.01	5.89	7.04	19.81	22.36	24.74	27.69	29.82
14	4.07	4.66	5.63	6.57	7.79	21.06	26.68	26.12	29.14	31.32
15	4.60	5.23	6.26	7.26	8.55	22.31	25.00	27.49	30.58	32.80
16	5.14	5.81	6.91	7.96	9.31	23.54	26.30	28.85	32.00	34.27
17	5.70	6.41	7.56	8.67	10.09	24.77	27.59	30.19	33.41	35.72
18	6.26	7.01	8.23	9.39	10.86	25.99	28.87	31.53	34.81	37.16
19	6.84	7.63	8.91	10.12	11.65	27.20	30.14	32.85	36.19	38.58
20	7.43	8.26	9.59	10.85	12.44	28.41	31.41	34.17	37.57	40.00
21	8.03	8.90	10.28	11.59	13.24	29.62	32.67	35.48	38.93	41.40
22	8.64	9.54	10.98	12.34	14.04	30.81	33.92	36.78	40.29	42.80
23	9.26	10.20	11.69	13.09	14.85	32.01	35.17	38.08	41.64	44.18
24	9.89	10.86	12.40	13.85	15.66	33.20	36.42	39.36	42.98	45.56
25	10.52	11.52	13.12	14.61	16.47	34.38	37.65	40.65	44.31	46.93
26	11.16	12.20	13.84	15.38	17.29	35.56	38.89	41.92	45.64	48.29
27	11.81	12.88	14.57	16.15	18.11	36.74	40.11	43.19	46.96	49.64
28	12.46	13.56	15.31	16.93	18.94	37.94	41.34	44.46	48.28	50.99
29	13.12	14.26	16.05	17.71	19.77	39.09	42.56	45.72	49.59	52.34
30	13.79	14.95	16.79	18.49	20.60	40.26	43.77	46.98	50.89	53.67
40	20.71	22.16	24.43	26.51	29.05	51.81	55.76	59.34	63.69	66.77
50	27.99	29.71	32.36	34.76	37.69	63.17	67.50	71.42	76.15	79.49
60	35.53	37.48	40.48	43.19	46.46	74.40	79.08	83.30	88.38	91.95
70	43.28	45.44	48.76	51.74	55.33	85.53	90.53	95.02	100.4	104.2
80	51.17	53.54	57.15	60.39	64.28	96.58	101.9	106.6	112.3	116.3
90	59.20	61.75	65.65	69.13	73.29	107.6	113.1	118.1	124.1	128.3
100	67.33	70.06	74.22	77.93	82.36	118.5	124.3	129.6	135.8	140.2

Source: Reprinted, with permission, from C. M. Thompson, "Table of Percentage Points of the Chi-Square Distribution," *Biometrika* 32 (1941), pp. 188–89.

Table A3.5 Random Numbers III

22 17 68 65 84	68 95 23 92 35	87 02 22 57 51	61 09 43 95 06	58 24 82 03 47
19 36 27 59 46	13 79 93 37 55	39 77 32 77 09	85 52 05 30 62	47 83 51 62 74
16 77 23 02 77	09 61 87 25 21	28 06 24 25 93	16 71 13 59 78	23 05 47 47 25
78 43 76 71 61	20 44 90 32 64	97 67 63 99 61	46 38 03 93 22	69 81 21 99 21
03 28 28 26 08	73 37 32 04 05	69 30 16 09 05	88 69 58 28 99	35 07 44 75 47
93 22 53 64 39	07 10 63 76 35	87 03 04 79 88	08 13 13 85 51	55 34 57 72 69
78 76 58 54 74	92 38 70 96 92	52 06 79 79 45	82 63 18 27 44	69 66 92 19 09
23 68 35 26 00	99 53 93 61 28	52 70 05 48 34	56 65 05 61 86	90 92 10 70 80
15 39 25 70 99	93 86 52 77 65	15 33 59 05 28	22 87 26 07 47	86 96 98 29 06
58 71 96 30 24	18 46 23 34 27	85 13 99 24 44	49 18 09 79 49	74 16 32 23 02
57 35 27 33 72	24 53 63 94 09	41 10 76 47 91	44 04 95 49 66	39 60 04 59 81
48 50 86 54 48	22 06 34 72 52	82 21 15 65 20	33 29 94 71 11	15 91 29 12 03
61 96 48 95 03	07 16 39 33 66	98 56 10 56 79	77 21 30 27 12	90 49 22 23 62
36 93 89 41 26	29 70 83 63 51	99 74 20 52 36	87 09 41 15 09	98 60 16 03 03
18 87 00 42 31	57 90 12 02 07	23 47 37 17 31	54 08 01 88 63	39 41 88 92 10
88 56 53 27 59	33 35 72 67 47	77 34 55 45 70	08 18 27 38 90	16 95 86 70 75
09 72 95 84 29	49 41 31 06 70	42 38 06 45 18	64 84 73 31 65	52 53 37 97 15
12 96 88 17 31	65 19 69 02 83	60 75 86 90 68	24 64 19 35 51	56 61 87 39 12
85 94 57 24 16	92 09 84 38 76	22 00 27 69 85	29 81 94 78 70	21 94 47 90 12
38 64 43 59 98	98 77 87 68 07	91 51 67 62 44	40 98 05 93 78	23 32 65 41 18
53 44 09 42 72	00 41 86 79 79	68 47 22 00 20	35 55 31 51 51	00 83 63 22 55
40 76 66 26 84	57 99 99 90 37	36 63 32 08 58	37 40 13 68 97	87 64 81 07 83
02 17 79 18 05	12 59 52 57 02	22 07 90 47 03	28 14 11 30 79	20 69 22 40 98
95 17 82 06 53	31 51 10 96 46	92 06 88 07 77	56 11 50 81 69	40 23 72 51 39
35 76 22 42 92	96 11 83 44 80	34 68 35 48 77	33 42 40 90 60	73 96 53 97 86
26 29 13 56 41	85 47 04 66 08	34 72 57 59 13	82 43 80 46 15	38 26 61 70 04
77 80 20 75 82	72 82 32 99 90	63 95 73 76 63	89 73 44 99 05	48 67 26 43 18
46 40 66 44 52	91 36 74 43 53	30 82 13 54 00	78 45 63 98 35	55 03 36 67 68
37 56 08 18 09	77 53 84 46 47	31 91 18 95 58	24 16 74 11 53	44 10 13 85 57
61 65 61 68 66	37 27 47 39 19	84 83 70 07 48	53 21 40 06 71	95 06 79 88 54
93 43 69 64 07	34 18 04 52 35	56 27 09 24 86	61 85 53 83 45	19 90 70 99 00
21 96 60 12 99	11 20 99 45 18	48 13 93 55 34	18 37 79 49 90	65 97 38 20 46
95 20 47 97 97	27 37 83 28 71	00 06 41 41 74	45 89 09 39 84	51 67 11 52 49
97 86 21 78 73	10 65 81 92 59	58 76 17 14 97	04 76 62 16 17	17 95 70 45 80
69 92 06 34 13	59 71 74 17 32	27 55 10 24 19	23 71 82 13 74	63 52 52 01 41
04 31 17 21 56	33 73 99 19 87	26 72 39 27 67	53 77 57 68 93	60 61 97 22 61
61 06 98 03 91	87 14 77 43 96	43 00 65 98 50	45 60 33 01 07	98 99 46 50 47
85 93 85 86 88	72 87 08 62 40	16 06 10 89 20	23 21 34 74 97	76 38 03 29 63
21 74 32 47 45	73 96 07 94 52	09 65 90 77 47	25 76 16 19 33	53 05 70 53 30
15 69 53 82 80	79 96 23 53 10	65 39 07 16 29	45 33 02 43 70	02 87 40 41 45
02 89 08 04 49	20 21 14 68 86	87 63 93 95 17	11 29 01 95 80	35 14 97 35 33
87 18 15 89 79	85 43 01 72 73	08 61 74 51 69	89 74 39 82 15	94 51 33 41 67
98 83 71 94 22	59 97 50 99 52	08 52 85 08 40	87 80 61 65 31	91 51 80 32 44
10 08 58 21 66	72 68 49 29 31	89 85 84 46 06	59 73 19 85 23	65 09 29 75 63
47 90 56 10 08	88 02 84 27 83	42 29 72 23 19	66 56 45 65 79	20 71 53 20 25
22 85 61 68 90	49 64 92 85 44	16 40 12 89 88	50 14 49 81 06	01 82 77 45 12
67 80 43 79 33	12 83 11 41 16	25 58 19 68 70	77 02 54 00 52	53 43 37 15 26
27 62 50 96 72	79 44 61 40 15	14 53 40 65 39	27 31 58 50 28	11 39 03 34 25
33 78 80 87 15	38 30 06 38 21	14 47 47 07 26	54 96 87 53 32	40 36 40 96 76
13 13 92 66 99	47 24 49 57 74	32 25 43 62 17	10 97 11 69 84	99 63 22 32 98
10 27 53 96 23	71 50 54 36 23	54 31 04 82 98	04 14 12 15 09	26 78 25 47 47
28 41 50 61 88	64 85 27 20 18	83 36 36 05 56	39 71 65 09 62	94 76 62 11 89

Source: Reprinted with permission from Random Numbers IV of Table XXXIII of R. A. Fisher and F. Yates, *Statistical Tables for Biological, Agricultural, and Medical Research* (Edinburgh: Oliver & Boyd, Ltd.).

Table A3.6 Useful Information

To find the circumference of a circle, multiply the diameter by 3.1416.

To find the diameter of a circle, multiply the circumference by 0.31831.

To find the area of a circle, multiply the square of the diameter by 0.7854.

The radius of a circle $\times$ 6.283185 = the circumference.

The square of the circumference of a circle $\times$ 0.07958 = the area.

One-half the circumference of a circle $\times$ one-half its diameter = the area.

The circumference of a circle $\times$ 0.159155 = the radius.

The square root of the area of a circle $\times$ 0.56419 = the radius.

The square root of the area of a circle $\times$ 1.12838 = the diameter.

To find the diameter of a circle equal in area to a given square, multiply a side of the square by 1.12838.

To find the side of a square equal in area to a given circle, multiply the diameter by 0.8862.

To find the side of a square inscribed in a circle, multiply the diameter by 0.7071.

To find the side of a hexagon inscribed in a circle, multiply the diameter of the circle by 0.500.

To find the diameter of a circle inscribed in a hexagon, multiply a side of the hexagon by 1.7321.

To find the side of an equilateral triangle inscribed in a circle, multiply the diameter of the circle by 0.866.

To find the diameter of a circle inscribed in an equilateral triangle, multiply a side of the triangle by 0.57735.

To find the area of the surface of a ball (sphere), multiply the square of the diameter by 3.1416.

To find the volume of a ball (sphere), multiply the cube of the diameter by 0.5236.

Doubling the diameter of a pipe increases its capacity four times.

To find the pressure in pounds per square inch at the base of a column of water, multiply the height of the column in feet by 0.433.

A gallon of water (U.S. standard) weighs 8.336 lb and contains 231 in^3. A cubic foot of water contains 7^1/$_2$ gal and 1,728 in^3 and weighs 62.425 lb at a temperature of about 39°F.

These weights change slightly above and below this temperature.

Table A3.7 15 Percent Compound Interest Factors

	Single Payment		Uniform Series			
n	Compound Amount Factor caf'	Present Worth Factor pwf'	Sinking Fund Factor sff	Capital Recovery Factor crf	Compound Amount Factor caf	Present Worth Factor pwf
	Given P To find S $(1+i)^n$	Given S To find P $\dfrac{1}{(1+i)^n}$	Given S To find R $\dfrac{i}{(1+i)^n-1}$	Given P To find R $\dfrac{i(1+i)^n}{(1+i)^n-1}$	Given R To find S $\dfrac{(1+i)^n-1}{i}$	Given R To find P $\dfrac{(1+i)^n-1}{i(1+i)^n}$
1	1.150	0.8696	1.00000	1.15000	1.000	0.870
2	1.322	0.7561	0.46512	0.61512	2.150	1.626
3	1.521	0.6575	0.28798	0.43798	3.472	2.283
4	1.749	0.5718	0.20026	0.35027	4.993	2.855
5	2.011	0.4972	0.14832	0.29832	6.742	3.352
6	2.313	0.4323	0.11424	0.26424	8.754	3.784
7	2.660	0.3759	0.09036	0.24036	11.067	4.160
8	3.059	0.3269	0.07285	0.22285	13.727	4.487
9	3.518	0.2843	0.05957	0.20957	16.786	4.772
10	4.046	0.2472	0.04925	0.19925	20.304	5.019
11	4.652	0.2149	0.04107	0.19107	24.349	5.234
12	5.350	0.1869	0.03448	0.18448	29.002	5.421
13	6.153	0.1625	0.02911	0.17911	34.352	5.583
14	7.076	0.1413	0.02469	0.17469	40.505	5.724
15	8.137	0.1229	0.02102	0.17102	47.580	5.847
16	9.358	0.1069	0.01795	0.16795	55.717	5.954
17	10.761	0.0929	0.01537	0.16537	65.075	6.047
18	12.375	0.0808	0.01319	0.16319	75.836	6.128
19	14.232	0.0703	0.01134	0.16134	88.212	6.198
20	16.367	0.0611	0.00976	0.15976	102.443	6.259
21	18.821	0.0531	0.00842	0.15842	118.810	6.312
22	21.645	0.0462	0.00727	0.15727	137.631	6.359
23	24.891	0.0402	0.00628	0.15628	159.276	6.399
24	28.625	0.0349	0.00543	0.15543	184.167	6.434
25	32.919	0.0304	0.00470	0.15470	212.793	6.464
26	37.857	0.0264	0.00407	0.15407	245.711	6.491
27	43.535	0.0230	0.00353	0.15353	283.568	6.514
28	50.065	0.0200	0.00306	0.15306	327.103	6.534
29	57.575	0.0174	0.00265	0.15265	377.169	6.551
30	66.212	0.0151	0.00230	0.15230	434.744	6.566
31	76.143	0.0131	0.00200	0.15200	500.956	6.579
32	87.565	0.0114	0.00173	0.15173	577.099	6.591
33	100.700	0.0099	0.00150	0.15150	664.664	6.600
34	115.805	0.0086	0.00131	0.15131	765.364	6.609
35	133.175	0.0075	0.00113	0.15113	881.168	6.617
40	267.862	0.0037	0.00056	0.15056	1,779.1	6.642
45	538.767	0.0019	0.00028	0.15028	3,585.1	6.654
50	1,083.652	0.0009	0.00014	0.15014	7,217.7	6.661
∞				0.15000		6.667

Table A3.8 Tables of Machine Interference Time *i* and Machine Running Time *m* for Selected Servicing Constants ($k = l/m$)
(Values expressed as percentages of total time, where $m + l + i = 100$ percent)

	(a)		(b)			(a)		(b)			(a)		(b)	
n	*i*	*m*	*i*	*m*	*n*	*i*	*m*	*i*	*m*	*n*	*i*	*m*	*i*	*m*
$k = 0.01$					**$k = 0.02$ (cont.)**					**$k = 0.03$ (cont.)**				
1	0.0	99.0	0.0	99.0	10	0.2	97.8	0.4	97.6	32			8.9	88.5
10	0.1	99.0	0.1	98.9	15	0.4	97.7	0.7	97.4	33			9.7	87.7
20	0.1	98.9	0.2	98.8	20	0.6	97.5	1.1	97.0	34			10.6	86.8
30	0.2	98.8	0.4	98.6	25	0.8	97.2	1.6	96.5	35			11.6	85.9
40			0.6	98.4	30	1.2	96.9	2.2	95.9	36			12.6	84.9
50			0.9	98.1	35			3.1	95.0	37			13.7	83.8
60			1.3	97.8	40			4.3	93.8	38			14.9	86.8
70			1.8	97.2	45			6.1	92.0	39			16.1	81.4
80			2.7	96.3	50			8.7	89.5	40			17.4	80.2
85			3.4	95.7	51			9.3	88.9	41			18.8	78.9
90			4.2	94.9	52			10.0	88.3	42			20.1	77.5
95			5.2	93.8	53			10.7	87.6	43			21.6	76.2
100			6.7	92.4	54			11.5	86.8	44			23.0	74.8
105			8.5	90.6	55			12.3	86.0	45			24.4	73.4
110			10.7	88.4	56			13.1	85.2	46			25.9	72.0
115			13.4	85.8	57			14.0	84.3	47			27.3	70.6
120			16.3	82.9	58			14.9	83.4	48			28.7	69.2
121			16.9	82.3	59			15.9	82.5					
122			17.5	81.7	60			16.8	81.5	**$k = 0.04$**				
123			18.1	81.1	61			17.9	80.5					
124			18.8	80.4	62			18.9	79.5	1	0.0	96.2	0.0	96.2
125			19.4	79.8	63			19.9	78.5	2	0.1	96.1	0.2	96.0
126			20.0	79.2	64			21.0	77.5	3	0.2	96.0	0.3	95.9
127			20.6	78.6	65			22.0	76.4	4	0.2	95.9	0.5	95.7
128			21.2	78.1	66			23.1	75.4	5	0.3	95.8	0.7	95.5
129			21.8	77.5	67			24.2	74.4	6	0.5	95.7	0.9	95.3
130			22.4	76.9	68			25.2	73.3	7	0.6	95.6	1.1	95.1
131			22.9	76.3	69			26.2	72.3	8	0.7	95.5	1.3	94.9
132			23.5	75.7	70			27.2	71.3	9	0.8	95.4	1.5	94.7
133			24.1	75.2	71			28.2	70.4	10	1.0	95.2	1.8	94.4
134			24.6	74.6	72			29.2	69.4	11	1.1	95.1	2.1	94.1
135			25.2	74.1						12	1.3	94.9	2.4	93.8
136			25.7	73.5	**$k = 0.03$**					13	1.5	94.7	2.8	93.5
137			26.3	73.0						14	1.8	94.5	3.2	93.1
138			26.8	72.5	1	0.0	97.1	0.0	97.1	15	2.0	94.2	3.6	92.7
139			27.3	71.9	5	0.2	96.9	0.4	96.7	16	2.3	94.0	4.0	92.3
140			27.9	71.4	10	0.5	96.6	1.0	96.2	17	2.6	93.6	4.5	91.8
141			28.4	70.9	15	1.0	96.2	1.8	95.4	18	3.0	93.3	5.1	91.3
142			28.9	70.4	20	1.6	95.5	3.0	94.2	19	3.4	92.9	5.7	90.7
143			29.4	69.9	25	2.8	94.4	4.7	92.5	20	3.9	92.4	6.4	90.0
144			29.9	69.4	26	3.1	94.1	5.2	92.1	21	4.5	91.8	7.1	89.3
					27	3.4	93.7	5.7	91.6	22	5.2	91.2	8.0	88.5
$k = 0.02$					28	3.8	93.4	6.2	91.1	23	6.0	90.4	8.9	87.6
					29	4.3	92.9	6.8	90.5	24	6.8	89.6	9.9	86.7
1	0.0	98.0	0.0	98.0	30	4.8	92.4	7.4	89.9	25	7.9	88.6	11.0	85.6
5	0.1	98.0	0.2	97.9	31			8.1	89.2	26	9.0	87.5	12.2	84.5

Table A3.8   (continued)

Left panel

	(a)		(b)	
n	i	m	i	m
k = 0.04 (cont.)				
27	10.4	86.2	13.4	83.2
28	11.9	84.7	14.8	81.9
29	13.6	83.0	16.3	80.5
30	15.5	81.3	17.9	79.0
31			19.6	77.4
32			21.3	75.7
33			23.0	74.0
34			24.8	72.3
35			26.6	70.6
36			28.4	68.9
37			30.1	67.2
k = 0.05				
1	0.0	95.2	0.0	95.2
2	0.1	95.1	0.2	95.0
3	0.2	95.0	0.5	94.8
4	0.4	94.9	0.7	94.5
5	0.5	94.7	1.0	94.3
6	0.7	94.6	1.4	94.0
7	0.9	94.4	1.7	93.6
8	1.1	94.2	2.1	93.3
9	1.4	93.9	2.5	92.9
10	1.6	93.7	3.0	92.4
11	2.0	93.4	3.5	91.9
12	2.3	93.0	4.1	91.4
13	2.7	92.6	4.7	90.8
14	3.2	92.2	5.4	90.1
15	3.8	91.7	6.2	89.3
16	4.4	91.0	7.1	88.5
17	5.2	90.3	8.1	87.6
18	6.1	89.5	9.1	86.5
19	7.1	88.5	10.4	85.4
20	8.4	87.3	11.7	84.1
21	9.8	85.9	13.1	82.7
22	11.5	84.3	14.7	81.2
23	13.4	82.5	16.5	79.6
24	15.5	80.5	18.3	77.8
25	17.8	78.2	20.2	76.0
26	20.3	75.9	22.2	74.1
27	22.8	73.6	24.3	72.1
28	25.3	71.2	26.4	70.1
29	27.9	68.8	28.5	68.1
k = 0.06				
1	0.0	94.3	0.0	94.3
2	0.2	94.2	0.3	94.0

Middle panel

	(a)		(b)	
n	i	m	i	m
k = 0.06 (cont.)				
3	0.4	94.0	0.7	93.7
4	0.6	93.8	1.1	93.3
5	0.8	93.6	1.5	92.9
6	1.1	93.3	2.0	92.5
7	1.4	93.1	2.5	92.0
8	1.7	92.7	3.1	91.4
9	2.1	92.4	3.7	90.8
10	2.6	91.9	4.5	90.1
11	3.1	91.4	5.3	89.4
12	3.8	90.8	6.2	88.5
13	4.5	90.1	7.3	87.5
14	5.4	89.2	8.4	86.4
15	6.5	88.2	9.7	85.2
16	7.8	87.0	11.2	83.8
17	9.3	85.6	12.8	82.3
18	11.1	83.9	14.6	80.6
19	13.2	81.9	16.5	78.8
20	15.6	79.7	18.6	76.8
21			20.8	74.7
22			23.1	72.5
23			25.5	70.3
24			27.9	68.0
25			30.3	65.8
k = 0.07				
1	0.0	93.5	0.0	93.5
2	0.2	93.2	0.4	93.1
3	0.5	93.0	0.9	92.6
4	0.8	92.7	1.4	92.1
5	1.1	92.4	2.0	91.6
6	1.5	92.1	2.7	91.0
7	1.9	91.7	3.4	90.3
8	2.4	91.2	4.3	89.5
9	3.1	90.6	5.2	88.6
10	3.8	89.9	6.3	87.6
11	4.7	89.1	7.5	86.4
12	5.7	88.1	8.9	85.1
13	7.0	86.9	10.4	83.7
14	8.6	85.4	12.2	82.1
15	10.4	83.7	14.1	80.3
16	12.6	81.6	16.2	78.3
17	15.2	79.3	18.5	76.2
18	18.1	76.6	21.0	73.9
19	21.1	73.7	23.5	71.5
20	24.4	70.7	26.2	69.0
21			28.9	66.5

Right panel

	(a)		(b)	
n	i	m	i	m
k = 0.08				
1	0.0	92.6	0.0	92.6
2	0.3	92.3	0.5	92.1
3	0.6	92.0	1.2	91.5
4	1.0	91.7	1.9	90.9
5	1.4	91.2	2.7	90.1
6	2.0	90.8	3.5	89.3
7	2.6	90.2	4.5	88.4
8	3.4	89.5	5.7	87.3
9	4.3	88.6	7.0	86.1
10	5.4	87.6	8.5	84.8
11	6.7	86.4	10.1	83.2
12	8.4	84.8	12.0	81.4
13	10.4	83.0	14.2	79.5
14	12.8	80.8	16.5	77.3
15	15.6	78.2	19.0	75.0
16	18.8	75.2	21.8	72.4
17	22.2	72.0	24.6	69.8
18	25.7	68.8	27.6	67.1
19	28.2	66.5	30.5	64.4
k = 0.09				
1	0.0	91.5	0.0	91.7
2	0.4	91.4	0.7	91.1
3	0.8	91.0	1.4	90.4
4	1.3	90.6	2.3	89.6
5	1.9	90.0	3.3	88.7
6	2.6	89.4	4.5	87.7
7	3.4	88.6	5.8	86.5
8	4.5	87.6	7.3	85.1
9	5.7	86.5	9.0	83.5
10	7.3	85.0	10.9	81.7
11	9.3	83.2	13.1	79.7
12	11.7	81.0	15.6	77.5
13	14.5	78.4	18.3	75.0
14	17.8	75.4	21.2	72.3
15	21.5	72.0	24.2	69.5
16	25.3	68.5	27.4	66.6
17	29.2	65.0	30.6	63.7
k = 0.10				
1	0.0	90.9	0.0	90.9
2	0.4	90.5	0.8	90.2
3	1.0	90.0	1.8	89.3
4	1.6	89.5	2.8	88.3
5	2.3	88.8	4.1	87.2
6	2.2	88.0	5.5	85.9

Table A3.8   (continued)

	(a)		(b)			(a)		(b)			(a)		(b)	
n	*i*	*m*	*i*	*m*	*n*	*i*	*m*	*i*	*m*	*n*	*i*	*m*	*i*	*m*
	k = 0.10 *(cont.)*					*k* = 0.15 *(cont.)*					*k* = 0.30			
7	4.4	86.9	7.1	84.4	6	8.0	80.0	11.8	76.7	1	0.0	76.9	0.0	76.9
8	5.8	85.7	9.0	82.7	7	11.2	72.2	15.4	73.5	2	3.0	74.6	5.1	73.0
9	7.5	84.1	11.2	80.8	8	15.2	73.7	19.5	70.0	3	7.4	71.3	11.1	68.4
10	9.7	82.1	13.6	78.5	9	20.1	69.5	23.8	66.2	4	13.3	66.7	18.0	63.1
11	12.4	79.8	16.3	76.1	10	25.5	64.8	28.4	62.3	5	21.1	60.7	25.4	57.4
12	15.6	76.8	19.3	73.4	11	31.0	60.0			6	29.9	53.9	33.0	51.6
13	19.2	73.4	22.5	70.4										
14	23.3	69.8	25.9	67.4		*k* = 0.20					*k* = 0.40			
15	27.4	66.0	29.4	64.2	1	0.0	83.3	0.0	83.3	1	0.0	71.4	0.0	71.4
16	31.5	62.0			2	1.5	82.0	2.7	81.1	2	4.8	68.0	7.5	66.0
					3	3.6	80.4	5.9	78.4	3	11.8	63.0	16.3	59.8
	k = 0.15				4	6.3	78.1	9.8	75.2	4	21.2	56.3	25.6	53.1
					5	10.0	75.0	14.2	71.5	5	31.9	48.6	34.9	46.5
1	0.0	87.0	0.0	87.0	6	14.7	71.1	19.2	67.4					
2	0.9	86.2	1.7	85.5	7	20.6	66.2	24.6	62.8					
3	2.1	85.1	3.6	83.8	8	27.3	60.6	30.3	58.1					
4	3.9	83.8	6.0	81.8	9	32.6	56.1							
5	5.5	82.2	8.7	79.4										

Note: All tables assume random calls for service. Column (a) is for constant servicing time and column (b) for an exponential distribution of servicing times. It is hoped that the missing values in column (a) can be secured by approximation in the near future. Where no entry appears in column, the figures were not available.

Table A3.9 Metric System Conversion Chart

U.S.   METRIC

1 inch = 25.4 millimeters
1 foot = 30.48 centimeters
1 yard = 0.914 meter

METRIC   U.S.

1 millimeter = 0.039 inch
1 centimeter = 0.394 inch
1 meter = 39.37 inches

LENGTH

METRIC (cm, meters)

ENGLISH (in-, ft., yds.)

THICKNESS

1 mil = .025 millimeter

1 millimeter = 39.37 mils

MM

MILS

AREA

1 sq. foot = 929.03 sq. centimeters

1 sq. centimeter = 0.155 sq. inch

CM²

SQ. FT.

VOLUME

1 gallon = 3.785 liters

1 liter = 0.264 gallon

LITERS

GALLONS

Table A3.9 (continued)

VOLUME

1 cu. inch = 16.387 cu. centimeters

1 cu. centimeter = 0.061 cu. inch

CC

CU. IN.

TEMPERATURE

$$°Fahrenheit = \frac{(°Celsius)\ 9}{5} + 32$$

$$°Celsius = \frac{(°Fahrenheit - 32)\ 5}{9}$$

°C

°F

WEIGHT

1 ounce (dry) = 28.35 grams

1 gram = .035 ounce

GRAMS

OUNCES

WEIGHT

1 pound = .454 kilogram

1 kilogram = 2.204 pounds

KILOGRAMS

POUNDS

PRESSURE

1 pound/sq. inch = 0.703 kilogram/sq. centimeter

1 kilogram/sq. centimeter = 14.22 pounds/sq. inch

KG/CM²

LBS./SQ. IN.

With the compliments of McGraw-Hill Book Company, College Division. Publishing for the engineer's diversity. 1221 Avenue of the Americas, New York, NY 10020.

INDEX

Page numbers followed by e, f, g, and t refer to examples, figures, glossary, and tables, respectively.